ADVANCES IN SECOND MESSENGER
AND PHOSPHOPROTEIN RESEARCH

Volume 33

Ion Channel Regulation

Advances in Second Messenger and
Phosphoprotein Research

ADVANCES IN SECOND MESSENGER
AND PHOSPHOPROTEIN RESEARCH

Volume 33

Ion Channel Regulation

Editors

David L. Armstrong
Laboratory of Signal Transduction
National Institute of Environmental Health Science
Research Triangle Park, North Carolina

Sandra Rossie
Department of Biochemistry
Purdue University
West Lafayette, Indiana

Academic Press

San Diego London Boston New York Sydney Tokyo Toronto

This book is printed on acid-free paper. ∞

Academic Press
a division of Harcourt Brace & Company
525 B Street, Suite 1900, San Diego, California 92101-4495, USA
http://www.apnet.com

Academic Press
24-28 Oval Road, London NW1 7DX, UK
http://www.hbuk.co.uk/ap/

International Standard Book Number: 0-12-036133-7

PRINTED IN THE UNITED STATES OF AMERICA
99 00 01 02 03 04 QW 9 8 7 6 5 4 3 2 1

Contents

Contributing Authors

Numbers in parentheses indicate the pages on which the authors' contributions begin.

Martin Biel (*231*), *Institut für Pharmakologie und Toxikologie der Technischen Universität München, 80802 Munich, Germany*

Kim Chan (*179*), *Laboratory of Cardiac/Membrane Physiology, The Rockefeller University, New York, New York 10021*

Annette C. Dolphin (*153*), *Department of Pharmacology, University College of London, London WC1E 6BT, England*

Kathleen Dunlap (*131*), *Departments of Physiology and Neuroscience, Tufts University School of Medicine, Boston, Massachusetts 02111*

David C. Gadsby (*79*), *Laboratory of Cardiac/Membrane Physiology, The Rockefeller University, New York, New York 10021*

Franz Hofmann (*231*), *Institut für Pharmakologie und Toxikologie der Technischen Universität München, 80802 Munich, Germany*

Richard L. Huganir (*49*), *Department of Neuroscience, Howard Hughes Medical Institute, The Johns Hopkins University Medical School, Baltimore, Maryland 21205*

Stephen R. Ikeda (*131*), *Laboratory of Molecular Physiology, Guthrie Research Institute, Sayre, Pennsylvania 18840*

Barry D. Johnson (*203*), *Department of Physiology and Neurobiology, University of Connecticut, Storrs, Connecticut 06269*

Marie-Noëlle Langan (*179*), *Department of Physiology and Biophysics, Mount Sinai School of Medicine, City University of New York, New York, New York 10029*

Irwin B. Levitan (*3*), *Biochemistry Department and Volen Center for Complex Systems, Brandeis University, Waltham, Massachusetts 02454*

Richard S. Lewis (*279*), *Department of Molecular and Cellular Physiology, Stanford University School of Medicine, Stanford, California 94305*

Diomedes E. Logothetis (*179*), *Department of Physiology and Biophysics, Mount Sinai School of Medicine, City University of New York, New York, New York 10029*

Stephen J. Moss (*49*), *The MRC Laboratory of Molecular Cell Biology and Department of Pharmacology, University College of London, London WC1E 6BT, England*

Angus C. Nairn (*79*), *Laboratory of Molecular and Cellular Neuroscience, The Rockefeller University, New York, New York 10021*

Stanley G. Rane (*107*), *Department of Biological Sciences, Purdue University, West Lafayette, Indiana 47907*

Lynn A. Raymond (*49*), *Department of Psychiatry, The Kinsmen Laboratories for Neurological Sciences, The University of British Columbia, Vancouver, British Columbia, Canada V6T 1Z3*

Sandra Rossie (*23*), *Department of Biochemistry, Purdue University, West Lafayette, Indiana 47907*

Jin-Liang Sui (*179*), *Cambridge Neuroscience, Cambridge, Massachusetts 02139*

Sheridan L. Swope (*49*), *Department of Neurology, Division of Neuroscience, Georgetown Institute for Cognitive and Computational Neuroscience, Georgetown University Medical Center, Washington, DC 20007*

Michel Vivaudou (*179*), *CEA, DBMS, Biophysique Moléculaire et Cellulaire (URA CNRS 520), 38054 Grenoble, France*

Richard E. White (*251*), *Department of Physiology and Biophysics, Wright State University School of Medicine, Dayton, Ohio 45435*

Xiangang Zong (*231*), *Institut für Pharmakologie und Toxikologie der Technischen Universität München, 80802 Munich, Germany*

Prologue

An entire volume of *Advances in Second Messenger and Phosphoprotein Research* dedicated to the subject of ion channel regulation signals a new threshold in our awareness of ion channels as signaling proteins. No longer simply the variable resistors in electrical equivalent circuits of biophysicists, ion channels have equally important roles in the development and plasticity of the central nervous system and in the growth regulation of nonexcitable cells. Even the genome of the simple roundworm, *Caenorhabditis elegans,* encodes approximately 80 distinct potassium-selective channels (1). Only a small subset of these channels are regulated by voltage. In humans the list of diseases that result from mutations in channel encoding genes continues to grow (2–5).

The chapters in this volume review many important advances in ion channel regulation by signal transduction pathways during the past decade. To put these chapters into perspective, we include here a brief summary of the origins of the field and the contents of this volume. We have annotated our discussion with a few references of historical interest and several more recent references to relevant areas that are not covered in the chapters. Both fields, signal transduction and ion channel physiology, are moving too fast for any book to remain topical for long, although *Neuromodulation,* the textbook edited by Irwin Levitan and Len Kaczmarek in 1987 (6), remains an outstanding and still instructive introduction. We hoped to make this volume a similarly enduring compendium of the classical problems and experiments for researchers in the field and an inviting introduction to nonelectrophysiologists considering ion channels as interesting and accessible substrates/effectors in their own signaling pathways of interest. Therefore, we chose authors who have been contributing to this field throughout their careers, and we encouraged them to give a historical and nontechnical introduction to their work. Nevertheless, the most thoroughly studied examples of ion channel regulation involve the channels that have been studied longest, the voltage and ligand-gated channels of Hodgkin, Huxley, and Katz (7). Consequently many of the data come from electrophysiological experiments. Even in the case of voltage-independent channels such as the CFTR gene product, which is mutated in cystic fibrosis, the patch clamp remains the technique of choice for studying ion channel regulation.

From one point of view, the patch clamp is the ultimate molecular technique. It allows investigators to observe in real time functional changes in the conformation of individual macromolecules *in situ.* Figure 1 illustrates the different configurations of the patch clamp, which allow one to measure single channel activity on intact cells or in cell-free patches. The inside-out configuration of the cell-free patch in particular provides experimental access to the cytoplasmic side of the membrane in the absence of much of the cell's complex metabolic machinery, including the nucleus. Neverthe-

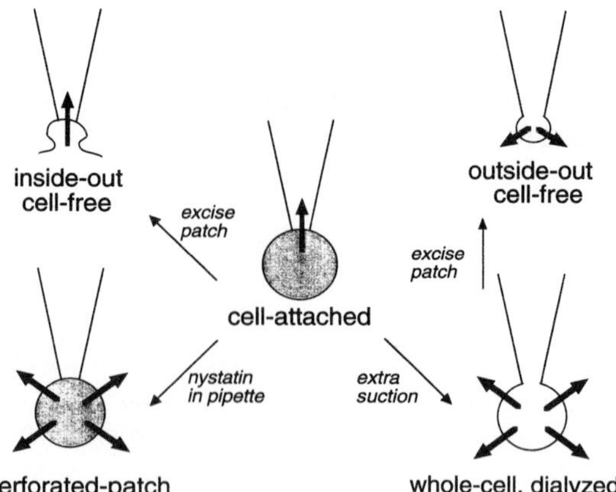

FIG. 1. Patch clamp configurations for recording single channel and whole cell currents. Bold arrows indicate the membrane that contributes to the current. Shaded cells are intact. All configurations begin with the formation of a cell-attached patch, which requires a tight or high-resistance (gigaohm) seal between the fire-polished tip of a glass micropipette and the plasma membrane. Longer, thinner arrows indicate how each configuration is derived from the initial cell-attached patch. When the seal is sufficiently tight, small patches of membrane can be excised from the cell in either orientation, with the former cytoplasmic side of the membrane exposed to the experimental chamber (inside-out) or to the inside of the pipette (outside-out). For simplicity the nystatin molecules that permeabilize the membrane under the pipette to small monovalent ions in the perforated patch configuration are not illustrated graphically.

less, such patches contain many more proteins than just a few ion channels. With minimum areas of 1–2 μm^2, there is room for thousands of integral membrane protein complexes in each patch, in addition to whatever other proteins, membranes, or organelles remain attached through cytoskeletal links to the plasma membrane.

Unfortunately, the simplest and most frequently employed configuration, whole-cell recording through ruptured patches, is also the least suited to studies of ion channel regulation. Rupturing the membrane under the patch pipette leads to dialysis of the cell with the pipette solution, the volume of which is infinite relative to the cell's volume. However, each molecule diffuses into the gel matrix of the cytoplasm at different but generally unmeasurable rates depending on their size and reactivity with cellular molecules. Maintaining cell viability while recording through ruptured patches also requires the addition of exogenous calcium chelators, which buffers many calcium transients and inhibits calcium-dependent signaling. The perforated patch technique introduced by Horn and Marty (8) is superior in several respects. It takes advantage of the pore-forming properties of certain antibiotics, such as nystatin, which kill microorganisms by dissipating the proton gradients they use to generate energy. In eukaryotic cells those energy-generating membranes are intracellular, but the nystatin molecules translocate across the bilayer so rarely that they can be added to the pipette solution at concentrations sufficiently high to permeabilize the membrane under the patch without altering the rest of the cell's membrane or reducing cell viability. In addition, the nystatin and amphotericin pores are only permeable to small monovalent ions, so calcium homeostasis and second messenger metabolism are not perturbed. For more detailed descriptions of the acquisition and interpretation of patch clamp data, there are several excellent collections of methods ranging from primers (9) to state of the art (10,11).

BRIEF HISTORY OF ION CHANNEL REGULATION

Figure 2 illustrates three different experimental views of ion channel proteins: as variable resistors in electrical equivalent circuits of excitable membranes, as physiological effectors regulated by signal transduction pathways, and as gene products. Much has been learned about the biophysical properties of ion channel pores in the years since Hodgkin, Huxley, and Katz initiated the modern study of ion channels (12), culminating recently in the first crystal structure of a potassium-selective pore (13). The initial studies were made possible by voltage-clamp amplifiers, which allowed investigators to measure directly the ionic currents flowing through the channels. Two additional experimental advances around 1980 led to the modern era. The patch-clamp technique (14) allowed the voltage clamp to record picoampere currents through individual channels and to record cellular currents from virtually any mammalian cell type. Gene cloning allowed scientists to identify the

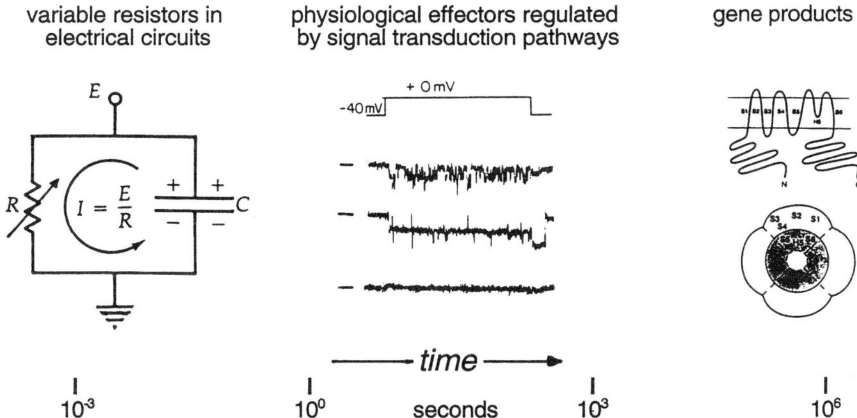

FIG. 2. Three views of ion channel proteins on different time scales: as variable resistors in electrical circuits on a time scale of milliseconds, as physiological effectors regulated by signal transduction pathways on a time scale of seconds, and as gene products on a time scale of hours. The middle view, which is the subject of this volume, shows three qualitatively different responses, separated by only a few seconds, of L-type calcium channels from the same patch to a constant stimulus. A single cell-attached patch on a rat pituitary tumor cell was depolarized repeatedly from −40 to 0 mV for 0.1 sec at 3 sec intervals. The dash at the beginning of each trace indicates the resting current level when the channels are closed. Inward current through the open channel appears as a downward deflection. The channels respond either with many brief openings, with a few very prolonged openings, or with no openings at all. Neurotransmitters, calcium-dependent enzymes and drugs used to treat heart disease alter the relative frequency of observing each *mode* of gating. See Reference 6 for a more detailed explanation.

proteins that form the pores (15–18). But neither of these advances by themselves provided a satisfying explanation for the physiological roles of ion channels in the development and plasticity of the nervous system or their roles in many nonexcitable cells outside the nervous system, such as lymphocytes (19).

Even in electrically excitable cells, the response of ion channels to depolarization is fleeting, subsiding within fractions of a second when the membrane potential returns to rest. On the other end of the biological time scale, transcriptional regulation of channel expression provides cells with a powerful mechanism for adapting to environmental changes, and the expression of recombinant channels in heterologous systems has taught molecular biologists much about the structural basis of the channels' biophysical properties. However, on the physiological time scale of seconds to hours, signal transduction pathways appear to provide the primary means of controlling ion channel activity. This book describes the evidence accumulated during the past 10 years, which supports that conclusion.

Table 1 provides an overview of the major signals that directly and indirectly control ion channels. Until recently, most channel physiologists would have reserved the term "regulation" for signals that directly activate the channels, through either depolarization or ligand binding at the extracellular or intracellular surface of the channel. "Modulation" would have been used to describe indirect pathways of channel regulation. But now we appreciate that many channels, such as the cystic fibrosis gene product described below, are not regulated directly by either voltage or ligands. Furthermore, protein phosphorylation can modulate the activity of voltage-dependent channels in an all-or-none fashion (20). Thus, we have chosen to call this volume *Ion Channel Regulation,* even though it focuses predominantly on the indirect mechanisms of ion channel regulation.

Paul Greengard was the first to articulate a general role for second messengers and phosphoproteins in the regulation of neuronal excitability (21), and Eric Kandel and his colleagues produced much of the earliest and most convincing evidence for

TABLE 1. *Ion channel regulation*

Activated directly by depolarization
For electrical signaling: Na, K channels
For chemical signaling: Ca channels
Activated directly by ligands
Extracellular ligands (neurotransmitters)
ACh, ATP, GABA, glycine, glutamate, serotonin
Intracellular ligands (second messengers)
Calcium, cyclic nucleotides, $G\beta\gamma$, inositol phosphates, lipids
Regulated indirectly through second messengers
Intracellular ligands (above)
Calmodulin
Cytoskeleton
Reversible protein phosphorylation
Regulated by transcription
Constitutively active channels
Alternate splicing of mRNA

channel regulation in their many studies of molluscan neurons, but by the end of the 1970s there were several "anomalous" effects of neurotransmitters in the vertebrate nervous system that appeared too slow to be explained by direct binding to traditional ligand-activated channels (22,23). "Too slow" was shown to be an irreducible delay of $>$100 ms between neurotransmitter binding at the postsynaptic membrane and the first detectable change in ionic currents across the postsynaptic membrane. Most notably there were two examples of such slow effects where the targets were clearly voltage-dependent calcium channels (24,25). Now, of course, we realize that these effects reflected the coupling of receptors to G proteins that initiate signal transduction cascades by controlling second messenger production.

Ironically, many electrophysiologists currently discount any response with subsecond kinetics as too impossibly fast to be explained by any signal transduction cascade that is more complicated than direct G-protein binding to the channels. They forget, however, that the delay between the arrival of the action potential in the terminals of mammalian motor nerves and the initiation of skeletal muscle shortening is less than 10 ms! During those 10 ms, calcium channels open, calcium influx triggers exocytosis, acetylcholine diffuses across the synaptic cleft and binds to nicotinic receptor channels, an action potential is initiated in the postsynaptic membrane, the action potential triggers calcium release from the sarcoplasmic reticulum, and calcium initiates the series of enzymatic reactions that lead to crossbridge formation and myofibril movement. Literally dozens of conformational changes, protein associations, and enzymatic reactions are completed in those 10 ms because all these components are highly organized at the membrane. We now know many cases where signaling proteins are assembled in preformed complexes, which allows rapid and selective signaling. The study of the proteins that organize these complexes is one of the most exciting recent advances in ion channel regulation, and there is a chapter describing those results later in the book.

There are several other exciting new areas of ion channel regulation that seemed too preliminary to include when we began planning this volume over two years ago. New channels have been discovered, and new categories of regulation have been established. The connexon channel proteins that form gap junctions are not new, but much has been learned about the properties of heteromultimers, their regulation, and their role in disease (26,27). Several physiologically important potassium channels have been cloned in the last few years, including M-channels (28), K_{ATP} channels (29,30), hyperpolarization-activated "pacemaking" channels (31), and the lower conductance, calcium-activated but voltage-independent potassium channels (32,33). Perhaps most surprising is the size of the novel two-pore potassium channel family that was identified first in the *C. elegans* genome project (34). These channels provide the background potassium permeability of many cell membranes; they are insensitive to voltage but respond to pH and other intracellular signals (35–38). One imagines that chloride channels might be equally diverse and numerous, but the identity, function, and regulation of most chloride channels remain to be determined (39). Perhaps the recent elucidation of the amino acids conferring anion selectivity in one chloride channel (40) will allow other family members to be identified more quickly. Although the gene mutated in cystic fibrosis clearly encodes a pore-forming chloride channel (41), ABC transporters in general do not appear to function as ion channels. Nevertheless, they do influence the activities of other channels through multiple mechanisms, either as subunits of the channel complex, as in the case of the sulfonylurea receptor (29,30), or as initiators of purinergic signaling (42). Other non-pore-forming proteins regulate channel activity by reversible (read regulatable) non-covalent associations with the channel proteins themselves (43,44). Finally, alternative splicing of mRNA transcripts has been recognized as a powerful mechanism for generating channel diversity (45–47), particularly during development (48,49) or in response to hormonal signals (50,51). These subjects will have to wait for the next volume devoted to ion channels.

BRIEF SUMMARY OF THE CONTENTS

In this volume on ion channel regulation by second messengers and phosphoproteins, the chapters are divided into three sections: protein phosphorylation, closely associated proteins, and second messengers. Like most other proteins involved in cell signaling, ion channels are regulated by reversible protein phosphorylation. Irwin Levitan, one of the original pioneers in this field, gives an authoritative overview of the role of protein phosphorylation in ion channel regulation and its significance for neuronal modulation. Sandra Rossie and Richard Huganir, two scientists who participated in some of the initial biochemical studies of ion channel regulation many years ago as postdoctorates in the Catterall (Rossie) and Greengard (Huganir) laboratories, review the effects of protein phosphorylation on the two traditional classes of ion channels, voltage and ligand-gated channels. David Gadsby and Angus Nairn review their experiments on chloride channel regulation in cardiac muscle. This is the same CFTR channel that is responsible for cystic fibrosis when mutated and that provides a wonderful example of how much insight one can gain from patch-clamp studies of individual channels in cell-free patches when biochemistry is combined with electrophysiology. Finally, Stanley Rane, who trained with Kathleen Dunlap, describes some of the ways growth factors regulate ion channel activity. Although there are now several examples of channel regulation by direct tyrosine phosphorylation (52,53), there are certain to be many other effects, like those discovered by Rane and his colleagues, that are mediated by receptor tyrosine kinase initiated changes in transcription and protein–protein interactions.

Whenever patch clampers present evidence of channel regulation by protein phosphorylation without the structural corroboration provided by ^{32}P incorporation and site-directed mutagenesis, they always qualify their conclusions by acknowledging the possibility of other "closely associated (i.e., unknown) proteins" in addition to the channel proteins themselves as potential substrates for phosphorylation. As mentioned earlier, we now know there are many such closely associated proteins: other channel proteins in heteromeric complexes, regulatory subunits, G proteins, and cytoskeletal scaffolding proteins, many of which are probably also regulated by protein phosphorylation. Barry Johnson, who discovered some of the first evidence for channel regulation by cytoskeleton as a graduate student with Lou Byerly and later characterized scaffold-dependent calcium channel regulation by protein kinases with Bill Catterall and John Scott, reviews the exciting new findings on channel regulation by structural proteins. However, in the remaining chapters of this section, we have chosen to focus on the most thoroughly studied example of such closely associated proteins, the heterotrimeric G proteins.

Many neurotransmitters and hormones reduce excitability by inhibiting calcium channels and stimulating potassium channels through receptors coupled to the pertussis toxin sensitive G proteins, G_i and G_o (54). These effects involve at least two distinct signaling pathways, one membrane-delimited pathway and another diffusible pathway that is calcium dependent (55). The G-protein $\beta\gamma$ subunits mimic many of the effects of the membrane delimited pathway on channel activity. Diomedes Logothetis, who participated in the first studies showing potassium channel regulation by $\beta\gamma$ subunits in David Clapham's laboratory, reviews with his colleagues the regulation of the inwardly rectifying potassium channels (K_{ir} 3.1/3.4, also known as K_{ACh} or GIRK) from atrial muscle cells. Kathleen Dunlap, who started the field of channel *modulation* by metabotropic receptors with her landmark paper as a postdoctorate with Gerald Fischbach (25), and Stephen Ikeda, who discovered the $\beta\gamma$ effect on calcium channels, review their work on N-type calcium channels in peripheral neurons. In contrast to the work described previously, there is no comparable evidence for α subunits binding directly to any channel (56). Such interactions have been postulated in many preparations, but G protein activation had no independent effect on channels in their native membranes when protein phosphorylation was prevented completely (57,58).

Nevertheless, G protein coupled receptors modulate many other channels. In that context Annette Dolphin reviews the G protein dependent regulation of the L-type calcium channels that provide the calcium for secretion and contraction in endocrine and muscle cells. Unlike the N-type calcium channels that are regulated by both the membrane-delimited and the diffusible second messenger signaling pathways, L-type channels are not regulated by the membrane-delimited signal (59), which is now presumed to be $\beta\gamma$.

Cyclic AMP was the original "second messenger," and this section begins with a review by Franz Hofmann and his colleagues of the channels that are activated directly by cyclic nucleotide binding. These channels are closely related to the more recently discovered class of hyperpolarization-activated potassium channels (60), which are critical for pacemaking cells that spontaneously spike at regular intervals. Cyclic GMP also binds directly to cationic channels, which it stimulates in photoreceptors, but most of its effects on excitability are mediated by the cyclic GMP-dependent protein kinases. Richard White, a former postdoctorate of Criss Hartzell and David Armstrong, with whom he provided some of the first evidence for magnesium ions (61) and for protein phosphatases (62) as intracellular regulators of channel activity, reviews the many indirect effects of the "other" cyclic nucleotide, cyclic GMP.

More recently it has become apparent that in addition to the cyclic nucleotides, there are literally dozens of second messengers, which have been implicated in one or more aspects of channel regulation, including calcium binding proteins, inositol phosphates and an uncounted number of lipid metabolites. Calcium appears to be the most pivotal intracellular signal, triggering exocytosis and contraction, binding directly to many channels and regulating many more through calmodulin or the calcium-dependent protein kinases and phosphatases (63). But calcium phosphate is insoluble, so the cell expends considerable energy to maintain resting levels of calcium below 200 nM. Voltage-activated calcium channels in the plasma membrane (64) and ligand-activated calcium channels in the endoplasmic reticulum (65) provide the main source of calcium for transient cytoplasmic increases. More recently new voltage-dependent (T-type) (66) and independent (TRP) (67,68) calcium channels have been identified. The latter channels are believed to represent the "capacitative" calcium entry pathway that is activated by the novel, but unidentified, signal transduction pathway resulting from depletion of the calcium stores in the endoplasmic reticulum (69; but see 70). Richard Lewis, who obtained some of the original evidence for ion channel regulation of lymphocyte function in Mike Cahalan's laboratory, reviews the regulation of these voltage-independent calcium channels in the plasma membrane.

In contrast to the mechanism of capacitative calcium entry, everyone is familiar with the molecular sequelae responsible for the early phase of calcium signaling through G protein coupled receptors. Lipid hydrolysis by G protein activated phospholipase C generates two second messengers: 1,4,5-inositol trisphosphate (IP_3) and diacylglycerol. IP_3 opens calcium channels in the endoplasmic reticulum (65), and diacylglycerol stimulates protein kinase C, which has many effects on ion channel activity (71). IP_3 is removed by further phosphorylation of inositol, but some of those metabolites may also regulate channel activity (72), either directly or indirectly by altering regulatory enzymes (73).

Phospholipase A_2 is also stimulated by G proteins. The arachidonic acid that is released can be metabolized into many signaling molecules by three distinct enzyme systems: the cyclooxygenases, lipoxygenases, and cytochrome P450s. Several of these metabolites regulate ion channel activity (74), but the mechanism(s) underlying these effects has been difficult to elucidate because the compounds are extremely labile and they have many potential side effects, including calcium release, protein kinase C stimulation, oxidation–reduction of the channel proteins, and bilayer deformation (75). Recent reports have implicated PIP_2, the precursor of IP_3 in potassium channel regulation (76,77), and growth factor stimulated increases in IP_3 kinase activity initiate complex signaling cascades regulating cell growth and death through many substrates, including channels (78). Other growth factors are believed to act through glycolipid metabolites, such as ceramide, sphingosine-1-phosphate, and ganglioside GM_1, but the mechanisms of their action in the few cases where they have been reported to alter ion channel activity have not yet been investigated thoroughly.

The only common theme among studies of ion channel regulation by lipid metabolites, other than the well-established PLC pathway, involves potassium channel stimulation by lipoxygenase metabolites of arachidonic acid (54). Like so many themes in ion channel regulation, this effect was first observed in studies of molluscan neurons by Eric Kandel, Steve Siegelbaum, Jimmy Schwartz, and their collaborators at Columbia University (79). In that work, PLA_2 was implicated as the effector for the pertussis toxin-sensitive G proteins mediating the action of an inhibitory neuropeptide, FMRFamide. Subsequently, lipoxygenase metabolites have been reported to stimulate potassium channels in many cell types, including atrial myocytes (80), cortical neurons (81,82), and pituitary tumor cells (83). Furthermore, lipoxygenase inhibitors prevent potassium channel stimulation by inhibitory neuropeptides. In general, these effects of lipoxygenase metabolites antagonize the effects of protein kinases, and there is evidence that they act through protein phosphatases (82–84), but the identity of the enzymes and the mechanism of their activation have not been determined conclusively. Now that the book is finally finished, the editors can return to their efforts to answer these questions.

ACKNOWLEDGMENTS

We are grateful to the editors of the series, Paul Greengard, Angus Nairn, and Shirish Shenolikar, for giving us this opportunity and to the editorial staffs of Lippincott-Raven and more recently Academic Press for their patience and their indulgence. We have learned much from attending the triennial meetings on phosphoproteins and second messengers and we hope this book partially repays those debts. We are particularly grateful to Paul Greengard and Bill Catterall for teaching us the importance of ion channel regulation in neuronal function and the power of combining electrophysiological, biochemical, and molecular approaches to these problems.

REFERENCES

1. Bargmannn CI: Neurobiology of the *Caenorhabditis elegans* genome. *Science* 1998;282:2028–2033.
2. Ackerman MJ, Clapham DE: Ion channels—basic science and clinical disease. *New England J Med.* 1997;336:1575–1585.
3. Keating MT, Sanguinetti MC: Pathophysiology of ion channel mutations. *Curr Opin Gen & Devel.* 1996;6:326–333.
4. Cannnon SC: Ion-channel defects and aberrant excitability in myotonia and periodic paralysis. *Trends Neurosci* 1996;19:3–10.
5. Doyle JL: Ataxia, arrhythmia and ion-channel gene defects. *Trends Genetics* 1998;14:92–98.
6. Kaczmarek LK, Levitan IB, eds: *Neuromodulation, the biochemical control of neuronal excitability.* New York: Oxford University Press, 1987.
7. Hille B. *Ionic channels of excitable membranes,* second edition. Sunderland, MA: Sinauer Assoc. Inc., 1992.
8. Horn R, Marty A: Muscarinic activation of ionic currents measured by a new whole-cell recording method. *J Gen Physiol* 1988;92:145–159.
9. Ogden, D, ed: *Microelectrode Techniques,* second edition. Cambridge, UK: The Company of Biologists, 1994.
10. Sakmann B, Neher E., eds: *Single channel recording,* second edition. New York: Plenum Press, 1995.
11. Rudy B, Iverson LE, eds: *Ion channels, Methods Enzymology,* vol 207. San Diego, Academic Press, 1992.
12. Armstrong CM, Hille B: Voltage-gated ion channels and electrical excitability. *Neuron* 1998;20: 371–380.

13. Moczydlowski E: Chemical basis for alkali cation selectivity in potassium-channel proteins. *Chem & Biol* 1998;5:R291–R301.
14. Hamill OP, Marty A, Neher E, Sakmann B, Sigworth FJ: Improved patch clamp techniques for high-resolution current recording from cells and cell-free membrane patches. *Pflugers Archiv* 1981;391: 85–100.
15. Ballivet M, Patrick J, Lee J, Heinemann S: Molecular cloning of cDNA coding for the gamma subunit of Torpedo acetylcholine receptor. *Proc Natl Acad Sci USA* 1982;79:4466–4470.
16. Noda M, Takahashi H, Tanabe T, Toyosato M, Fururani Y, Hirose T, Asai M, Inayama S, Miyata T, Numa S: Primary structure of alpha-subunit precursor of *Torpedo Californica* acetylcholine receptor deduced from cDNA sequence. *Nature* 1982;299:793–797.
17. Noda M, Shimizu S, Tsutomu T, Takai T, Kayano T, Ikeda T, Takahashi H, Hakayama H, Yuichi K, Minamino N, Kangawa, Matsuo H, Raftery MA, Hirose T, Inayama S, Hayashida H, Miyata T, Numa S: Primary structure of *electrophorus electricus* sodium channel deduced from cDNA sequence. *Nature* 1984;312:121–127.
18. Jan LY, Jan YN: Voltage-gated and inwardly rectifying potassium channels. *J Physiol* 1997;505.2: 267–282.
19. Lewis RS, Cahalan MD: Ion channels and signal transduction in lymphocytes. *Ann Rev Physiol* 1990; 52:415–430.
20. Levitan IB: Modulation of ion channels by protein phosphorylation and dephosphorylation. *Ann Rev Physiol* 1994;56:193–212.
21. Greengard P: Possible role for cyclic nucleotides and phosphorylated membrane proteins in postsynaptic actions of neurotransmitters. *Nature* 1976;260:101–108.
22. Kehoe J, Marty A: Certain slow synaptic responses: their properties and possible underlying mechanisms. *Ann Rev Biophys Bioeng* 1980;9:437–465.
23. Hartzell HC: Mechanisms of slow postsynaptic potentials. *Nature* 1981;291:539–544.
24. Reuter H: Properties of two inward membrane currents in the heart. *Ann Rev Physiol* 1979;41: 413–424.
25. Dunlap K, Fischbach GD: Neurotransmitters decrease the calcium component of sensory neuron action potentials. *Nature* 1978;276:837–839.
26. Alexander MS, Goodenough DA: Diverse functions of vertebrate gap junctions. *Cell Biol* 1998;8: 477–483.
27. Steel KP: One connexin, two diseases. *Nature Genetics* 1998;20:319–320.
28. Wang H-S, Pan Z, Shi W, Brown BS, Wymore RS, Cohen IS, Dixon JE, McKinnon D: KCNG2 and KCNQ3 potassium channel subunits: Molecular correlates of the M-channel. *Science* 1998;282: 1890–1893.
29. Inagaki N, Gonoi T, Clement IV JP, Namba N, Inazawa J, Gonzalez G, Aguilar-Bryan L, Seino S, Bryan J: Reconstitution of I_{KATP}: An inward rectifier subunit plus the sulfonylurea receptor. *Science* 1995;270:1166–1170.
30. Clement IV JP, Kunjilwar K, Gonzalez G, Schwanstecher M, Panten U, Aguilar-Bryan L, Bryan J: Association and stoichiometry of K_{ATP} channel subunits. *Neuron* 1997;19:827–838.
31. Santoro B, Liu DT, Yau H, Bartsch D, Kandel ER, Siegelbaum SA, Tibbs G, R.: Identification of a gene encoding a hyperpolarization-activated pacemaker channel of brain. *Cell* 1998;93:717–729.
32. Kohler M, Hirschbrg B, Bond CT, Kinzie JM, Marrion NV, Maylie J, Adelman JP: Small-conductance, calcium-activated potassium channels from mammalian brain. *Science* 1996;273: 1709–1714.
33. Ishii TM, Silvia C, Hirschberg B, Bond CT, Adelman JP, Maylie J: A human intermediate conductance calcium-activated potassium channel. *Proc Natl Acad Sci USA* 1997;94:11651–11656.
34. Wei A, Jegla T, Salkoff L: Eight potassium channel families revealed by the *C. elegans* genome project. *Neuropharmacol* 1996;35:805–829.
35. Lesage F, Guillemare E, Fink M, Duprat F, Lazdunski M, Romey G, Barhanin J: TWIK-1, a ubiquitous human weakly inward rectifying K^+ channel with a novel structure. *EMBO J* 1996;15:1004–1011.
36. Fink M, Duprat F, Lesage F, Reyes R, Romey G, Heurteaux C, Lazdunski M: Cloning, functional expression and brain localization of a novel unconventional outward rectifier K^+ channel. *EMBO J* 1996;15:6854–6862.
37. Duprat F, Lesage F, FInk M, Reyes R, Heurteaus C, Lazdunski M: TASK, a human background K+ channel to sense external pH variations near physiological pH. *EMBO J* 1997;16:5464–5471.
38. Patel AJ, Honore E, Maingret F, Lesage F, Fink M, Duprat F, Lazdunski M: A mammalian two pore domain mechano-gated S-like K^+ channel. *EMBO J* 1998;17:4283–4290.
39. Jentsch TJ: Molecular physiology of anion channels. *Curr Opin Cell Biol* 1994;6:600–606.

40. Fahlke C, Yu HT, Beck CL, Rhodes TH, George Jr. AL: Pore-forming segments in voltage-gated chloride channels. *Nature* 1997;390:529–532.
41. Welsh MJ: Cystic fibrosis transmembrane conductance regulator: A chloride channel with novel regulation. *Neuron* 1992;8:821–829.
42. Schwiebert EM, Egan ME, Hwang TH, Fulmer SB, Allen SS, Cutting GR, Guggino WB: CFTR regulates outwardly rectifying chloride channels through an autocrine mechanism involving ATP. *Cell* 1995;8:1063–1073.
43. Adelman JP: Proteins that interact with the pore-forming subunits of voltage-gated ion channels. *Curr Biol* 1995;5:286–295.
44. Schopperle WM, Holmqvist MH, Zhou Y, Wang J, Wang Z, Griffith LC, Keselman I, Kusinitz F, Dagan D, Levitan IB: Slob, a novel protein that interacts with the slowpoke calcium-dependent potassium channel. *Neuron* 1998;20:565–573.
45. Lagrutta A, Shen K-Z, North RA, Adelman JP: Functional differences among alternately spliced variants of Slowpoke, a Drosophila calcium-activated potassium channel. *J Biol Chem* 1994;269: 20347–20351
46. Seeburg PH: The role of RNA editing in controlling glutamate receptor channel properties. *J Neurochem* 1996;66:1–5.
47. Patton DE, Silva T, Bezanilla F: RNA editing generates a diverse array of transcripts encoding squid Kv2 K^+ channels with altered functional properties. *Neuron* 1997;19:711–722.
48. Rosenblatt KP, Sun ZP, Heller S, Hudspeth AJ: Distribution of Ca^{2+}-activated K channel isoforms along the tonotopic gradient of the chicken's cochlea. *Neuron* 1997;19:1061–1075.
49. Navaratnam DS, Bell TJ, Tu TD, Cohen EL, Oberholtzer JC: Differential distribution of Ca^{2+}-activated K channel splice variants among hair cells along the tonotopic axis of the chick cochlea. *Neuron* 1997;19:1077–1085.
50. Liu SJ, Kaczmarek LK: The expression of two splice variants of the Kv3.1 potassium channel gene is requlated by different signaling pathways. *J Neurosci* 1998;18:2881–2890.
51. Xie J, McCobb DP: Control of alternative splicing of potassium channels by stress hormones. *Science* 1998;280:443–446.
52. Siegelbaum SA: Ion channel control by tyrosine phosphorylation. *Curr Biol* 1994;4:242–245.
53. Boxall AR, Lancaster B: Tyrosine kinases and synaptic transmission. *Euro J Neuroscience* 1998;10: 2–7.
54. Armstrong DL, White RE: An enzymatic mechanism for potassium channel stimulation through pertussis toxin-sensitive G proteins. *Trends Neurosci* 1992;15:403–408.
55. Bernheim L, Beech DJ, Hille B: A diffusible second messenger mediates one of the pathways coupling receptors to calcium channels in rat sympathetic neurons. *Neuron* 1991;6:859–867.
56. Clapham DE: Direct G protein activation of ion channels? *Annu Rev Neurosci* 1994;17:441–464.
57. Hwang TC, Horie M, Nairn AC, Gadsby DC: Role of GTP-binding proteins in the regulation of mammalian cardiac chloride conductance. *J Gen Physiol* 1992;99,465–489.
58. Hartzell HC, Fischmeister R: Direct regulation of cardiac Ca^{2+} channels by G proteins: Neither proven nor necessary? *Trends Pharmacol Sci* 1992;13:380–385.
59. Mathie A, Bernheim L, Hille B: Inhibition of N- and L-type calcium channels by muscarinic receptor activation in rat sympathetic neurons. *Neuron* 1992;8:907–914.
60. Clapham DE: Not so funny anymore: Pacing channels are cloned. *Neuron* 1998;21:5–7.
61. White RE, Hartzell HC: Effects of intracellular free magnesium on calcium current in isolated cardiac myocytes. *Science* 1988;239:778–780.
62. White RE, Schonbrunn A, Armstrong DL: Somatostatin stimulates Ca^{2+}-activated K channels through protein dephosphorylation. *Nature* 1991;351:570–573.
63. Verkhratsky A, Toescu EC, eds: *Integrative aspects of calcium signaling.* New York: Plenum Press, 1998.
64. Perez-Reyes E, Schneider T: Molecular biology of calcium channels. *Kidney Intnat'l* 1995;48:1111–1124.
65. Mikoshiba K: The insP_3 receptor and intracellular Ca^{2+} signaling. *Curr Opin Neurobiol* 1997;7:339–345.
66. Huguenard JR: Low-voltage-activated (T-type) calcium-channel genes identified. *Trends Neurosci* 1998;21:451–452.
67. Montell C: New light on TRP and TRPL. *Molec Pharm* 1997;52:755–763.
68. Philipp S, Hambrecht J, Braslavski L, Schroth G, Freichel M, Murakami M, Cavalie A, Flockerzie V: A novel capacitative calcium entry channel expressed in excitable cells. *EMBO J* 1998;17:4274–4282.
69. Parekh AB, Penner R: Store depletion and calcium influx. *Physiol Rev* 1997;77:901–930.

70. Scott K, Sun Y, Beckingham K, Zuker CS: Calmodulin regulation of Drosophila light-activated channels and receptor function mediates termination of the light response *in vivo*. *Cell* 1997;91: 375–383.
71. Jonas EA, Kaczmarek LK: Regulation of potassium channels by protein kinases. *Curr Opin Neurobiol* 1996;6:318–323.
72. Shears SB: The versatility of inositol phosphates as cellular signals. *Biochim Biophys Acta* 1998; 1436:49–67.
73. Xie W, Solomons KR, Freeman S, Kaetzel MA, Bruzik KS, Nelson DJ, Shears SB: Regulation of Ca^{2+}-dependent Cl^- conductance in a human colonic epithelial cell line (T84): Cross talk between Ins(3,4,5,6)IP4 and protein phosphatases. *J Physiol* 1998;510.3:661–673.
74. Meves H: Modulation of ion channels by arachidonic acid. *Prog Neurobiol* 1994;43:175–186.
75. Casado M, Ascher P: Opposite modulation of NMDA receptors by lysophospholipids and arachidonic acid. *J Physiol* 1998;513.2:317–330.
76. Hilgemann DW: Cytoplasmic ATP-dependent regulation of ion transporters and channels. *Annu Rev Physiol* 1997;59:193–220.
77. Ashcroft FM: Exciting times for PIP2. *Science* 1998;282:1059–1060.
78. Blair LA, Marshall J: IGF-1 modulates N and L calcium channels in a PI3-kinase dependent manner. *Neuron* 1997;19:421–429.
79. Piomelli D, Volterra A, Dale N, Diegelbaum SA, Kandel ER, Schwartz JH, Belardetti F: Lipoxygenase metabolites of arachidonic acid as econd messengers for presynaptic inhibition of Aplysia sensory cells. *Nature* 1987;328:38–43.
80. Scherer RW, Lo CF, Breitwieser GE: Leukotriene C4 modulation of muscarinic K current activation in bull frog atrial myocytes. *J Gen Physiol* 1993;102:125–141.
81. Schweitzer P, Madamba S, Champagnat J, Siggins GR: Somatostatin inhibition of hippocampal CA1 pyramidal neurons: Mediation by arachidonic acid and its metabolites. *J Neurosci* 1993;13:2033–2049.
82. Zhu M, Gelband CH, Moore JM, Posner P, Sumners C: Angiotensin II type 2 receptor stimulation of neuronal delayed rectifier potassium current involves phospholipase A2 and arachidonic acid. *J Neurosci* 1998;18:679–686.
83. Duerson K, White RE, Jiang F, Schonbrunn A, Armstrong DL: Somatostatin stimulates BKca channels in rat pituitary tumor cells through lipoxygenase metabolites of arachidonic acid. *Neuropharmacol* 1996;35:949–961.
84. Petit-Jacques J, Hartzell HC: Effect of arachidonic acid on the L type calcium current in frog cardiac myocytes. *J Physiol* 1996;493.1:67–81.

ADVANCES IN SECOND MESSENGER
AND PHOSPHOPROTEIN RESEARCH

Volume 33

Ion Channel Regulation

Part I

Protein Phosphorylation

Advances in Second Messenger and Phosphoprotein Research, Vol. 33

1040-7952/99 $30.00

1

Modulation of Ion Channels by Protein Phosphorylation

How the Brain Works

Irwin B. Levitan

Biochemistry Department and Volen Center for Complex Systems, Brandeis University, Waltham, Massachusetts 02454

INTRODUCTION

The subtitle of this chapter is not entirely facetious. The electrical activity of individual nerve cells, and communication among them, are determined by the movement of ions across the plasma membrane through the highly specialized intrinsic membrane proteins known as ion channels. Because the essence of brain function is information transfer, mediated by ion channels, within and between nerve cells, it is safe to say that detailed knowledge about how ion channels work and how they are regulated is necessary for even the most rudimentary understanding of how the brain controls behavior. Of course, ion channels are not restricted to nerve cells, or even to excitable cells, and many of the chapters in this volume will describe the fundamental role of ion channels and their modulation in the functioning of cells of many kinds.

It is instructive to recall that the term "ion channel modulation" was not long ago regarded as something of an oxymoron. The classic studies of Alan Hodgkin and Andrew Huxley defined the roles of voltage-dependent sodium and potassium currents in the generation and propagation of squid axon action potentials as "all-or-none." Although Hodgkin and Huxley themselves never explicitly excluded the possibility that these ionic currents might be subject to cellular regulation, the very brilliance of their work and the completeness of their description of action potential properties led others to conclude that the properties of the ion channels responsible for these currents might be immutable. Similarly, the pioneering studies of Bernard Katz and his colleagues on synaptic transmission at the frog neuromuscular junction led to the belief that all synapses behave like the neuromuscular junction, and

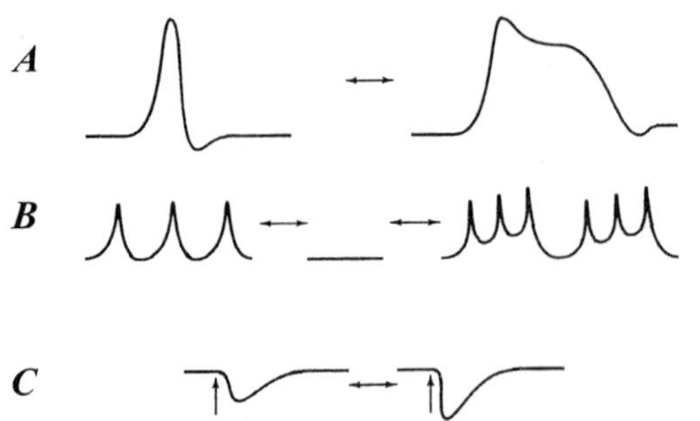

FIG. 1. Modulation of the electrical properties of excitable cells. Common modulatory targets include the amplitude and/or duration of action potentials **(A),** endogenous action potential pattern and firing rate **(B),** and the strength of synaptic transmission **(C).** (Modified from Ref. 89.)

engendered considerable skepticism about early claims for a role for intracellular second messengers in some forms of synaptic transmission.

It is now widely accepted, however, that patterns of neuronal electrical activity and intercellular communication vary greatly. Even closely neighboring neurons can exhibit distinct electrical properties, reflecting their distinct complement of ion channels that are functional under a given set of conditions. At least as important for brain function, and more important from the perspective of this volume, is the fact that these patterns of electrical activity are not fixed, but can be modulated by the actions of neurotransmitters, hormones, or sensory inputs. The medley of modulatory changes observed in excitable cells matches the diversity of endogenous electrical properties themselves, but some patterns can be discerned: modulation generally results in a change in action potential amplitude or duration, in endogenous action potential firing, or in the efficacy of synaptic transmission (Fig. 1). These may not be mutually exclusive—for example, a change in action potential duration will influence neurotransmitter release, hence will modulate synaptic transmission indirectly. As this volume illustrates, cells make use of an enormous assortment of molecular mechanisms to produce these changes. I will focus my discussion on the ubiquitous mechanism that has been most thoroughly studied, modulation by protein phosphorylation.

IONIC CURRENTS ARE MODULATED BY PROTEIN PHOSPHORYLATION AND DEPHOSPHORYLATION

One of the earliest and best understood examples of modulation by protein phosphorylation comes from studies of the prolongation of the cardiac action potential by β-adrenergic agonists. In principle, such an increase in action potential duration could come about either by an increase in the calcium current that underlies the depolarizing phase of the cardiac action potential or by a decrease in the potassium current(s) responsible for repolarization (or both). By the early 1970s, work from the laboratories of Harald Reuter, Wolfgang Trautwein, Richard Tsien, and others had established that voltage-dependent calcium currents in the heart can be enhanced by stimulation of β-adrenergic receptors, and that this modulation is probably mediated by that archetypal intracellular second messenger, cyclic AMP (1). This was followed by the demonstration of a role for cyclic AMP in the modulation of the electrical properties of several different molluscan neurons (2–4). It soon became evident that these actions of cyclic AMP are mediated by protein phosphorylation, by the cyclic-AMP-dependent protein kinase (PKA). For example, injection of the active catalytic subunit of PKA into both cardiac muscle cells and molluscan neurons can mimic the effects of neurotransmitters and cyclic AMP on ionic currents (5–8), and intracellular injection of a specific protein inhibitor of PKA can block them (9,10); some of these early studies are reviewed by Kennedy (11). Subsequent experiments demonstrated that other kinds of protein kinase, as well as phosphoprotein phosphatases, are involved in the modulation of ionic currents in a variety of cell types (12–15).

MOLECULAR MECHANISMS OF MODULATION OF IONIC CURRENTS

Studies such as those just described made it evident that protein phosphorylation is both necessary and sufficient for certain actions of hormones and neurotransmitters on cellular electrical properties, but they did not speak to the question of which proteins are phosphorylated by the kinases. The ion channel proteins, responsible for the ionic currents that are subject to modulation, were obvious candidates, but only with the advent of single channel recording and molecular cloning approaches did it become possible to address this issue directly.

In this chapter I will summarize biophysical, biochemical, and molecular evidence that potassium channels are phosphorylated directly by protein kinases and that this phosphorylation results in channel modulation. It is clear that ion channels other than potassium channels also are subject to this kind of regulation, and I will refer here to several notable examples that help to illustrate some of the points I wish to convey. However, work with voltage-dependent sodium and calcium channels and ligand-gated ion channels is described in more detail in Chapters 2 and 3, and my detailed discussion is restricted to the modulation of several different classes of potassium channels by phosphorylation. Because this literature is becoming rather extensive, I will be selective rather than comprehensive in this summary and will focus on the historical development of the field. I then will review the emerging evidence that some modulatable ion channels exist as part of a regulatory protein complex, which includes the protein kinase and phosphoprotein phosphatase activities that participate in channel modulation. Finally, I will return from this reductionist excursion to a brief consideration of the cellular physiological consequences of channel phosphorylation, and of the intimate association of channels with regulatory enzymes.

POTASSIUM CHANNEL PROTEINS ARE SUBSTRATES FOR PROTEIN KINASES AND PHOSPHATASES

Biochemical Measurements of Potassium Channel Phosphorylation

Because there are tissue sources containing high concentrations of nicotinic acetylcholine receptor/channels and voltage-gated sodium and calcium channels, and specific affinity reagents that can be used for their purification and assay, it has been known for a long time from biochemical measurements that these channels are substrates for a variety of protein kinases. For example, the nicotinic acetylcholine receptor/channel purified from *Torpedo* is a substrate for both PKA (16) and an endogenous tyrosine kinase (17), and both sodium channels (18) and calcium channels (19) from various tissues can be phosphorylated by PKA. However, no rich tissue sources of potassium channels exist, and thus in only a few cases have comparable protein biochemistry experiments been done on potassium channels. One of the best studied examples is the $K_{v1.3}$ potassium channel, which has been cloned from several mammalian tissues and is the major voltage-gated potassium channel expressed in T lymphocytes (where it is known as the type *n* potassium channel) (20,21). Cai and Douglass (22) prepared antibodies against fusion proteins encoding $K_{v1.3}$ sequences and demonstrated using the human Jurkat T lymphocyte cell line that the type *n* potassium channel is phosphorylated on serine and threonine residues by both PKA and protein kinase C (PKC). The same antibodies were used to demonstrate that cloned $K_{v1.3}$, expressed heterologously in a human embryonic kidney (HEK) 293 cell line, can be phosphorylated on tyrosine residues by either endogenous or cotransfected exogenous tyrosine kinases (23).

Potassium Channel Phosphorylation Inferred from Modulation of Function

Because of the difficulty of such biochemical experiments with low abundance membrane proteins, a key approach to determining whether ion channels might be direct targets for protein kinases and phosphatases is to apply the enzymes directly to detached membrane patches containing the channel of interest, and ask whether such treatment modulates channel function. Because soluble intracellular proteins and diffusible small molecules are left behind when membrane patches are detached from the host cell, there are fewer cellular components that might intervene between protein phosphorylation and channel modulation. This was done first with two different kinds of potassium channel from molluscan neurons. Direct application of the catalytic subunit of PKA, together with ATP, to the cytoplasmic surface of membrane patches detached from *Aplysia* sensory neurons, decreases the number of functional S potassium channels in the patch without changing the gating properties of those channels that remain (24). By contrast, PKA application increases the open probability of calcium-dependent potassium (K_{Ca}) channels in detached patches from *Helix* neurons, without affecting the number of functional channels (25). These experiments emphasize that the functional consequences of phosphorylation may be very different for different channels. In addition, they demonstrate that channel modulation does not always require cytoplasmic components, but can result from the phosphorylation of some protein, perhaps the ion channel itself, that comes away with the detached patch.

An even more reductionist approach involves the application of protein kinase or phosphatase to ion channels reconstituted in artificial phospholipid bilayers. K_{Ca} channels from *Helix* neurons and rat brain can be modulated by PKA, PKC, or protein phosphatase 2A added to the bilayer (25–27). Because proteins are effectively at infinite dilution in the vast ocean of bilayer lipid (28), it can be concluded that the phosphorylation target that mediates functional modulation is either the ion channel protein itself or some regulatory protein so intimately associated with the ion channel protein that it swims with it in the bilayer.

Other more recent experiments, using mutational analysis of cloned potassium channels expressed in heterologous host cells, have demonstrated a requirement for specific serine, threonine, or tyrosine residues in ion channel proteins for functional modulation to occur. For example, when *Shaker* voltage-gated potassium channels are expressed in *Xenopus* oocytes and their activity is examined in detached membrane patches, application of a phosphoprotein phosphatase to the cytoplasmic face of the patch slows the rapid N-type inactivation characteristic of these channels (29). This modulation can be reversed by subsequent application of the purified catalytic subunit of PKA, together with ATP, to the cytoplasmic face of the patch. Interestingly, although it is the amino terminal region of *Shaker* that mediates N-type inactivation (30,31), mutation of a PKA consensus phosphorylation site near the amino terminal of the channel does not influence the modulation; instead, removal of another PKA consensus site, near the carboxyl terminal, eliminates the effect of phosphatase on N-type inactivation kinetics (29).

The inactivation of potassium channels can also be modulated by PKC phosphorylation. $K_{v3.4}$ is a cloned mammalian voltage-gated potassium channel that exhibits rapid N-type inactivation, comparable to that of *Shaker*, mediated by the channel's amino terminal region (32). When $K_{v3.4}$ is expressed in *Xenopus* oocytes, treatment of the oocytes with activators of PKC leads to a substantial slowing of the inactivation (32). In addition, application of purified PKC to the cytoplasmic surface of detached membrane patches increases the average mean open time of individual $K_{v3.4}$ channels in these patches. In contrast to the results described earlier for the actions of PKA on *Shaker*, mutational analysis demonstrates that several PKC consensus sites within the amino terminal inactivation domain of $K_{v3.4}$ participate in this modulation (32). Since it is known that electrostatic interactions, mediated by positively charged amino acids within the amino terminal region, are critical for N-type inactivation (33), it is possible that negatively charged phosphate groups inserted in this region disrupt the required electrostatic interactions and slow inactivation. In support of this idea is the finding that replacement of one of the PKC phosphorylation sites in $K_{v3.4}$ with a negatively charged aspartic acid residue also slows inactivation of the channel (32).

Modulation by Tyrosine Phosphorylation

Although the roles of protein tyrosine kinases and phosphatases in such phenomena as growth control and cellular differentiation are becoming well understood, it is only recently that ion channel modulation by tyrosine phosphorylation has begun to be studied in detail (see Ref. 34 for a recent review). For example, a cation channel in *Aplysia* bag cell neurons is modulated by tyrosine phosphorylation/

dephosphorylation, via a complex pathway that involves PKA-mediated serine/threonine phosphorylation as well (35). The voltage-gated potassium channel $K_{v1.2}$, which is distributed widely in mammalian brain and heart, is suppressed by receptor-mediated stimulation of phospholipase C, both in native cells and when it is expressed in *Xenopus* oocytes (36). This suppression of $K_{v1.2}$ requires the involvement of PKC, which is activated as a result of phospholipase C stimulation, but direct phosphorylation of the channel by PKC is not sufficient (and may not be necessary) to account for the modulation. Instead, PKC appears to participate in the activation of the PYK2 tyrosine kinase (37), which in turn phosphorylates the channel at a tyrosine residue in the presumed intracellular amino terminal region prior to the first predicted membrane-spanning domain (36). Mutation of this tyrosine residue to phenylalanine decreases the tyrosine phosphorylation of the channel, measured biochemically, and also decreases (but does not completely eliminate) the modulation of $K_{v1.2}$ current that results from receptor stimulation (36).

The cloned $K_{v1.3}$ potassium channel is also modulated by tyrosine phosphorylation when it is expressed heterologously in HEK 293 cells. Treatment of HEK 293 cells with the membrane permeant tyrosine phosphatase inhibitor pervanadate results in a rapid and robust tyrosine phosphorylation of $K_{v1.3}$, which is accompanied by suppression of the channel current (23). The phosphorylation and suppression presumably result from the actions of constitutively active tyrosine kinase(s) endogenous to the HEK 293 cells. The suppression by pervanadate can be eliminated by mutation of one tyrosine residue, in the presumed intracellular carboxyl terminal domain of the channel, to phenylalanine (23). Biochemical analysis confirms that this tyrosine residue is a major target for phosphorylation following pervanadate treatment (23). However, there are several other tyrosine residues in $K_{v1.3}$ that lie within good consensus sequences for tyrosine kinases and that are phosphorylated when the channel is coexpressed with the constitutively active tyrosine kinase v-Src (23,38) or the receptor tyrosine kinase epidermal growth factor (EGF) receptor (23,39). These tyrosine kinases also produce suppression of $K_{v1.3}$ current, as well as changes in channel inactivation and deactivation kinetics that may require phosphorylation of multiple tyrosine residues (38,39). Thus phosphorylation of different tyrosine residues on the same channel, by different tyrosine kinases, produces different functional consequences. This finding has interesting implications for integration of cell signaling at the level of a single ion channel. Phosphorylation of multiple serine/threonine residues by different protein kinases has also been observed with voltage-dependent sodium channels (40).

REGULATORY ENZYME COMPLEXES ASSOCIATED WITH ION CHANNELS: FUNCTIONAL EVIDENCE

There are now many studies, illustrated by those discussed in the preceding sections, that demonstrate unequivocally that potassium (and other) channels are indeed substrates for protein kinases and phosphatases, and that phosphorylation of channels can modulate their function. In fact, modulability by phosphorylation is so wide-

FIG. 2. The legend of the lonesome ion channel. Ion channel biophysicists have often preferred to think of potassium channels as sitting alone in the plasma membrane, functioning in isolation. Much work, some of which is summarized in this chapter, demonstrates that this view is incorrect.

spread that it can be thought of as a fundamental feature of ion channels, as intrinsic as such properties as voltage dependence, conduction, and selectivity. More recently, the idea has arisen that, in many instances, modulation may be mediated via a protein complex that includes an ion channel closely associated with the regulatory enzymes that are involved in its modulation. I will first summarize functional electrophysiological studies that support this idea and then discuss direct biochemical evidence for channel/kinase association.

The Legend of the Lonesome Channel

Protein–protein interactions are ubiquitous in biology. This statement will come as no surprise to those who study such phenomena as DNA transcription, which is dependent on the formation of a complex of interacting proteins. Similarly, it is well established that growth factor signaling involves a series of specific protein–protein interactions mediated by modular binding domains. It is also evident that ion channels of some kinds must interact with other proteins in order to function properly. For example, the clustering of nicotinic acetylcholine receptor/channels at the neuromuscular junction, which is essential for synapse formation, requires the formation of a multiprotein complex (41). In spite of this, there has been a tendency for workers in the potassium channel field, particularly biophysicists, to think of their favorite channels as loners (Fig. 2). Indeed, much effort has been expended and Nobel Prizes have been awarded for success in recording the activity of individual ion channel proteins. This is entirely appropriate, because the unique information provided by single channel recording has revolutionized our understanding of how potassium and other ion channels work. Nevertheless it is becoming clear that this picture is incomplete, that the legend of the lonesome potassium channel (Fig. 2) is a myth, and that potassium channels in fact are intimately associated with other proteins that are important for channel function (see Table 1). Because many of these other proteins—for example, auxiliary subunits and guanyl nucleotide binding proteins—are treated elsewhere in this volume, I will focus on modulatory complexes that include ion channels together with protein kinases and phosphoprotein phosphatases.

A Channel/Protein Kinase Regulatory Complex

Application of protein kinases or phosphatases to potassium channels in detached membrane patches or artificial phospholipid bilayers has indeed provided evidence

TABLE 1. *Some examples of proteins that are tightly associated with potassium channel α subunits*

Protein	Representative reference(s)
Other α subunits	86
β subunits	57
G protein subunits	87,88
PSD-95	67,68
Protein kinases and phosphatases	27,43,48,50,85

that K_{Ca} channels are substrates for these enzymes. However, the modulation of at least one K_{Ca} channel, the slowly gating type II K_{Ca} channel from rat brain (42), turns out to be both more complex and more interesting, because the activity of this channel can be increased in the bilayer system by application of intracellular ATP without the addition of an exogenous protein kinase (43). ATP-sensitive potassium channels have been identified in a number of mammalian tissues, most notably pancreas and brain (44). These are ligand-gated ion channels, whose activity is generally inhibited by the reversible binding of the ligand ATP to an intracellular site on the channel. Although at first it seemed possible that the type II K_{Ca} channel from rat brain is such an ATP-gated channel, that is not the case. Nonhydrolyzable ATP analogs such as adenylylimidodiphosphate (AMPPNP), which can substitute for ATP as a ligand for the ATP-sensitive potassium channels, do not modulate the type II K_{Ca} channel (43). On the other hand, channel activity is modulated very effectively by ATPγS (43), another ATP analog that is hydrolyzable and can act as a substrate for protein kinases (45). These data suggested that the modulation of K_{Ca} channel activity by ATP might involve protein phosphorylation, and confirmation was provided by the demonstration that application of purified phosphoprotein phosphatase 1 to the channel can reverse the effects of ATP (43).

Since no exogenous protein kinase was added in these experiments, protein phosphorylation and channel modulation must result from the action of some endogenous protein kinase activity that accompanies the K_{Ca} channel in the bilayer. One possibility is that the membrane vesicles contain membrane-associated protein kinase activity, independent of the K_{Ca} channel, that enters the bilayer when a vesicle fuses. The protein kinase might then collide randomly with the channel and, in the presence of ATP, phosphorylate it and modulate its activity (Fig. 3A). However, measurements of protein kinase activities in these vesicle preparations (43) make it evident that this is not the case. The number of kinase molecules inserted into the bilayer upon vesicle fusion is so small that, based on diffusion considerations, it would take *several days* on average for the channel and kinase to find one another under the infinite dilution conditions of the bilayer. Since modulation of K_{Ca} channel activity is seen routinely within seconds after the addition of ATP (27,43), the random collision model (Fig. 3A) cannot be correct. More likely is the hypothesis that the channel and kinase are so intimately associated that they swim together in the bilayer protein phosphorylation and provide evidence that the endogenous protein kinase activity that mediates the modulation is PKC-like.

A Phosphoprotein Phosphatase Activity Is Also Part of the Regulatory Complex

The pseudo-substrate PKC inhibitor peptide was also used to determine whether persistent protein kinase activity is required for the modulation by ATP. This was tested by first treating a K_{Ca} channel with ATP to increase channel activity, and then adding the PKC inhibitor in the continued presence of ATP. Addition of the PKC inhibitor reverses the modulation (27), suggesting that there is a phosphoprotein phosphatase activity that also remains intimately associated with the channel in the bilayer (Fig. 3C). This leads to the hypothesis that, in the presence of ATP, channel activity is regulated by the balance between endogenous kinase and phosphatase activities; such a hypothesis predicts that the modulation by ATP should be greater in the presence of phosphatase inhibitors. Indeed, the effect of ATP is enhanced (27) by treatment with the relatively nonspecific phosphatase inhibitor microcystine, as well as by addition of inhibitor 1, a highly specific protein inhibitor of phosphoprotein phosphatase 1 (47). Thus the endogenous phosphatase activity that participates in the K_{Ca} channel regulatory complex is phosphoprotein phosphatase 1-like (27).

How Common Is Channel Association with Regulatory Enzymes?

Is the functional association of an ion channel with protein kinase and/or phosphatase activities a widespread phenomenon, or is it restricted to the type II K_{Ca} channel from brain. This question is especially relevant in view of the finding that the closely related brain type I K_{Ca} channel (42) does not appear to be modulated by ATP, although its activity can be regulated by exogenous PKA and phosphatase 2A (26). In fact, K_{Ca} channels from a number of different tissues and cell types, including pituitary tumor cells (48,49), peptidergic nerve terminals from the posterior pituitary (50,51), cortical neurons (52), and arterial (53) and gastric (54) smooth muscle cells, can be modulated in detached membrane patches by protein phosphorylation/dephosphorylation without the addition of exogenous protein kinase or phosphoprotein phosphatase. Although the detached patch experiments do not address the issue of direct *physical* association between the channel and the regulatory enzymes, because there are many other proteins present in such patches that could act as mediators, certainly they demonstrate that the *functional* association of K_{Ca} channels with membrane-delimited protein kinase and phosphatase activities occurs in cells of several different kinds. The extent to which this is true for other kinds of potassium channel is not clear, although a protein kinase activity copurifies with and modulates a dendrotoxin-binding protein that is probably a voltage-gated potassium channel (55). In the case of ligand-gated ion channels, it has been shown that modulation of glutamate receptor/channels in hippocampal neurons requires the tight association of PKA with the channels via a specific PKA anchoring protein (56). In addition, as I

A. Channel and kinase associate randomly

B. Channel/kinase complex

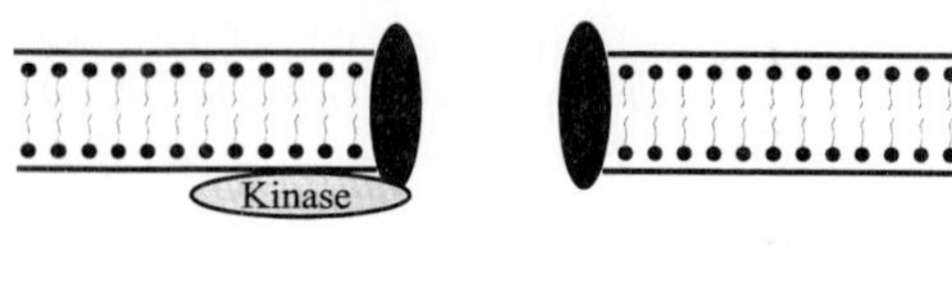

C. Channel/kinase/phosphatase modulatory complex

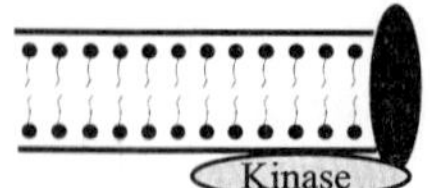

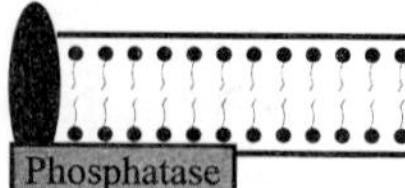

FIG. 3. A modulatory enzyme comp associated with the type II K(Ca) cha nel from rat brain. **(A)** Random intera tion, by lateral diffusion in the plane the plasma membrane, of an ion cha nel and a protein kinase that exist ind pendently of one another in th membrane. This scheme cannot ac count for the results with the type K(Ca) channel. **(B)** Tight binding be tween the ion channel and the protei kinase keeps them together, even un der conditions of infinite dilution of pro teins in artificial bilayers. **(C)** A phosphoprotein phosphatase is also part of the channel/kinase modulatory complex. (See Refs. 27,43).

(Fig. 3B), allowing channel phosphorylation and modulation immediately upon the addition of ATP.

The Endogenous Protein Kinase Activity Is PKC-like

Since the activity of the type II K_{Ca} channel from brain is decreased by additio of the catalytic subunit of PKA to the bilayer (26), the endogenous protein kinas activity that increases the activity of the channel cannot be PKA-like. In contras application of PKC causes an increase in channel activity (27), consistent with th possibility that the modulation by ATP involves an endogenous PKC-like kina activity. To test this further, a specific peptide inhibitor of PKC was used. This 1 amino-acid peptide, whose sequence corresponds to that of the pseudo-substr autoinhibitory domain of brain PKC (46), blocks the modulation of the K_{Ca} chan by ATP (27). In contrast, an analog peptide, in which a critical arginine required inhibition of PKC (46) is changed to glutamate, does not inhibit the modulation ATP (27). These results confirm further that the modulation by ATP results fi will summarize, many cloned potassium channels (as well as ligand-gated ion channels) contain specific amino acid sequence motifs that could bind to modular binding domains that are common in protein kinases, phosphatases, and other regulatory proteins.

A Separate Enzyme Protein or Intrinsic Enzymatic Activity?

One possible explanation for the findings just described is that protein kinases and phosphoprotein phosphatases constitute tightly associated subunits of at least some modulable ion channels. It is becoming evident that the pore-forming α subunits of many ion channels bind to auxiliary (β, γ, δ, etc.) subunits (57) and scaffolding proteins (58), which play critical roles in channel localization and activity (see Table 1). The possibility cannot be excluded, however, that regulatory enzymatic activities are intrinsic to the potassium channel α subunit itself (this is the reason I have taken pains to use the terms kinase *activity* and phosphatase *activity,* rather than simply kinase and phosphatase). This possibility might be tested by identifying potential kinase or phosphatase domains within channel amino acid sequences and employing mutagenesis strategies to determine whether these sequences participate in channel modulation. The first K_{Ca} channel to be cloned, that encoded by the *Drosophila Slopoke* gene (59,60), in fact contains within its primary amino acid sequence an ATP-binding domain of the sort found in protein kinases. We have described modulation of this *dSlo* channel, expressed in *Xenopus* oocytes, by ATPγS in the absence of exogenous protein kinase (61). However, the remarkable variability in the basal activity of heterologously expressed *dSlo* (62,63) led us to reassess this finding (63), and thus the intriguing idea that regulatory enzyme activity may be intrinsic to the channel α subunit remains to be tested.

REGULATORY ENZYME COMPLEXES ASSOCIATED WITH ION CHANNELS: BIOCHEMICAL EVIDENCE FOR DIRECT INTERACTION BETWEEN CHANNELS AND REGULATORY PROTEINS

All the studies summarized thus far involve electrophysiological measurements of ion channel activity and its modulation, and the conclusions about closely associated regulatory enzyme activities are inferred from these functional measurements. An alternative approach is to use biochemical and genetic methods to look for the direct interaction of regulatory proteins with ion channels. Two complementary strategies are becoming widely used for this purpose: the yeast two-hybrid screen and copurification or coprecipitation.

Strategies for Identifying Proteins That Interact with Ion Channels

The yeast two-hybrid screen is a powerful genetic tool to screen for protein–protein interactions (64). It is based on the fact that some yeast transcription factors are modular proteins consisting of separable DNA-binding and transcriptional-

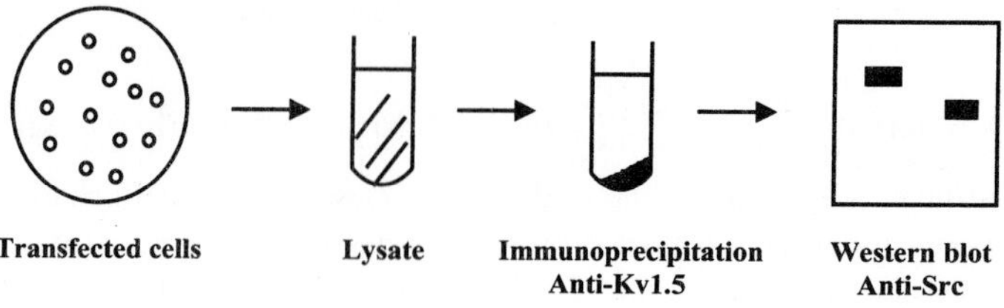

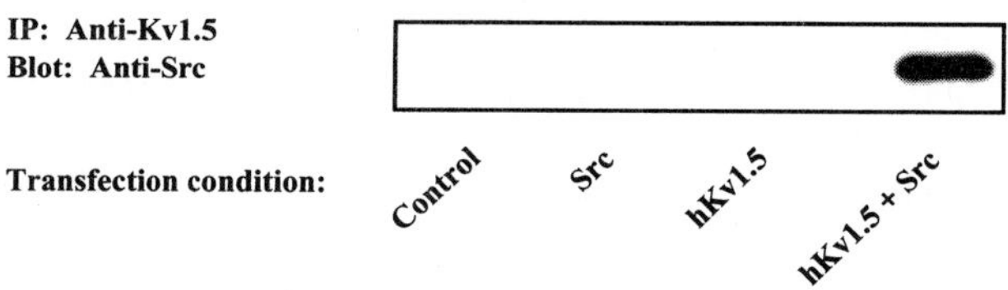

FIG. 4. Biochemical evidence for direct channel/kinase binding. **(A)** Coimmunoprecipitation allows testing for interactions between two proteins for which specific antibodies are available. **(B)** When HEK 293 cells are cotransfected with cDNA constructs for human $K_{v1.5}$ (hKv1.5) and Src, the channel and the kinase coimmunoprecipitate. Modified with permission from Ref. 85, Holmes *et al.*, 1996. Copyright 1996 American Association for the Advancement of Science.

activation domains. The screen is set up so that when these domains are expressed as separate proteins in the yeast, transcription of some gene(s) critical for yeast survival cannot occur. However, when the separate domains are expressed as fusion proteins with portions of other proteins, and these other proteins interact with one another, hence bringing the two modular transcription factor domains together, the yeast will survive and the interacting proteins can be cloned and identified. The two-hybrid screen is relatively nonstringent, and thus is capable of detecting relatively weak and transient interactions between proteins. A major advantage is that no *a priori* information about the interacting proteins is required, and thus the screen can be used to identify novel binding partners for ion channels (or other proteins).

The most spectacular success to date of the two-hybrid screen in the realm of ion channels has come in defining the roles of several abundant synaptic proteins, including PSD-95 (65) and SAP-102 (66), in the binding and clustering of both voltage-gated potassium channels and glutamate receptor/channels (66–71). Although a discussion of this emerging (and fascinating) story (58) is well beyond the scope of this chapter, it is of interest that channel binding is mediated by well-defined modular protein interaction domains. More specifically, a short conserved amino acid sequence present at the carboxyl termini of many different channels (67) can bind to specific domains, called PDZ domains (72), several copies of which are present in the various synaptic proteins. Thus far the yeast two-hybrid screen has not identified protein kinases or phosphatases that bind to channels; but it certainly is conceivable that some kinases, phosphatases, or other regulatory enzymes will be found to contain PDZ domains or other modular binding domains (to be discussed) that can interact with specific sequences in ion channel proteins. In any event, phosphorylation can influence channel binding to PDZ domains; it has been found that the interaction of PSD-95 with an inward rectifier potassium channel can be disrupted by PKA phosphorylation of a serine residue, located in the carboxyl terminal region of the channel that is critical for binding to the PDZ domain of PSD-95 (73).

In its most commonly used form, coimmunoprecipitation, the coprecipitation strategy is more stringent but also more limited than the yeast two-hybrid screen. It involves immunoprecipitation of one protein—for example, an ion channel—with one specific antibody, followed by probing of the immunoprecipitate on a Western blot with another specific antibody that recognizes another protein—for example, a protein kinase or phosphatase (Fig. 4A). The approach is more stringent because the immunoprecipitation is often done in detergent-containing solution (necessary when dealing with intrinsic membrane proteins), conditions under which only tight-binding partners are likely to stay together. It is more limited because it is restricted to known proteins for which specific antibodies are available. In spite of this limitation, however, coimmunoprecipitation (together with the related strategies copurification and fusion protein precipitation, discussed later) is providing novel information about the direct association of ion channels with protein kinases.

Protein Kinase Activity Copurifies with Some Ion Channels

Although potassium channels are rare membrane proteins, making biochemical purification of channel protein from native tissues notoriously difficult, toxins that bind tightly to the channel have been used in several cases as reagents for channel purification and assay. As mentioned earlier, a dendrotoxin-binding protein that might be a subunit of a voltage-gated potassium channel has been isolated, and a protein kinase activity that may modulate channel activity copurifies with this dendrotoxin-binding protein (55). Another kind of ion channel, the inositol 1,4,5-trisphosphate (IP_3) receptor/calcium channel, also copurifies with a protein kinase activity that autophosphorylates the channel (74). The IP_3 receptor/channel cannot be separated from the kinase activity through several independent purification strategies, and the phosphorylating activity remains associated with the receptor/channel even after SDS gel electrophoresis (74), suggesting that this is a true autophosphorylation and that the kinase activity is intrinsic to the receptor/channel protein itself.

Coimmunoprecipitation and Fusion Protein Precipitation Strategies Demonstrate That Tyrosine Kinases Bind to Some Ion Channels

A fundamental theme in cell biology is that protein–protein interactions mediated by modular binding domains play important roles in cell regulation and signaling. An example of particular relevance to ion channels is the PDZ domain interaction with different channels, discussed earlier, which appears to be critical for the molecular organization of the synapse (58). Among the most thoroughly studied of these modular binding domains are the so-called Src homology 2 (SH2) and Src homology 3 (SH3) domains, first defined in the Src tyrosine kinase, and subsequently found in a large number of tyrosine kinases and other signaling proteins (75). Indeed, PSD-95 (65) and SAP-102 (66) are among the proteins that contain SH3 domains, in addition to their PDZ domains already noted.

SH2 and SH3 domains have different binding specificities (75,76). SH2 domains bind to short sequences surrounding tyrosine residues, but only if the tyrosine is phosphorylated. In contrast, SH3 domains bind to short proline-rich regions in target proteins, with no requirement for phosphorylation, and the specific proline-rich sequences preferred by different SH3 domain containing proteins are beginning to be defined (77–79). These different binding specificities can give rise to sequential regulation of a target protein by two different tyrosine kinases (75). The first kinase may bind, via its SH3 domain, to a proline-rich region in the target protein, and once bound may phosphorylate the target on one or more tyrosine residues. The SH2 domain of the second kinase may then bind to the sequence around this newly phosphorylated tyrosine, subsequently phosphorylating another tyrosine residue in the target protein. Such sequential recruitment of kinases and multiple phosphorylations are common features of growth factor signaling (75).

One example of the direct interaction of ion channels with binding domains in protein kinases comes from work with the nicotinic acetylcholine receptor/channel from *Torpedo* electric organ and muscle. Two members of the Src family of tyrosine kinases, Fyn and Fyk, coimmunoprecipitate with the phosphorylated nicotinic acetylcholine receptor/channel from *Torpedo* electric organ, apparently via an SH2 domain interaction (80,81). Furthermore, a combined fusion protein precipitation and coimmunoprecipitation strategy demonstrates that the large cytoplasmic loop of the β subunit of the receptor/channel from mammalian myotubes specifically binds Src (82). This binding is not dependent on tyrosine phosphorylation, hence is not likely to involve an SH2 domain interaction; nor does there appear to be an appropriate proline-rich motif in this region of the β subunit that could bind to the Src SH3 domain. Hence the molecular basis for the binding of Src to the muscle nicotinic acetycholine receptor/channel remains to be determined (82).

Src also binds directly to a human voltage-gated potassium channel. The human isoform of the $K_{v1.5}$ channel (hKv1.5) (83) contains, near its amino terminal, two repeats of the proline-rich amino acid sequence motif that is most preferred by the SH3 domain of Src (78). Interestingly, the rat $K_{v1.5}$ does not contain this preferred sequence (84). When Src and hKv1.5 are expressed together in HEK 293 cells and coimmunoprecipitation experiments are carried out (Fig. 4), the kinase and the channel coimmunoprecipitate (85). A similar tight interaction between native Src and native hKv1.5 is seen in human myocardial tissue (85). In contrast, the homologous rat $K_{v1.5}$ does not coimmunoprecipitate with Src, suggesting a role for the specific hKv1.5 proline-rich motif in the interaction of Src with the human channel.

The sequence requirements for the binding were explored further by means of fusion protein precipitation. A GST fusion protein encoding the Src SH3 domain is able to bind to and precipitate hKv1.5 (but not rat $K_{v1.5}$), and this binding is disrupted if the fusion protein is preabsorbed with a peptide corresponding in sequence to the proline-rich motif of hKv1.5 (85). These findings provide strong evidence that the binding results from interaction of the SH3 domain of Src with the proline-rich motif of hKv1.5. No intermediary proteins are required for the interaction, because the GST–Src–SH3 fusion protein can bind directly to the hKv1.5 channel protein when the latter is separated from other proteins by SDS gel electrophoresis. Finally, the hKv1.5 channel is robustly tyrosine phosphorylated, and its activity is modulated, when it is coexpressed with Src (85).

How rare is this kind of direct binding of channel and kinase? Is it restricted to one species variant of one particular voltage-gated potassium channel? In fact, a search of the sequences of cloned ion channels reveals that proline-rich motifs show up with astonishing frequency, in potassium channels of various families as well as in glutamate receptor/channels (Table 2, ref. 85). Although few of these proline-rich motifs are identical to the preferred Src-binding sequences, SH3 domains that prefer other proline-rich motifs are a common feature of signaling proteins (78,79). Thus it seems likely that channel interaction with signaling proteins via modular binding domains is a widespread phenomenon.

PHYSIOLOGICAL CONSEQUENCES OF CHANNEL PHOSPHORYLATION AND CHANNEL/KINASE BINDING

For more than two decades the focus of research on modulation of ion channels has been to demonstrate, often in the face of skepticism, that modulation does in fact exist. As illustrated by this volume, the evidence is now overwhelming that virtually all ion channels are subject to modulation, and that there exists a stunning diversity of modulatory mechanisms. Indeed, even when one considers only protein phosphorylation as a modulatory mechanism, the diversity of the phenomenology observed remains astonishing. Different ion channels may have their activities increased or decreased by phosphorylation, a variety of biophysical properties are subject to modulation, and the same channel may even be modulated in different directions by phosphorylation on different residues by different protein kinases.

TABLE 2. *Proline-rich motifs in mammalian ion channels*

Channels	Proline-rich motifs[a]
Potassium channels	
Shaker-related subfamily	
Kv1.2 (human)	^{16}PGHPQDTYDP, ^{436}PKIPSSP
Kv1.3 (rat)	^{35}RYEPLPPALP
Kv1.3 (human)	^{43}RYEPLPPSLP
Kv1.5 (rat)	^{65}RPLPPMA
Kv1.5 (human)	^{65}RPLPPLPDPGVRPLPPLPEELPRRP
Kv1.6 (human)	^{149}KPLPSQP
Shab-related subfamily	
Kv2.1 (rat)	^{13}PPEPMEIV, ^{562}PSPVAPLP
	^{576}PTPLLP, ^{596}PLPTSPKFRP
Kv2.2 (rat)	^{21}PPEPVEII, ^{619}PLTPVP
Shaw-related subfamily	
Kv3.1 (rat)	^{461}PRPPQLGSP
Kv3.2 (rat)	^{56}PLPPPLSPPPRPPPLSPVP
	^{498}PPAPLASSP, ^{550}PPLSPPERLP
Kv3.3 (rat)	^{33}PAPTPQPPESSPPPLLPP
	^{565}PRPPQPGSPNYCKPDPPPPPPHP
	^{597}PPPPITPP, ^{615}PPGPHTHP
	^{638}PPLPAPGEPCP
Kv3.4 (human)	^{551}PQWPREFPNGP
Shal-related subfamily	
Kv4.1 (mouse)	^{25}QPLPPAP, ^{606}PTPPANTPDESQPSSP
Kv4.2 (rat)	^{21}PAVSPMPAPP
Kv4.3 (rat)	^{21}PVANCPMPLAP
Small K(Ca) channel subfamily	
hSK1 (human)	^{69}PARPSPGSPRGQP, ^{526}RPPPPPLPPRPGPGP
Glutamate receptor/channels	
NMDA receptor/channel family	
NMDA2C (mouse)	^{942}PGPPGQPSPSGWRPP
	^{962}PLARRAPQPPARPQP
NMDA2D (rat)	^{900}PPPAKPPPPPQPLPRPPPGPAP
	^{926}PLSPPTTQPPQKPPPPGFPSPPAPP

[a]Proline-rich motifs in ion channels obtained from sequence databases (see note 9 in Ref. 85 for details).

Source: Modified from Ref. 85.

The only unifying theme to date is that a cell will use any and all of the tricks at its disposal to regulate the properties of its ion channels. Why is this the case? What are the physiological advantages of ready reconfiguration of neuronal electrical properties? The honest answer to this question is that we do not really know; but one can guess that modulation has evolved because it provides a way for a nervous system to respond rapidly and flexibly to a rich and varied environment. Although I emphasized early on that modulatory changes fall into just a few simple categories (Fig. 1), subtle differences in response that are afforded by a diversity of mechanisms must be critical for survival.

In much of this chapter, I have focused on studies that emphasize that at least some (and perhaps many) ion channels bind very tightly to signaling proteins, and remain bound even under such harsh experimental conditions as detergent extraction or preparation of isolated membranes for bilayer reconstitution. Why do channels go to such lengths to associate with partner proteins? What are the physiological consequences of such association? This question becomes all the more pressing with the realization that tight binding may not be an absolute requirement for channel phosphorylation and modulation. For example, the $K_{v1.3}$ potassium channel, which contains a proline-rich motif (Table 2) that does not appear to be preferred by Src (78), nevertheless is phosphorylated robustly on tyrosine residues when it is coexpressed with Src (23). Furthermore, this tyrosine phosphorylation is accompanied by a profound suppression of $K_{v1.3}$ current, as well as by modulation of the channel's inactivation and deactivation kinetics (38). Thus, tight binding may not be *necessary* for channel phosphorylation and modulation, but it may nevertheless influence such fundamental physiological features of signaling as target specificity and time course. Elucidation of the physiological significance of channel association with signaling proteins is an important task for the future.

ACKNOWLEDGMENT

Work from my laboratory described in this chapter was supported by grants from the U.S. National Institutes of Health.

REFERENCES

1. Tsien RW: Cyclic AMP and contractile activity in heart. *Adv Cyclic Nucl Res* 1977;8:363–420.
2. Brunelli M, Castellucci V, Kandel ER: Synaptic facilitation and behavioral sensitization in *Aplysia:* possible role of serotonin and cyclic AMP. *Science* 1976;194:1176–1181.
3. Treistman SN, Levitan IB: Alteration of electrical activity in molluscan neurones by cyclic nucleotides and peptide factors. *Nature* 1976;261:62–64.
4. Kaczmarek LK, Strumwasser F: A voltage-clamp analysis of currents underlying cyclic AMP-induced membrane modulation in isolated peptidergic neurons of *Aplysia. J Neurophysiol* 1984;52: 340–349.
5. Castellucci VF, Kandel ER, Schwartz JH, Wilson FD, Nairn AC, Greengard P: Intracellular injection of the catalytic subunit of cyclic AMP-dependent protein kinase simulates facilitation of transmitter release underlying behavioral sensitization in *Aplysia. Proc Natl Acad Sci U S A* 1980;77:7492–7496.
6. Kaczmarek LK, Jennings KR, Strumwasser F, Nairn AC, Walter U, Wilson FD, Greengard P: Microinjection of catalytic subunit of cyclic AMP-dependent protein kinase enhances calcium action potentials of bag cell neurons in cell culture. *Proc Natl Acad Sci U S A* 1980;77:7487–7491.
7. DePeyer JE, Cachelin AB, Levitan IB, Reuter H: Ca^{2+}-activated K^+ conductance in internally perfused snail neurons is enhanced by protein phosphorylation. *Proc Natl Acad Sci U S A* 1982;79: 4207–4211.
8. Brum G, Flockerzi V, Hofmann F, Osterrieder W, Trautwein W: Injection of catalytic subunit of cAMP-dependent protein kinase into isolated cardiac myocytes. *Pfluegers Arch* 1983;398:147–154.
9. Adams WB, Levitan IB: Intracellular injection of protein kinase inhibitor blocks the serotonin-induced increase in K^+ conductance in *Aplysia* neuron R15. *Proc Natl Acad Sci U S A* 1982;79: 3877–3880.
10. Castellucci VF, Nairn A, Greengard P, Schwartz JH, Kandel ER: Inhibitor of adenosine 3':5'-monophosphate-dependent protein kinase blocks presynaptic facilitation in *Aplysia. J Neurosci* 1982;2: 1673–1681.
11. Kennedy MB: Experimental approaches to understanding the role of protein phosphorylation in the regulation of neuronal function. *Annu Rev Neurosci* 1983;6:493–525.
12. DeRiemer SA, Strong JA, Albert KA, Greengard P, Kaczmarek LK: Enhancement of calcium current in *Aplysia* neurones by phorbol ester and protein kinase C. *Nature* 1985;313:313–316.

13. Huganir RL, Delcour AH, Greengard P, Hess GP: Phosphorylation of the nicotinic acetylcholine receptor regulates its rate of desensitization. *Nature* 1986;321:774–776.
14. Armstrong D, Eckert R: Voltage-activated calcium channels that must be phosphorylated to respond to membrane depolarization. *Proc Natl Acad Sci U S A* 1987;84:2518–2522.
15. Hopfield JF, Tank DW, Greengard P, Huganir RL: Functional modulation of the nicotinic acetylcholine receptor by tyrosine phosphorylation. *Nature* 1988;336:677–680.
16. Huganir RL, Greengard P: cAMP-dependent protein kinase phosphorylates the nicotinic acetylcholine receptor. *Proc Natl Acad Sci U S A* 1983;80:1130–1134.
17. Huganir RL, Miles K, Greengard P: Phosphorylation of the nicotinic acetylcholine receptor by an endogenous tyrosine-specific protein kinase. *Proc Natl Acad Sci U S A* 1984;81:6968–6972.
18. Rossie S, Catterall WA: Cyclic AMP-dependent phosphorylation of voltage-sensitive sodium channels in primary cultures of rat brain neurons. *J Biol Chem* 1987;262:12735–12744.
19. Flockerzi V, Oeken HJ, Hofmann F, Pelzer D, Cavalie A, Trautwein W: Purified dihydropyridine-binding site from skeletal muscle t-tubules is a functional calcium channel. *Nature* 1986;323:66–68.
20. Matteson DR, Deutsch C: K channels in T lymphocytes: a patch clamp study using monoclonal antibody adhesion. *Nature* 1984;307:468–471.
21. Cahalan MD, Chandy KG, DeCoursey TE, Gupta S: A voltage-gated potassium channel in human T lymphocytes. *J Physiol* 1985;358:197–237.
22. Cai Y-C, Douglass J: *In vivo* and *in vitro* phosphorylation of the T lymphocyte type *n* (Kv1.3) potassium channel. *J Biol Chem* 1993;268:23720–23727.
23. Holmes TC, Fadool DA, Levitan IB: Tyrosine phosphorylation of the Kv1.3 potassium channel. *J Neurosci* 1996;16:1581–1590.
24. Shuster MJ, Camardo JS, Siegelbaum SA, Kandel ER: Cyclic AMP-dependent protein kinase closes the serotonin-sensitive K^+ channels of *Aplysia* sensory neurones in cell-free membrane patches. *Nature* 1985;313:392–395.
25. Ewald D, Williams A, Levitan IB: Modulation of single Ca^{2+}-dependent K^+ channel activity by protein phosphorylation. *Nature* 1985;315:503–506.
26. Reinhart PH, Chung SK, Martin BL, Brautigan DL, Levitan IB: Modulation of calcium-activated potassium channels from rat brain by protein kinase A and phosphatase 2A. *J Neurosci* 1991;11:1627–1635.
27. Reinhart PH, Levitan IB: Kinase and phosphatase activities intimately associated with a reconstituted calcium-dependent potassium channel. *J Neurosci* 1995;15:4572–4579.
28. Miller C. *Ion channel reconstitution.* New York: Plenum, 1986.
29. Drain P, Dubin AE, Aldrich RW: Regulation of *Shaker* K^+ channel inactivation gating by the cAMP-dependent protein kinase. *Neuron* 1994;12:1097–1109.
30. Hoshi T, Zagotta WN, Aldrich RW: Biophysical and molecular mechanisms of *Shaker* potassium channel inactivation. *Science* 1990;250:533–538.
31. Zagotta WN, Hoshi T, Aldrich RW: Restoration of inactivation in mutants of *Shaker* potassium channels by a peptide derived from ShB. *Science* 1990;250:568–571.
32. Covarrubias M, Wei A, Salkoff L, Vyas TB: Elimination of rapid potassium channel inactivation by phosphorylation of the inactivation gate. *Neuron* 1994;13:1403–1412.
33. Murrell-Lagnado RD, Aldrich RW: Interactions of amino terminal domains of *Shaker* K channels with a pore blocking site studied with synthetic peptides. *J Gen Physiol* 1993;102:949–975.
34. Jonas EA, Kaczmarek LK. Regulation of potassium channels by protein kinases. *Curr. Opin. Neurobiol.* 1996;6:318–323.
35. Wilson GF, Kaczmarek LK: Mode-switching of a voltage-gated cation channel is mediated by a protein kinase A-regulated tyrosine phosphatase. *Nature* 1993;366:433–438.
36. Huang X-Y, Morielli AD, Peralta EG: Tyrosine kinase-dependent suppression of a potassium channel by the G protein-coupled m1 muscarinic acetylcholine receptor. *Cell* 1993;75:1145–1156.
37. Lev S, Moreno H, Martinez R et al: Protein tyrosine kinase PYK2 involved in Ca^{2+}-induced regulation of ion channel and MAP kinase functions. *Nature* 1995;376:737–745.
38. Fadool DA, Holmes TC, Berman K, Dagan D, Levitan IB: Tyrosine phosphorylation modulates current amplitude and kinetics of a neuronal voltage-gated potassium channel. *J Neurophysiol* 1997; 78:1563–1573.
39. Bowlby MR, Fadool DA, Holmes TC, Levitan IB: Modulation of the Kv1.3 potassium channel by receptor tyrosine kinases. *J Gen Physiol* 1997;110:601–610.
40. Li M, West JW, Numann R, Murphy BJ, Scheuer T, Catterall WA: Convergent regulation of sodium channels by protein kinase C and cAMP-dependent protein kinase. *Science* 1993;261:1439–1442.
41. Apel ED, Roberds SL, Campbell KP, Merlie JP: Rapsyn may function as a link between the acetylcholine receptor and the agrin-binding dystrophin-associated glycoprotein complex. *Neuron* 1995;15: 115–126.

42. Reinhart PH, Chung S, Levitan IB: A family of calcium-dependent potassium channels from rat brain. *Neuron* 1989;2:1031–1041.
43. Chung SK, Reinhart PH, Martin BL, Brautigan D, Levitan IB: Protein kinase activity closely associated with a reconstituted calcium-activated potassium channel. *Science* 1991;253:560–562.
44. Ashcroft FM: Adenosine 5'-triphosphate-sensitive potassium channels. *Annu Rev Neurosci* 1988;11: 97–118.
45. Gratecos D, Fischer EH: Adenosine 5'-*O*(3-thiotriphosphate) in the control of phosphorylase activity. *Biochem Biophys Res Commun* 1974;58:960–967.
46. House C, Kemp BE: Protein kinase C contains a pseudosubstrate prototype in its regulatory domain. *Science* 1987;238:1726–1728.
47. Nairn AC, Shenolikar S: The role of protein phosphatases in synaptic transmission, plasticity and neuronal development. *Curr Opin Neurobiol* 1992;2:296–301.
48. White RE, Schonbrunn A, Armstrong DL: Somatostatin stimulates Ca^{2+}-activated K^+ channels through protein dephosphorylation. *Nature* 1991;351:570–573.
49. White RE, Lee AB, Shcherbatko AD, Lincoln TM, Schonbrunn A, Armstrong DL: Potassium channel stimulation by natriuretic peptides through cGMP-dependent dephosphorylation. *Nature* 1993;361: 263–266.
50. Bielefeldt K, Jackson MB: Intramolecular and intermolecular enzymatic modulation of ion channels in excised membrane patches. *Biophys J* 1994;66:1904–1914.
51. Bielefeldt K, Jackson MB: Phosphorylation and dephosphorylation modulate a Ca^{2+}-activated K^+ channel in rat peptidergic nerve terminals. *J Physiol* (Lond) 1994;475:241–254.
52. Lee K, Rowe ICM, Ashford MLJ: Characterization of an ATP-modulated large conductance Ca^{2+}-activated K^+ channel present in rat cortical neurones. *J Physiol* (Lond) 1995;488:319–337.
53. Hartley SA, Kozlowski RZ: ATP increases Ca^{2+}-activated K^+ channel activity in isolated rat arterial smooth muscle cells. *Biochim Biophys Acta* 1996;1283:192–198.
54. Lee M-Y, Bang H-W, Lim I-J, Uhm D-Y, Rhee S-D: Modulation of large conductance calcium-activated K^+ channel by membrane-delimited protein kinase and phosphatase activities. *Pfluegers Arch* 1994;429:150–152.
55. Rehm H, Pelzer S, Cochet C et al: Dendrotoxin-binding brain membrane protein displays a K^+ channel activity that is stimulated by both cAMP-dependent and endogenous phosphorylations. *Biochemistry* 1989;28:6455–6460.
56. Rosenmund C, Carr DW, Bergeson SE, Nilaver G, Scott JD, Westbrook GL: Anchoring of protein kinase A is required for modulation of AMPA/kainate receptors on hippocampal neurons. *Nature* 1994;368:853–856.
57. Isom LL, De Jongh KS, Catterall WA: Auxiliary subunits of voltage-gated ion channels. *Neuron* 1994;12:1183–1194.
58. Sheng M: PDZs and receptor/channel clustering: rounding up the latest suspects. *Neuron* 1996;17: 575–578.
59. Atkinson NS, Robertson GA, Ganetzky B: A component of calcium-activated potassium channels encoded by the *Drosophila slo* locus. *Science* 1991;253:551–555.
60. Adelman JP, Shen K, Kavanaugh MP et al: Calcium-activated potassium channels expressed from cloned complementary DNAs. *Neuron* 1992;9:209–216.
61. Esguerra M, Wang J, Foster CD, Adelman JP, North RA, Levitan IB: Phosphorylation at a specific serine residue by a functionally associated protein kinase modulates a cloned calcium-dependent potassium channel. *Nature* 1994;369:563–565.
62. Silberberg SD, Lagrutta A, Adelman JP, Magleby KL: Wanderlust kinetics and variable Ca^{2+}-sensitivity of Drosophila, a large conductance Ca^{2+}-activated K^+ channel, expressed in oocytes. *Biophys J* 1996;70:2640–2651.
63. Bowlby MR, Levitan IB: Kinetic variability and modulation of *dSlo,* a cloned calcium-dependent potassium channel. *Neuropharmacology* 1996;35:867–875.
64. Fields S, Song O: A novel genetic system to detect protein–protein interactions. *Nature* 1989;340: 245–246.
65. Cho K-O, Hunt CA, Kennedy MB: The rat brain postsynaptic density fraction contains a homolog of the *Drosophila* discs—large tumor suppressor protein. *Neuron* 1992;9:929–942.
66. Müller BM, Kistner U, Kindler S et al: SAP102, a novel postsynaptic protein that interacts with NMDA receptor complexes *in vivo. Neuron* 1996;17:255–265.
67. Kornau H-C, Schenker LT, Kennedy MB, Seeburg PH: Domain interaction between NMDA receptor subunits and the postsynaptic density protein PSD-95. *Science* 1995;269:1737–1740.
68. Kim E, Neithammer M, Rothschild A, Jan YN, Sheng M. Clustering of *Shaker*-type K^+ channels by interaction with a family of membrane-associated guanylate kinases. *Nature* 1995;378:85–88.

69. Niethammer M, Kim E, Sheng M: Interaction between the C terminus of NMDA receptor subunits and multiple members of the PSD-95 family of membrane-associated guanylate kinases. *J Neurosci* 1996;16:2157–2163.
70. Kim E, Cho K-O, Rothschild A, Sheng M: Heteromultimerization and NMDA receptor-clustering activity of chapsyn-110, a member of the PSD-95 family of proteins. *Neuron* 1996;17:103–113.
71. Lau L-F, Mammen A, Ehlers MD, Kindler S, Chung WJ, Garner CC, Huganir RL: Interaction of the *N*-methyl-D-aspartate receptor complex with a novel synapse-associated protein, SAP102. *J Biol Chem* 1996;271:21622–21628.
72. Doyle DA, Lee A, Lewis J, Kim E, Sheng M, MacKinnon R: Crystal structures of a complexed and peptide-free membrane protein-binding domain: Molecular basis of peptide recognition by PDZ. *Cell* 1996;85:1067–1076.
73. Cohen NA, Brenman JE, Snyder SH, Bredt DS: Binding of the inward rectifier K^+ channel Kir 2.3 to PSD-95 is regulated by protein kinase A phosphorylation. *Neuron* 1996;17:759–767.
74. Ferris CD, Cameron AM, Bredt DS, Huganir RL, Snyder SH: Autophosphorylation of inositol 1,4,5-trisphosphate receptors. *J Biol Chem* 1992;267:7036–7041.
75. Pawson T: Protein modules and signalling networks. *Nature* 1995;373:573–580.
76. Cohen GB, Ren R, Baltimore D: Modular binding domains in signal transduction proteins. *Cell* 1995; 80:237–248.
77. Ren R, Mayer BJ, Cicchetti P, Baltimore D: Identification of a ten-amino acid proline-rich SH3 binding site. *Science* 1993;259:1157–1161.
78. Rickles RJ, Botfield MC, Weng Z, Taylor JA, Green OM, Brugge JS, Zoller MJ: Identification of Src, Fyn, Lyn, P13K and Abl SH3 domain ligands using phage display libraries. *EMBO J* 1994;13:5598–5604.
79. Alexandropoulos K, Cheng G, Baltimore D: Proline-rich sequences that bind to src homology 3 domains with individual specificities. *Proc Natl Acad Sci U S A* 1995;92:3110–3114.
80. Swope SL, Huganir RL: Binding of the nicotinic acetylcholine receptor to SH2 domains of Fyn and Fyk protein tyrosine kinases. *J Biol Chem* 1994;269:29817–29824.
81. Swope SL, Qu Z, Huganir RL: Phosphorylation of the nicotinic acetylcholine receptor by protein tyrosine kinases. *Ann N Y Acad Sci* 1995;757:197–214.
82. Fuhrer C, Hall ZW: Functional interaction of Src family kinases with the acetylcholine receptor in C2 myotubes. *J Biol Chem* 1996;271:32474–32481.
83. Tamkun MM, Knoth KM, Walbridge JA, Kroemer H, Roden DM, Glover DM: Molecular cloning and characterization of two voltage-gated K^+ channel cDNAs from human ventricle. *FASEB J* 1991; 5:331–337.
84. Swanson R, Marshall J, Smith JS et al: Cloning and expression of cDNA and genomic clones encoding three delayed rectifier potassium channels in rat brain. *Neuron* 1990;4:929–939.
85. Holmes TC, Fadool DA, Ren R, Levitan IB: Direct association of Src tyrosine kinase with a human potassium channel mediated by SH3 domain. *Science* 1996;274:2089–2091.
86. Liao YJ, Jan YN, Jan LY: Heteromultimerization of G-protein-gated inwardly rectifying K^+ channel proteins GIRK1 and GIRK2 and their altered expression in *weaver* brain. *J Neurosci* 1996;16:7137–7150.
87. Huang C-L, Slesinger PA, Casey PJ, Jan YN, Jan LY: Evidence that direct binding of $G_{\beta\Gamma}$ to the GIRK1 G protein-gated inwardly rectifying K^+ channel is important for channel activation. *Neuron* 1995;15:1133–1143.
88. Krapivinsky G, Krapivinsky L, Wickman K, Clapham DE: $G_{\beta\Gamma}$ binds directly to the G protein-gated K^+ channel, I_{KACh}. *J Biol Chem* 1995;270:29059–29062.
89. Levitan IB, Kaczmarek LK: *The neuron: Cell and molecular biology.* 2nd ed. New York: Oxford University Press, 1997.

Advances in Second Messenger and Phosphoprotein Research, Vol. 33

1040-7952/99 $30.00

2

Regulation of Voltage-Sensitive Sodium and Calcium Channels by Phosphorylation

Sandra Rossie

Department of Biochemistry, Purdue University, West Lafayette, Indiana 47907

INTRODUCTION

Voltage-sensitive ion channels are a superfamily of structurally related integral membrane proteins that control the selective flow of ions down their electrochemical gradient in response to changes in membrane potential. These channels are responsible for the initiation and propagation of regenerative action potentials in neurons and in skeletal and cardiac muscle, and they control the increase in intracellular calcium that couples electrical excitation to contraction or secretion. Thus, changes in the individual response properties of voltage-sensitive ion channels can profoundly affect vital physiological processes such as the transmission of neural and hormone signals and muscle contraction.

In its simplest view, a voltage-sensitive ion channel can be envisioned as existing in three conformations: closed, open, and inactivated (Fig. 1). These conformations exist in dynamic equilibrium, and transitions between states are controlled by changes in membrane potential. Many voltage-sensitive channels are also modulated by direct phosphorylation and by the direct binding of G protein subunits or calcium. These modulatory events often alter the rate at which a channel undergoes a conformational change from one state to another, or the sensitivity of these transitions to membrane potential. The time scale for a multistep, second messenger–controlled phosphorylation event is typically longer than the time required for a single opening and closing of many voltage-sensitive ion channels. Thus, phosphorylation is not thought to cause individual opening or closing events. By analogy to the more familiar paradigm of enzyme kinetics, a change in the peak current is similar to changing an enzyme's V_{max}, whereas shifting the voltage dependence of activation to more negative or more positive membrane potentials is similar to decreasing or increasing the K_m of an enzyme, respectively. Fundamental properties of channels such as ion selectivity and unitary conductance are not usually altered by modulatory events.

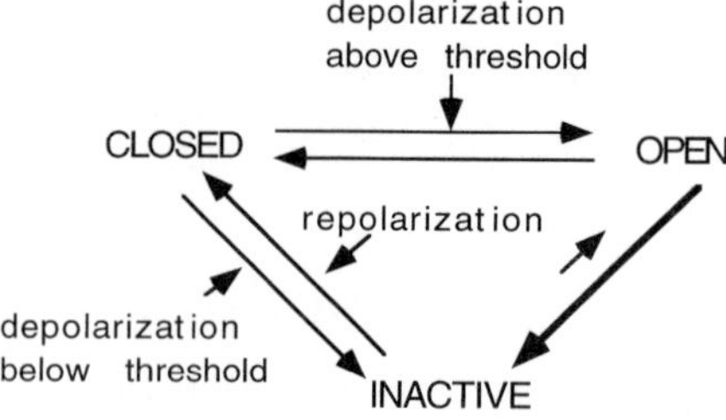

FIG. 1. Transitions between the closed, open, and inactive conformations of voltage-sensitive sodium channels as a function of membrane potential.

This chapter focuses on the modulation of voltage-sensitive sodium channels and calcium channels by direct phosphorylation events. Among the topics considered are the regulation of sodium channels from brain and skeletal and cardiac muscle by cAMP-dependent protein kinase (PKA) and protein kinase C (PKC)-mediated phosphorylation. Recent efforts to identify phosphatases and physiologic stimuli controlling sodium channel phosphorylation also are reviewed. In the case of voltage-sensitive calcium channels, regulation of L-type channels from cardiac and skeletal muscle by PKA-mediated phosphorylation is discussed. These channels currently represent the best understood examples of voltage-sensitive channels controlled by direct phosphorylation. Several general reviews on the structure and regulation of these channels have been published recently (1–4).

THE STRUCTURE OF VOLTAGE-SENSITIVE ION CHANNELS

The voltage-sensitive sodium channel contains a large α subunit that forms a pore and a voltage-sensing element (reviewed in 5). The α subunit contains four homologous domains, I–IV (Fig. 2). Each homologous domain is thought to contain six transmembrane α helices and one additional membrane-associated loop that partially spans the membrane and forms one wall of the channel pore. The fourth transmembrane segment in each homologous domain contains charged residues that are responsible for the voltage-sensitive conformational change that takes place during channel activation. The α1 subunit of voltage-sensitive calcium channels is thought to have a similar overall structure, whereas voltage-dependent potassium channels are composed of four separate subunits, each subunit being the equivalent of one homologous domain. In addition to their major pore-forming subunits, voltage-sensitive ion channels have auxiliary subunits that influence their response properties.

VOLTAGE-SENSITIVE SODIUM CHANNELS

Voltage-sensitive sodium channels are responsible for the rising phase of the action potential in most neurons and in skeletal and cardiac muscle. Thus, altering their response to membrane depolarization will affect the excitability of these tissues. At

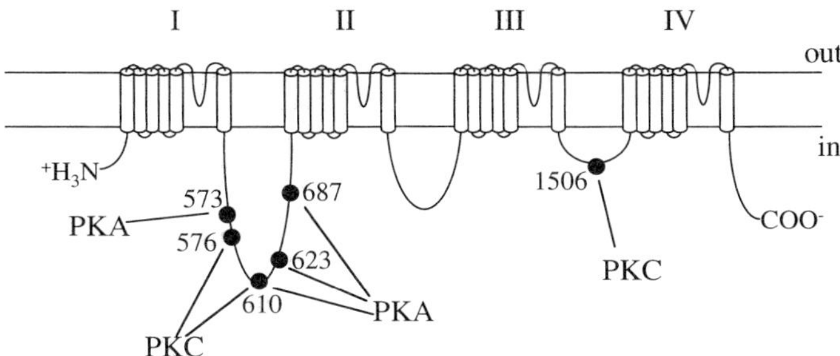

FIG. 2. A schematic representation of the α subunit of the voltage-sensitive sodium channel, illustrating four homologous domains, I–IV, each of which is thought to contain six α helices and one membrane-associated loop thought to form the channel pore. The fourth α helix within each homologous domain contains charged residues and represents the voltage sensor. Identified sites of phosphorylation by PKA (21) and PKC (28), for rat brain type IIA α subunits, are depicted.

a very negative membrane holding potential, voltage-sensitive sodium channels are closed. A stimulus that depolarizes the membrane potential beyond a critical threshold causes the channels to undergo a conformational change, opening the sodium selective pore (Fig. 1). After opening, sodium channels rapidly become inactivated, a state in which they are closed and resistant to reopening even though the membrane remains depolarized. Repolarization of the membrane permits sodium channels to return to the closed state, from which they can again be activated. At less negative membrane potentials, a portion of the closed channels will be inactivated, and consequently fewer channels will be available to open when a depolarizing stimulus beyond threshold occurs, resulting in less current. At typical resting potentials for neurons and muscle cells, some portion of the sodium channels are inactive.

Cardiac and skeletal muscle cells and brain neurons express distinct forms of the sodium channel α subunit, which differ in their electrophysiological characteristics, auxiliary subunits, and specific cellular localization (6). The α subunit is phosphorylated by PKA and PKC. Cyclic AMP dependent phosphorylation sites and the effects of PKA phosphorylation are different for brain and cardiac and skeletal muscle sodium channels. In the case of PKC, specific phosphorylation sites have been defined only in brain α subunits. Although one of these sites is conserved in α subunits from all three tissues, it does not appear to serve the same role in all types of sodium channel.

RAT BRAIN SODIUM CHANNELS

Several types of α subunit are expressed in brain in a developmentally and spatially distinct manner (6). They are highly similar in sequence and contain multiple phosphorylation sites for PKA and PKC that are conserved in all the brain forms, types I, II, IIA, and III (7–9). Brain sodium channels also contain auxiliary β1 and β2 subunits that influence the electrophysiological properties of the α subunit in coexpression studies (10–12). The β subunits do not appear to be targets of phosphorylation.

Phosphorylation by PKA

Effect of PKA Phosphorylation on Channel Function

Sodium channels in brain were demonstrated to be targets for phosphorylation by second messenger–controlled protein kinases (13–15) long before it was appreciated that their electrophysiological responses were altered by phosphorylation or by neurotransmitter-controlled signal transduction pathways. This was due in part to a bias in the field—namely, that sodium channels performed a set task, action potential initiation, that was not thought to be modulated—and in part because the effects of phosphorylation were not easily observed, possibly because of high levels of channel phosphorylation in resting neurons (16). In early studies using rat brain synaptosomes, treatment with 8-bromo-cAMP decreased veratridine-induced sodium influx, suggesting that cAMP-dependent phosphorylation inhibits sodium current (14). Electrophysiological studies later demonstrated that in excised membrane patches from cultured rat brain neurons and from Chinese hamster ovary (CHO) cells expressing type IIA α subunits, treatment with the catalytic subunit of PKA decreased sodium current in response to membrane depolarization (17). Single channel analysis showed that this resulted from a decrease in the probability of channel opening. Similar results were found for rat brain type IIA α subunits expressed in *Xenopus* oocytes during treatment with a cAMP analog or stimulation of coexpressed β-adrenergic receptors (18,19). When much higher levels of β-adrenergic receptors were expressed together with rat brain type IIA α subunits in oocytes, a paradoxical increase in sodium current was observed upon β-adrenergic stimulation or the addition of a cAMP analog (20). The reason for this response is not yet understood, but it does not require phosphorylation of the identified sites on the channel protein described below (19). Protein kinase A mediated phosphorylation had no effect on the voltage dependence of channel activation or inactivation (17–20).

Phosphorylation Sites for PKA and Their Role in Controlling Channel Function

Early studies demonstrated that the sodium channel α subunit was phosphorylated by PKA at multiple sites *in vitro* and *in vivo* (13,14,16). In the rat brain type IIA α subunit, PKA phosphorylates Ser-573, Ser-610, Ser-623, and Ser-687 *in vitro* and *in vivo* (21). These sites are all present within the intracellular loop linking homologous domains I and II (linker I–II) of the channel protein (Fig. 2). Site-directed mutagenesis studies have shown that Ser-573 is both necessary and sufficient for PKA-mediated inhibition of sodium current in *Xenopus* oocytes (22) or in human embryonic kidney (HEK) cells (23) expressing type IIA α subunits. Ser-610, Ser-623, and Ser-687 are not required for the inhibition of sodium current by cAMP-dependent phosphorylation (22). It is not known whether these sites play other cAMP-regulated roles for sodium channels. In cultured rat brain neurons, all four sites are phosphorylated to some extent in the absence of added stimuli, and their phosphorylation is increased in a cAMP-dependent manner (16). Immature α subunits that have not yet reached the cell surface are also phosphorylated on all four sites in a cAMP-dependent manner. The possibility that cAMP-dependent phosphorylation plays a role for sodium channels before they reach their functional state at the cell surface has not been explored.

Phosphorylation by PKC

Effect of PKC Phosphorylation on Channel Function

In *Xenopus* oocytes injected with chick brain RNA, stimuli that activate PKC inhibited voltage-sensitive sodium current; this effect was blocked by tamoxifen, a specific PKC inhibitor (24). In this study, activation of PKC had no effect on the voltage dependence of activation or inactivation. Others also observed that activation of PKC decreased the sodium current resulting from expression of chick brain RNA or RNA encoding the rat brain type IIA sodium channel α subunit in *Xenopus* oocytes (25). In this study, however, the decreased current appeared to result from a shift in the voltage dependence of activation to more positive potentials. In CHO cells expressing type IIA sodium channel α subunit and in cultured embryonic rat brain cells, activation of PKC with phorbol ester or diacylglycerol decreased sodium current and slowed current inactivation (26). No change in the voltage dependence of activation or inactivation was observed. Single channel analysis showed that an increase in channel open time and an increase in reopenings during depolarization accounted for the slowed inactivation. The slowing of inactivation was observed at a lower level of PKC activation than that required to inhibit current, suggesting that phosphorylation at two separate sites may control these distinct effects on channel function.

Phosphorylation Sites for PKC and Their Role in Controlling Channel Function

Protein kinase C phosphorylates sodium channels in lysed synaptosomal membranes (15) and purified channels reconstituted in phospholipid vesicles (27). Three sites of phosphorylation *in vitro* have been identified in the rat brain type IIA channel: Ser-576, Ser-610, and Ser-1506 (28). Ser-576 and Ser-610 are located in linker I–II and Ser-1506 is located in the intracellular loop linking domains III and IV (linker III–IV), which is highly conserved and mediates channel inactivation (5) (Fig. 2). When Ser-1506 was replaced with Ala, stimulation of PKC no longer slowed channel inactivation or decreased current (29), indicating that this phosphorylation site may be required for both effects. In preliminary studies, mutations within linker I–II led to the selective loss of PKC-induced current inhibition (30); however, it has not yet been determined whether phosphorylation of Ser-576, Ser-610, or both is required for this effect.

Voltage-Dependent Convergent Regulation of Sodium Channels by PKC and PKA

Although PKA inhibited sodium current in excised membrane patches from CHO cells expressing rat brain type IIA α subunits, in cell-attached patches PKA-mediated inhibition of current was observed only following activation of PKC (31). Mutation of Ser-1506 to Ala prevented the PKA-mediated reduction in sodium current, indicating that phosphorylation of Ser-1506, and not other PKC sites, was responsible for this effect. If Ser-1506 was mutated to Asp, or the surrounding residues changed to resemble a consensus site for PKA phosphorylation, 8-bromo-cAMP inhibited sodium current and slowed inactivation. Thus, the effect is due to the presence of a negative charge at position 1506, and no other PKC site is essential for this effect. The requirement for prior phosphorylation of Ser-1506 to see the effect of cAMP-dependent phosphorylation has been found to depend on the membrane holding potential prior to depolarization (23). At very negative membrane holding potentials, prior phosphorylation by PKC at Ser-1506 is necessary to see the effect of cAMP-dependent phosphorylation. At less negative membrane potentials, cAMP-dependent phosphorylation can inhibit sodium current during a depolarizing test pulse without prior activation of PKC. Thus, the ability of stimuli acting through cAMP to influence sodium channels will depend on the resting membrane potential and on the level of PKC activity, in addition to the competing action of Ser/Thr protein phosphatases. It will be of interest to learn whether the requirement for Ser-1506 phosphorylation to observe PKC-mediated current inhibition (29) also depends on membrane potential.

Dephosphorylation of Rat Brain Sodium Channels

Calcineurin and the catalytic subunit of protein phosphatase 2A (PP2A) can dephosphorylate cAMP-dependent phosphorylation sites of sodium channels *in vitro* (21). Treatment of synaptosomes with cyclosporin A or okadaic acid increased phosphorylation of cAMP-dependent phosphorylation sites on sodium channels, suggesting that both calcineurin and PP2A or a related phosphatase are also active toward sodium channels *in vivo* (32). In soluble extracts prepared from rat brain, two forms of PP2A, $PP2A_0$ and $PP2A_1$, were active toward sodium channels phosphorylated in cAMP-dependent phosphorylation sites. These forms contain distinct B regulatory subunits that may control the substrate specificity and the subcellular localization of PP2A (33). It is not yet known whether one or both of these forms are responsible for dephosphorylating sodium channels in intact neurons.

In *in vitro* studies, sodium channels phosphorylated by both PKC and the catalytic subunit of PKA were poor substrates for calcineurin or PP2A compared with channels phosphorylated by the catalytic subunit of PKA alone (34). This suggests that phosphorylation by PKC may somehow protect PKA sites from dephosphorylation. This is one of a few examples of phosphorylation of a substrate at one site altering dephosphorylation at another site (35–38), and it may provide a biochemical mechanism for convergent regulation of sodium channels by PKC and PKA. However, it remains to be demonstrated that PKC phosphorylation of sodium channels in intact neurons slows or prevents dephosphorylation. The dephosphorylation of PKC-phosphorylated sodium channels has not been studied.

Physiological Signals Controlling Reversible Phosphorylation

Several studies have identified physiologic stimuli that can alter neuronal sodium channel phosphorylation and function. In rat brain striatal neurons, stimulation of D1 dopamine receptors inhibited sodium current; this was blocked by D1 antagonists (39,40). In one study, the response to D1 dopaminergic stimulation was mimicked by intracellular application of the catalytic subunit of PKA and blocked by the specific PKA inhibitor peptide PKI_{5-24}, indicating that the response to D1 agonist was mediated by PKA (40). In isolated rat brain hippocampal neurons, treatment with a D1 agonist also inhibited sodium current, and this response was blocked by PKI_{5-24} (23). In all cases, D1 agonists caused a decrease in peak sodium current, which is consistent with the direct effect of cAMP-dependent phosphorylation to decrease channel opening during depolarization (17). In one study, D1 stimulation also shifted the steady state inactivation curve to more negative membrane potentials (39); however, no change in the voltage dependence or kinetics of activation or inactivation was observed in other studies (23,40). Thus, dopamine may promote cAMP-dependent phosphorylation of sodium channels and sodium current inhibition in central neurons of the neostriatum and hippocampus expressing D1 receptors. The inhibition of sodium current by dopamine would be expected to increase the threshold for initiating action potentials (23). This is consistent with the effect of dopamine in suppressing excitability in neostriatal neurons (41). Dopaminergic inhibition of hippocampal neurons may also result from inhibition of sodium channels, together with effects on other ion channels (42,43).

Muscarinic stimulation, via PKC activation, inhibited sodium current in hippocampal neurons acutely isolated from rat brain (44). In this study, the slowing of channel inactivation was much less pronounced than that observed during PKC activation in cultured rat brain cells or CHO cells expressing type IIA α subunits (26). Since the slowing of inactivation occurs at low levels of PKC stimulation that do not inhibit peak sodium current, it is possible that, under the conditions used in this study, channels were already extensively phosphorylated at Ser-1506 and no further slowing of inactivation would be observed. Alternatively, if the requirement for Ser-1506 phosphorylation is voltage dependent, as was found in the case of cAMP-dependent current inhibition (23), then at the holding potential used in this study PKC-induced current inhibition may occur in the absence of PKC phosphorylation of Ser-1506 and the resulting slowing of inactivation.

In rat brain synaptosomes, depolarization in the presence of calcium increased the level of phosphorylation on all four cAMP-dependent phosphorylation sites of sodium channels (34). Surprisingly, this effect was not blocked by inhibition of PKA, but was prevented by inhibition of PKC and mimicked by activation of PKC with phorbol ester. The mechanism by which the increase occurs is not known. One possible explanation is that PKC may attack all PKA sites *in situ,* even though PKC phosphorylates only one of these sites, Ser-610, *in vitro* (27,28). Alternatively, PKC may decrease dephosphorylation of PKA sites on sodium channels by altering the susceptibility of channels to phosphatase action, as suggested by *in vitro* studies of channel dephosphorylation (34), or by inhibiting phosphatases. This study suggests that repetitive or prolonged depolarization may lead to enhanced phosphorylation and, consequently, a decrease in sodium channel activity. This may represent a mechanism for inhibiting neuronal excitation following periods of increased activity to prevent the accumulation of high levels of calcium.

RAT SKELETAL MUSCLE SODIUM CHANNELS

Skeletal muscle sodium channels contain an α subunit and a β1 subunit (reviewed in Ref. 45). Two distinct types of sodium channel are expressed in skeletal muscle, a tetrodotoxin-resistant channel that is expressed in embryonic and denervated muscle, and a tetrodotoxin-sensitive channel expressed both in embryonic and adult muscle. These arise from the expression of distinct α subunits, SkM1 (46), which encodes the tetrodotoxin-sensitive α subunit and SkM2 (47), which encodes the tetrodotoxin-resistant α subunit. SkM1 lacks the PKA phosphorylation sites present in brain α subunits but contains other potential PKA phosphorylation sites. The consensus sequence for PKC phosphorylation in linker III–IV is conserved in SkM1; however, the PKC phosphorylation sites found in linker I–II of the brain α subunits are not conserved. Other potential PKC phosphorylation sites are present. SkM2 is the dominant form of sodium channel α subunit expressed in cardiac muscle (47,48), and is discussed in the section on cardiac sodium channels.

Phosphorylation by PKA

Sodium channels from adult skeletal muscle can be phosphorylated by PKA *in vitro* (49) to the extent of 0.5 mol of phosphate per mol of channel, suggesting the presence of a single phosphorylation site. Sodium channels have also been shown to be phosphorylated in a cAMP-dependent manner in cultured rat myotubes (49) and in HEK cells expressing SkM1 α subunits (50). The electrophysiological properties of SkM1 channels were unaffected by cAMP-dependent phosphorylation when expressed either in HEK cells (51) or in *Xenopus* oocytes (19). The site of cAMP-dependent phosphorylation is not known, although limited proteolysis studies have mapped it to the N-terminal half of the α subunit encompassing homologous domains I and II (49). The physiological significance of cAMP-dependent phosphorylation of skeletal muscle sodium channels has not yet been defined.

Phosphorylation by PKC

In the mouse satellite cell line MM14, which expresses only SkM1-type sodium channels, treatment with the diacylglycerol analog 1-oleoyl-2-acetyl-*sn*-glycerol slowed channel inactivation and decreased peak sodium current (52). When SkM1 α subunits were expressed in HEK cells, treatment with diacylglycerol inhibited sodium current, shifted the voltage dependence of inactivation to more negative potentials, and slightly accelerated the development of inactivation (51). This response was not altered by coexpression of the β1 subunit or by mutation of Ser-1321, the equivalent of Ser-1506 in linker III–IV of the rat brain type IIA α subunit. While the differences in observed effects may be attributed to differences in electrophysiological measurements or the properties of the cells used in these studies, the mutagenesis results (51) indicate that at least one PKC phosphorylation event other than phosphorylation of Ser-1321 is required for current inhibition. Further work will be required to determine whether Ser-1321 or other sites in SkM1 α are phosphorylated by PKC and whether direct phosphorylation of any site on the α subunit itself is responsible for the electrophysiological response to PKC activation, as well as to define the physiological signals regulating skeletal muscle sodium channels via PKC.

CARDIAC SODIUM CHANNELS

The major form of sodium channel in cardiac cells is encoded by the H1 α subunit gene (53). The rat form of this subunit, rH1 α (48) is identical to the SkM2 α subunit encoding the tetrodotoxin-resistant sodium channel in embryonic and denervated skeletal muscle (47). Sodium channel β1 subunits may also be present in heart (53,54). The H1 α sequence lacks the PKA phosphorylation sites present in linker I–II of brain α subunits, but contains other potential PKA sites in linker I–II and elsewhere. The H1 α subunit contains the PKC phosphorylation site, Ser-1505 in the rH1 α subunit, within the highly conserved linker III–IV region involved in inactivation. Although the surrounding sequence is not identical, Ser-529 of SkM2 α, the equivalent of Ser-576 of the rat brain type IIA α subunit in linker I–II, is still a potential target for PKC. Other potential PKC phosphorylation sites are also present.

Phosphorylation by PKA

Effect of cAMP-Dependent Phosphorylation on Channel Function

Numerous observations indicate that cAMP-dependent phosphorylation modulates cardiac sodium channels, but the specific effects reported are often contradictory. These differences may arise from technical differences in experiments and from the complex voltage-dependent nature of the changes observed. In studies examining β-adrenergic regulation, both direct binding of G_s and cAMP-dependent phosphorylation have been proposed to control sodium channel function, further complicating the analysis of phosphorylation effects.

In several studies, β-adrenergic stimulation or treatments that elevate intracellular cAMP reduced sodium current in rat (55) or guinea pig ventricular myocytes (56). This effect was voltage dependent and was more pronounced at less negative membrane holding potentials. In rat ventricular myocytes, β-adrenergic-induced inhibition still occurred in a membrane-delimited patch in the absence of added ATP, and the response was mimicked by application of preactivated G_s, but not G_k (55). This suggests that β-adrenergic receptors control sodium channels by two mechanisms, direct coupling to G_s and indirect control by cAMP-dependent phosphorylation. In contrast to sodium current inhibition, others observed that β-adrenergic stimulation or elevation of cAMP increased sodium current in rabbit cardiac myocytes (57). This effect was also dependent on the membrane holding potential and appeared to involve direct coupling to G_s and indirect activation through cAMP-dependent phosphorylation, since the effect of forskolin, but not isoproterenol, could be completely blocked by kinase inhibition.

One possible explanation for these discrepancies is that the response of sodium current to β-adrenergic stimulation depends on the membrane potential. In ventricular myocytes held at very negative membrane potentials, β-adrenergic stimulation or treatment with cAMP or a cAMP analog enhanced sodium current elicited by a depolarizing stimulus, whereas in cells held at less negative membrane holding potentials, the same treatments inhibited sodium current (58,59). In their study, Muramatsu and colleagues (59) found that the response to β-adrenergic stimulation was fully inhibited by kinase blockade; thus all measured responses could be attributed to phosphorylation, not to direct G protein action.

Finally, in studies in which β-adrenergic receptors were coexpressed in *Xenopus* oocytes together with rat SkM2 (60) or human H1 α subunits (61), cAMP-dependent phosphorylation caused an increase in peak current, with no alteration of the voltage dependence of channel function. These findings are reminiscent of the observations of Smith and Goldin (19,20), who found that in *Xenopus* oocytes coexpressing rat brain type IIA α subunits and β-adrenergic receptors, PKA could increase or decrease sodium channel activity, depending on the level of β-adrenergic receptor expression. In both cases, studies with chimeric channel constructs showed that linker I/II is required for the PKA-induced increase (19,61). Although linker I/II contains all identified cAMP-dependent phosphorylation sites of the rat brain type IIA α subunit, the PKA-induced increase in activity did not require direct phosphorylation of any of these sites (19). In the case of cardiac sodium channels expressed in *Xenopus* oocytes (60,61), it remains to be determined whether these effects are mediated by direct phosphorylation of the channel protein itself.

Phosphorylation Sites for PKA and Their Role in Controlling Channel Function

Cardiac sodium channels can be phosphorylated by the catalytic subunit of PKA *in vitro* (62–64) and by elevation of cAMP *in situ* (64). Two sites, Ser-526 and Ser-529 in linker I–II, were phosphorylated *in vitro* and in Chinese hamster lung cells expressing rH1 sodium channels (64). Future site-directed mutagenesis studies should reveal whether PKA-mediated effects on channel function require either or both of these sites and will also help distinguish the effects of direct channel phosphorylation from those caused by direct G protein interaction or phosphorylation of other proteins that may modulate channel function.

Phosphorylation by PKC

In neonatal rat ventricular myocytes and in Chinese hamster lung cells expressing rH1 α subunits, treatment with 1-oleoyl-2-*sn*-glycerol inhibited sodium current (65);

this effect was blocked by a specific peptide inhibitor of PKC, indicating that the response is mediated by PKC and is not a direct effect of lipid on channel function. The response to PKC activation was due to a shift in the voltage dependence of inactivation to more negative membrane potentials and a decrease in peak current. In ventricular myocytes, activation of PKC also slowed channel inactivation. Mutation of the conserved PKC phosphorylation site in linker III/IV, Ser-1505 in the rH1 α subunit, prevented both the reduction in peak current and the negative shift in the voltage dependence of inactivation (66). When the human H1 α subunit was expressed in *Xenopus* oocytes, activation of PKC decreased peak sodium current but had no effect on the voltage dependence of inactivation (67). The inhibition of peak current was reduced, but not completely prevented, by mutation of Ser-1503, the equivalent of Ser-1505 in the rH1 α subunit and Ser-1506 in the rat type IIA α subunit, suggesting that in addition to Ser-1503, another phosphorylation site for PKC may exist that mediates the remaining inhibition of current. As with the SkM1 α subunit, biochemical studies will be required to verify phosphorylation of Ser-1503 in linker III–IV and to identify additional PKC phosphorylation sites for mutagenesis studies.

SUMMARY OF SODIUM CHANNEL PHOSPHORYLATION

Voltage-sensitive sodium channels are subject to complex regulation by second messenger–controlled phosphorylation. Specific phosphorylation sites for PKA and PKC, and the consequences of phosphorylation are different for sodium channels present in neurons, skeletal muscle, and cardiac muscle. Cyclic AMP dependent phosphorylation of brain sodium channel α subunits on Ser-573 in linker I–II decreases peak sodium current because it serves to decrease channel opening during depolarization. This would decrease neuronal excitability. Phosphorylation by PKC at Ser-1506 in linker III–IV slows channel inactivation, which would prolong action potential duration and alter the frequency of action potential firing. The prolonged membrane depolarization may also affect calcium influx. At higher levels of activation PKC also inhibits peak current; this involves an unidentified site within linker I–II, as well as phosphorylation at Ser-1506.

Sodium channels containing SkM1 α subunits found in both embryonic and adult skeletal muscle can be phosphorylated by PKA *in situ,* but this does not alter their electrophysiological response. In the case of tetrodotoxin-resistant sodium channels, which contain SkM2 and are expressed in cardiac muscle and in embryonic or denervated skeletal muscle, cAMP-dependent phosphorylation has been reported to both decrease and increase sodium current. Two phosphorylation sites identified within linker I–II, Ser-526 and Ser-529, are candidates for mediating the effect of cAMP-dependent phosphorylation on cardiac channels; however, their roles in controlling the response of these channels to cAMP have not yet been defined.

For SkM1 and SkM2 α subunits, PKC activation inhibits sodium current, but the slowing of inactivation is not observed in all cases. For both types of α subunit, the response to PKC requires the conserved PKC phosphorylation site in linker III–IV; however, additional unidentified PKC phosphorylation events may also be involved.

A complete understanding of how direct phosphorylation controls sodium channel function in each of these tissues will require the identification of remaining phosphorylation sites and analysis of their role in channel function through site-directed mutagenesis. The role of Ser/Thr phosphatases and additional physiological signals in controlling the state of channel phosphorylation, and the role(s) of phosphorylation sites that do not appear to control electrophysiological responses, are important topics that have only begun to be addressed. Continued investigation of these topics is likely to reveal important insights into the control of sodium channel function.

VOLTAGE-SENSITIVE L-TYPE CALCIUM CHANNELS

Voltage-sensitive calcium channels, present in neurons, muscle cells, and secretory cells, provide a voltage-regulated path for the influx of calcium that directly or indirectly controls diverse responses such as contraction, secretion, and gene expression, depending on the cell type (4). In contrast to voltage-sensitive sodium channels, it has long been recognized that voltage-dependent calcium channels are subject to regulation by hormone-induced phosphorylation events (68). However, the identification of physiologically relevant phosphorylation sites has been difficult, owing to the diversity of calcium channel structure, the presence of multiple phosphorylated subunits, the low levels of protein available for biochemical analysis, and the susceptibility of channels to degradation. Five classes of voltage-sensitive calcium channels have been identified on the basis of electrophysiological and pharmacological characteristics and, more recently, cloning studies (reviewed in Ref. 4). This discussion focuses on the regulation of L-type calcium channels in cardiac and skeletal muscle by cAMP-dependent phosphorylation. A large body of literature documenting the effects of various physiological stimuli on L-type calcium channel responses in cardiac and skeletal muscle was reviewed in 1994 (1) and is not covered here. Instead, work on the identification of phosphorylation sites, the role of anchored PKA in controlling channel function, and the role of phosphorylation in voltage-dependent potentiation is emphasized.

Like voltage-sensitive sodium channels, L-type calcium channels open during membrane depolarization and inactivate despite continued depolarization. L-type calcium channel responses are slower than those of sodium channels, and inactivation of these channels is controlled by at least two mechanisms, one voltage dependent and one calcium dependent (1). L-type channels have at least two open states, a short opening and a prolonged opening (69). Inactivation decreases the probability of channel opening, but does not alter the duration of short or prolonged openings. In addition to these basic responses, L-type calcium channels may also be controlled by direct interaction with G proteins (see Chapter 7) and are subject to voltage-dependent potentiation, a process by which a period of high frequency depolarizing prepulses results in increased calcium current during a test depolarization (70). Phosphorylation modulates channel activation and inactivation, and has been implicated in voltage-dependent potentiation of skeletal muscle and cardiac L-type calcium channels. Other types of voltage-dependent calcium channel also exhibit these complex behaviors. The extent to which phosphorylation by PKA and other kinases mediates these responses in L-type and other voltage-sensitive calcium channels is a matter of active debate.

L-type and other voltage-sensitive calcium channels contain a large $\alpha1$ subunit that is structurally similar to the sodium channel α subunit and is a target for phosphorylation (2). Cloning studies have identified many forms of the L-type $\alpha1$ subunit (4) that differ in their putative phosphorylation sites. In addition, full-length $\alpha1$ subunits can be proteolyzed to generate a C-terminally truncated version (71,72). The significance of the proteolyzed form of $\alpha1$ and the extent to which it occurs naturally in different tissues are not well understood. However, since the C terminus contains PKA phosphorylation sites that are lost upon proteolysis (72–74), the presence of full-length and proteolyzed forms has complicated attempts to identify phosphorylation sites in the channel itself that may alter function. Skeletal muscle L-type calcium channels contain three additional subunits: an intracellular β subunit, and transmembrane γ and $\alpha2/\delta$ subunits (4). Cardiac L-type channels contain β and $\alpha2/\delta$ subunits, in addition to $\alpha1$ (4). The β subunit can also be phosphorylated (75–79), and cloning studies have revealed that many forms of this subunit (4) exist which, like $\alpha1$ subunits, vary in their putative phosphorylation sites.

SKELETAL MUSCLE L-TYPE CALCIUM CHANNELS

In vertebrate skeletal muscle, L-type calcium channels are present in T tubules, where they play two roles: a voltage sensor coupling membrane depolarization to calcium release from the sarcoplasmic reticulum through a direct interaction with the calcium release complex of the sarcoplasmic reticular membrane, and a calcium channel permitting the influx of extracellular calcium required for the eventual restoration of sarcoplasmic reticulum calcium pools (1). As mentioned earlier, two forms of $\alpha1$ have been observed in skeletal muscle, a minor full-length 212-kDa form and a major 165- to 190-kDa form that is C-terminally truncated (71). The truncated form appears to be present in native tissue and is not a result of proteolysis during purification (80). Recombinant truncated calcium channel $\alpha1$ subunits can function both as calcium channels and as voltage sensors when expressed in dysgenic myotubes, indicating that the C terminus is not essential for these functions (81).

Effect of PKA Phosphorylation on Channel Function

Skeletal muscle L-type calcium current is increased by stimuli that elevate or mimic cAMP, and by the direct application of the catalytic subunit of PKA in the presence of ATP (reviewed in Ref. 1). Phosphorylation increases the probability of channel opening and shifts the voltage dependence of activation to more negative membrane potentials. The activity of purified reconstituted calcium channels is also increased by phosphorylation, suggesting that phosphorylation of the channel itself is responsible for increased function (82–84).

Cyclic AMP-Dependent Phosphorylation Sites and Their Contribution to Altered Function

The α1 and β subunits of skeletal muscle L-type calcium channels are both phosphorylated *in vitro* by PKA (75,77,78,85,86). The majority of α1 subunits isolated from adult rabbit skeletal muscle are a 165- to 190-kDa truncated form that is phosphorylated primarily on Ser-687 (in linker II–III) by PKA *in vitro* (74,87). A small fraction of L-type calcium channels in this tissue contains the full-length 212-kDa α1 subunit (71,88), which is phosphorylated primarily on Ser-1757 and Ser-1854 by PKA *in vitro* (74,89), sites that are lost by truncation. In cultured rabbit skeletal muscle myotubes, most of the α1 subunits were present as the full-length form and were phosphorylated *in situ* on Ser-1757 and Ser-1854 in a cAMP-dependent manner (89); Ser-687 represented a minor phosphorylation site whose state of phosphorylation did not change significantly upon elevation of cAMP. These studies suggest that Ser-1757 and Ser-1854 in the C-terminus, and not Ser-687, are the major sites of cAMP-dependent phosphorylation in full-length α1 subunits.

The significance of individual phosphorylation sites in controlling channel function has not yet been determined. In early reconstitution studies, purified channels exhibited increased activity upon phosphorylation that was accompanied by phosphorylation of both α1 and β subunits (82,84). Studies in which changes in channel function and subunit phosphorylation were directly compared indicate that phosphorylation of the α1 subunit is correlated with changes in channel activity (90,91). However, most of these early studies were performed with preparations containing primarily truncated α1 subunits lacking Ser-1757 and Ser-1854. In 1997, skeletal muscle L-type calcium channel α1, β, and α2/δ subunits were coexpressed in HEK cells (92). Although expressed current was low and treatment with a dihydropyridine agonist was required for its measurement, functional calcium channels that responded to PKA were obtained. This heterologous expression system may permit the identification of specific phosphorylation sites controlling channel function and a comparison of channels containing full-length or truncated α1 subunits, although the use of a dihydropyridine channel activator could prevent measurement of subtle PKA-induced changes in channel responses.

The Role of Phosphorylation and Anchored PKA in Voltage-Dependent Potentiation

L-type calcium channels are subject to activity and voltage-dependent potentiation (reviewed in Refs. 1,70). In skeletal muscle, potentiation may contribute to the enhanced calcium influx and increased contractile force seen following high frequency stimulation (93). Experimentally, voltage-dependent potentiation can be generated by a series of depolarizing prepulses or by a single high amplitude depolarizing prepulse. Repetitive depolarization of frog or rat skeletal muscle cells was shown to potentiate L-type calcium current (94–96). In rat skeletal muscle cells, repetitive stimulation shifted the voltage dependence of channel activation to more negative membrane potentials and slowed inactivation (96). When barium was substituted for calcium as the charge carrier, a single high amplitude prepulse promoted a similar response. Potentiation was blocked by the specific PKA inhibitor PKI, indicating that basal activity of PKA was required for this response. Intracellular application of a peptide that disrupts the binding of PKA to A-kinase anchoring proteins (AKAPs) also prevented the development of potentiation, suggesting that this process requires strategic anchoring of PKA near the calcium channels to be modified (97). A 15-kDa AKAP that has been recently isolated from rabbit skeletal muscle and shown to bind both L-type calcium channels and PKA is a candidate for docking PKA near to calcium channels (98).

These studies led to the hypothesis that depolarization induces a conformational change that makes a site on the channel protein available for phosphorylation (96). The effect of potentiation and of PKA stimulation on calcium channel function are similar; however, it is not known whether the same phosphorylation sites are involved in controlling channel responses to cAMP elevation or to repetitive depolarization (92). Heterologously expressed, PKA-stimulated skeletal muscle L-type channels also undergo potentiation that requires the presence of anchored PKA (92), and may be useful in defining phosphorylation sites required for potentiation and cAMP-dependent stimulation.

CARDIAC L-TYPE CALCIUM CHANNELS

In cardiac muscle, extracellular calcium entering the cell through L-type calcium channels during depolarization promotes muscle contraction and, by stimulating calcium release from the sarcoplasmic reticulum, triggers further elevation of intracellular calcium. Cloning studies revealed that the cardiac $\alpha 1$ subunit has distinct C-terminal PKA consensus sequences from those contained in the skeletal muscle $\alpha 1$ subunit and is missing Ser-687 in linker II–III (99). The β subunit of cardiac calcium channels also differs in structure and potential phosphorylation sites from that of the skeletal muscle channel (100).

Effect of PKA Phosphorylation on Channel Function

β-Adrenergic stimulation of cardiac cells increases L-type calcium current. This response is mimicked by treatment with cAMP analogs, agents that stimulate cAMP synthesis, and by the intracellular application of the purified catalytic subunit of PKA (reviewed in Ref. 101). At the single channel level, PKA increases the probability of channel opening and promotes long channel openings (101,102).

Recordings of L-type calcium channels from cardiac myocytes showed that β-adrenergic stimulation still activated calcium channels in the presence of PKI or in the absence of ATP (103). Stimulation under these conditions required GTP and was mimicked by direct application of the α subunit of G_s. This led to the proposal that β-adrenergic stimulation activates L-type calcium channels by direct binding of $G\alpha_s$, as well as by cAMP-dependent phosphorylation. However, under physiologic conditions cAMP-dependent phosphorylation has been shown to be responsible for most, if not all, of the changes in calcium current during β-adrenergic stimulation (104,105), and the contribution of direct $G\alpha_s$ coupling does not appear to be significant (106).

Cyclic AMP-Dependent Phosphorylation Sites and Their Contribution to Altered Function

Attempts to demonstrate the structural basis for cardiac calcium channel regulation by PKA were complicated by contradictory findings on two fronts: biochemical analysis of phosphorylation, and the reconstitution of calcium channels that were responsive to cAMP-dependent phosphorylation. In early studies of cardiac L-type calcium channels from various species, partially purified channel preparations contained a major α1-like subunit of 200 kDa or less that was not phosphorylated by PKA (107–109). In later biochemical studies, antiserum specific for the cardiac α1 subunit immunopurified two polypeptides, a 240- to 250-kDa polypeptide that could be phosphorylated by PKA *in vitro* and a smaller one (200–210 kDA) that was not phosphorylated (72,73); antiserum recognizing a C-terminal epitope of α1 detected only the 240- to 250-kDa polypeptide (72,73). This suggested that one or more sites in the C terminus of α1 were phosphorylated and that this region of the polypeptide may have been missing in early studies in which smaller polypeptides were isolated (107–109). Phosphopeptide mapping studies, together with sequence analysis of phosphopeptides isolated from recombinant α1 and site-directed mutagenesis studies, demonstrated that Ser-1928 in the C terminus was phosphorylated *in vitro* (73,110). Cardiac α1 subunits expressed in CHO cells were also phosphorylated in a cAMP-dependent manner (72) on Ser-1928 (73). Cardiac β subunits have also been shown to be phosphorylated *in vivo* in a cAMP-dependent manner (79,111,112); however, expression studies suggest that α1 phosphorylation is responsible for PKA-induced changes in calcium current. The physiological significance of β subunit phosphorylation by PKA is not yet known.

In expression studies, PKA-mediated changes in cardiac L-type calcium channel function have been observed in some cases, but not in others (72,112–117). The basis for these discrepancies is not known, but it may be attributable in part to differences in resting levels of cAMP and in levels or localization of endogenous PKA and Ser/Thr phosphatases. In one study using stably transfected CHO cells expressing the cardiac α1 subunit, treatment with cAMP analogs increased calcium current approximately two-fold (72). However, when cells were perfused with the catalytic subunit of PKA, calcium current was enhanced 5- to 10-fold (113). This suggested that low levels of phosphorylation by endogenous kinase may have limited the response to cAMP in earlier studies (72) and demonstrated that the α1 subunit alone can produce the PKA-induced increase in calcium current.

In several expression studies using HEK 293 cells (114) or *Xenopus* oocytes (115), treatment intended to activate PKA failed to increase calcium current; PKA inhibitors decreased it, however, suggesting that high levels of endogenous cAMP may cause extensive phosphorylation of expressed channels in the absence of added stimuli.

When this decrease in current during treatment with PKA inhibitors was used as an assay, mutation of Ser-1928 was shown to prevent the PKA-sensitive change in current (118). In an expression study demonstrating PKA-sensitive calcium current, mutation of Ser-1928 prevented both phosphorylation of the α1 subunit and cAMP-induced current increase, demonstrating that phosphorylation of Ser-1928 on the α1 subunit contributes to cAMP-dependent regulation of cardiac calcium channels (112). In this study, regulation of calcium channels expressed in HEK 293 cells by PKA required coexpression of a PKA anchoring protein, AKAP 79. This suggests that without the aid of an anchoring protein, endogenous PKA levels near the channel protein in these cells are insufficient to promote the phosphorylated state. The magnitude of the cAMP-induced current increase observed, 1.5- to 1.8-fold, was modest compared to the 3- to 4-fold elevation in current seen during cAMP-mediated stimulation of cardiac cells (119), and required the presence of phosphatase inhibitors. Since earlier expression studies demonstrated that the α1 subunit expressed alone generates a calcium channel able to fully respond to cAMP in the presence of sufficient PKA (113), this modest response may reflect a failure of the localization and concentration of endogenous kinases and phosphatases in this heterologous expression system fully to mimic the components responsible for the cAMP regulation of L-type calcium current in cardiac myocytes. It will be important to assess the consequences of Ser-1928 mutation under conditions in which recombinant channels demonstrate a robust response to cAMP-dependent phosphorylation.

The Role of Phosphorylation in Activity and Voltage-Dependent Potentiation

In cardiac cells, L-type calcium channels also undergo voltage- and activity-dependent potentiation (1,70). As in skeletal muscle, this can be generated by a series of depolarizing prepulses or by a single high amplitude depolarizing prepulse. The role of calcium in promoting potentiation of cardiac L-type calcium channels has been explored. Because these channels are inactivated by elevation of intracellular calcium, potentiation can be observed only in the presence of an intracellular calcium buffer. Calcium influx appears to be required for potentiation induced by repetitive depolarization (120–122), but not in the case of potentiation produced by a single high amplitude depolarizing prepulse (113,122,123). When repetitive depolarization was used to induce potentiation in rat (122) or guinea pig ventricular myocytes (120,121), the development of potentiation was prevented by substitution of another charge-carrying ion, barium, for calcium in the extracellular solution. Increasing calcium by flash photolysis of intracellular caged calcium also potentiated calcium current (124–126). Specific blockade of calcium/calmodulin-dependent protein kinase (CaM kinase) prevented the potentiation produced by release of caged calcium (126), as well as potentiation produced by repeated depolarization (122). These studies suggest that calcium influx during repeated depolarization promotes a CaM kinase mediated phosphorylation event that can potentiate calcium current in cardiac cells.

In studies using rat ventricular myocytes (122,123) or CHO cells expressing cardiac α1 subunits (113), voltage-dependent potentiation was also observed after a single high magnitude depolarizing prepulse using barium as the charge-carrying cation. Xiao and colleagues (122) demonstrated that specific blockade of CaM kinase also prevented potentiation generated by this protocol. Thus, basal activity of CaM kinase can mediate channel potentiation even in the absence of calcium influx. The study of Sculptoreanu and colleagues (113) showed that potentiation induced by a single high amplitude depolarizing prepulse was due to a negative shift in the voltage dependence of channel activation and slowed inactivation, and that the α1 subunit alone was sufficient for this response. In this investigation, the development of potentiation required intracellular perfusion with the catalytic subunit of PKA and the reversal of potentiation was blocked by okadaic acid, suggesting that voltage-dependent potentiation is due to a conformation-dependent phosphorylation of the channel's α1 subunit. When channels were maximally activated by PKA alone, a subsequent depolarizing prepulse produced no further potentiation. This result is similar to the observation that in cardiac cells, following maximal β-adrenergic stimulation, no further calcium-dependent, depolarization-induced potentiation could be achieved (121,125). The observation that PKA was required for the development of potentiation of heterologously expressed cardiac α1 subunits suggests that PKA phosphorylation may be required for this response (113). It is not clear whether PKA and CaM kinase both phosphorylate a common site to produce voltage-dependent potentiation, phosphorylate two different sites that are both essential for this response, or phosphorylate two different sites that cause a similar response.

Further work will be required to identify the site or sites responsible for voltage-dependent phosphorylation and potentiation, and to determine whether phosphorylation of the same site is responsible for potentiation caused by the two different depolarization protocols. Studies of the effects of various phosphatase inhibitors on L-type calcium current in cardiac cells suggest the presence of two distinct phosphorylation sites that are attacked by different kinases (127–130). Although it is not clear whether the phosphorylation sites revealed by these studies play a role in potentiation or PKA activation, these sites appear to control similar aspects of channel responses.

The Role of Ser/Thr Phosphatases in Controlling L-Type Calcium Channel Function

L-type calcium channels in a number of preparations are subject to reversible and irreversible loss of activity during patch–clamp analysis, referred to as rundown (1). Irreversible loss of activity appears to be due to proteolysis, since it could be blocked by protease inhibitors and mimicked by protease treatment (1). In some cases, reversible rundown has been attributed to dephosphorylation (131). In rabbit ventricular myocytes, rundown in isolated membrane patches was slowed by treatment with the Ser/Thr phosphatase inhibitor okadaic acid and could be reversed by treatment with the catalytic subunit of PKA and ATP (132). This study, together with work demonstrating functional effects of phosphatase inhibitors on calcium channels (127–130), highlights the important role of phosphatases in controlling cardiac L-type calcium channel activity. The ability of okadaic acid, calyculin, and microcystin to alter L-type calcium current in cardiac cells suggests that protein phosphatase 1,2A, or a related enzyme may dephosphorylate cardiac L-type calcium channels (127–130), although the inhibitor concentrations used in these studies are too high to permit analysts to distinguish which phosphatase may be involved.

Far less is known about the control of Ser/Thr phosphatases than is known about the regulation of kinases acting on ion channels. However, a growing body of evidence suggests that Ser/Thr phosphatases are more diverse than previously appreciated, and that they are subject to regulation and can be specifically targeted to distinct subcellular compartments (33). The selective activation or inhibition of the relevant Ser/Thr phosphatase(s) can be expected to have profound influence on calcium channel function. It will be important to identify the enzymes that dephosphorylate L-type calcium channels and to determine their contribution to controlling the state of calcium channel phosphorylation.

SUMMARY OF L-TYPE CALCIUM CHANNEL PHOSPHORYLATION

L-type calcium channels in skeletal and cardiac muscle differ in the roles they serve in muscle contraction, in their response kinetics, in their isoforms of $\alpha 1$ and β subunits, and in the phosphorylation sites contained within these subunits. Despite these differences, several general statements can be made about their responses to cAMP-dependent phosphorylation. Phosphorylation increases channel availability by increasing the probability of opening and shifting the voltage sensitivity of activation to more negative membrane potentials; this results in an increase in brief openings during membrane depolarization. Phosphorylation by PKA may also increases long channel openings. Voltage-dependent potentiation increases calcium current by slowing inactivation and causing a negative shift in the voltage dependence of activation. Depolarization may induce a conformational change that increases the accessibility of a site of phosphorylation by PKA or another kinase. The basal activity of PKA appears to be sufficient for this response in skeletal muscle cells. In the case of cardiac L-type calcium channels, both PKA and CaM kinase have been implicated. The phosphorylation site or sites required for voltage and activity-dependent potentiation have not been identified.

For both cardiac and skeletal muscle channels, phosphorylation of the $\alpha 1$ subunit is best correlated with the cAMP-dependent increase in channel activity, although more work is required to ascertain which biochemically identified phosphorylation sites fully account for PKA-mediated changes in channel function. The biological significance of $\alpha 1$ subunits that are truncated by proteolysis, resulting in the loss of cAMP-dependent phosphorylation sites, requires further study. In addition, phosphorylation of the $\alpha 1$ subunit by other second messenger–regulated kinases, and of the β subunit by PKA and other kinases, has been reported. Specific phosphorylation sites have not been identified, however, and their physiological significance is not known. Even the presence of the same phosphorylation sites in two different forms of $\alpha 1$ subunit may not control channel responses in the same way. Smooth muscle L-type calcium channels, whose $\alpha 1$ subunits contain the same putative PKA phosphorylation sites present in the cardiac α1 subunit (133), are not activated by cAMP (1). The presence of accessory subunits and other interacting proteins, and the proximity and abundance of relevant kinases and phosphatases near the channel complex, all contribute to the functional response of the channel pore and may explain why the presence of the same phosphorylation site does not serve the same function in two different channels expressed in different environments. Future work is likely to reveal additional roles for phosphorylation in controlling protein–protein interactions and cross talk between different signal transduction pathways. Several such examples have already been described for neuronal voltage-dependent calcium channels (134,135). In addition, as more is learned about the localization and regulation of Ser/Thr phosphatases, a more dynamic role for these enzymes in controlling the function of calcium and other ion channels may emerge.

ACKNOWLEDGMENTS

The author thanks Drs. David Armstrong, William Catterall, Aaron Fox, Amber Pond, Stanley Rane, and Todd Scheuer for their helpful input, and Drs. Angela Cantrell, Alan Goldin, and Marlene Hosey for providing reprints and preprints of their work. The author apologizes for omitting discussion of some literature for reasons of space. Research in the author's laboratory is supported by the National Institutes of Health (NINDS) and the American Heart Association.

REFERENCES

1. McDonald TF, Pelzer S, Trautwein W, Pelzer DJ: Regulation and modulation of calcium channels in cardiac, skeletal, and smooth muscle cells. *Physiol Rev* 1994;74:365–507.
2. Catterall WA: Structure and function of voltage-gated ion channels. *Annu Rev Biochem* 1995;64: 493–531.
3. Cukierman S: Regulation of voltage-dependent sodium channels. *J Membr Biol* 1996;151:203–214.
4. De Waard M, Gurnett CA, Campbell KP: Structural and functional diversity of voltage-activated calcium channels. *Ion Channels* 1996;4:41–87.
5. Catterall WA: Molecular properties of sodium and calcium channels. *J Bioenerg Biomembr* 1996; 28:219–230.
6. Catterall WA: Cellular and molecular biology of voltage-gated sodium channels. *Physiol Rev* 1992; 72:S15–48.
7. Noda M, Ikeda T, Kayano T, Suzuki H, Takeshima H, Kurasaki M, Takahashi H, Numa S: Existence of distinct sodium channel messenger RNAs in rat brain. *Nature* 1986;320:188–192.
8. Kayano T, Noda M, Flockerzi V, Takahashi H, Numa S: Primary structure of rat brain sodium channel III deduced from the cDNA sequence. *FEBS Lett* 1988;228:187–194.
9. Auld VJ, Goldin AL, Krafte DS, Marshall J, Dunn JM, Catterall WA, Lester HA, Davidson N, Dunn RJ: A rat brain Na^+ channel α subunit with novel gating properties. *Neuron* 1988;1:449–461.
10. Isom LL, De Jongh KS, Patton DE, Reber BF, Offord J, Charbonneau H, Walsh K, Goldin AL, Catterall WA: Primary structure and functional expression of the β 1 subunit of the rat brain sodium channel. *Science* 1992;256:839–842.
11. Isom LL, Scheuer T, Brownstein AB, Ragsdale DS, Murphy BJ, Catterall WA: Functional co-expression of the β1 and type IIA α subunits of sodium channels in a mammalian cell line. *J Biol Chem* 1995;270:3306–3312.
12. Isom LL, Ragsdale DS, De Jongh KS, Westenbroek RE, Reber BF, Scheuer T, Catterall WA: Structure and function of the β 2 subunit of brain sodium channels, a transmembrane glycoprotein with a CAM motif. *Cell* 1995;83:433–442.

13. Costa MR, Casnellie JE, Catterall WA: Selective phosphorylation of the α subunit of the sodium channel by cAMP-dependent protein kinase. *J Biol Chem* 1982;257:7918–7921.
14. Costa MR, Catterall WA: Cyclic AMP-dependent phosphorylation of the α subunit of the sodium channel in synaptic nerve ending particles. *J Biol Chem* 1984;259:8210–8218.
15. Costa MR, Catterall WA: Phosphorylation of the α subunit of the sodium channel by protein kinase C. *Cell Mol Neurobiol* 1984;4:291–297.
16. Rossie S, Catterall WA: Cyclic-AMP-dependent phosphorylation of voltage-sensitive sodium channels in primary cultures of rat brain neurons. *J Biol Chem* 1987;262:12735–12744.
17. Li M, West JW, Lai Y, Scheuer T, Catterall WA: Functional modulation of brain sodium channels by cAMP-dependent phosphorylation. *Neuron* 1992;8:1151–1159.
18. Gershon E, Weigl L, Lotan I, Schreibmayer W, Dascal N: Protein kinase A reduces voltage-dependent Na^+ current in *Xenopus* oocytes. *J Neurosci* 1992;12:3743–3752.
19. Smith RD, Goldin AL: Phosphorylation of brain sodium channels in the I–II linker modulates channel function in *Xenopus* oocytes. *J Neurosci* 1996;16:1965–1974.
20. Smith RD, Goldin AL: Protein kinase A phosphorylation enhances sodium channel currents in *Xenopus* oocytes. *Am J Physiol* 1992;263:C660–666.
21. Murphy BJ, Rossie S, De Jongh KS, Catterall WA: Identification of the sites of selective phosphorylation and dephosphorylation of the rat brain Na^+ channel α subunit by cAMP- dependent protein kinase and phosphoprotein phosphatases. *J Biol Chem* 1993;268:27355–27362.
22. Smith RD, Goldin AL: Phosphorylation at a single site in the rat brain sodium channel is necessary and sufficient for current reduction by protein kinase A. *J Neurosci* 1997;17:6086–6093.
23. Cantrell AR, Smith RD, Goldin AL, Scheuer T, Catterall WA: Dopaminergic modulation of sodium current in hippocampal neurons via cAMP-dependent phosphorylation of specific sites in the sodium channel α subunit. *J Neurosci* 1997;17:7330–7338.
24. Sigel E, Baur R: Activation of protein kinase C differentially modulates neuronal Na^+, Ca^{2+}, and gamma-aminobutyrate type A channels. *Proc Natl Acad Sci U S A* 1988;85:6192–6196.
25. Dascal N, Lotan I: Activation of protein kinase C alters voltage dependence of a Na^+ channel. *Neuron* 1991;6:165–175.
26. Numann R, Catterall WA, Scheuer T: Functional modulation of brain sodium channels by protein kinase C phosphorylation. *Science* 1991;254:115–118.
27. Murphy BJ, Catterall WA: Phosphorylation of purified rat brain Na^+ channel reconstituted into phospholipid vesicles by protein kinase C. *J Biol Chem* 1992;267:16129–16134.
28. Murphy BJ, Catterall WA: Identification of three sites of selective phosphorylation of the rat brain Na^+ channel α subunit by protein kinase C. *J. Neurosci* 1994;20:719.
29. West JW, Numann R, Murphy BJ, Scheuer T, Catterall WA: A phosphorylation site in the Na^+ channel required for modulation by protein kinase C. *Science* 1991;254:866–868.
30. West JW, Numann R, Murphy BJ, Scheuer T, Catterall WA: Phosphorylation of a conserved protein kinase C site is required for modulation of Na^+ currents in transfected Chinese hamster ovary cells. *Biophys J* 1992;62:31–33.
31. Li M, West JW, Numann R, Murphy BJ, Scheuer T, Catterall WA: Convergent regulation of sodium channels by protein kinase C and cAMP- dependent protein kinase. *Science* 1993;261: 1439–1442.
32. Chen TC, Law B, Kondratyuk T, Rossie S: Identification of soluble protein phosphatases that dephosphorylate voltage-sensitive sodium channels in rat brain. *J Biol Chem* 1995;270:7750–7756.
33. Wera S, Hemmings BA: Serine/threonine protein phosphatases. *Biochem J* 1995;311:17–29.
34. Kondratyuk T, Rossie S: Depolarization of rat brain synaptosomes increases phosphorylation of voltage-sensitive sodium channels. *J Biol Chem* 1997;272:16978–16983.
35. Desdouits F, Siciliano JC, Greengard P, Girault JA: Dopamine- and cAMP-regulated phosphoprotein DARPP-32: phosphorylation of Ser-137 by casein kinase I inhibits dephosphorylation of Thr-34 by calcineurin. *Proc Natl Acad Sci U S A* 1995;92:2682–2685.
36. Guan KL, Butch E: Isolation and characterization of a novel dual specific phosphatase, HVH2, which selectively dephosphorylates the mitogen-activated protein kinase. *J Biol Chem* 1995;270: 7197–7203.
37. Denu JM, Zhou G, Wu L, Zhao R, Yuvaniyama J, Saper MA, Dixon JE: The purification and characterization of a human dual-specific protein tyrosine phosphatase. *J Biol Chem* 1995;270: 3796–3803. [Erratum in *J Biol Chem* 1995;270(17):10358.]
38. Bornancin F, Parker PJ: Phosphorylation of threonine 638 critically controls the dephosphorylation and inactivation of protein kinase Cα. *Curr Biol* 1996;6:1114–1123.

39. Surmeier DJ, Eberwine J, Wilson CJ, Cao Y, Stefani A, Kitai ST: Dopamine receptor subtypes colocalize in rat striatonigral neurons. *Proc Natl Acad Sci U S A* 1992;89:10178–10182.
40. Schiffmann SN, Lledo PM, Vincent JD: Dopamine D1 receptor modulates the voltage-gated sodium current in rat striatal neurones through a protein kinase A. *J Physiol (Lond)* 1995;483:95–107.
41. Calabresi P, Mercuri N, Stanzione P, Stefani A, Bernardi G: Intracellular studies on the dopamine-induced firing inhibition of neostriatal neurons *in vitro:* evidence for D1 receptor involvement. *Neuroscience* 1987;20:757–771.
42. Benardo LS, Prince DA: Dopamine modulates a Ca^{2+}-activated potassium conductance in mammalian hippocampal pyramidal cells. *Nature* 1982;297:76–79.
43. Malenka RC, Nicoll RA: Dopamine decreases the calcium-activated afterhyperpolarization in hippocampal CA1 pyramidal cells. *Brain Res* 1986;379:210–215.
44. Cantrell AR, Ma JY, Scheuer T, Catterall WA: Muscarinic modulation of sodium current by activation of protein kinase C in rat hippocampal neurons. *Neuron* 1996;16:1019–1026.
45. Barchi RL: Molecular pathology of the skeletal muscle sodium channel. *Annu Rev Physiol* 1995;57: 355–385.
46. Trimmer JS, Cooperman SS, Tomiko SA, Zhou JY, Crean SM, Boyle MB, Kallen RG, Sheng ZH, Barchi RL, Sigworth FJ, Goodman RH, Agnew WS, Mandel G: Primary structure and functional expression of a mammalian skeletal muscle sodium channel. *Neuron* 1989;3:33–49.
47. Kallen RG, Sheng ZH, Yang J, Chen LQ, Rogart RB, Barchi RL: Primary structure and expression of a sodium channel characteristic of denervated and immature rat skeletal muscle. *Neuron* 1990;4: 233–242.
48. Rogart RB, Cribbs LL, Muglia LK, Kephart DD, Kaiser MW: Molecular cloning of a putative tetrodotoxin-resistant rat heart Na^+ channel isoform. *Proc Natl Acad Sci U S A* 1989;86:8170–8174.
49. Yang J, Barchi R: Phosphorylation of the rat skeletal muscle sodium channel by cyclic AMP-dependent protein kinase. *J Neurochem* 1990;54:954–962.
50. Ukomadu C, Zhou J, Sigworth FJ, Agnew WS: μI Na^+ channels expressed transiently in human embryonic kidney cells: biochemical and biophysical properties. *Neuron* 1992;8:663–676.
51. Bendahhou S, Cummins TR, Potts JF, Tong J, Agnew WS: Serine-1321-independent regulation of the μ1 adult skeletal muscle Na^+ channel by protein kinase C. *Proc Natl Acad Sci U S A* 1995;92: 12003–12007.
52. Numann R, Hauschka SD, Catterall WA, Scheuer T: Modulation of skeletal muscle sodium channels in a satellite cell line by protein kinase C. *J Neurosci* 1994;14:4226–4236.
53. Fozzard HA, Hanck DA: Structure and function of voltage-dependent sodium channels: comparison of brain II and cardiac isoforms. *Physiol Rev* 1996;76:887–926.
54. Sutkowski EM, Catterall WA: β1 subunits of sodium channels. Studies with subunit-specific antibodies. *J Biol Chem* 1990;265:12393–12399.
55. Schubert B, VanDongen AM, Kirsch GE, Brown AM: β-Adrenergic inhibition of cardiac sodium channels by dual G-protein pathways. *Science* 1989;245:516–519.
56. Ono K, Kiyosue T, Arita M: Isoproterenol, DBcAMP, and forskolin inhibit cardiac sodium current. *Am J Physiol* 1989;256:C1131–C1137.
57. Matsuda JJ, Lee H, Shibata EF: Enhancement of rabbit cardiac sodium channels by β-adrenergic stimulation. *Circ Res* 1992;70:199–207.
58. Ono K, Fozzard HA, Hanck DA: Mechanism of cAMP-dependent modulation of cardiac sodium channel current kinetics. *Circ Res* 1993;72:807–815.
59. Muramatsu H, Kiyosue T, Arita M, Ishikawa T, Hidaka H: Modification of cardiac sodium current by intracellular application of cAMP. *Pfluegers Arch* 1994;426:146–154.
60. Schreibmayer W, Frohnwieser B, Dascal N, Platzer D, Spreitzer B, Zechner R, Kallen RG, Lester HA: β-Adrenergic modulation of currents produced by rat cardiac Na^+ channels expressed in *Xenopus laevis* oocytes. *Receptors Channels* 1994;2:339–350.
61. Frohnwieser B, Chen LQ, Schreibmayer W, Kallen RG: Modulation of the human cardiac sodium channel α-subunit by cAMP- dependent protein kinase and the responsible sequence domain. *J Physiol (Lond)* 1997;498:309–318.
62. Gordon D, Merrick D, Wollner DA, Catterall WA: Biochemical properties of sodium channels in a wide range of excitable tissues studied with site-directed antibodies. Biochemistry 1988;27:7032–7038.
63. Cohen SA, Levitt LK: Partial characterization of the rH1 sodium channel protein from rat heart using subtype-specific antibodies. *Circ Res* 1993;73:735–742.
64. Murphy BJ, Rogers J, Perdichizzi AP, Colvin AA, Catterall WA: cAMP-dependent phosphorylation of two sites in the α subunit of the cardiac sodium channel. *J Biol Chem* 1996;271:28837–28843.

65. Qu Y, Rogers J, Tanada T, Scheuer T, Catterall WA: Modulation of cardiac Na^+ channels expressed in a mammalian cell line and in ventricular myocytes by protein kinase C. *Proc Natl Acad Sci U S A* 1994;91:3289–3293.
66. Qu Y, Rogers JC, Tanada TN, Catterall WA, Scheuer T: Phosphorylation of S1505 in the cardiac Na^+ channel inactivation gate is required for modulation by protein kinase C. *J Gen Physiol* 1996; 108:375–379.
67. Murray KT, Hu NN, Daw JR, Shin HG, Watson MT, Mashburn AB, George AL Jr: Functional effects of protein kinase C activation on the human cardiac Na^+ channel. *Circ Res* 1997;80:370–376.
68. Reuter H: Calcium channel modulation by neurotransmitters, enzymes and drugs. *Nature* 1983;301: 569–574.
69. Bean B: Calcium channels. Gating for the physiologist. *Nature* 1990;348:192–193.
70. Dolphin AC: Facilitation of Ca^{2+} current in excitable cells [see comments]. *Trends Neurosci* 1996; 19:35–43.
71. De Jongh KS, Merrick DK, Catterall WA: Subunits of purified calcium channels: a 212-kDa form of α 1 and partial amino acid sequence of a phosphorylation site of an independent β subunit. *Proc Natl Acad Sci U S A* 1989;86:8585–8589.
72. Yoshida A, Takahashi M, Nishimura S, Takeshima H, Kokubun S: Cyclic AMP-dependent phosphorylation and regulation of the cardiac dihydropyridine-sensitive Ca channel. *FEBS Lett* 1992; 309:343–349.
73. De Jongh KS, Murphy BJ, Colvin AA, Hell JW, Takahashi M, Catterall WA: Specific phosphorylation of a site in the full-length form of the α1 subunit of the cardiac L-type calcium channel by adenosine 3',5'- cyclic monophosphate-dependent protein kinase. *Biochemistry* 1996;35:10392–10402.
74. Rotman EI, De Jongh KS, Florio V, Lai Y, Catterall WA: Specific phosphorylation of a COOH-terminal site on the full-length form of the α1 subunit of the skeletal muscle calcium channel by cAMP-dependent protein kinase. *J Biol Chem* 1992;267:16100–16105.
75. Curtis BM, Catterall WA: Phosphorylation of the calcium antagonist receptor of the voltage-sensitive calcium channel by cAMP-dependent protein kinase. *Proc Natl Acad Sci U S A* 1985;82: 2528–2532.
76. Nastainczyk W, Rohrkasten A, Sieber M, Rudolph C, Schachtele C, Marme D, Hofmann F: Phosphorylation of the purified receptor for calcium channel blockers by cAMP kinase and protein kinase C. *Eur J Biochem* 1987;169:137–142.
77. Jahn H, Nastainczyk W, Rohrkasten A, Schneider T, Hofmann F: Site-specific phosphorylation of the purified receptor for calcium-channel blockers by cAMP- and cGMP-dependent protein kinases, protein kinase C, calmodulin-dependent protein kinase II and casein kinase II. *Eur J Biochem* 1988; 178:535–542.
78. O'Callahan CM, Hosey MM: Multiple phosphorylation sites in the 165-kilodalton peptide associated with dihydropyridine-sensitive calcium channels. *Biochemistry* 1988;27:6071–6077.
79. Haase H, Karczewski P, Beckert R, Krause EG: Phosphorylation of the L-type calcium channel β subunit is involved in β-adrenergic signal transduction in canine myocardium. *FEBS Lett* 1993;335: 217–222.
80. Lai Y, Seagar MJ, Takahashi M, Catterall WA: Cyclic AMP-dependent phosphorylation of two size forms of α 1 subunits of L-type calcium channels in rat skeletal muscle cells. *J Biol Chem* 1990; 265:20839–20848.
81. Beam KG, Adams BA, Niidome T, Numa S, Tanabe T: Function of a truncated dihydropyridine receptor as both voltage sensor and calcium channel. *Nature* 1992;360:169–171.
82. Flockerzi V, Oeken HJ, Hofmann F, Pelzer D, Cavalie A, Trautwein W: Purified dihydropyridine-binding site from skeletal muscle T-tubules is a functional calcium channel. *Nature* 1986;323: 66–68.
83. Hymel L, Striessnig J, Glossmann H, Schindler H: Purified skeletal muscle 1,4-dihydropyridine receptor forms phosphorylation-dependent oligomeric calcium channels in planar bilayers. *Proc Natl Acad Sci U S A* 1988;85:4290–4294.
84. Nunoki K, Florio V, Catterall WA: Activation of purified calcium channels by stoichiometric protein phosphorylation. *Proc Natl Acad Sci U S A* 1989;86:6816–6820.
85. Flockerzi V, Oeken HJ, Hofmann F: Purification of a functional receptor for calcium-channel blockers from rabbit skeletal–muscle microsomes. *Eur J Biochem* 1986;161:217–224.
86. Takahashi M, Seagar MJ, Jones JF, Reber BF, Catterall WA: Subunit structure of dihydropyridine-sensitive calcium channels from skeletal muscle. *Proc Natl Acad Sci U S A* 1987;84:5478–5482.

87. Rohrkasten A, Meyer HE, Nastainczyk W, Sieber M, Hofmann F: cAMP-dependent protein kinase rapidly phosphorylates serine-687 of the skeletal muscle receptor for calcium channel blockers. *J Biol Chem* 1988;263:15325–15329.
88. De Jongh KS, Warner C, Colvin AA, Catterall WA: Characterization of the two size forms of the α 1 subunit of skeletal muscle L-type calcium channels. *Proc Natl Acad Sci U S A* 1991;88:10778–10782.
89. Rotman EI, Murphy BJ, Catterall WA: Sites of selective cAMP-dependent phosphorylation of the L-type calcium channel α1 subunit from intact rabbit skeletal muscle myotubes. *J Biol Chem* 1995; 270:16371–16377.
90. Chang CF, Gutierrez LM, Mundina-Weilenmann C, Hosey MM: Dihydropyridine-sensitive calcium channels from skeletal muscle. II. Functional effects of differential phosphorylation of channel subunits. *J Biol Chem* 1991;266:16395–16400.
91. Mundina-Weilenmann C, Chang CF, Gutierrez LM, Hosey MM: Demonstration of the phosphorylation of dihydropyridine-sensitive calcium channels in chick skeletal muscle and the resultant activation of the channels after reconstitution. *J Biol Chem* 1991;266:4067–4073.
92. Johnson BD, Brousal JP, Peterson BZ, Gallombardo PA, Hockerman GH, Lai Y, Scheuer T, Catterall WA: Modulation of the cloned skeletal muscle L-type Ca^{2+} channel by anchored cAMP-dependent protein kinase. *J Neurosci* 1997;17:1243–1255.
93. Kernell D, Eerbeek O, Verhey BA: Relation between isometric force and stimulus rate in cat's hindlimb motor units of different twitch contraction time. *Exp Brain Res* 1983;50:220–227.
94. Feldmeyer D, Melzer W, Pohl B, Zollner P: Fast gating kinetics of the slow Ca^{2+} current in cut skeletal muscle fibres of the frog. *J Physiol (Lond)* 1990;425:347–367.
95. Garcia J, Avila-Sakar AJ, Stefani E: Repetitive stimulation increases the activation rate of skeletal muscle Ca^{2+} currents. *Pfluegers Arch* 1990;416:210–212.
96. Sculptoreanu A, Scheuer T, Catterall WA: Voltage-dependent potentiation of L-type Ca^{2+} channels due to phosphorylation by cAMP-dependent protein kinase. *Nature* 1993;364:240–243.
97. Johnson BD, Scheuer T, Catterall WA: Voltage-dependent potentiation of L-type Ca^{2+} channels in skeletal muscle cells requires anchored cAMP-dependent protein kinase. *Proc Natl Acad Sci U S A* 1994;91:11492–11496.
98. Gray PC, Tibbs VC, Catterall WA, Murphy BJ: Identification of a 15-kDa cAMP-dependent protein kinase-anchoring protein associated with skeletal muscle L-type calcium channels. *J Biol Chem* 1997;272:6297–6302.
99. Mikami A, Imoto K, Tanabe T, Niidome T, Mori Y, Takeshima H, Narumiya S, Numa S: Primary structure and functional expression of the cardiac dihydropyridine-sensitive calcium channel. *Nature* 1989;340:230–233.
100. Hullin R, Singer-Lahat D, Freichel M, Biel M, Dascal N, Hofmann F, Flockerzi V: Calcium channel β subunit heterogeneity: functional expression of cloned cDNA from heart, aorta and brain. *EMBO J* 1992;11:885–890.
101. Tsien RW, Bean BP, Hess P, Lansman JB, Nilius B, Nowycky MC: Mechanisms of calcium channel modulation by β-adrenergic agents and dihydropyridine calcium agonists. *J Mol Cell Cardiol* 1986; 18:691–710.
102. Yue DT, Herzig S, Marban E: β-Adrenergic stimulation of calcium channels occurs by potentiation of high-activity gating modes. *Proc Natl Acad Sci U S A* 1990;87:753–757.
103. Yatani A, Codina J, Imoto Y, Reeves JP, Birnbaumer L, Brown AM: A G protein directly regulates mammalian cardiac calcium channels. *Science* 1987;238:1288–1292.
104. Kameyama M, Hofmann F, Trautwein W: On the mechanism of β-adrenergic regulation of the Ca channel in the guinea-pig heart. *Pfluegers Arch* 1985;405:285–293.
105. Hartzell HC, Mery PF, Fischmeister R, Szabo G: Sympathetic regulation of cardiac calcium current is due exclusively to cAMP-dependent phosphorylation. *Nature* 1991;351:573–576.
106. Hartzell HC, Fischmeister R: Direct regulation of cardiac Ca^{2+} channels by G proteins: neither proven nor necessary? *Trends Pharmacol Sci* 1992;13:380–385.
107. Chang FC, Hosey MM: Dihydropyridine and phenylalkylamine receptors associated with cardiac and skeletal muscle calcium channels are structurally different. *J Biol Chem* 1988;263:18929–18937.
108. Schneider T, Hofmann F: The bovine cardiac receptor for calcium channel blockers is a 195-kDa protein. *Eur J Biochem* 1988;174:369–375.
109. Yoshida A, Takahashi M, Fujimoto Y, Takisawa H, Nakamura T: Molecular characterization of 1,4-dihydropyridine-sensitive calcium channels of chick heart and skeletal muscle. *J Biochem (Tokyo)* 1990;107:608–612.

110. Mitterdorfer J, Froschmayr M, Grabner M, Moebius FF, Glossmann H, Striessnig J: Identification of PK-A phosphorylation sites in the carboxyl terminus of L-type calcium channel α1 subunits. *Biochemistry* 1996;35:9400–9406.
111. Haase H, Bartel S, Karczewski P, Morano I, Krause EG: *In-vivo* phosphorylation of the cardiac L-type calcium channel β- subunit in response to catecholamines. *Mol Cell Biochem* 1996;163: 99–106.
112. Gao T, Yatani A, Dell'Acqua ML, Sako H, Green SA, Dascal N, Scott JD, Hosey MM: cAMP-dependent regulation of cardiac L-type Ca^{2+} channels requires membrane targeting of PKA and phosphorylation of channel subunits. *Neuron* 1997;19:185–196.
113. Sculptoreanu A, Rotman E, Takahashi M, Scheuer T, Catterall WA: Voltage-dependent potentiation of the activity of cardiac L-type calcium channel α1 subunits due to phosphorylation by cAMP-dependent protein kinase. *Proc Natl Acad Sci U S A* 1993;90:10135–10139.
114. Perez-Reyes E, Yuan W, Wei X, Bers DM: Regulation of the cloned L-type cardiac calcium channel by cyclic-AMP-dependent protein kinase. *FEBS Lett* 1994;342:119–123.
115. Singer-Lahat D, Lotan I, Biel M, Flockerzi V, Hofmann F, Dascal N: Cardiac calcium channels expressed in *Xenopus* oocytes are modulated by dephosphorylation but not by cAMP-dependent phosphorylation. *Receptors Channels* 1994;2:215–226.
116. Zong X, Schreieck J, Mehrke G, Welling A, Schuster A, Bosse E, Flockerzi V, Hofmann F: On the regulation of the expressed L-type calcium channel by cAMP-dependent phosphorylation. *Pfluegers Arch* 1995;430:340–347.
117. Yatani A, Wakamori M, Niidome T, Yamamoto S, Tanaka I, Mori Y, Katayama K, Green S: Stable expression and coupling of cardiac L-type Ca^{2+} channels with β1-adrenoceptors. *Circ Res* 1995;76: 335–342.
118. Perets T, Blumenstein Y, Shistik E, Lotan I, Dascal N: A potential site of functional modulation by protein kinase A in the cardiac Ca^{2+} channel α 1C subunit. *FEBS Lett* 1996;384:189–192.
119. Trautwein W, Hescheler J: Regulation of cardiac L-type calcium current by phosphorylation and G proteins. *Annu Rev Physiol* 1990;52:257–274.
120. Fedida D, Noble D, Spindler AJ: Mechanism of the use dependence of Ca^{2+} current in guinea-pig myocytes. *J Physiol (Lond)* 1988;405:461–475.
121. Zygmunt AC, Maylie J: Stimulation-dependent facilitation of the high threshold calcium current in guinea-pig ventricular myocytes. *J Physiol (Lond)* 1990;428:653–671.
122. Xiao RP, Cheng H, Lederer WJ, Suzuki T, Lakatta EG: Dual regulation of Ca^{2+}/calmodulin-dependent kinase II activity by membrane voltage and by calcium influx. *Proc Natl Acad Sci U S A* 1994;91:9659–9663.
123. Pietrobon D, Hess P: Novel mechanism of voltage-dependent gating in L-type calcium channels. *Nature* 1990;346:651–655.
124. Gurney AM, Charnet P, Pye JM, Nargeot J: Augmentation of cardiac calcium current by flash photolysis of intracellular caged-Ca^{2+} molecules. *Nature* 1989;341:65–68.
125. Bates SE, Gurney AM: Ca^{2+}-dependent block and potentiation of L-type calcium current in guinea-pig ventricular myocytes. *J Physiol (Lond)* 1993;466:345–365.
126. Anderson ME, Braun AP, Schulman H, Premack BA: Multifunctional Ca^{2+}/calmodulin-dependent protein kinase mediates Ca^{2+}-induced enhancement of the L-type Ca^{2+} current in rabbit ventricular myocytes. *Circ Res* 1994;75:854–861.
127. Ono K, Fozzard HA: Two phosphatase sites on the Ca^{2+} channel affecting different kinetic functions. *J Physiol (Lond)* 1993;470:73–84.
128. Frace AM, Hartzell HC: Opposite effects of phosphatase inhibitors on L-type calcium and delayed rectifier currents in frog cardiac myocytes. *J Physiol (Lond)* 1993;472:305–326.
129. Hartzell HC, Hirayama Y, Petit-Jacques J: Effects of protein phosphatase and kinase inhibitors on the cardiac L-type Ca current suggest two sites are phosphorylated by protein kinase A and another protein kinase. *J Gen Physiol* 1995;106:393–414.
130. Wiechen K, Yue DT, Herzig S: Two distinct functional effects of protein phosphatase inhibitors on guinea-pig cardiac L-type Ca^{2+} channels. *J Physiol (Lond)* 1995;484:583–592.
131. Armstrong DL: Calcium channel regulation by calcineurin, a Ca^{2+}-activated phosphatase in mammalian brain. *Trends Neurosci* 1989;12:117–122.
132. Ono K, Fozzard HA: Phosphorylation restores activity of L-type calcium channels after rundown in inside-out patches from rabbit cardiac cells. *J Physiol (Lond)* 1992;454:673–688.
133. Biel M, Ruth P, Bosse E, Hullin R, Stuhmer W, Flockerzi V, Hofmann F: Primary structure and functional expression of a high voltage activated calcium channel from rabbit lung. *FEBS Lett* 1990; 269:409–412.

134. Zamponi GW, Bourinet E, Nelson D, Nargeot J, Snutch TP: Crosstalk between G proteins and protein kinase C mediated by the calcium channel α_1 subunit [see comments]. *Nature* 1997;385: 442–446.
135. Yokoyama CT, Sheng ZH, Catterall WA: Phosphorylation of the synaptic protein interaction site on N-type calcium channels inhibits interactions with SNARE proteins. *J Neurosci* 1997;17:6929–6938.

Advances in Second Messenger and Phosphoprotein Research, Vol. 33

1040-7952/99 $30.00

3

Regulation of Ligand-Gated Ion Channels by Protein Phosphorylation

Sheridan L. Swope,* Stephen J. Moss,[†] Lynn A. Raymond,[‡] and Richard L. Huganir[§]

**Department of Neurology, Division of Neuroscience, Georgetown Institute for Cognitive and Computational Neuroscience, Georgetown University Medical Center, Washington, DC 20007, and [†]The MRC Laboratory of Molecular Cell Biology and Department of Pharmacology, University College of London, London WC1E 6BT, England, and [‡]Department of Psychiatry, The Kinsmen Laboratories for Neurological Sciences, The University of British Columbia, Vancouver, British Columbia, Canada V6T 1Z3, and [§]Department of Neuroscience, Howard Hughes Medical Institute, The Johns Hopkins University Medical School, Baltimore, Maryland 21205*

INTRODUCTION

Neurotransmitter receptors transduce presynaptic information into a postsynaptic response via the binding of a chemical messenger. One class of receptors includes the ligand-gated ion channels, which, upon binding of neurotransmitter, bring about a rapid change in electrical potential across the membrane by opening an intrinsic ion channel. The predominant ligand-gated ion channels in the mammalian central and peripheral nervous systems are the nicotinic acetylcholine receptors (AChR), the glutamate receptors, the γ-aminobutyric acid (GABA) receptors, and the glycine receptors. Because of the pivotal role of ligand gated ion channels in synaptic transmission, their regulation provides a molecular mechanism to bring about changes in the efficiency of synaptic transmission.

Protein phosphorylation is an important post-translational modification known to regulate essentially all cellular processes. Based on their substrate specificity, protein kinases are broadly classified as either protein serine/threonine or protein tyrosine kinases. The most common protein serine/threonine kinases are those that are regulated by intracellular second messengers such as the cAMP-dependent protein kinase (PKA), the phospholipid/calcium-dependent protein kinase (PKC), and the calcium/calmodulin-dependent protein kinases (CaMK). The two major classes of protein tyrosine kinases are the receptor protein tyrosine kinases, including growth factor and neurotrophin receptors, and the nonreceptor tyrosine kinases, for which the major subclass is the Src family. Many protein kinases are most highly expressed in the central nervous system, suggesting their importance in neuronal function.

Accumulating evidence supports phosphorylation as an important mechanism in the regulation of ligand-gated ion channels. This regulation includes alterations in channel properties as well as control of receptor expression, mobility, or localization. In some instances, regulation is mediated by a direct phosphorylation of the receptor. In this chapter we review the role of protein phosphorylation in the modulation of ligand-gated ion channels. The neuromuscular junction and the AChR are the most thoroughly characterized synapse and ligand-gated ion channel, respectively. We begin by describing what is known about how phosphorylation regulates the AChR at the neuromuscular junction. We will then go on to review the evidence that phosphorylation modulates ligand gated ion channels in the central nervous system, including $GABA_A$ receptors and NMDA and non-NMDA glutamate receptors. In addition, we discuss the role of phosphorylation of neurotransmitter receptors in synaptic plasticity.

THE NICOTINIC ACETYLCHOLINE RECEPTOR

At the neuromuscular junction, the AChR is the ligand-gated ion channel that mediates rapid depolarization of the postsynaptic muscle cell membrane in response to binding of acetylcholine released from the motor neuron (1). Because nature has provided the electric organs of *Torpedo californica* as an enriched source and α-bungarotoxin as an irreversible antagonist, the AChR has historically served as a model for study of the structure, function, and regulation of ligand-gated ion channels. The AChR is a pentameric protein (see Fig. 1A) comprised of four homologous subunits in a stoichiometry of $\alpha_2\beta\gamma\delta$ (2). Each subunit has a large extracellular N-terminus, which in the α subunits binds acetylcholine, four putative transmembrane domains (M1–M4), a large intracellular loop between transmembrane domains M3 and M4, and an extracellular C-terminus (see Fig. 1B). The five subunits form a ring around a central pore which is lined by the second transmembrane domain of each subunit. Upon binding of acetylcholine, the channel opens allowing influx of cations and depolarization of the muscle cell plasma membrane (3).

Serine and Tyrosine Phosphorylation of the AChR

The AChR from *Torpedo,* the first receptor or ion channel shown to be phosphorylated, is phosphorylated on both serine and tyrosine residues. All the phosphorylation sites are located on the major intracellular loop between M3 and M4 (see Fig. 1B and Table 1). Serine phosphorylation is catalyzed by both PKA (4)and PKC (5,6). As demonstrated by sequencing phosphorylated peptide fragments of the receptor,

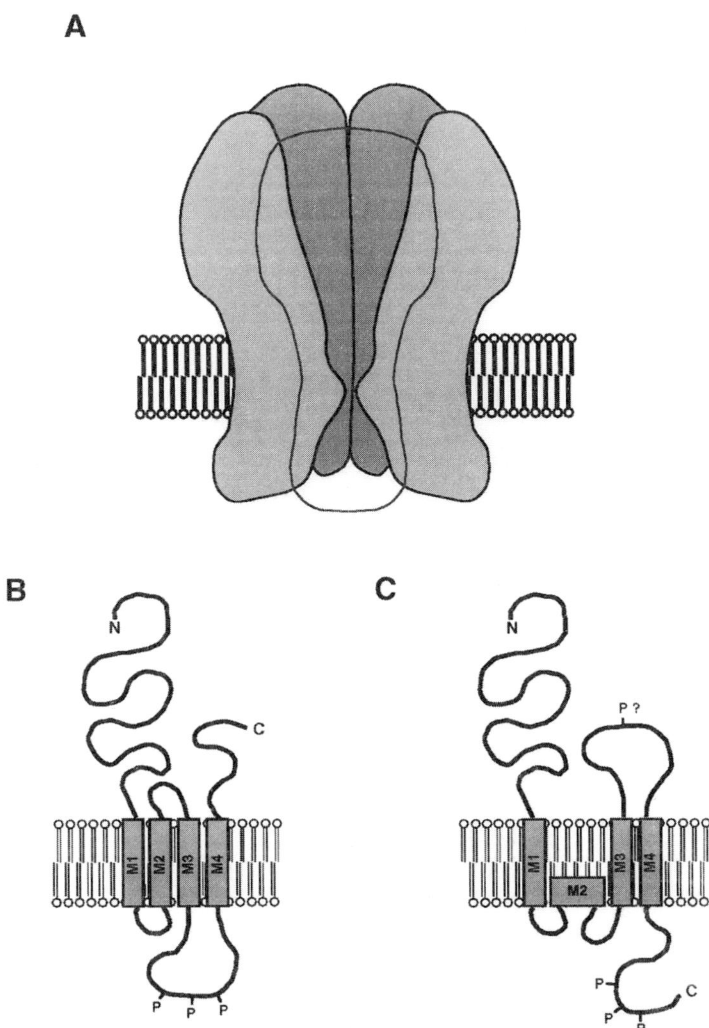

FIG. 1. Schematic structure of ligand gated ion channels. **(A)** Oligomeric structure of these receptors viewed through the plane of the membrane. Each receptor is believed to be a pentameric complex of homologous subunits. **(B)** Transmembrane topology of AChR and $GABA_A$ receptors. Each subunit has a large extracellular N-terminal domain, four transmembrane domains, and a major intracellular loop between M3 and M4. All identified phosphorylation sites have been localized to the major intracellular loop region (P). **(C)** Transmembrane topology of the glutamate receptors. Each subunit has a large extracellular N-terminal region and four transmembrane domains. However, the M2 region is thought to form a hairpin structure resulting in an intracellular C-terminal domain. Many of the identified phosphorylation sites are located on the C-terminal intracellular domain, although some have been reported to be located on the loop between M3 and M4.

TABLE 1. *Amino acid sequence of identified phosphorylation sites on ligand-gated ion channels*

Receptor/ Subunit	Sequence	Protein kinase
AChR/β	348IleSerArgAlaAsnAspGlu**Tyr**PheIleArgLys359	PTK
AChR/γ	346LysProGlnProArgArgArg**Ser**SerPheGlyIle357	PKA
	359IleLysAlaGluGlu**Tyr**IleLeuLys367	PTK
AChR/δ	355LeuLysLeuArgArgSer**Ser**SerValGlyTyr365	PKA
	355LeuLysLeuArgArgSerSer**Ser**ValGlyTyr365	PKC
	373PheAsnIleLys**Ser**Arg**Ser**GluLeuMetPhe383	PKC
	368LysAlaGlnGlu**Tyr**PheAsnIleLysSerArg378	PTK
$GluR_1$	841LeuProArgAsn**Ser**GlyAlaGly848	PKA
	826LeuIleProGlnGln**Ser**IleAsnGluAlaIleArg837	PKC
	622ValGluArgMetVal**Ser**ProIleGlu630	CaMKII
$GluR_2$	690ValAlaArgValArgLys**Ser**LysGlyLysTyrAla701	PKC
$GluR_6$	681ArgArgGln**Ser**ValLeuValLys688	PKA
NR_1	886ThrLeuAlaSer**Ser**PheLysArgArgArg**Ser**SerLysAsp899	PKC
	886ThrLeuAlaSerSerPheLysArgArgArgSer**Ser**LysAsp899	PKA
$GABA_A/\beta_1/\beta_2/\beta_3$	401LysGlyArgIleArgArgArgAla**Ser**GlnLeuLys413	PKA/PKC
$GABA_A/\gamma_{2L}$	338LeuLeuArgMetPhe**Ser**PheLys345	PKC
$GABA_A/\gamma_{2S}$	323AsnArgLysPro**Ser**LysGlnLysGlnLys332	PKC
$GABA_A/\gamma_2$	359GlnGlnArgGlnAspAsp**Tyr**Gly**Tyr**GlnLysLysLeuAsp372	PTK

PKA phosphorylation occurs on Ser-353 of the γ subunit and Ser-361 of the δ subunit (7). PKC phosphorylates the δ subunit on Ser-362 and Ser-377/379, as indicated by peptide sequencing and the use of synthetic peptide substrates as well as anti-peptide antibodies (6,8). In addition to serine phosphorylation, the AChR is phosphorylated to a high stoichiometry by endogenous protein tyrosine kinases in isolated postsynaptic membranes (9). In *Torpedo,* the sites for tyrosine phosphorylation of the AChR are Tyr-355 of the β subunit, Tyr-364 of the γ subunit, and Tyr-372 of the δ subunit (10). It was demonstrated that the AChR can be phosphorylated *in vitro* on tyrosine residues by two Src-like protein tyrosine kinases, Fyn and Fyk, or in transfected cells via the activation of the receptor protein tyrosine kinase MuSK (11,12). Although the specific sites for phosphorylation of the AChR by Fyn and Fyk have not been determined, activation of MuSK results in phosphorylation of Tyr-390 on the mouse β subunit, which is analogous to Tyr-355 of the *Torpedo* β subunit (12). Except for the PKC and protein tyrosine kinase sites on the γ subunit, the consensus sites for both serine and tyrosine phosphorylation of the *Torpedo* AChR are conserved in mammalian receptors, supporting the importance of AChR phosphorylation (13).

The phosphorylation of the AChR by both serine and tyrosine specific protein kinases is well documented. In contrast, serine and tyrosine phosphatases involved in dephosphorylation of the AChR have not been as intensely investigated. A novel tyrosine-specific protein phosphatase that dephosphorylates the AChR has been identified and purified (14), however, serine-specific protein phosphatases that dephosphorylate the AChR have not been characterized.

Functional Effects of Phosphorylation of the AChR

Regulation of Desensitization of the AChR

The functional roles for phosphorylation of the AChR have been investigated. In the continued presence of acetylcholine, the AChR ion channel inactivates, a process known as desensitization. Serine phosphorylation by PKA and PKC, as well as tyrosine phosphorylation of the AChR, regulates the rate of receptor desensitization. The effect of PKA has been most carefully characterized. When PKA-phosphorylated AChR of *Torpedo* is reconstituted into liposomes, the rapid phase of desensitization is eightfold faster for the phosphorylated AChR than for the nonphosphorylated receptor (15). These results agree with experiments examining the effect of PKA phosphorylation of the mammalian AChR. In cultured mammalian muscle cells, both cAMP analogs and forskolin, an activator of adenylate cyclase, increase AChR δ subunit phosphorylation (16,17). Analogs of cAMP (18,19) and forskolin (18,20,21) also increases the rate of AChR desensitization. The effects of forskolin on AChR phosphorylation and desensitization are strikingly similar in time course, concentration dependence, and pharmacology. In addition, as discussed shortly, a neuropeptide, calcitonin gene related peptide (CGRP), that induces receptor-mediated activation of adenylate cyclase also stimulates AChR phosphorylation (22) and an increase in the rate of desensitization (19).

Phosphorylation of the AChR, mediated by PKC and protein tyrosine kinase, also appears to regulate the rate of receptor desensitization. Phorbol esters, activators of PKC, stimulate serine phosphorylation of the chick δ subunit (23) and rat δ and γ subunits of the AChR (24). In addition, phorbol esters increase the rate of receptor desensitization and decrease the sensitivity to acetylcholine (25). Thus, pharmacological evidence suggests that an increase in the desensitization rate is a result of AChR phosphorylation by PKC. The effect of AChR tyrosine phosphorylation was examined by patch–clamp analysis of *Torpedo* receptor reconstituted into phospholipid vesicles. The rapid rate of AChR desensitization shows a striking correlation with the stoichiometry of tyrosine phosphorylation (26). Therefore, tyrosine phosphorylation, analogous to serine phosphorylation by PKA and PKC, appears to modulate the desensitization rate of the AChR.

Regulation of the Synaptic Targeting of the AChR

Intense interest is focused on the role of tyrosine phosphorylation in clustering of AChRs during synaptogenesis. Clustering of postsynaptic components at the nerve–muscle contact seems to be regulated by factors contained in the synaptic extracellular matrix, including the neuronally derived protein agrin (27,28). Agrin-induced AChR clustering in muscle appears to be mediated by tyrosine phosphorylation. In fact, a role for protein tyrosine kinase activation has been implicated in the clustering of postsynaptic components induced by a variety of neuronal and nonneuronal stimuli including nerve (29,30), agrin (30,31), basic fibroblast growth factor (32), heparin binding, growth-associated molecule (33), expression of exogenous rapsyn (34), electric fields (35), and polymer beads (36). Innervation of muscle regulates both the clustering and tyrosine phosphorylation of the AChR (29,30).

Agrin is an extracellular matrix protein that mediates the effect of the neuron to induce AChR phosphorylation and clustering (30,31,37,38,39). The action of agrin to stimulate AChR clustering is blocked by protein tyrosine kinase inhibitors (37,40). In addition, agrin-induced tyrosine phosphorylation precedes AChR aggregation, suggesting that the phosphorylation is not a result of clustering but may mediate clustering (31). It is not clear whether the inhibition of agrin-stimulated AChR clustering by protein tyrosine kinase inhibitors is due to block of receptor phosphorylation or, alternatively, the phosphorylation of other synaptic components involved in AChR clustering. In addition, although agrin-deficient mice show dramatic disorganization of the neuromuscular junction, focal AChR-rich aggregates are observed, suggesting the existence of additional factors that regulate AChR clustering (39).

Mutated AChRs lacking the tyrosine phosphorylation site have been used to examine the ability of another postsynaptic component, rapsyn, to mediate receptor clustering. Rapsyn, formerly known as the 43-kDa protein, is an intracellular peripheral membrane protein thought to anchor the AChR to the synaptic cytoskeleton (1,41). When rapsyn is coexpressed with the AChR in heterologous systems, coclustering of the receptor with rapsyn is seen (42–44). These rapsyn-induced AChR clusters contain phosphotyrosine, and coexpression of rapsyn with the AChR results in the tyrosine phosphorylation of the AChR (34). Furthermore, rapsyn-deficient knockout mice do not develop postsynaptic specializations including AChR clusters (45). Interestingly, myotubes cultured from rapsyn-minus embryos do not aggregate AChR in response to agrin (45). Taken together, these data suggest that rapsyn may be important for the effective tyrosine phosphorylation and clustering of the AChR. However, mutational analysis demonstrated that the tyrosine phosphorylation sites of the AChR are not required for rapsyn-induced clustering of the AChR in heterologous cells (34,44). Thus, if rapsyn does mediate the effect of agrin, these data argue that, at least in these heterologous expression systems, tyrosine phosphorylation of the AChR is not required for aggregation.

Signal Transduction Cascades Regulating AChR Phosphorylation

The signal transduction cascades for phosphorylation of the AChR by both serine- and tyrosine-specific kinases have been investigated. The most likely candidate for an extracellular factor that regulates AChR phosphorylation by PKA is CGRP. This neuropeptide is contained in the presynaptic vesicles at the neuromuscular junction (46). It is known that CGRP receptors are linked to adenylate cyclase via a guanine nucleotide binding protein (47) and treatment of rat myotubes with CGRP results in cAMP accumulation (22). CGRP also stimulates AChR phosphorylation on the α and δ subunits (22) and enhances the rapid phase of receptor desensitization (19). As described earlier, cAMP analogs and forskolin, activators of the PKA-mediated pathway, stimulate AChR phosphorylation and increase the rate of receptor desensitization. Taken together, these data indicate that CGRP is released from the motor neuron and binds to its postsynaptic plasma membrane receptor, thereby activating adenylyte cyclase via a G protein mediated process, resulting in the generation of cAMP and activation of PKA. The subsequent phosphorylation of the AChR results in an increase in the rate of receptor desensitization.

Another extracellular signal that may regulate PKA-mediated phosphorylation of the AChR is adenosine. Stimulation of cultured rat muscle with carbachol leads to the secretion of adenosine, while treatment with either adenosine or carbachol leads to elevated cAMP levels. In addition, an adenosine receptor antagonist partially blocks carbachol-induced desensitization of the AChR (48). These data suggest an autoregulatory mechanism by which binding of acetylcholine to the AChR leads to secretion of adenosine, which activates adenylyte cyclase. The subsequent rise in cAMP and stimulation of PKA would result in AChR phosphorylation and regulation of channel desensitization.

Several studies support acetylcholine as the extracellular signal that stimulates PKC-mediated phosphorylation and regulation of the AChR. Carbachol and phorbol ester stimulate the phosphorylation of AChR subunits in both chick (δ) and rat (δ and γ) cultured muscle cells (23,24). Carbachol-induced phosphorylation is blocked by curare, demonstrating the involvement of the nicotinic AChR. In addition, the effect of carbachol is dependent on extracellular calcium, blocked by channel blockers, and mimicked by calcium ionophore, supporting the importance of calcium influx (24). Furthermore, carbachol and phorbol ester both regulate AChR desensitization (25,49). Taken together, these studies suggest that activation of the nicotinic AChR by acetylcholine leads to an influx of calcium and activation of PKC, which phosphorylates the receptor, resulting in an increase in the rate of AChR desensitization. However, more recent observations argue against this model. Down regulation of PKC with phorbol ester does not block the effect of carbachol to stimulate AChR phosphorylation (24). These data suggest that a novel calcium-dependent signal transduction pathway mediates autoregulation of the AChR.

As described earlier, agrin is now believed to be the extracellular factor that stimulates AChR tyrosine phosphorylation and clustering. Agrin is synthesized by the motor neuron and is deposited in the synaptic basal lamina, where it stably associates with the extracellular matrix (50). Treatment of cultured chick myotubes with agrin induces phosphorylation of the AChR on the β and δ subunits (30,31).

The identity of the agrin receptor is an area of intense investigation. High affinity binding of agrin to α-dystroglycan, an extracellular component of the dystrophin-associated glycoprotein complex, has been demonstrated (51–54), and a monoclonal antibody to a-dystroglycan inhibits agrin-induced AChR aggregation, suggesting that the dystrophin-associated glycoprotein complex is the functional agrin receptor (52). Dystrophin is the gene product defective in Duchenne muscular dystrophy (55), and the transmembrane dystrophin-associated glycoprotein complex is thought to link the extracellular matrix with the actin cytoskeleton, thus stabilizing skeletal muscle structure (56). Although agrin binds directly to α-dystroglycan, the dystrophin-associated glycoprotein complex does not appear to be a functional agrin receptor, since the relative binding of agrin splice isoforms and fragments to α-dystroglycan does not correlate with stimulation of AChR clustering (54,57).

MuSK, a receptor tyrosine kinase specifically expressed in skeletal muscle (58), is an excellent candidate for the functional agrin receptor. MuSK expression parallels expression of the AChR. During development, MuSK is up-regulated upon myotube fusion, while in adult muscle expression is restricted to the postsynaptic membrane but is up-regulated by denervation (58). MuSK-deficient knockout mice die perinatally (59). Neuromuscular junctions including AChR aggregates do not form in the mutant mice, and death results from an inability to breathe. Myotubes cultured from these mice express normal levels of AChR, but the AChRs do not cluster in response to agrin treatment (60). When control myotubes are treated with agrin, a rapid induction of MuSK autophosphorylation occurs, and the relative effectiveness of agrin splice isoforms to activate MuSK correlates with their ability to induce AChR clustering (60). In addition, agrin forms a complex with MuSK, although the binding is dependent on a myotube-specific accessory (60). These data support MuSK as a component of the functional agrin receptor.

MuSK autophosphorylation can also be stimulated upon coexpression with the cytoskeletal protein rapsyn (12). As described earlier, postsynaptic specializations including AChR clusters are absent in rapsyn-deficient knockout mice, and the cultured myotubes do not aggregate AChR in response to agrin (45). In addition, agrin induces rapsyn clustering (28). These data suggest that that rapsyn mediates the effect of agrin on AChR clustering. Coexpression of MuSK with rapsyn also induces the coclustering of MuSK and the AChR. Furthermore, MuSK activation by rapsyn results in phosphorylation of the AChR β subunit (12), the same subunit that is phosphorylated in response to innervation and agrin treatment of muscle (30). Although the exact molecular mechanism has not been clarified, these data suggest that MuSK and rapsyn are intimately involved in the signal transduction pathway for agrin-stimulated AChR phosphorylation and aggregation.

In addition to tyrosine phosphorylation of the AChR by MuSK activation, the AChR is a substrate *in vitro* for Fyn and Fyk, two Src-like protein tyrosine kinases (11). These two kinases are very abundant in *Torpedo* electric organ, together comprising almost half the total protein tyrosine kinase activity of the postsynaptic membranes. In addition, Fyn and Fyk stably associate with the AChR via binding of the tyrosine-phosphorylated receptor δ subunit with the SH2 domains of the kinases (11,60a). Complex formation between Fyn and Fyk and the AChR may either prevent dephosphorylation or promote the stabilization of the postsynaptic receptor aggregates. Fyn and Fyk are cytoplasmic kinases lacking extracellular domains. The signal transduction pathways by which these kinases are activated to directly phosphorylate the AChR are unknown. Perhaps these pathways involve the agrin–MuSK–rapsyn cascade or other novel molecular mechanisms.

THE GABA$_A$ RECEPTORS

GABA$_A$ receptors are the major sites of fast synaptic inhibition in the brain. Activation of these receptors leads to chloride flux, which usually results in hyperpolarization of the neuronal membrane. Molecular cloning has revealed that $GABA_A$ receptors are members of a channel superfamily that includes AChRs and glycine receptors (61). A large number of $GABA_A$ receptor subunits have been identified which can be divided into subunit classes based on sequence homology: α_{1-6}, β_{1-4}, γ_{1-4}, δ_1 (62,63). Like the AChR, the $GABA_A$ receptor is proposed to be a pentamer of these subunits (Fig. 1A). Each subunit is thought to have a large N-terminal extracellular domain and four transmembrane domains (M1–M4), with a large intracellular loop between the third and fourth transmembrane domain (Fig. 1B). All the identified phosphorylation sites on $GABA_A$ receptors have been localized to this region (Table 1). The repertoire of $GABA_A$ receptor subunits is further enhanced by alternative splicing of the mRNAs of the α_6, β_2, β_4, and γ_2 subunits in a number of species (62,63). In the case of the γ_2 subunit an insertion of eight amino acids within the M3/M4 loop distinguishes two forms of this subunit termed γ_2L and γ_2S (64,65). This insertion contains a serine residue, which fits the consensus for phosphorylation by a number of protein kinases including PKC.

In situ hybridization and immunohistochemical methodologies utilizing subunit-specific antisera have demonstrated a large developmental and regional heterogeneity in the expression of $GABA_A$ receptor subunits in the central nervous system, with neurons of many types expressing multiple subunits (62,63). Expression of receptor cDNAs in *Xenopus* oocytes and mammalian cell lines has been used to determine the minimal subunit requirement for the production of GABA-gated chloride channels. Expression of α and β subunits produces GABA-gated currents, which are modulated by barbiturates and inhibited by GABA antagonists, channel blockers, and zinc ions, but are not enhanced by benzodiazepines (62,63). Coexpression of α, β, and either γ_2 or γ_3 subunits is crucial in conferring benzodiazepine sensitivity and relative insensitivity to inhibition by zinc (62,63). These data suggest that $GABA_A$ receptors *in vivo* consist of α, β, and γ subunits.

Serine and Tyrosine Phosphorylation of the $GABA_A$ Receptor

Within the major intracellular domains of many $GABA_A$ receptor subunits there are consensus sites for a number of both serine/threonine and tyrosine protein kinases. Most notably, all β subunit cDNAs isolated to date encode a conserved site for phosphorylation by a number of second messenger–dependent protein kinases, including PKA. The γ_2L subunit encodes a site for phosphorylation that conforms to the consensus of a number of protein kinases including PKC. Both the γ_1 and γ_2 subunits contain conserved consensus sites for tyrosine phosphorylation.

Affinity-purified preparations of $GABA_A$ receptors have been shown to be phosphorylated by a number of different protein kinases. PKA and PKC both appear to phosphorylate "β subunits" based on apparent molecular masses (53–57 kDa) observed on SDS-PAGE (66–68). In addition, a receptor-associated kinase has been reported to phosphorylate an "α subunit" based on its apparent molecular mass of 51 kDa (69,70). Recent studies have demonstrated that purified receptors are also substrates of v-Src *in vitro,* which phosphorylates "β" and "γ" subunits as determined by migration on SDS-PAGE (71). Because of the heterogeneity of affinity-purified receptor preparations and the low abundance of $GABA_A$ receptors in the brain, the precise identity of the subunits phosphorylated in these studies remains uncertain.

To circumvent these problems, phosphorylation studies of the intracellular domains of $GABA_A$ receptor subunits expressed in *E. coli* have been performed. This approach has permitted the identification of high affinity kinase substrates in the intracellular domains of some $GABA_A$ receptor subunits. The murine β_1 subunit intracellular domain is phosphorylated to high stoichiometry, with high affinity, by cGMP-dependent protein kinase (PKG), PKC, CaMKII, and PKA on Ser-409 (72,73). This conserved residue is also phosphorylated in the β_2 and β_3 subunits by these four kinases *in vitro* (74). Additional sites in the β_3 and β_1 subunits are also phosphorylated by (73,74), and the β_1 subunit intracellular domain can also be phosphorylated by v-Src (71). Phosphorylation of the γ_2 subunit has been analyzed using similar methodologies (64,65). These studies reveal that Ser-343 within the eight amino acid insertion that differentiates the γ_2S and γ_2L forms of the γ_2 subunit is a high affinity substrate of both PKC and CaMKII (64,73,75,76). As well, both the γ_2L and γ_2S intracellular domains are phosphorylated on additional residues by PKC and CaMKII. PKC phosphorylates Ser-327, while CaMKII phosphorylates both Ser-348 and Thr-350 (73,75). The γ_2L intracellular domain can also be phosphorylated by v-Src (71).

The phosphorylation of $GABA_A$ receptor subunits expressed in human embryonic kidney cells (HEK 293) cells has also been examined. $GABA_A$ receptors composed of either $\alpha_1\beta_1$ or $\alpha_1\beta_1\gamma_2$S subunits expressed in HEK 293 cells are phosphorylated specifically by PKA on Ser-409 within the M3–M4 intracellular domain of the β subunit (72). Receptors composed of $\alpha_1\beta_1$, $\alpha_1\beta_1\gamma_2$S, and $\alpha_1\beta_1\gamma_2$L subunits are also phosphorylated by PKC, which specifically phosphorylates Ser-409 in the β_1 subunit (77), Ser-327 in the γ_2S subunit, and serines 327 and 343 in the γ_2L subunit (77).

Tyrosine phosphorylation of receptors expressed in HEK 293 cells has also been examined. This cell line has very low steady state levels of phosphotyrosine. This allows $GABA_A$ receptor tyrosine phosphorylation to be controlled by coexpressing activated tyrosine kinases (78). Coexpression of $GABA_A$ receptors consisting of α_1, β_1, and γ_2L subunits with vSrc results in tyrosine phosphorylation of the γ_2L subunit on residues Tyr-365 and Tyr-367. The β_1 subunit is also phosphorylated but to a much lower stoichiometry on residues Tyr-370 and Tyr-372. These sites for tyrosine phosphorylation are found within the predicted intracellular domain of these two subunits (78).

Functional Effects of $GABA_A$ Receptor Phosphorylation

PKA-Mediated Phosphorylation

PKA-mediated phosphorylation appears to regulate $GABA_A$ receptor desensitization and to inhibit receptor activation in cortical neurons and cortical synaptoneurosomes (79–83) but not in spinal cord neurons (84). However, in the retina and in cerebellar Purkinje cells, vasointestinal peptide (VIP) and noradrenaline enhance

$GABA_A$ receptor responses (85–87). These results suggests that PKA phosphorylation may inhibit, have no effect on, or enhance $GABA_A$ receptor function in these differing neuronal systems. To determine the molecular basis for this differential regulation, the phosphorylation of recombinant receptors of defined subunit composition has been examined.

The functional effects of PKA-mediated phosphorylation have been studied on $GABA_A$ receptors expressed in HEK 293 cells. Phosphorylation of S409 in the β_1 subunit mediated either by intracellular dialysis with cAMP or with the catalytic subunit of PKA results in a time-dependent decrease of GABA-evoked currents (72). The inhibition is greater in magnitude for receptors composed of $\alpha_1\beta_1\gamma_2S$ subunits than for receptors composed of only α_1 and β_1 subunits. Receptor desensitization is also modulated in receptors composed of $\alpha_1\beta_1$ subunits. All these functional effects could be abolished by mutation of Ser-409 in the β_1 subunit, the site of PKA phosphorylation in these receptors (72). The catalytic subunit of PKA has been used to demonstrate similar modulation of $GABA_A$ receptor function in cultured superior cervical ganglia (SCG), spinal cord neurons, and cerebellar granule cells (72,82,88). GABA-induced chloride flux from brain microsacs is also reduced by inclusion of the catalytic subunit of PKA (89). In addition to effects on channel function, chronic activation of PKA has been reported to enhance the assembly of $GABA_A$ receptors (90).

These findings with recombinant receptors do not explain the enhancements of $GABA_A$ receptor function seen in the retina and cerebellar Purkinje cells, where short-term exposure to VIP and noradrenaline has been shown enhance $GABA_A$ receptor currents (85,87). In cerebellar Purkinje cells, the effect of noradrenaline can be mimicked by cAMP analogs and blocked by the specific PKA inhibitor peptide (PKI) (91). This enhancement of $GABA_A$ receptor function can be induced by glutamate receptor activation and elevated levels of intracellular calcium (92). Elucidation of the reasons for the differences in apparent effects of PKA activation in these neuron types compared with SCG, spinal cord, and cerebellar granule neurons will require further experimentation. Receptor heterogeneity may underlie these differing modes of regulation, and this possibility can be most conveniently addressed by means of recombinant methodologies.

PKC-Mediated Phosphorylation

The first indication that $GABA_A$ receptor function could be modulated by PKC activity came from expression studies using rat or chick brain mRNA in *Xenopus* oocytes. Activation of PKC by means of phorbol ester treatment resulted in a significant inhibition of GABA-induced whole-cell currents (93,94). This observation has been further examined by utilizing heterologous expression of receptor cDNAs. These initial studies have shown that PKC activity can inhibit the function of a range of receptors constructed from α_{1-5}, β_{1-2}, and γ_2 subunits utilizing phorbol ester treatment for PKC activation (95–97). In the case of receptors composed of $\alpha_1\beta_1\gamma_2L$ subunits the effects of phorbol esters can be blocked by PKC inhibitory peptide

(PKCI) (96). The role of specific phosphorylation sites for PKC within defined receptor subunits, as determined by biochemical methodologies, has been examined by site-specific mutagenesis. PKC exerts its inhibitory effect on receptors composed of $\alpha_1\beta_1$ or β_2 and either γ_2L or γ_2S subunits via phosphorylation of Ser-409 within the β_1 subunit, Ser-327 in both the γ_2L and γ_2S subunits, and Ser-343 within the γ_2L subunit (77,98).

The effects of phosphorylation are related to agonist concentration-greater inhibition was seen at high concentrations of GABA, with receptors incorporating the γ_2L subunit. Selective mutagenesis revealed that phosphorylation at any of the sites on the β_1 or γ_2 subunits is sufficient to produce negative modulation. However, phosphorylation at Ser-343, which is contained within the eight extra amino acids within the γ_2L subunit, produces the largest effect, suggesting that the phosphorylation sites are not functionally equivalent (77).

In addition to the effects on GABA-evoked current amplitude, PKC phosphorylation of S409 in the β_1 subunit modulates the rate of desensitization for receptors composed of the α_1 and β_1 subunits. Phorbol ester treatment has also been used to examine the regulation of $GABA_A$ receptors by PKC in neurons. Responses in SCG neurons are inhibited by phorbol ester treatment but not by α-phorbols (77). Likewise, GABA-induced chloride flux from cerebellar microsacs can be selectively inhibited by PKC activators (96). Inclusion of the specific peptide inhibitor PKCI in recording pipettes enhances $GABA_A$-mediated inhibitory postsynaptic potentials (99). In contrast to the results already described, intracellular dialysis of trypsin-activated PKC into L929 cells transiently expressing $GABA_A$ receptors composed of α_1, β_1 and γ_2L subunits has been reported to enhance receptor function (100). This enhancement could be blocked by the PKCI peptide. Whether these effects are due to direct receptor phosphorylation by PKC, however, has not been determined.

PKG- and CaMKII-Mediated Phosphorylation

The functional effects of PKG and CaMKII on $GABA_A$ receptor function are poorly understood. However, $GABA_A$-mediated currents from the nucleus of the tractus solitarius neurons of the rat are inhibited by cGMP (101). Calcium ions can inhibit or activate $GABA_A$-mediated currents in differing neuronal populations (102). The role of CaMKII in these actions of calcium is not well characterized. However, recent experiments have shown that CaMKII can activate $GABA_A$-mediated whole-cell currents in cultured rat spinal cord neurons (103). Whether these effects were mediated by direct receptor phosphorylation by CaMKII was not determined.

Protein Tyrosine Kinase Mediated Phosphorylation

The effects of Src-mediated phosphorylation on the functional properties of $GABA_A$ receptors composed of $\alpha_1\beta_1$ and γ_2L subunits expressed in HEK 293 cells was analyzed in 1995. Tyrosine phosphorylation of Tyr-365 and Tyr-367 in the γ_2L subunit by Src enhances GABA-induced currents (78). In agreement with this, Valenzuela et al. (71) have shown that tyrosine kinase inhibitors reduce GABA-gated currents recorded from *Xenopus* oocytes expressing $GABA_A$ receptors. Neuronal $GABA_A$ receptors can also be modulated by tyrosine phosphorylation. Whole-cell currents from cultured SCG neurons are reduced by exposure to tyrosine kinase inhibitors and enhanced by tyrosine phosphatase inhibitors (78). Furthermore, GABA-mediated chloride flux from brain microsacs is decreased by tyrosine kinase inhibitors (71). These results suggest that tyrosine phosphorylation may be a means of enhancing or maintaining $GABA_A$ receptor function. Interestingly, phosphorylation by unidentified kinases (not PKA or PKC) or "phosphorylation factors" has been implicated in preventing the washout or rundown of $GABA_A$ receptor responses in a variety of neuron types (104–106).

Multiple Effects of Receptor Phosphorylation Activation and Inhibition?

There is general agreement that $GABA_A$ receptors are the substrates of multiple protein kinases. The β and γ_2 subunits appear to be the major substrates for phosphorylation by PKA, PKC, PKG, CaMKII, and the prototypic tyrosine kinase Src, based on both *in vitro* and *in vivo* analysis. Recombinant studies have demonstrated that direct phosphorylation by either PKA or PKC inhibits receptor function. PKA phosphorylation may also be important in receptor assembly. Similar regulation of neuronal receptors by PKA and PKC phosphorylation also has been reported. In some neuronal preparations, however, PKA-mediated phosphorylation increases receptor function. These discrepancies may be due to differences in receptor structure or to the activation of other signaling pathways. The functional effects of PKC phosphorylation on both neuronal and recombinant receptors are almost exclusively inhibitory when PKC is activated by means of phorbol ester treatment. However, the use of PKC activated by trypsin appears to activate $GABA_A$ receptor function.

In contrast, tyrosine phosphorylation of sites on the γ_2 subunit appears to enhance the function of recombinant receptors composed of the α_1, β_1, and γ_2L subunits. Similar modulation of receptor function is also seen in cultured SCG neurons. These results suggest that this process may be important in enhancing receptor function. Recent observations with tyrosine kinases are also important in light of the varying effects reported on receptor function by PKC. A novel protein tyrosine kinase homologous to focal adhesion kinase, called PYK_2 or CAK_β, which is rapidly activated by elevations in intracellular calcium and also by PKC, has been identified (107,108). This kinase, widely expressed throughout the brain and also in a number cell lines (108), has been shown to phosphorylate and regulate inwardly rectifying potassium channels (107). Given that tyrosine phosphorylation enhances receptor function (71,78), some of the variable effects seen on $GABA_A$ receptor function on activation of PKC may be mediated by a PYK-like tyrosine kinase.

THE GLUTAMATE RECEPTORS

Fast excitatory neurotransmission in the mammalian central nervous system is mediated largely by ionotropic glutamate receptors. A role for these receptors has been implicated in processes ranging from synaptogenesis, learning, memory, and sensorimotor processing to acute and chronic neuropathological conditions (109–113). Therefore, elucidation of mechanisms by which the function of these receptors may be modulated will enhance understanding of synaptic plasticity and may also suggest new therapeutic targets for drug development in neurological disease.

Three main classes of ionotropic glutamate receptors—*N*-methyl-*D*-aspartate (NMDA), kainate, and α-amino-3-hydroxy-5-methyl-4-isoxazole propionate (AMPA)—have been differentiated on the basis of physiological and pharmacological properties (114,115). AMPA and kainate receptors, also known as "non-NMDA" receptors, activate and desensitize rapidly in response to glutamate application and are permeable mainly to monovalent cations, thus mediating fast, reliable depolarization of the postsynaptic membrane (116). NMDA receptors, on the other hand, exhibit high calcium permeability and also show voltage-dependent magnesium block, such that receptor activation occurs only under conditions of concurrent membrane depolarization (114). These properties are critical to the role of NMDA receptors in synaptic plasticity, allowing them to act as "coincidence" detectors and to trigger a cascade of second messenger systems (110).

The recent cloning of 16 different subunits within the family of ionotropic glutamate receptors has greatly facilitated studies of these receptors' structure, function, and modulation. Each of the glutamate receptor subtypes has been shown to be composed of oligomeric complexes, most likely pentamers (see Fig. 1A) of homologous subunits. Combinations of NR_1, along with one or more of the NR2A–D subunits form NMDA receptors; GluR5–7 and KA1,2 form kainate receptors; and GluR1–4(A–D) form AMPA receptors (117–119). In addition, variants of many of these subunits exist as a result of alternative splicing of exons.

Based on hydrophobicity plots and analogy to nicotinic AChRs, ionotropic glutamate receptors were originally proposed to have four transmembrane domains (M1–M4), and a large extracellular N-terminal and smaller extracellular C-terminal region; the large loop between M3 and M4 (referred to here as the M3–M4 loop) was thought to be intracellular and was noted to contain consensus sites for a variety of protein kinases (120) (Fig. 1B). More recent evidence suggests a new model for membrane topology of these receptors (Fig. 1C). Notable features include an intracellular C-terminus, an M2 region that forms a hairpin structure within the membrane, and an entirely extracellular M3–M4 loop (119, 121–128). M2 has been shown to line the ion channel, and agonist-binding domains have been localized to two regions: one just N-terminal to M1 and the other within the M3–M4 loop (118,119,129–131). Studies combining site-directed mutagenesis with electrophysiological and biochemical analysis of recombinant glutamate receptors have provided strong evidence for direct phosphorylation and modulation of these receptors by a variety of protein kinases. These data are reviewed next, along with the results of studies suggesting a role for this process in mediating synaptic plasticity.

Serine and Tyrosine Phosphorylation of AMPA and Kainate Receptors

Recent studies indicate that AMPA and kainate receptors may be directly phosphorylated by both serine and tyrosine kinases. By combining a variety of biochemical techniques with site-directed mutagenesis, sites for phosphorylation by PKA, PKC, CaMKII, and PTKs have been identified. In most cases, electrophysiological studies support a role for phosphorylation of these sites in regulating ion channel function, as discussed later.

Forskolin-induced PKA phosphorylation of the AMPA receptor $GluR_1$ has been demonstrated both in transfected HEK 293 cells and in cultured rat cortical neurons (132). Phosphopeptide map analysis of ^{32}P-labeled mutant and wild-type $GluR_1$ expressed in HEK 293 cells has shown that the major site for PKA phosphorylation is Ser-845, within the C-terminal region (133). Similar evidence has identified Ser-684, within the M3–M4 loop, as the major site for PKA phosphorylation of the kainate receptor GluR6 (134).

Sites of PKC and CaMKII phosphorylation have also been localized to both the M3–M4 loop and the C-terminal region. GluR1-containing AMPA receptors can be phosphorylated by activated CaMKII or PKC in cultured rat hippocampal neurons as well as rat brain synaptosomes *in vitro* (135). Site-directed mutagenesis and expression of recombinant AMPA receptors identified the major site for phosphorylation of $GluR_1$ by CaMKII as Ser-627, within the M3–M4 loop (136), and by PKC as Ser-831, within the C-terminal region (133). In addition, Nakazawa *et al.* (136a) used a phosphospecific antibody raised against a putative PKC site (serine 696) within the M3–M4 loop of $GluR_2$ to demonstrate phosphorylation of this site on AMPA receptor subunits in Purkinje cell dendritic membranes in rat brain slice preparations.

Membrane Topology of AMPA/Kainate Receptors and Localization of Phosphorylation Sites

As discussed earlier, in the current membrane topology model for ionotropic glutamate receptors the M3–M4 loop is located entirely extracellularly, whereas the C-terminal region is intracellular (Fig. 1C). Although several phosphorylation sites have been localized to the C-terminal region of GluR1 (Table 1) (133), some of the sites within GluR1–4 and GluR6 have been reported to be within the N-terminal half of the M3–M4 loop, suggesting that this region is accessible to protein kinases on the intracellular side of the membrane (Fig. 1C and Table 1). One alternative hypothesis is that the topology of the M3–M4 loop is dynamic, and in fact intracellular access to M3–M4 loop phosphorylation sites may depend on the agonist (or on use). This intriguing possibility requires further testing.

Functional Effects of AMPA and Kainate Receptor Phosphorylation

PKA-Mediated Phosphorylation

Substantial evidence has accumulated to suggest that kainate and AMPA receptor function can be modulated by protein phosphorylation. Initially, much of this work had centered on regulation of these receptors by PKA activation, since in their pioneering studies, Dowling and colleagues had used the whole-cell patch–clamp technique to show that teleost retinal kainate receptors could be potentiated by dopamine receptor agonists through activation of PKA (137,138). Subsequently, activation of PKA by dopamine was found to potentiate AMPA/kainate receptor-mediated currents in the goldfish Mauthner cells and in chick motor neurons (139,140). Furthermore, extracellular application of PKA activators or intracellular perfusion of purified PKA enhanced these currents in cultured rat hippocampal neurons (141,142). In both teleost retinal cells and rat hippocampal neurons, an increase in AMPA/kainate receptor channel opening frequency and open time were found to underlie this PKA-mediated potentiation (141,143).

In addition to these studies of neuronal receptors, a variety of studies of recombinant AMPA and kainate receptors in heterologous expression systems have demonstrated functional regulation by phosphorylation. Currents mediated by $GluR_1$ and/or $GluR_3$ expressed in *Xenopus* oocytes were enhanced following extracellular application of membrane-permeable cAMP analogs (144). Furthermore, in whole-cell patch–clamp recordings from transfected HEK 293 cells, current responses mediated by $GluR_6$ and $GluR_1$ were potentiated by intracellular perfusion of purified PKA, and potentiation was abolished in recordings from mutant receptors lacking the PKA phosphorylation sites described earlier (133,134,145). Indeed, potentiation of GluR6 whole-cell currents by PKA phosphorylation may be mediated by an increase in channel open probability (Wohl P, Mott D, Traynelis SF: *Abstr Soc Neurosci* 1996: 22:1540). Taken together, these results suggest that PKA activation via a variety of G protein-coupled neurotransmitter receptors in neurons may regulate AMPA/kainate receptor responsiveness.

CaMKII- and PKC-Mediated Phosphorylation

Data from the mid-1990s indicate that AMPA receptor function can be modulated by phosphorylation by CaMKII and PKC, two protein kinases implicated in mechanisms underlying hippocampal LTP/LTD, as well as cerebellar LTD (for PKC) (long-term potentiation and depression are discussed in the section preceding the chapter summary). Intracellular perfusion of the purified catalytic fragment of PKC or of autophosphorylated CaMKII during whole-cell patch–clamp recordings from cultured hippocampal neurons resulted in potentiation of AMPA receptor-mediated current responses (135,146,147). The molecular mechanism(s) by which CaMKII- or PKC-mediated phosphorylation regulates AMPA receptor-mediated currents has not yet been determined. However, data indicating that previously "silent" AMPA receptors may be activated following hippocampal LTP induction (148,149) suggest that phosphorylation by these protein kinases may, in part, mediate membrane insertion or redistribution of AMPA receptors.

Serine and Tyrosine Phosphorylation of NMDA Receptors

Results of several studies indicate that NMDA receptors may be directly phosphorylated by PKC and PTKs. Using GST-fusion proteins of the NR1 C-terminal domain (i.e., from the end of M4 to the carboxy terminus—see Fig. 1C), this segment has been shown to be an excellent substrate for phosphorylation by PKC (122). This result has been confirmed in *in situ* studies; immunoprecipitation of NR_1 from ^{32}P-prelabeled primary cultures of rat cortical neurons and NR_1-transfected HEK 293 cells demonstrated phorbol ester induced phosphorylation of four serines (889,890,896,897) within the C_1 splice cassette of NR1 (122). Biochemical evidence for direct phosphorylation of NMDA receptors by PTKs is also strong. Data from three different studies have demonstrated tyrosine phosphorylation of NR2A and/or NR2B subunits in synaptic membranes of rat hippocampal and forebrain neurons (150–152). Furthermore, tyrosine phosphorylation was shown to be specific for NR_2 subunits and does not occur on the NR_1 subunit (151).

Functional Effects of Phosphorylation of NMDA Receptors

PKC-Mediated Phosphorylation

There is strong evidence to suggest that NMDA receptors can be functionally modulated by a variety of protein kinases, including PKC, PKA, and PTKs. Especially compelling are data suggesting that PKC activation regulates NMDA receptor function. In the thin-slice preparation from rat medulla, NMDA receptor mediated currents recorded from spinal trigeminal neurons were potentiated by activation of μ-opioid receptors, an effect mimicked by intracellular perfusion of purified PKC through the patch pipette (153). Recordings from these neurons after dissociation showed that PKC potentiation was due to an increased probability of NMDA receptor channel opening, as well as decreased Mg^{2+} block of these channels (154). The effect of PKC on Mg^{2+} block (hence on NMDA receptor subunit composition) may be brain region-specific, since NMDA receptor mediated currents recorded from dissociated hippocampal neurons can be potentiated by intracellular PKC perfusion without a reduction in Mg^{2+} block (146; J.F. Macdonald personal communication). In the hippocampal slice, application of PKC activators such as phorbol esters or metabotropic receptor agonists has resulted in conflicting results; the former produced suppression and the latter potentiation of NMDA-evoked current responses (155,156). This discrepancy may be due to effects other than PKC activation, which may be mediated by phorbol esters and/or metabotropic receptor agonists, or to additional pre- or polysynaptic effects of these drugs that indirectly affect postsynaptic NMDA receptor activation. Alternatively, the net effect of PKC activation on NMDA receptor currents may depend on the preceding history of afferent stimulation, which may in itself affect levels of NMDA receptor phosphorylation (157).

Studies investigating the effect of PKC activation on recombinant NMDA receptor function in heterologous expression systems have uniformly demonstrated potentiation of NMDA receptor mediated currents. Intracellular perfusion of purified PKC during whole-cell patch–clamp recording from NR_{1A}/NR_{2A}-transfected HEK 293 cells resulted in 2 to 3-fold potentiation of NMDA receptor current responses (158), whereas these currents were potentiated up to 20-fold by extracellular application of phorbol esters to *Xenopus* oocytes expressing certain splice variants of the NR_1 subunit alone or in combination with NR_{2A} or NR_{2B} (159–163). Whether these effects are due to direct PKC phosphorylation of NMDA receptors remains unclear.

Electrophysiological studies of recombinant NMDA receptors expressed in oocytes have indicated that PKC phosphorylation of serines 889, 890, 896, and 897 in the C-terminal region of NR_{1A} (see preceding section) is not responsible for the phorbol ester mediated potentiation of NMDA receptor currents in oocytes (164,165). In fact, none of the serine or threonine residues within the C-terminal region or the M3–M4 loop of NR_1 appear to contribute to this effect (164,165). On the other hand, phorbol ester induced potentiation of heteromeric NMDA receptor mediated currents expressed in oocytes is observed for receptors composed of NR1 (ζ_1) subunits in combination with NR_{2A} or NR_{2B} (ε_1 and ε_2) subunits but not NR2C (ε_3) subunits (163) and appears to depend on the C-terminal region of the NR2 subunits (166). Taken together with results of studies described next, these data suggest the possibility that PKC-mediated potentiation of NMDA receptors may be due to phosphorylation of interacting protein(s) rather than the receptor itself.

Effects of Phosphatases and PKA- and PTK-Mediated Phosphorylation

PKA, protein phosphatases, and PTKs may play an important role in modulating NMDA receptor function. One study has suggested that synaptic NMDA receptors in the resting state (in the absence of transmitter) may be phosphorylated by PKA, and then become dephosphorylated during synaptic activity upon activation of the calcium-dependent serine/threonine phosphatase calcineurin by NMDA receptor mediated calcium influx. PKA phosphorylation of the receptor appears to correlate with a decreased rate of receptor inactivation during synaptic transmission (157). Prior to this study, calcineurin had been shown to inhibit receptor function in outside-out patch recordings from cultured rat hippocampal neurons and acutely dissociated adult rat dentate granule cells (167,168). In addition, exposure of hippocampal neuronal inside-out patches to phosphatases 1 and 2A has been shown to decrease NMDA receptor channel open probability (169). Finally, a role for modulation of NMDA receptor channel function by PTKs has been suggested, as well. In whole-cell recordings from spinal dorsal horn neurons, inhibitors of protein tyrosine kinases suppressed, whereas intracellular application of the PTK pp60 c-*src* or inhibitors of protein tyrosine phosphatases enhanced, NMDA receptor currents (142). Considering the variety of neurotransmitter, growth factor, and hormone receptors capable of regulating PKA, PKC, PTK, and protein phosphatase activities, a highly diverse set of signaling pathways may be involved in phosphorylation-mediated modulation of NMDA receptor function.

Cytoskeletal Protein Interactions and Phosphorylation of NMDA Receptors

Data suggest that NMDA receptors interact with cytoskeletal proteins and that this interaction may be regulated by phosphorylation of the receptor. One study showed that NR_1 subunits expressed in mammalian cell lines exhibit a clustered distribution at the plasma membrane. The C_1 splice cassette was shown to be required for clustering, and PKC phosphorylation of serine 890 within this cassette resulted in dispersion of these clusters (170,171).

Protein Phosphorylation and Synaptic Plasticity

Among the best studied examples of synaptic plasticity in the mammalian brain are the long-term changes in synaptic strength encoded by stimulus frequency at the glutamatergic Schaffer collateral–CA_1 neuron synapse in the hippocampus. Enhancement of synaptic responsiveness is observed for hours following a high frequency (50–100 Hz, 1–2 s) stimulus train, whereas persistent suppression is seen after ~10 minutes of low frequency (1–2 Hz) stimulation; the former condition is known as long-term potentiation (LTP) and the latter as long-term depression (LTD) (172,173). Both processes require NMDA receptor activation and calcium influx (172,173), and expression appears to involve, at least in part, an alteration in postsynaptic glutamate receptor responsiveness (174–181). These results suggest that processes that regulate glutamate receptor function may contribute importantly to mechanisms underlying synaptic plasticity.

A variety of data suggest a critical role for protein kinases and phosphatases in mediating the changes in synaptic strength underlying LTP and LTD in the hippocampus. Induction of hippocampal LTP in the CA_1 region has been shown to involve activation of PKC, CaMKII, and PTKs (112,182,183). Results of electrophysiological studies have shown that direct injection of CaMKII or PKC inhibitors into the postsynaptic CA_1 neuron blocked induction, and in some cases expression, of LTP (184–186). Further, hippocampal slice neurons infected with a virus containing the constitutively active form of CaMKII exhibited enhanced synaptic transmission and occlusion of LTP by standard induction protocols (187). Similarly, direct intracellular injection of calcium and calmodulin postsynaptically resulted in potentiation of CA_1 synaptic responses and also occluded LTP induction in rat hippocampal slices (103). Tyrosine kinase inhibitors applied either extra- or intracellularly (into CA_1 neurons) also blocked LTP induction (188).

In addition to these electrophysiological studies, biochemical data support a role for PKC in the expression of LTP. One study demonstrated persistent activation of PKC during the maintenance phase of LTP (189), while another study showed that LTP expression was associated with high levels of constitutively active PKCζ (190).

Moreover, induction of LTP by standard protocols at the Schaffer collateral–CA_1 synapse was severely impaired in transgenic mice lacking either PKCγ, α-CaMKII, or the tyrosine kinase *fyn,* providing further evidence that activation of these kinases is required for long-lasting enhancement of synaptic strength in this paradigm (191–193).

On the other hand, induction of hippocampal LTD appears to be favored by activation of protein phosphatases. Protein phosphatase inhibitors applied either extra- or intracellularly (into CA_1 neurons) have been shown to impair or eliminate induction of NMDA receptor dependent LTD (194,195) and in some cases, to enhance AMPA receptor mediated synaptic transmission (196). Further, in an NMDA receptor independent paradigm in which a voltage pulse protocol was used to induce potentiation of CA_1 neuronal responses in slices, bath application or intracellular injection of protein kinase or phosphatase inhibitors demonstrated that AMPA receptor mediated postsynaptic currents were enhanced by a calcium-sensitive kinase (likely CaMKII) and that phosphatase activation antagonized this effect (197). However, in contrast to these data, transgenic mice overexpressing an equivalent of the autophosphorylated form of CaMKII demonstrated a shift in favor of LTD induction even at frequencies that would usually produce LTP (198).

Since LTD and LTP induction both require postsynaptic calcium influx, a popular theory has been that spatiotemporal differences in postsynaptic intracellular calcium accumulation, in response to afferent stimulation frequency, determine whether specific protein kinases vs. phosphatases are activated, which in turn trigger LTP vs. LTD. More recently, however, a study using calcium imaging suggested that dendritic calcium levels resulting from both LTP and LTD induction protocols are quite similar (199). Although subtle differences in spatial distribution of intracellular calcium could still account for selective activation of kinases and phosphatases, there may be other factors involved as well.

Synaptic plasticity in response to specific patterns of afferent stimulation has also been described in brain regions outside the hippocampus. One well-studied example is long-term depression of the parallel fiber (PF)–Purkinje cell (PC) synapse in the cerebellum, induced by coincident stimulation of parallel fiber and climbing fiber (CF) afferents to the PC (200). The decreased PF–PC synaptic responsiveness is due to inhibition of AMPA receptor mediated current, and PKC activation is required for induction of LTD, suggesting that PKC phosphorylation and modulation of AMPA receptors may play a role in the mechanisms underlying this process (201).

The data outlined here strongly suggest that plasticity in each of these paradigms is due, at least in part, to changes in postsynaptic responsiveness as a result of phosphorylation and functional modulation of ionotropic glutamate receptors. Direct tests of this hypothesis are difficult in the brain slice. However, now that specific sites for phosphorylation by a variety of protein kinases have been identified within glutamate receptors, phosphospecific antibodies for these sites can be developed. These tools could be applied to experimental paradigms that produce long-lasting changes in synaptic efficacy in the brain slice to help determine the role of phosphorylation of GluRs in mediating various forms of synaptic plasticity in the brain.

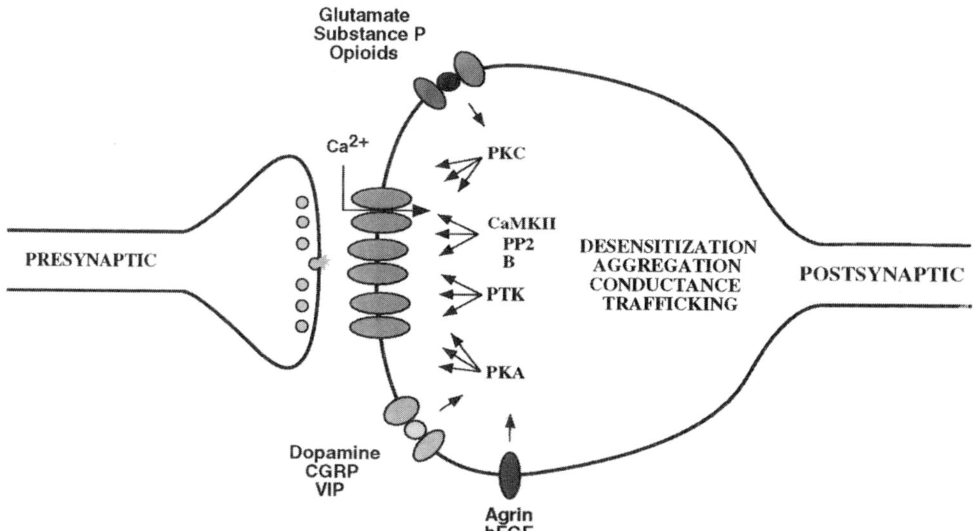

FIG. 2. Regulation of ligand-gated ion channels by protein phosphorylation. A variety of extracellular factors such as neurotransmitters (glutamate and dopamine), neuropeptides (substance P, CGRP, VIP, and opioids), extracellular matrix proteins (agrin), and growth factors (FGF) activate various protein kinases, such as CaM-KII, PKC and PKA, and PTK, or protein phosphatases, such as calcineurin (PP2B). These enzymes can then regulate the phosphorylation state of ligand-gated ion channels in the postsynaptic membrane. The resulting changes in receptor function may depend on the specific subunit composition of the receptors and alternative splicing of the receptor subunits.

SUMMARY

The studies discussed in this review demonstrate that phosphorylation is an important mechanism for the regulation of ligand-gated ion channels. Structurally, ligand-gated ion channels are heteromeric proteins comprised of homologous subunits. For both the AChR and the $GABA_A$ receptor, each subunit has a large extracellular N-terminal domain, four transmembrane domains, a large intracellular loop between transmembrane domains M3 and M4, and an extracellular C-terminal domain (Fig. 1B). All the phosphorylation sites on these receptors have been mapped to the major intracellular loop between M3 and M4 (Table 1). In contrast, glutamate receptors appear to have a very large extracellular N-terminal domain, one membrane hairpin loop, three transmembrane domains, a large extracellular loop between transmembrane domains M3 and M4, and an intracellular C-terminal domain (Fig. 1C). Most phosphorylation sites on glutamate receptors have been shown to be on the intracellular C-terminal domain, although some have been suggested to be on the putative extracellular loop between M3 and M4 (Table 1).

A variety of extracellular factors and intracellular signal transduction cascades are involved in regulating phosphorylation of these ligand-gated ion channels (Fig. 2).

Once again, the AChR at the neuromuscular junction is the most fully understood system. Phosphorylation of the AChR by PKA is stimulated synaptically by the neuropeptide CGRP and in an autocrine fashion by adenosine released from the muscle in response to acetylcholine. In addition, acetylcholine, via calcium influx through the AChR, appears to activate calcium-dependent kinases including PKC to stimulate serine phosphorylation of the receptor. Presently, agrin is the only extracellular factor known to stimulate phosphorylation of the AChR on tyrosine residues. For glutamate receptors, non-NMDA receptor phosphorylation by PKA is stimulated by dopamine, while NMDA receptor phosphorylation by PKA and PKC can be induced via the activation of β-adrenergic receptors, and metabotropic glutamate or opioid receptors, respectively. In addition, Ca^{2+} influx through the NMDA receptor has been shown to activate PKC, CaMKII, and calcineurin, resulting in phosphorylation of AMPA receptors (by CaMKII) and inactivation of NMDA receptors (at least in part through calcineurin). In contrast to the AChR and glutamate receptors, no information is presently available regarding the identities of the extracellular factors and intracellular signal transduction cascades that regulate phosphorylation of the $GABA_A$ receptor. Surely, future studies will be aimed at further clarifying the molecular mechanisms by which the central receptors are regulated.

The presently understood functional effects of ligand-gated ion channel phosphorylation are diverse. At the neuromuscular junction, a regulation of the AChR desensitization rate by both serine and tyrosine phosphorylation has been demonstrated. In addition, tyrosine phosphorylation of the AChR or other synaptic components appears to play a role in AChR clustering during synaptogenesis. For the $GABA_A$ receptor, the data are complex. Both activation and inhibition of $GABA_A$ receptor currents as a result of PKA and PKC phosphorylation have been reported, while phosphorylation by PTK enhances function. The predominant effect of glutamate receptor phosphorylation by a variety of kinases is a potentiation of the peak current response. However, PKC also modulates clustering of NMDA receptors. This complexity in the regulation of ligand-gated ion channels by phosphorylation provides diverse mechanisms for mediating synaptic plasticity. In fact, accumulating evidence supports the involvement of protein phosphorylation and dephosphorylation of AMPA receptors in LTP and LTD respectively.

There has been a dramatic increase in our understanding of the nature by which phosphorylation regulates ligand-gated ion channels. However, many questions remain unanswered. It is still unclear exactly what intramolecular changes occur upon phosphorylation of a receptor and how those changes modulate channel properties. It is also not understood how phosphorylation regulates receptor aggregation and density at the postsynaptic membrane to mediate synaptogenesis and possibly synaptic plasticity. Furthermore, the signal transduction pathways utilized for the regulation of receptor clustering, expression, and channel properties need to be elucidated. However, a picture is emerging in which phosphorylation plays a pivotal role in the dynamic regulation of ligand-gated ion channels resulting in short- and long-term changes in synaptic transmission.

REFERENCES

1. Hall Z, Sanes J: Synaptic structure and development: the neuromuscular junction. *Neuron /Cell* 1993;72/10(Suppl):99–121.
2. Galzi J, Revah F, Bessis A, Changeux J: Functional architecture of the nicotinic acetylcholine receptor: from electric organ to brain. *Annu Rev Pharmacol* 1991;31:37–72.
3. Changeux, JP: The TIPS Lecture: the nicotinic acetylcholine receptor: an allosteric protein prototype of ligand-gated ion channels. *Trends Phamacol Sci* 1990;11:485–492.
4. Huganir RL, Greengard, P: cAMP-Dependent protein kinase phosphorylates the nicotinic acetylcholine receptor. *Proc Natl Acad Sci U S A* 1983;80:1130–1134.
5. Huganir RL: Regulation of the nicotinic acetylcholine receptor by protein phosphorylation. *J Recept Res* 1987;7:241–256.
6. Safran A, Sagi-Eisenberg R, Neumann D, Fuchs S: Phosphorylation of the acetylcholine receptor by protein kinase C and identification of the phosphorylation site within the receptor subunit. *J Biol Chem* 1987;262:10506–10510.
7. Yee GH, Huganir RL: Determination of the sites of cAMP-dependent phosphorylation on the nicotinic acetylcholine receptor. *J Biol Chem* 1987;262:16748–16753.
8. Schroeder W, Mayer HE, Buchner K, Bayer H, Hucho F: Phosphorylation sites of the nicotinic acetylcholine receptors. A novel site detected in position S362. *Biochemistry* 1991;30:3583–3588.
9. Huganir RL, Miles K, Greengard P: Phosphorylation of the nicotinic acetylcholine receptor by an endogenous tyrosine-specific protein kinase. *Proc Natl Acad Sci U S A* 1984;81:6968–6972.
10. Wagner K, Edson K, Heginbotham L, Post M, Huganir RL, Czernik AJ: Determination of the tyrosine phosphorylation sites of the nicotinic acetylcholine receptor. *J Biol Chem* 1991;266:23784–23789.
11. Swope SL, Huganir RL: Molecular cloning of two abundant protein tyrosine kinases in *Torpedo* electric organ that associate with the acetylcholine receptor. *J Biol Chem* 1993;268:25152–25161.
12. Gillespie S, Balasubramanian S, Fung E, Huganir R: Rapsyn clusters and activates the synapse-specific receptor tyrosine kinase MuSK. *Neuron* 1996;16:953–962.
13. Huganir RL, Miles K: Protein phosphorylation of nicotinic acetylcholine receptors. *Crit Rev Biochem Mol Biol* 1989;24:183–215.
14. Mei L, Huganir R: Purification and characterization of a protein tyrosine phosphatase which dephosphorylates the nicotinic acetylcholine receptor. *J. Biol Chem* 1991;266:16063–16072.
15. Huganir RL, Delcour AH, Greengard P, Hess GP: Phosphorylation of the nicotinic acetylcholine receptor regulates its rate of desensitization. *Nature* 1986;321:774–776.
16. Miles K, Anthony DT, Rubin LL, Greengard P, Huganir RL: Regulation of nicotinic acetylcholine receptor phosphorylation in rat myotubes by forskolin and cAMP. *Proc Natl Acad Sci U S A* 1987; 84:6591–6595.
17. Smith M, Merlie J, Lawrence JJ: Regulation of phosphorylation of nicotinic acetylcholine receptors in mouse BC3H1 myocytes. *Proc Natl Acad Sci U S A* 1987;84:6601–6605.
18. Middleton P, Rubin L, Schuetze S: Desensitization of acetylcholine receptors in rat myotubes is enhanced by agents that elevate intracellular cAMP. *J Neurosci* 1988;8:3405–3412.
19. Mulle C, Benoit P, Pinset C, Roa M, Changeux J-P: Calcitonin gene-related peptide enhances the rate of desensitization of the nicotinic acetylcholine receptor in cultured mouse muscle cells. *Proc Natl Acad Sci U S A* 1988;85:5728–5732.
20. Middleton P, Jaramillo F, Schuetze S: Forskolin increases the rate of acetylcholine receptor desensitization at rat soleus endplates. *Proc Natl Acad Sci U S A* 1986;83:4967–4971.
21. Albuquerque EX, Deshpande SS, Aracava Y, Alkondon M, Daly JW: A possible involvement of cyclic AMP in the expression of desensitization of the nicotinic acetylcholine receptor: a study with forskolin and its analogs. *FEBS Let* 1986;199:113–120.
22. Miles K, Greengard P, Huganir RL: Calcitonin gene-related peptide regulates phosphorylation of the nicotinic acetylcholine receptor in rat myotubes. *Neuron* 1989;2:1517–1524.
23. Ross A, Rapuano M, Prives J: Induction of phosphorylation and cell surface redistribution of acetylcholine receptors by phorbol ester and carbamylcholine in cultured chick muscle cells. *J Cell Biol* 1988;107:1139–1145.
24. Miles K, Audigier SS, Greengard P, Huganir RL: Autoregulation of phosphorylation of the nicotinic acetylcholine receptor. *J Neurosci* 1994;14:3271–3279.
25. Eusebi F, Molinaro M, Zani B: Agents that activate protein kinase C reduce acetylcholine sensitivity in cultured myotubes. *J Cell Biol* 1985;100:1339–1342.

26. Hopfield JF, Tank DW, Greengard P, Huganir RL: Functional modulation of the nicotinic acetylcholine receptor by tyrosine phosphorylation. *Nature* 1988;336:677–680.
27. Magill-Solc C, McMahan U: Synthesis and transport of agrin-like molecules in motor neurons. *J Exp Biol* 1990;153:1–10.
28. Wallace B: Agrin-induced specializations contain cytoplasmic, membrane, and extracellular matrix-associated components of the postsynaptaic apparatus. *J Neurosci* 1989;9:1294–1302.
29. Qu ZC, Moritz E, Huganir RL: Regulation of tyrosine phosphorylation of the nicotinic acetylcholine receptor at the rat neuromuscular junction. *Neuron* 1990;4:367–378.
30. Qu Z, Huganir RL: Comparison of innervation and agrin-induced tyrosine phosphorylation of the nicotinic acetylcholine receptor. *J Neurosci* 1994;14:6834–6841.
31. Wallace BG, Qu Z, Huganir RL: Agrin induces phosphorylation of the nicotinic acetylcholine receptor. *Neuron* 1991;6:869–878.
32. Peng HB, Baker LP, Chen Q: Induction of synaptic development in cultured muscle cells by basic fibroblast growth factor. *Neuron* 1991;6:237–246.
33. Peng BH, Ali AA, Dai Z, Daggett DF, Raulo E, Rauvala, H: The role of heparin-binding growth-associated molecule (HB-GAM) in the postsynaptic induction in cultured muscle cells. *J Neurosci* 1995;15:3027–3038.
34. Qu Z, Apel ED, Doherty CA, Hoffman PW, Merlie JP, Huganir RL: The synapse-associated protein rapsyn regulates tyrosine phosphorylation of proteins colocalized at nicotinic acetylcholine receptor clusters. *Mol Cell Neurosci* 1996;8:171–84.
35. Peng HB, Baker LP, Dai Z: A role of tyrosine phosphorylation in the formation of acetylcholine receptor clusters induced by electric fields in cultured *Xenopus* muscle cells. *J Cell Biol* 1993;120: 197–204.
36. Baker LP, Chen Q, Peng HB: Induction of acetylcholine receptor clustering by native polystyrene beads. Implication of an endogenous muscle-derived signalling system. *J Cell Sci* 1992;102:543–555.
37. Wallace B: Staurosporine inhibits agrin-induced acetylcholine receptor phosphorylation and aggregation. *J Cell Biol* 1994;125:661–668.
38. McMahan U: The agrin hypothesis. *Cold Spring Harbor Symp Quant Biol* 1990;55:407–418.
39. Gautam M, Noakes P, Moscoso L et al: Defective neuromuscular synaptogenesis in agrin-deficient mutant mice. *Cell* 1996;85:525–535.
40. Ferns M, Deiner M, Hall Z: Agrin-induced acetylcholine receptor clustering in mammalian muscle requires tyrosine phosphorylation. *J Cell Biol* 1996;132:937–944.
41. Froehner SC: The submembrane machinery for nicotinic acetylcholine receptor clustering. *J Cell Biol* 1991;114:1–7.
42. Froehner S, Luetje C, Scotland P, Patrick J: The postsynaptic 43K protein clusters muscle nicotinic acetylcholine receptors in *Xenopus* oocytes. *Neuron* 1990;5:403–410.
43. Phillips WD, Kopta C, Blount P, Gardner PD, Steinbach JH, Merlie JP: ACh receptor-rich membrane domains organized in fibroblasts by recombinant 43-kilodalton protein. *Science* 1991;251:568–570.
44. Yu X, Hall Z: The role of the cytoplasmic domains of indiviual subunits of the acetylcholine receptor in 43 kDa protein-induced clustering in COS cells. *J Neurosci* 1994;14:785–795.
45. Gautam M, Noakes P, Mudd J et al: Failure of postsynaptic specialization to develop at neuromuscular junctions of rapsyn-deficient mice. *Nature* 1995;377:232–236.
46. Matteoli M, Haimann C, Torri-Tarelli F, Polak JM, Carelli B, De Camilli P: Differential effect of alpha-1 atrotoxin on exocytosis from small synaptic vesicles and from large dense-core vesicles containing calcitonin gene-related peptide at the frog neuromuscular junction *Proc Natl Acad Sci U S A* 1988;85:7366–7370.
47. Yamaguchi A, Chiba T, Yamatani T et al: Calcitonin gene-related peptide stimulates adenylate cyclase activation via a guanine nucleotide-dependent process in rat liver plasma membranes. *Endocrinology* 1988;123:2591–2596.
48. Pitchford S, Day JW, Gordon A, Mochly-Rosen D: Nicotinic acetylcholine receptor desensitization is regulated by activation-induced extracellular adenosine accumulation. *J Neurosci* 1992;12:4540–4544.
49. Eusebi F, Grassi F, Molinaro M, Zani B: Acetylcholine regulation of nicotinic receptor channels through a putative G protein in chick myotubes. *J Physiol (Lond)* 1987;393:635–645.
50. Magill-Solc C, McMahan U: Motor neurons contain agrin-like molecules. *J Cell Biol* 1988;107: 1825–1833.
51. Campanelli J, Roberds S, Campbell K, Scheller R: A role for dystrophin-associated glycoproteins and utrophin in agrin-induced AChR clustering. *Cell* 1994;77:663–674.

52. Gee SH, Montanaro F, Lindenbaum MH, Carbonetto S: Dystroglycan-alpha, a dystrophin-associated glycoprotein, is a functional agrin receptor. *Cell* 1994;77:675–686.
53. Bowe M, Deyst K, Leszyk J, Fallon J: Identification and purification of an agrin receptor from *Torpedo* postsynaptic membranes: a heteromeric complex related to the dystroglycans. *Neuron* 1994; 12:1173–1180.
54. Sugiyama J, Bowen D, Hall Z: Dystroglycan binds nerve and muscle agrin. *Neuron* 1995;13: 103–115.
55. Hoffman E, Brown R, Kunkel L: The protein product of the Duchenne muscular dystrophy locus. *Cell* 1987;51:919–928.
56. Ervasti J, Campbell K: A role for the dystrophin–glycoprotein complex as a transmembrane linker between laminin and actin. *J Cell Biol* 1993;122:809–823.
57. Gesemann M, Denzer A, Cavalli V et al: Alternative splicing of agrin alters its binding to heparin, dystroglycan, and the putative agrin receptor. *Neuron* 1996;16:755–767.
58. Valenzuela DM, Stitt TN, DiStefano PS, Rojas E, Mattson K, Compton DL, Nunez L, Park JS, Stark JL, Geis DR, Thomas S, LeBeau MM, Fernald AA, Copeland NG, Jenkins NA, Burden SJ, Glass DJ, Vancopoulos GD: Receptor tyrosine kinase specific for the skeletal muscle lineage: Expression in embryonic muscle, at the neuromuscular junction, and after injury. *Neuron* 1995;15:573–584.
59. DeChiara T, Bowen D, Valenzuela D et al: The receptor tyrosine kinase MuSK is required for neuromuscular junction formation *in vivo*. *Cell* 1996;85:1–20.
60. Glass D, Bowen D, Stitt T et al: Agrin acts via a MuSK receptor complex. *Cell* 1996;85:513–523.
60a. Swope SL, Huganir RL: Binding of the nicotinic acetylcholine receptor to SH2 domans of fyn and fyk protein tyrosine kinases. *J Biol Chem* 1994;269:29817–29824.
61. Unwin N. Neurotransmitter action: opening of ligand-gated ion channels. *Cell* 1993;72 (Suppl): 31–41.
62. Macdonald RL, Olsen RW: $GABA_A$ receptor channels. *Annu Rev Neurosci* 1994;17:569–602.
63. Rabow LE, Russek SJ, Farb DH: From ion currents to genomic analysis: recent advances in $GABA_A$ receptor research. *Synapse* 1995;21:189–274.
64. Whiting P, McKernan RM, Iversen LL: Another mechanism for creating diversity in gamma-aminobutyrate type A receptors: RNA splicing directs expression of two forms of gamma 2 phosphorylation site. *Proc Natl Acad Sci U S A* 1990;87:9966–9970.
65. Kofuji P, Wang JB, Moss SJ, Huganir RL, Burt DR: Generation of two forms of the gamma-aminobutyric $acid_A$ receptor gamma 2-subunit in mice by alternative splicing. *J Neurochem* 1991; 56:713–715.
66. Kirkness EF, Bovenkerk CF, Ueda T, Turner AJ: Phosphorylation of gamma-aminobutyrate (GABA)/ benzodiazepine receptors by cyclic AMP-dependent protein kinase. *Biochem J* 1989;259:613–616.
67. Browning MD, Bureau M, Dudek EM, Olsen RW: Protein kinase C and cAMP-dependent protein kinase phosphorylate the beta subunit of the purified gamma-aminobutyric acid A receptor. *Proc Natl Acad Sci U S A* 1990;87:1315–1318.
68. Tehrani MH, Barnes EM Jr: $GABA_A$ receptors in mouse cortical homogenates are phosphorylated by endogenous protein kinase A. *Brain Res Mol Brain Res* 1994;24:55–64.
69. Sweetnam PM, Lloyd J, Gallombardo P et al: Phosphorylation of the $GABA_A$/benzodiazepine receptor α subunit by a receptor-associated protein kinase. *J Neurochem* 1988;51:1274–1284.
70. Bureau MH, Laschet JJ: Endogenous phosphorylation of distinct gamma-aminobutyric acid type A receptor polypeptides by Ser/Thr and Tyr kinase activities associated with the purified receptor. *J. Biol Chem* 1995;270:26482–26487.
71. Valenzuela CF, Machu TK, McKernan RM et al: Tyrosine kinase phosphorylation of $GABA_A$ receptors. *Brain Res Mol Brain Res* 1995;31:165–172.
72. Moss SJ, Smart TG, Blackstone CD, Huganir RL: Functional modulation of $GABA_A$ receptors by cAMP-dependent protein phosphorylation. Science 1992;257:661–665.
73. McDonald BJ, Moss SJ: Differential phosphorylation of intracellular domains of gamma-aminobutyric acid type A receptor subunits by calcium/calmodulin type 2-dependent protein kinase and cGMP-dependent protein kinase. *J. Biol Chem* 1994;269:18111–18117.
74. McDonald KA, Lakonishok M, Horwitz AF: Alpha-v and alpha-3 integrin subunits are associated with myofibrils during myofibrillogenesis. *J Cell Sci* 1995;108:975–983.
75. Moss SJ, Doherty CA, Huganir RL: Identification of the cAMP-dependent protein kinase and protein kinase C phosphorylation sites within the major intracellular domains of the beta 1, gamma 2S, and gamma 2L subunits of the gamma-aminobutyric acid type A receptor. *J. Biol Chem* 1992;267: 14470–14476.

76. Machu TK, Firestone JA, Browning MD: Ca^{2+}/calmodulin-dependent protein kinase II and protein kinase C phosphorylate a synthetic peptide corresponding to a sequence that is specific for the gamma 2L subunit of the $GABA_A$ receptor. *J Neurochem* 1993;61:375–377.
77. Krishek BJ, Xie X, Blackstone C, Huganir RL, Moss SJ, Smart TG: Regulation of $GABA_A$ receptor function by protein kinase C phosphorylation. *Neuron* 1994;12:1081–1095.
78. Moss SJ, Gorrie GH, Amato A, Smart TG: Modulation of $GABA_A$ receptors by tyrosine phosphorylation. *Nature* 1995;377:344–348.
79. Tehrani MH, Hablitz JJ, Barnes EM Jr: cAMP increases the rate of $GABA_A$ receptor desensitization in chick cortical neurons. *Synapse* 1989;4:126–131.
80. Heuschneider G, Schwartz RD: cAMP and forskolin decrease gamma-aminobutyric acid–gated chloride flux in rat brain synaptoneurosomes. *Proc Natl Acad Sci U S A* 1989;86:2938–2942.
81. Schwartz RD, Heuschneider G, Edgar PP, Cohn JA: cAMP analogs inhibit gamma-aminobutyric acid–gated chloride flux and activate protein kinase A in brain synaptoneurosomes. *Mol Pharmacol* 1991;39:370–375.
82. Porter NM, Twyman RE, Uhler MD, Macdonald RL: Cyclic AMP-dependent protein kinase decreases $GABA_A$ receptor current in mouse spinal neurons. *Neuron* 1990;5:789–796.
83. White G, Li C, Ishac E: 1,9-Dideoxyforskolin does not mimic all cAMP and protein kinase A independent effects of forskolin on GABA activated ion currents in adult rat sensory neurons. *Brain Res* 1992;586:157–161.
84. Ticku MK, Mehta, AK: gamma-Aminobutyric acidA receptor desensitization in mice spinal cord cultured neurons: lack of involvement of protein kinases A and C. *Mol Pharmacol* 1990;38: 719–724.
85. Veruki ML, Yeh HH: Vasoactive intestinal polypeptide modulates $GABA_A$ receptor function in bipolar cells and ganglion cells of the rat retina. *J Neurophysiol* 1992;67:791–797.
86. Veruki ML, Yeh HH: Vasoactive intestinal polypeptide modulates $GABA_A$ receptor function through activation of cyclic AMP. *Vis Neurosci* 1994;11:899–908.
87. Parfitt KD, Hoffer BJ, Bickford-Wimer PC: Potentiation of gamma-aminobutyric acid-mediated inhibition by isoproterenol in the cerebellar cortex: receptor specificity. *Neuropharmacology* 1990; 29:909–916.
88. Robello M, Amico C, Cupello A: Regulation of $GABA_A$ receptor in cerebellar granule cells in culture: differential involvement of kinase activities. *Neuroscience* 1993;53:131–138.
89. Leidenheimer NJ, Browning MD, Harris RA: $GABA_A$ receptor phosphorylation: multiple sites, actions and artifacts. *Trends Pharmacol Sci* 1991;12:84–87.
90. Angelotti TP, Uhler MD, Macdonald RL: Enhancement of recombinant gamma-aminobutyric acid type A receptor currents by chronic activation of cAMP-dependent protein kinase. *Mol Pharmacol* 1993;44:1202–1210.
91. Kano M, Konnerth A: Potentiation of GABA-mediated currents by cAMP-dependent protein kinase. *Neuroreport* 1992;3:563–566.
92. Kano M, Rexhausen U, Dreessen J, Konnerth A: Synaptic excitation produces a long-lasting rebound potentiation of inhibitory synaptic signals in cerebellar Purkinje cells. *Nature* 1992;356: 601–604.
93. Sigel E, Baur R: Activation of protein kinase C differentially modulates neuronal Na^+, Ca^{2+}, and gamma-aminobutyrate type A channels. *Proc Natl Acad Sci U S A* 1988;85:6192–6196.
94. Moran O, Dascal N: Protein kinase C modulates neurotransmitter responses in *Xenopus* oocytes injected with rat brain RNA. *Brain Res Mol Brain Res* 1989;5:193–202.
95. Sigel E, Baur R, Malherbe P: Activation of protein kinase C results in down-modulation of different recombinant $GABA_A$-channels. *FEBS Lett* 1991;291:150–152.
96. Leidenheimer NJ, McQuilkin SJ, Hahner LD, Whiting P, Harris RA: Activation of protein kinase C selectively inhibits the gamma-aminobutyric acidA receptor: role of desensitization. *Mol Pharmacol* 1992;41:1116–1123.
97. Leidenheimer NJ, Whiting PJ, Harris RA: Activation of calcium-phospholipid-dependent protein kinase enhances benzodiazepine and barbiturate potentiation of the $GABA_A$ receptor. J Neurochem 1993;60:1972–1975.
98. Kellenberger S, Malherbe P, Sigel E: Function of the alpha 1 beta 2 gamma 2S gamma-aminobutyric acid type A receptor is modulated by protein kinase C via multiple phosphorylation sites. *J. Biol Chem* 1992;267:25660–25663.
99. Weiner JL, Zhang L, Carlen PL: Potentiation of $GABA_A$-mediated synaptic current by ethanol in

hippocampal CA1 neurons: possible role of protein kinase C. *J Pharmacol Exp Ther* 1994;268: 1388–1395.
100. Lin YF, Browning MD, Dudek EM, Macdonald RL: Protein kinase C enhances recombinant bovine alpha 1 beta 1 gamma 2L $GABA_A$ receptor whole-cell currents expressed in L929 fibroblasts. *Neuron* 1994;13:1421–1431.
101. Glaum SR, Miller RJ: Activation of metabotropic glutamate receptors produces reciprocal regulation of ionotropic glutamate and GABA responses in the nucleus of the tractus solitarius of the rat. *J Neurosci* 1993;13:1636–1641.
102. Stelzer A: Intracellular regulation of $GABA_A$-receptor function. *Ion Channels* 1992;3:83–136.
103. Wang JH, Kelly PT: Postsynaptic injection of CA^{2+}/CaM induces synaptic potentiation requiring CaMKII and PKC activity. *Neuron* 1995;15:443–452.
104. Stelzer A, Kay AR, Wong RK: $GABA_A$-receptor function in hippocampal cells is maintained by phosphorylation factors. *Science* 1988;241:339–341.
105. Gyenes M, Farrant M, Farb DH: "Run-down" of gamma-aminobutyric $acid_A$ receptor function during whole-cell recording: a possible role for phosphorylation. *Mol Pharmacol* 1988;34: 719–723.
106. Gyenes M, Wang Q, Gibbs TT, Farb DH: Phosphorylation factors control neurotransmitter and neuromodulator actions at the gamma-aminobutyric acid type A receptor. *Mol Pharmacol* 1994;46: 542–549.
107. Lev S, Moreno H, Martinez R et al: Protein tyrosine kinase PYK2 involved in Ca^{2+}-induced regulation of ion channel and MAP kinase functions [see comments]. *Nature* 1995;376:737–745.
108. Sasaki H, Nagura K, Ishino M, Tobioka H, Kotani K, Sasaki T: Cloning and characterization of cell adhesion kinase beta, a novel protein-tyrosine kinase of the focal adhesion kinase subfamily. *J. Biol Chem* 1995;270:21206–21219.
109. Choi DW: Glutamate neurotoxicity and diseases of the nervous system. *Neuron* 1988;1:623–634.
110. Collingridge GL, Singer W: Excitatory amino acid receptors and synaptic plasticity. *Trends Pharmacol Sci* 1990;11:290–296.
111. Albin RL, Greenamyre JT: Alternative excitotoxic hypotheses. *Neurology* 1992;42:733–738.
112. Bliss TV, Collingridge GL: A synaptic model of memory: long-term potentiation in the hippocampus. *Nature* 1993;361:31–39.
113. Coyle JT, Puttfarcken P: Oxidative stress, glutamate, and neurodegenerative disorders. *Science* 1993;262:689–695.
114. Mayer ML, Westbrook GL: The physiology of excitatory amino acids in the vertebrate central nervous system. *Prog Neurobiol* 1987;28:197–276.
115. Monaghan DT, Bridges RJ, Cotman CW: The excitatory amino acid receptors: their classes, pharmacology, and distinct properties in the function of the central nervous system. *Annu Rev Pharmacol Toxicol* 1989;29:365–402.
116. Westbrook GL: Glutamate receptor update. *Curr Opin Neurobiol* 1994;4:337–346.
117. Nakanishi S: Molecular diversity of glutamate receptors and implications for brain function. *Science* 1992;258:597–603.
118. Seeburg PH: The *TINS/TiPS* Lecture. The molecular biology of mammalian glutamate receptor channels. *Trends Neurosci* 1993;16:359–365.
119. Hollmann M, Heinemann S: Cloned glutamate receptors. *Annu Rev Neurosci* 1994;17:31–108.
120. Gasic GP, Hollmann M: Molecular neurobiology of glutamate receptors. *Annu Rev Physiol* 1992; 54:507–536.
121. Petralia RS, Wenthold RJ: Light and electron immunocytochemical localization of AMPA-selective glutamate receptors in the rat brain. *J Comp Neurol* 1992;318:329–354.
122. Tingley WG, Roche KW, Thompson AK, Huganir RL: Regulation of NMDA receptor phosphorylation by alternative splicing of the C-terminal domain. *Nature* 1993;364:70–73.
123. Molnar E, McIlhinney RA, Baude A, Nusser Z, Somogyi P: Membrane topology of the GluR1 glutamate receptor subunit: epitope mapping by site-directed antipeptide antibodies. *J Neurochem* 1994;63:683–693.
124. Roche KW, Tingley WG, Huganir RL: Glutamate receptor phosphorylation and synaptic plasticity. *Curr Opin Neurosci* 1994;4:383–388.
125. Taverna FA, Wang LY, Macdonald JF, Hampson DR: A transmembrane model for an ionotropic glutamate receptor predicted on the basis of the location of asparagine-linked oligosaccharides. *J. Biol Chem* 1994;269:14159–14164.

126. Wo ZG, Oswald RE: Transmembrane topology of two kainate receptor subunits revealed by N-glycosylation. *Proc Natl Acad Sci U S A* 1994;91:7154–7158.
127. Bennett JA, Dingledine R: Topology profile for a glutamate receptor: three transmembrane domains and a channel-lining reentrant membrane loop. *Neuron* 1995;14:373–384.
128. Wood MW, VanDongen HM, VanDongen AM: Structural conservation of ion conduction pathways in K channels and glutamate receptors. *Proc Natl Acad Sci U S A* 1995;92:4882–4886.
129. Stern-Bach Y, Bettler B, Hartley M, Sheppard PO, O'Hara PJ, Heinemann SF: Agonist selectivity of glutamate receptors is specified by two domains structurally related to bacterial amino acid-binding proteins. *Neuron* 1994;13:1345–1357.
130. Kuryatov A, Laube B, Betz H, Kuhse J: Mutational analysis of the glycine-binding site of the NMDA receptor: structural similarity with bacterial amino acid-binding proteins. *Neuron* 1994;12:1291–1300.
131. Wafford KA, Kathoria M, Bain CJ et al: Identification of amino acids in the *N*-methyl-D-aspartate receptor NR1 subunit that contribute to the glycine binding site. *Mol Pharmacol* 1995;47:374–380.
132. Blackstone C, Murphy TH, Moss SJ, Baraban JM, Huganir RL: Cyclic AMP and synaptic activity-dependent phosphorylation of AMPA-preferring glutamate receptors. *J Neurosci* 1994;14:7585–7593.
133. Roche KW, O'Brien RJ, Mammen AL, Bernhardt J, Huganir RL: Characterization of multiple phosphorylation sites on the AMPA receptor GluR1 subunit. *Neuron* 1996;16:1179–1188.
134. Raymond LA, Blackstone CD, Huganir RL: Phosphorylation and modulation of recombinant GluR6 glutamate receptors by cAMP-dependent protein kinase. *Nature* 1993;361:637–641.
135. McGlade-McCulloh E, Yamamoto H, Tan S-E, Brickey DA, Soderling TR: Phosphorylation and regulation of glutamate receptors by calcium/calmodulin-dependent protein kinase II. *Nature* 1993;362:640–642.
136. Yakel JL, Vissavajjhala P, Derkach VA, Brickey DA, Soderling TR: Identification of a Ca^{2+}/calmodulin-dependent protein kinase II regulatory phosphorylation site in non-*N*-methyl-D-aspartate glutamate receptors. *Proc Natl Acad Sci U S A* 1995;92:1376–1380.
136a. Nakazawa K, Mikawa S, Hashikawa T, Ito M: Transient and persistent phosphorylation of AMPA-type receptors subunits in cerebellar Purkinje cells. *Neuron* 1995;15:697–709.
137. Knapp AG, Dowling JE: Dopamine enhances excitatory amino acid-gated conductances in cultured retinal horizontal cells. *Nature* 1987;325:437–439.
138. Liman ER, Knapp AG, Dowling JE: Enhancement of kainate-gated currents in retinal horizontal cells by cyclic AMP-dependent protein kinase. *Brain Res* 1989;481:399–402.
139. Pereda AE, Nairn AC, Wolszon LR, Faber DS: Postsynaptic modulation of synaptic efficacy at mixed synapses on the Mauthner cell. *J Neurosci* 1994;14:3704–3712.
140. Smith DO, Lowe D, Temkin R, Jensen P, Hatt, H: Dopamine enhances glutamate-activated currents in spinal motoneurons. *J Neurosci* 1995;15:3905–3912.
141. Greengard P, Jen J, Nairn AC, Stevens CF: Enhancement of the glutamate response by cAMP dependent protein kinase in hippocampal neurons. *Science* 1991;253:1135–1138.
142. Wang L-Y, Salter MW, Macdonald JF: Regulation of kainate receptors by cAMP-dependent protein kinase and phosphatases. *Science* 1991;253:1132–1135.
143. Knapp AG, Schmidt, KF, Dowling JE: Dopamine modulates the kinetics of ion channels gated by excitatory amino acids in retinal horizontal cells. *Proc Natl Acad Sci U S A* 1990;87:767–771.
144. Keller BU, Hollmann M, Heinemann S, Konnerth A: Calcium influx through subunits GluR1/GluR3 of kainate/AMPA receptor channels is regulated by cAMP dependent protein kinase. *EMBO J* 1992; 11:891–896.
145. Wang L-Y, Taverna FA, Huang XD, Macdonald JF, Hampson DR: Phosphorylation and modulation of a kainate receptor (GluR6) by cAMP-dependent protein kinase. *Science* 1993;259:1173–1175.
146. Wang L-Y, Dudek EM, Browning MD, Macdonald JF: Modulation of AMPA/kainate receptors in cultured murine hippocampal neurones by protein kinase C. *J Physiol* 1994;475:431–437.
147. Lledo PM, Jhelmstad GO, Mukherji S, Soderling TR, Malenka RC, Nicoll RA: Calcium/calmodulin-dependent kinase II and long-term potentiation enhance synaptic transmission by the same mechanism. *Proc Natl Acad Sci U S A* 1995;92:11175–11179.
148. Liao D, Hessler NA, Malinow R: Activation of postsynaptically silent synapses during pairing-induced LTP in CA1 region of hippocampal slice. *Nature* 1995;375:400–404.
149. Isaac JT, Nicoll RA, Malenka RC: Evidence for silent synapses: implications for the expression of LTP. *Neuron* 1995;15:427–434.
150. Moon IS, Apperson ML, Kennedy MB: The major tyrosine-phosphorylated protein in the postsynaptic density fraction is *N*-methyl-D-aspartate receptor subunit 2B. *Proc Natl Acad Sci U S A* 1994; 91:3954–3958.

151. Lau LF, Huganir RL: Differential tyrosine phosphorylation of *N*-methyl-D-aspartate receptor subunits. *J. Biol Chem* 1995;270:20036–20041.
152. Suzuki T, Okumura-Noji K: NMDA receptor subunits epsilon 1 (NR2A) and epsilon 2 (NR2B) are substrates for Fyn in the postsynaptic density fraction isolated from the rat brain. *Biochem Biophys Res Commun* 1995;216:582–588.
153. Chen L, Huang LY: Sustained potentiation of NMDA receptor-mediated glutamate responses through activation of protein kinase C by a mu opioid. *Neuron* 1991;7:319–326.
154. Chen L, Huang LY: Protein kinase C reduces Mg^{2+} block of NMDA-receptor channels as a mechanism of modulation. *Nature* 1992;356:521–523.
155. Markram H, Segal M: Activation of protein kinase C suppresses responses to NMDA in rat CA1 hippocampal neurones. *J Physiol (Lond)* 1992;457:491–501.
156. Aniksztejn L, Otani S, Ben-Ari Y: Quisqualate metabotropic receptors modulate NMDA currents and facilitate induction of long-term potentiation through protein kinase C. *Eur J Neurosci* 1992;4: 500–505.
157. Raman IM, Tong G, Jahr CE: beta-Adrenergic regulation of synaptic NMDA receptors by cAMP-dependent protein kinase. *Neuron* 1996;16:415–421.
158. Raymond LA, Tingley WG, Blackstone CD, Roche KW, Huganir RL: Glutamate receptor modulation by protein phosphorylation. *J Physiol (Paris)* 1994;88:181–192.
159. Kelso SR, Nelson TE, Leonard JP: Protein kinase C-mediated enhancement of NMDA currents by metabotropic glutamate receptors in *Xenopus* oocytes. *J Physiol (Lond)* 1992;449:705–718.
160. Urushihara H, Tohda M, Nomura Y: Selective potentiation of *N*-methyl-D-aspartate-induced current by protein kinase C in *Xenopus* oocytes injected with rat brain RNA. *J. Biol Chem* 1992;267:11697–11700.
161. Yamazaki M, Mori H, Araki K, Mori KJ, Mishina M: Cloning, expression and modulation of a mouse NMDA receptor subunit. *FEBS Lett* 1992;300:39–45.
162. Durand GM, Bennett MV, Zukin RS: Splice variants of the *N*-methyl-D-aspartate receptor NR1 identify domains involved in regulation by polyamines and protein kinase C. *Proc Natl Acad Sci U S A* 1993;90:6731–6735. [Erratum in *Proc Natl Acad Sci U S A* 1993;90(20):9739.]
163. Kutsuwada T, Kashiwabuchi N, Mori H et al: Molecular diversity of the NMDA receptor channel [see comments]. *Nature* 1992;358:36–41.
164. Yamakura, T, Mori H, Shimoji K, Mishina M: Phosphorylation of the carboxyl-terminal domain of the zeta 1 subunit is not responsible for potentiation by TPA of the NMDA receptor channel. *Biochem Biophys Res Commun* 1993;196:1537–1544.
165. Sigel E, Baur R, Malherbe P: Protein kinase C transiently activated heteromeric *N*-methyl-D-aspartate receptor channels independent of the phosphorylatable C-terminal splice domain and of consensus phosphorylation sites. *J. Biol Chem* 1994;269:8204–8208.
166. Mori H, Yamakura T, Masaki H, Mishina M: Involvement of the carboxyl-terminal region in modulation by TPA of the NMDA receptor channel. *Neuroreport* 1993;4:519–522.
167. Lieberman DN, Mody I: Regulation of NMDA channel function by endogenous Ca^{2+}-dependent phosphatase. *Nature* 1994;369:235–239.
168. Tong G, Jahr CE: Regulation of glycine-insensitive desensitization of the NMDA receptor in outside-out patches. *J Neurophysiol* 1994;72:754–761.
169. Wang LY, Orser BA, Brautigan DL, Macdonald JF: Regulation of NMDA receptors in cultured hippocampal neurons by protein phosphatases 1 and 2A. *Nature* 1994;369:230–232.
170. Ehlers MD, Tingley WG, Huganir RL: Regulated subcellular distribution of the NR1 subunit of the NMDA receptor. *Science* 1995;269:1734–1737.
171. Tingley WG, Ehlers MD, Kameyama K et al: Characterization of protein kinase A and protein kinase C phosphorylation of the *N*-methyl-D-aspartate receptor NR1 subunit using phosphorylation site-specific antibodies. *J. Biol Chem* 1997;272:5157–5166.
172. Dudek SM, Bear, MF: Homosynaptic long-term depression in area CA1 of hippocampus and effects of *N*-methyl-D-aspartate receptor blockade. *Proc Natl Acad Sci U S A* 1992;89:4363–4367.
173. Mulkey, RM, Malenka RC: Mechanisms underlying induction of homosynaptic long-term depression in area CA1 of the hippocampus. *Neuron* 1992;9:967–975.
174. Kauer JA, Malenka RC, Nicoll RA: A persistent postsynaptic modification mediates long-term potentiation in the hippocampus. *Neuron* 1988;1:911–917.
175. Davies SN, Lester RAJ, Reymann KG, Collingridge GL: Temporally distinct pre- and postsynaptic mechanisms maintain long-term potentiation. *Nature* 1989;338:500–503.

176. Bashir ZI, Alford S, Davies SN, Randall AD, Collingridge GL: Long-term potentiation of NMDA receptor-mediated synaptic transmission in the hippocampus. *Nature* 1991;349:156–158.
177. Kullmann DM, Nicoll RA: Long-term potentiation is associated with increases in quantal content and quantal amplitude. *Nature* 1992;357:240–244.
178. Manabe T, Renner P, Nicoll RA: Postsynaptic contribution to long-term potentiation revealed by the analysis of miniature synaptic currents. *Nature* 1992;355:50–55.
179. O'Connor JJ, Rowan MJ, Anwyl R: Tetanically induced LTP involves a similar increase in the AMPA and NMDA receptor components of the excitatory postsynaptic current: investigations of the involvement of mGlu receptors. *J Neurosci* 1995;15:2013–2020.
180. Nicoll RA, Malenka RC: Contrasting properties of two forms of long-term potentiation in the hippocampus. *Nature* 1995;377:115–118.
181. Selig DK, Hjelmstad GO, Herron C, Nicoll RA, Malenka RC: Independent mechanisms for long-term depression of AMPA and NMDA responses. *Neuron* 1995;15:417–426.
182. Madison DV, Malenka RC, Nicoll RA: Mechanisms underlying long-term potentiation of synaptic transmission. *Annu Rev Neurosci* 1991;14:379–397.
183. Ben-Ari Y, Aniksztejn L, Bregestovski P: Protein kinase C modulation of NMDA currents: an important link for LTP induction. Trends Neurosci 1992;15:333–339.
184. Malenka RC, Kauer JA, Perkel DJ et al: An essential role for postsynaptic calmodulin and protein kinase activity in long-term potentiation. *Nature* 1989;340:554–557.
185. Malinow R, Schulman H, Tsien RW: Inhibition of postsynaptic PKC or CaM-KII blocks induction but not expression of LTP. *Science* 1989;245:862–866.
186. Wang JH, Feng D-P: Postsynaptic protein kinase C essential to induction and maintenance of long-term potentiation in the hippocampal CA1 region. *Proc Natl Acad Sci U S A* 1992;89:2576–2580.
187. Pettit DL, Perlman S, Malinow R: Potentiated transmission and prevention of further LTP by increased CaMKII activity in postsynaptic hippocampal slice neurons. *Science* 1994;266:1881–1885.
188. O'Dell TJ, Kandel ER, Grant SGN: Long-term potentiation in the hippocampus is blocked by tyrosine kinase inhibitors. *Nature* 1991;353:558–560.
189. Klann E, Chen SJ, Sweatt JD: Mechanism of protein kinase C activation during the induction and maintenance of long-term potentiation probed using a selective peptide substrate [see comments]. *Proc Natl Acad Sci U S A* 1993;90:8337–8341.
190. Sacktor TC, Osten P, Valsamis H, Jiang X, Naik MU, Sublette, E: Persistent activation of the zeta isoform of protein kinase C in the maintenance of long-term potentiation [see comments]. *Proc Natl Acad Sci U S A* 1993;90:8342–8346.
191. Grant SGN, O'Dell TJ, Karl KA, Stein PL, Soriano P, Kandel ER: Impaired long-term potentiation, spatial learning, and hippocampal development in *fyn* mutant mice. *Science* 1992;258:1903–1910.
192. Silva AJ, Paylor R, Wehner JM, Tonegawa S: Deficient hippocampal long-term potentiation in α-calcium-calmodulin kinase II mutant mice. *Science* 1992;257:206–209.
193. Abeliovich A, Chen C, Goda Y, Silva AJ, Stevens, CF, Tonegawa S: Modified hippocampal long-term potentiation in PKC gamma mutant mice. *Cell* 1993;75:1253–1262.
194. Mulkey MM, Herron CE, Malenka RC: An essential role for protein phosphatases in hippocampal long-term depression. *Science* 1993;261:1051–1055.
195. Xiao MY, Karpefors M, Gustafsson B, Wigstrom H: On the linkage between AMPA and NMDA receptor-mediated EPSPs in homosynaptic long-term depression in the hippocampal CA1 region of young rats. *J Neurosci* 1995;15:4496–4506.
196. Figurov A, Boddeke H, Muller D: Enhancement of AMPA-mediated synaptic transmission by the protein phosphatase inhibitor calyculin A in rat hippocampal slices. *Eur J Neurosci* 1993;5:1035–1041.
197. Wyllie DJ, Nicoll RA: A role for protein kinases and phosphatases in the Ca^{2+}-induced enhancement of hippocampal AMPA receptor-mediated synaptic responses. *Neuron* 1994;13:635–643.
198. Mayford M, Wang J, Kandel ER, O'Dell TJ: CaMKII regulates the frequency–response function of hippocampal synapses for the production of both LTD and LTP. *Cell* 1995;81:891–904.
199. Neveu D, Zucker RS: Postsynaptic levels of $[Ca^{2+}]_i$ needed to trigger LTD and LTP. *Neuron* 1996;16:619–629.
200. Ito M: Long-term depression. *Annu Rev Neurosci* 1989;12:85–102.
201. Linden DJ: Long-term synaptic depression in the mammalian brain. *Neuron* 1994;12:457–472.

Advances in Second Messenger and Phosphoprotein Research, Vol. 33

1040-7952/99 $30.00

4

Regulation of CFTR Cl^- Ion Channels by Phosphorylation and Dephosphorylation

David C. Gadsby* and Angus C. Nairn[†]

*Laboratories of *Cardiac/Membrane Physiology and †Molecular and Cellular Neuroscience, The Rockefeller University, New York, New York 10021*

INTRODUCTION: THE CYSTIC FIBROSIS GENE PRODUCT IS A Cl^- CHANNEL WITH COMPLEX REGULATION

The disease cystic fibrosis stems from mutations in the gene that encodes the cystic fibrosis transmembrane conductance regulator (CFTR; Refs. 1,2). CFTR's predicted primary structure immediately identified it as a member of the superfamily of ATP-binding cassette (ABC) transporters, which includes the yeast *a* mating factor exporter (STE6) and the mammalian P-glycoprotein linked to multidrug resistance (3). Like these other ABC transporters, CFTR comprises two roughly similar halves, each incorporating six putative transmembrane α helices followed by an intracellular nucleotide binding domain (NBD; Fig. 1). Unique to CFTR is a large intracellular regulatory (R) domain containing multiple sites for phosphorylation by cyclic AMP dependent protein kinase (PKA) and protein kinase C (PKC; Ref. 1). Despite this superficial structural resemblance to molecules widely supposed to engage in active transport, there is now overwhelming evidence that CFTR is a Cl^- channel whose gating is regulated by kinase-mediated phosphorylation and by hydrolysis of ATP (4–8; for review see Ref. 9). In contrast, there is as yet no compelling evidence that any other ABC transporter forms an ion channel (an earlier suggestion not withstanding: Refs. 10,11). There is, however, mounting evidence that certain members of the ABC family, including CFTR and the sulfonylurea receptor (SUR; Ref. 12), can somehow modulate, or even control, the behavior of other ion channels (13), possibly through direct protein–protein interaction (14).

The evidence that CFTR itself comprises a Cl^- ion pore comes from several sources. First, expression of CFTR in a wide variety of cell types consistently yields the same, small conductance (roughly 10 pS), ohmic Cl^- channels that require phosphorylation by PKA before they will open (reviewed in Refs. 15–17). Second, and perhaps most straightforward, recombinant CFTR overexpressed in Sf9 cells and

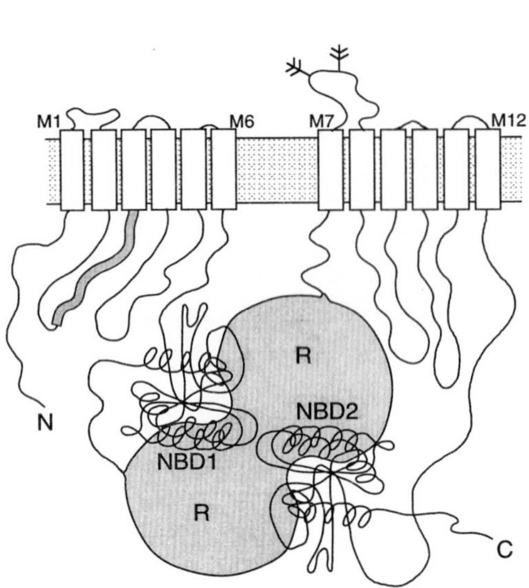

FIG. 1. Topological model of CFTR. The cytosolic amino (N) and carboxyl (C) termini, nucleotide binding domains (NBD1, NBD2), regulatory (R) domain, and predicted (1) membrane-spanning α-helices (M1–M12), and glycosylation sites in the M7–M8 extracellular loop are shown. The thickened section of the M2–M3 cytoplasmic loop represents the 30 amino acids encoded by exon 5, believed to be spliced out of the cardiac isoform of CFTR (123). Lengths of the extracellular and intracellular loops are drawn in approximate proportion to the number of amino acids they contain. NBD1 and NBD2 are drawn with the fold of p21-*ras* (124) to emphasize our proposal (9) of structural and functional homology between CFTR's NBDs and the catalytic sites of G proteins (cf. Refs. 112,113). The NBDs are drawn in close proximity to each other and in contact with the R domain to emphasize the likelihood that the two NBDs directly interact with each other and with the R domain.

extracted with detergent, purified to a single band on a silver-stained gel, and then renatured and reconstituted into artifical liposomes and incorporated into lipid bilayers, gave rise to typical Cl^- channels of small ohmic conductance that were activated by PKA catalytic subunit plus ATP (18). Third, mutations introduced into the putative transmembrane α helices M1 and M6 of CFTR (Fig. 1) can change the permeation characteristics of the resulting channels in several ways. For instance, the charge-reversing mutations K95D (in M1) or K335E (in M6) altered the anion permeability sequence of CFTR from $Br^- > Cl^- > I^-$ to $I^- > Br^- > Cl^-$ (19), the mutation S341A (in M6) modified the affinity and voltage dependence of open channel block by diphenylamine-2-carboxylate (DPC) (20), and the cysteine scanning accessibility method has shown that water-soluble reagents can reach residues in M1 and M6 arranged as if on one face of α helices that thus seem to line the pore (21,22); comparisons of reaction rates for the anionic and cationic forms of these reagents further suggest that the anion selectivity filter resides close to the cytoplasmic end of a wide water-filled pore (22). And, charge-neutralizing or reversing mutations at R347 in M6 greatly reduced average channel conductance, and altered interactions between permeant anions within CFTR channels (23). Indeed, with R347 replaced by histidine, which is positively charged (like arginine) at low pH but neutral at high pH, channel conductance and ionic interactions could be toggled back and forth between wild-type and mutant behavior by switching cytoplasmic pH between 5.5 and 8.7 (23); and detailed single channel analysis of the pH-dependent dwell times in high and low conductance states has revealed the voltage-dependent kinetics of protonation of the side chain of residue 347 within the ion pore (24).

There is now widespread agreement that CFTR channels will not open unless they are phosphorylated (5,4,25), but also that phosphorylation per se, although necessary for channel opening (but cf. Refs. 26,27), is not sufficient. After phosphorylation, CFTR channels require exposure to Mg^{2+} ions and ATP, or another hydrolyzable nucleoside triphosphate, to allow them to open and close (4,25). In fact, present evidence suggests that, during each gating cycle, a single phosphorylated CFTR channel hydrolyzes one molecule of ATP at its N-terminal nucleotide binding domain (NBD1, Fig. 1) to open, and then hydrolyzes a second ATP at the C-terminal NBD (NBD2, Fig. 1) to permit the channel to close (6–9,28,29). But it is also clear that activation of CFTR channels by phosphorylation with PKA is not simply an all-or-none event, because biochemical measurements have shown that PKA phosphorylates the R domain of CFTR to a stoichiometry of >5 mol/mol (30,31), and electrophysiological measurements have identified at least three functionally distinct phosphorylated states of individual CFTR channels (32,33). Our aim in this chapter is to review the evidence for CFTR's complex gating mechanism, with particular emphasis on the role of incremental phosphorylation and dephosphorylation of the R domain in orchestrating the cycles of ATP binding and hydrolysis at CFTR's two NBDs.

PHOSPHORYLATION AND REGULATION OF CFTR BY PKA

Phosphorylation by PKA: Biochemical Analysis

The R domain of human epithelial CFTR was predicted (1) to contain multiple sites for phosphorylation by PKA. A variety of subsequent studies have determined that at least 10 sites in CFTR can be phosphorylated by PKA *in vitro* (30,34–36) (Table 1) and most likely *in vivo.* To date, only seryl residues in CFTR have been found to be phosphorylated, although evidence for regulation of CFTR by agents that modulate tyrosine phosphorylation has recently been obtained. Nine of the sites identified are classical dibasic PKA consensus sites (37), with the amino acid sequence R-R/K-X-S/T (X represents any amino acid). In human CFTR, one of the sites (Ser-753) is found within a monobasic R-X-S sequence, the most common (twice as frequent as the classical dibasic) consensus sequence for PKA substrates (37). All but one of the sites phosphorylated by PKA are contained within the R domain; the exception is Ser-422, present just N-terminal to NBD1 (Fig. 2).

An important feature of CFTR phosphorylation by PKA is that although the phosphorylated serines lie within classical consensus sites, their relative levels of phosphorylation vary tremendously. *In vitro,* a variety of methods, including direct amino acid sequencing, site-directed mutagenesis together with two-dimensional peptide mapping, and mass spectrometry (30,34–36,38), have identified serines 660, 700, 737, 768, 795, and 813 as being phosphorylated rapidly, while serines 422, 670, 712, and 753 are phosphorylated at lower rates. The relatively slow phosphorylation of some sites explains why recombinant R-domain peptides are phosphorylated only to a stoichiometry of approximately 5–6 mol/mol after relatively brief exposure to PKA

TABLE 1. *Sites in CFTR phosphorylated by protein kinase A*

Residue number	Amino acid sequence	Phosphorylation		Effect of site-directed mutagenesis		
		In vitro PKA	*In vivo* PKA	PKA[a]	Stim[b]	Inh
S422	RKTS	+	–	+ (10SA)		
S660	RRNS	++	++	++++ (4SA)	++	
S670	RFS	–	–	null (12SA)	+	
S686	KKQSFK	–	–	+ (8SA)	Null	
T690	KQT	–	–	Null (12SA)		
S700	KRKNS	+++	++	Null (6SA)	+	
S712	RKFS	++	–	Null (6SA)	Null	
S728	EDSDE	–	–	Null (11SA)		
S737	RRLS	+++	+++	++++ (4SA)		+++
S753	RIS	+	–	+ (11SA)		
S756	VIST	–	–	Null (11SA)		
S768	RRQS	++++	–	+ (8SA)		++++
T787	RKT	–	–	Null (12SA)		
T788	RKTT	–	–	Null (9SA)		
S790	STRK	–	–	Null (12SA)		
S795	RKVS	++++	+++	++++ (4SA)	++	
S813	RRLS	++	++++	++++ (4SA)	++++	

The *in vitro* and *in vivo* tabulation summarizes biochemical results from a number of studies (see Refs. 30,34,36,38). The *in vitro* information takes into account results from amino acid sequencing, mass spectrometry, and kinetic analysis of phosphorylation of an R-domain peptide and short peptides derived from the R-domain sequence. The *in vivo* information takes into account results from ^{32}P-labeling followed by immunoprecipitation and two-dimensional peptide mapping. The relative level of phosphorylation reflects ^{32}P incorporation into a particular site during an acute stimulation of PKA activity (5–15 min). In intact cells, the level of phosphorylation will reflect both the activity of PKA toward a particular site and the activity of one or more protein phosphatases.

Stim: stimulatory influence reflected by rightward shift of IBMX dose–response curve in mutant; Inh: inhibitory influence reflected by leftward shift of dose–response curve in mutant; Null, no effect; +, positive; –, negative; blank spaces indicate sites not included in study.

[a] Summary of studies by Rich et al. (43), Chang et al. (44), and Seibert et al. (35,41) examining the gross ability of cAMP and/or PKA to activate CFTR mutants. The minimal CFTR mutant yielding a discernible effect is shown in parentheses. It was not possible to assign functional roles to individual serines in the 4SA mutant in the studies carried out by Rich et al. (43) and Chang et al. (44) because the coarse assay employed revealed no obvious effect of mutating any one of Ser-660, Ser-737, Ser-795, and Ser-813. Similarly, no particular role could be assigned to any single residue additionally mutated in the 8SA mutant. However, in the 10SA and 11SA mutants, a contribution can be assigned to Ser422 and Ser753, since a single amino acid substitution was made and results compared to the 9SA and 10SA mutants, respectively.

[b] Summary of studies performed by Wilkinson et al. (42) in which sensitivity to activation by IBMX was examined in CFTR mutants expressed in oocytes.

(30,31). Phosphorylation with higher levels of PKA or for much longer incubation times would presumably eventually lead to stoichiometric phosphorylation of all PKA sites. Indeed, phosphorylation of poorly phosphorylated sites can be detected only by using highly sensitive techniques. Phosphorylation of Ser-753, for example, was first detected in a mutant CFTR molecule in which all 10 serines in dibasic sites had been mutated to alanines (35), or by mass spectrometry of phosphorylated R-domain peptide (38).

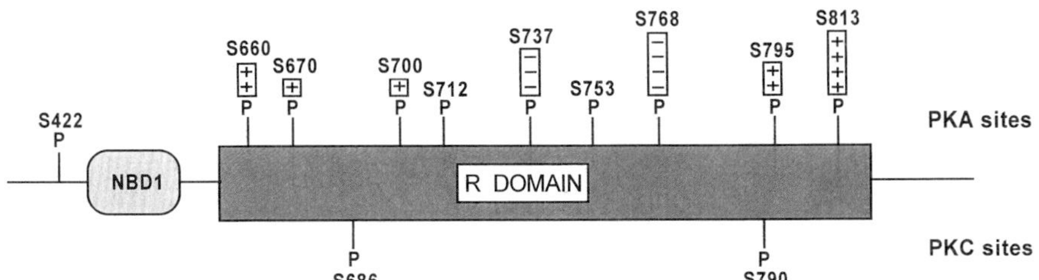

FIG. 2. Diagrammatic summary of CFTR phosphorylation by PKA, PKG, and PKC. Seryl residues phosphorylated by PKA are shown pointing upward; sites phosphorylated by PKC are shown pointing down. For PKA, the diagram attempts to integrate the relative level of phosphorylation of individual sites observed in intact cells following a brief stimulation of PKA (except Ser-768), as well as the relative contribution each site appears to make to regulation of CFTR channel activity as inferred from site-directed mutagenesis: note difference between stimulatory (+) sites and inhibitory (−) sites. An important issue not yet resolved is the apparent lack of phosphorylation of Ser-768 in intact cells. For PKC, the diagram jointly summarizes *in vitro* and *in vivo* phosphorylation studies. All *in vitro* biochemical results to date suggest that PKG phosphorylates the same sites as PKA, but no *in vivo* analysis of phosphorylation sites or site-directed mutagenesis has been carried out yet.

There are likely to be several reasons for the differences in the rates of phosphorylation of the various sites by PKA. Kinetic analysis of the phosphorylation of short synthetic peptides encompassing several of the CFTR serine residues has indicated more than 10-fold differences in the catalytic efficiencies of PKA toward those serines (30), reflecting intrinsic differences in the ability of PKA to bind to and phosphorylate specific sites. In intact CFTR, secondary and tertiary structure is also likely to influence the rate of phosphorylation of individual sites. It is possible that the pattern of phosphorylation is ordered, with changes in structure of CFTR due to phosphorylation of one site influencing the phosphorylation of another. In this regard, it is of interest that phosphorylation by PKA appears to induce conformational changes in the R domain as assessed from analysis of circular dichroism (CD) spectra (31,39), or from the several discrete mobility shifts seen with SDS–polyacrylamide gel electrophoresis (30,31). Recent studies have indicated that the largest of these mobility shifts occurs upon phosphorylation of Ser-737 (38,40), although mutation of Ser-737 to alanine did not appear to affect the phosphorylation of other R-domain serines (40). The likelihood that phosphorylation by PKA induces structural changes in the R domain, and the possibility that phosphorylation may occur in an ordered manner, must be borne in mind when results from site-directed mutagenesis studies are analyzed.

Biochemical studies and site-directed mutagenesis have been used to investigate which sites in CFTR are phosphorylated following activation of PKA in intact cells (Table 1), a presumption being that if mutation of a serine residue (e.g., to alanine) affects CFTR function, then that site is normally phosphorylated. Clear biochemical evidence has been obtained that serines 660, 700, 737, 795, and 813 are phosphorylated *in vivo.* Site-directed mutagenesis studies imply that, in addition, serines 422, 686, 712, 753, and 768 are likely to be phosphorylated. However, the relative levels of phosphorylation of different sites observed in intact cells do not always follow their relative rates of phosphorylation by PKA *in vitro.* Ser-813, for example, is phosphorylated relatively highly in intact cells but at a relatively low rate *in vitro.* Even more striking, phosphorylation of Ser-768 has not yet been demonstrated biochemically in intact cells, even though it is the best *in vitro* site. Because the steady state level of phosphorylation of a site *in vivo* is a function both of its rate of phosphorylation by the protein kinase and of its rate of dephosphorylation by cellular protein phosphatases, the failure to observe *in vivo* phosphorylation of Ser-768 likely reflects the fact that this site is also a very good substrate for one or more protein phosphatases. This observation underlines the important role played by protein dephosphorylation in the regulation of CFTR function in cells. In addition, results obtained from the biochemical studies to date emphasize the need for more sensitive methods, such as mass spectrometry, that can be used to assess quantitatively the level of phosphorylation of individual sites in CFTR in intact cells.

Do the 10 serines in CFTR identified as being phosphorylated by PKA constitute the complete picture, or are there sites still to be identified? Seibert et al. (35,41) were able to detect residual phosphorylation (correlated with residual channel activity) in a mutant CFTR channel (termed 11SA) in which these 10 serine residues, plus the threonine (Thr-788) in the tenth dibasic consensus site, were mutated to alanine. Also, two-dimensional phosphopeptide mapping of CFTR phosphorylated in intact cells indicated one prominent peptide that was not found in CFTR phosphorylated by PKA *in vitro* (30); but that (those) site(s) need not necessarily reside in the R domain and need not necessarily be directly phosphorylated by PKA (i.e., PKA phosphorylation of some other protein could stimulate another kinase, or inhibit a phosphatase, that impinges on CFTR). One of the relatively poorly phosphorylated sites, Ser-753, lies within a monobasic R-X-S motif. Interestingly, the human CFTR R domain (1) includes another R-X-S site at Ser-670 [a site recently shown to be phosphorylated at low levels *in vitro* (38) and also shown, by mutation, to contribute to channel regulation (42)], which might therefore contribute to the labeling and channel activity seen in the 11SA mutant. Curiously, in all other species examined, an additional Arg residue confers a classical dibasic R-R-X-S consensus motif on the Ser-670 site.

With the exception of Ser-422, all the well-characterized phosphorylation sites occur within the R domain. However, it is possible that some of the residual phosphorylation observed in the 11SA mutant, and/or the unidentified site(s) phosphorylated in intact cells, do not reside in the R domain. For instance, a cluster of serine, arginine and lysine residues at the C-terminus of CFTR (residues 1453–1457 of human CFTR) constitute potential phosphorylation sites for PKA (and possibly for PKC). In addition, the sequence R-X-S-S-K/R at that location is conserved in CFTR from different species, and preliminary studies show that a synthetic peptide including these amino acids is a substrate for PKA (K. Chan and A. C. Nairn, unpublished observations).

Phosphorylation by PKA: Functional Analysis

Following the biochemical identification of what appeared to be the principal physiological phosphorylation sites (R-domain serines 660, 700, 737, 795, and 813), site-directed mutagenesis was employed to assess the functional importance of each one. Hopes of a simple answer vanished, however, with the demonstrations that individual mutation of any one of those five serines (or of serines 686, 712, or 768) to alanine (34), or even simultaneous mutation of two or three of them in a host of different combinations (43), seemed to have little or no effect on gross measurements of cAMP-stimulated anion efflux. The first convincing indication that loss of PKA phosphorylation sites impairs CFTR channel function was seen with a "quad" mutant (4SA) in which four of the principal serines, 660, 737, 795, and 813, were simultaneously mutated (Table 1). In the 4SA mutant both cAMP-stimulated iodide efflux and CFTR channel P_o (open probability: the average fraction of time that a channel spends open) were reduced, but unexpectedly only to approximately 50% that of wild-type CFTR (43,44). Moreover, further mutation of up to all eight dibasic R-domain serines (8SA: the 4SA mutant with serines 686, 700, 712, and 768 mutated to alanine) caused only a slight additional reduction of PKA-activated channel function, measured either by iodide efflux, or by channel P_o (44,43).

Phosphorylation of the various CFTR mutants was also examined in intact cells by ^{32}P-labeling and immunoprecipitation (43,44) or by immunoprecipitation followed by phosphorylation *in vitro* (35,41,43). In the 4SA mutant, cAMP-stimulated phosphorylation was found to be substantially reduced from wild type, but still detectable. A slight further reduction in phosphorylation was observed for the 8SA mutant, but additional mutation of Ser-422 (outside the R domain) and of Thr-788, to yield a 10SA mutant, reduced the level of phosphorylation to below the resolution of the techniques then in use (44). The mutation of Ser-422 also slightly reduced channel function as assayed by iodide efflux. But it is important to note that these mutations of the major phosphorylation sites resulted in no obvious relative increase in phosphorylation levels of minor sites (43), raising the apparent paradox that PKA-mediated activation of mutant CFTR channels could occur without discernible phosphorylation. However, as already mentioned, higher resolution techniques were subsequently able to detect phosphorylation within the R domain in the 10SA mutant (35). Additional mutation of Ser-753 to alanine (the 11SA mutant) further reduced R-domain phosphorylation, and both cAMP-stimulated iodide efflux and channel P_o, by about another 40%. And both residual phosphorylation and PKA-dependent channel activity could be detected even in a 16SA mutant, in which several remaining R-domain serines and threonines, with a basic residue nearby, were mutated (35,41). These results, as well as the apparent contribution of Ser-422 to CFTR regulation, suggest that some site(s) outside the R domain might help regulate CFTR channels under certain circumstances.

In terms of functional mechanisms, the results of these initial residue-specific mutagenesis studies raised some vexing questions that we mention here but discuss in detail later. They indicated no specific roles for the sites (serines 660, 737, 795, and 813) most prominently phosphorylated in intact cells in response to PKA activation, and yet they suggested that sites (serines 422 and 753) presumably phosphorylated to lower levels in normal cells could also contribute to CFTR regulation, at least in channels lacking the major sites. The findings led to the idea of functional redundancy, or degeneracy, of individual phosphorylation sites (43,44), and they raised the question of the significance of small functional effects observed in a background of multiple mutations of major phosphorylation sites. Could, perhaps, phosphorylation of minor, kinetically less favorable, sites contribute to regulation of wild-type CFTR when cellular PKA activity is unusually high, or when phosphatases are relatively inactive? Although there is *in vitro* evidence for specific phosphorylation-mediated alterations in R-domain structure, an important general consideration in these mutagenesis studies is whether CFTR structure is affected in a nonspecific manner (39). If nothing else, these studies established the need for a careful evaluation of the contribution of each phosphorylation site to channel function. Eventually, this must include detailed functional analysis at the single-channel level.

A recent study (42), examining dose–response curves for activation of CFTR channels expressed in *Xenopus* oocytes by IBMX (isobutylmethylxanthine; which most likely predominantly inhibits cAMP phosphodiesterase; ref. 45) in the presence of forskolin (to activate PKA), provides a step in the right direction. As expected for "stimulatory" phosphorylation sites, substitution of serine 660, 670, 700, 795, or 813 individually with alanine increased the [IBMX] required ($K_{0.5}$) for half-maximal activation of CFTR (Table 1). Generally consistent with the phosphorylation levels observed in intact cells (Table 1), mutation of Ser-813 had the largest effect, while mutation of Ser-660 or Ser-795 had lesser effects, mutation of Ser-670 or Ser-700 had small effects, and mutation of Ser-712 or Ser-686 (a PKC site, as discussed later) had no measurable effect. Most importantly, mutation of Ser-737 (smaller effect) or Ser-768 (larger effect) *decreased* the $K_{0.5}$ for IBMX, demonstrating that these are "inhibitory" sites.

Interestingly, mutations of various combinations of stimulatory sites, or of combinations of stimulatory and inhibitory sites, yielded approximately additive effects. For example, mutation of Ser-768 (the major inhibitory site) together with Ser-813 (the major stimulatory site) effectively canceled each other's influence (42). These findings emphasize the quantitatively and qualitatively distinct contributions of individual PKA phosphorylation sites to the regulation of CFTR channel activity, and they further cloud the interpretation of earlier mutagenesis studies employing coarser assays and multiple mutations (cf. Refs. 43,44). The study of Wilkinson et al. (42) also highlights the potentially important role of phosphorylation of Ser-768, a site found to be rapidly phosphorylated *in vitro,* but not yet shown to be phosphorylated in intact cells.

PHOSPHORYLATION AND REGULATION OF CFTR BY OTHER KINASES

Protein Kinase C

While phosphorylation of CFTR by PKA appears to be the major mechanism by which channel gating is acutely modulated, it is now certain that CFTR can be phosphorylated, and potentially regulated, by other protein kinases. *In vitro,* PKC, with or without Ca^{2+}, phosphorylated CFTR on the R domain to a stoichiometry of ~2 mol/mol (30,31,46). PKC phosphorylated predominantly on Ser-686 and Ser-790, although some of the sites phosphorylated by PKA, including Ser-660 and Ser-700, could also be phosphorylated, though at lower rates, by PKC (30). Ser-686 occurs in the sequence K-K-Q-S-F-K, which conforms to the consensus found in many good substrates for PKC, such as the MARCKS family of proteins (47). However, in contrast to findings with PKA, phosphorylation of the R domain by PKC caused no conformational change discernible in CD spectra (31) nor, correspondingly, any substantial mobility shift on SDS–polyacrylamide gels (30,31). In intact cells, stimulation of PKC with phorbol ester increased phosphorylation of Ser-686, but not of Ser-790, and also slightly increased phosphorylation of several PKA sites (30). Since PKC phosphorylates Ser-660 and Ser-700 *in vitro,* it might directly phosphorylate those and other PKA sites *in vivo,* albeit slowly. But it is also possible that, in cells, phosphorylation of Ser-686 by PKC could facilitate subsequent phosphorylation of other sites by basally active PKA. Indeed, preliminary *in vitro* biochemical tests seemed to suggest that phosphorylation by PKA was enhanced after prephosphorylation by PKC (44; but see discussion in Ref. 48).

Functional studies have shown that direct application of PKC weakly activated recombinant CFTR channels in patches excised from Chinese hamster ovary (CHO) or NIH 3T3 cells (5,44,46). Stimulation of PKC with phorbol ester also activated Cl^- efflux in C127i mammary epithelial cells expressing CFTR, but not in cells expressing ΔF508-CFTR (49), and activated CFTR channels in cell-attached patches on colonic HT-29cl.19A cells (50) and on cardiac myocytes (51). Significantly, activation, or application, of PKC markedly potentiated the subsequent activation by PKA of CFTR channels in cell-attached or in excised patches (5,50), even of 10SA mutant channels (44).

A recent detailed investigation found that following excision of membrane patches from CFTR-transfected CHO cells, PKA-activated channels ran down over a period of 10 minutes or less; the rundown could be prevented by addition of PKC, but not of PKA (48). PKC apparently did not affect the number of active channels in a patch, but enhanced the ability of PKA to subsequently increase channel P_o. Preliminary analysis suggested that PKC shortened mean closed time without affecting the mean open time (open burst duration). The results suggest that phosphorylation of CFTR by constitutively active PKC in cells prior to excision of membrane patches is essential for subsequent acute regulation by PKA; this interpretation is supported by the reduced PKA-dependent channel activity observed after short-term treatment of transfected cells with PKC inhibitors (48).

Together, the various studies strongly suggest a modulatory role for PKC in the regulation of CFTR. But whether that modulation reflects direct phosphorylation of CFTR by PKC, and whether it is essential for subsequent channel activation by PKA, remains to be clarified. Relevant to this last question, it has been found that following extensive purification, reconstitution, and incorporation of CFTR into a lipid bilayer, channels can be activated by addition of only PKA and MgATP (18). But, even in that case, permissive constitutive phosphorylation by PKC cannot be ruled out, because a study of the ATPase activity of similarly purified CFTR suggested that it can remain partially phosphorylated (at sites sensitive to PP2A; see later) throughout the purification procedure (52). Unfortunately, the opportunity was not taken to test the ability of PKA alone (i.e., without PKC) to rephosphorylate and reactivate the purified CFTR after it had been dephosphorylated by treatment with PP2A. Remarkably, potentiation of channel activity by PKC was not abolished by mutation, in the 10SA channel, of Ser-686 (44), the major *in vitro* and *in vivo* site of phosphorylation by PKC, and mutation of Ser-686 did not affect PKA-dependent activation of macroscopic CFTR current in *Xenopus* oocytes (42); the 10SA mutant does retain Ser-790, however, an *in vitro* PKC site that might therefore mediate the potentiation by PKC. Alternatively, the effects of PKC might reflect an indirect mechanism, such as an influence on the cytoskeleton, or cell membrane area as suggested by Winpenny et al. (53). It is not clear, then, whether the effects of PKC on CFTR channels in cells or excised patches are due to PKC phosphorylation of CFTR or of some other target, nor whether the effects depend on prior or ongoing phosphorylation of CFTR by PKA (either basally active or stimulated). It is hoped that appropriate tests with selective inhibitors of PKA and/or of PKC will help answer these questions.

Many isoforms of PKC have been characterized and several, but not all, depend on Ca^{2+} for activity (54). A mixture of Ca^{2+}-dependent isoforms has been used in most of the studies with CFTR, although some evidence suggests that CFTR can also be phosphorylated by PKC in a Ca^{2+}-independent manner (30,46). It will be interesting to learn whether Ca^{2+}-independent, perhaps basally active, PKC isoforms might be involved in regulating CFTR in cells. In some studies, stimulation of PKC had a small and/or variable effect, possibly attributable to variations in PKC isoform distribution and activity in different cell and cell-free preparations. Given the synergistic effects of PKC and PKA on CFTR activation (48), variable PKC-dependent phosphorylation of CFTR, or other factor, could even underlie variable responses to PKA and might also complicate interpretation of studies using protein phosphatase inhibitors. Whatever its mechanism(s), CFTR channel regulation by PKC seems important and so warrants further detailed investigation.

Finally, PKC has been found to regulate the expression and degradation of CFTR in a manner that seems independent of direct phosphorylation of CFTR. Treatment of epithelial cells for several hours with phorbol ester leads to a significant reduction in transcription of CFTR mRNA (55–57), and Ca^{2+} ionophores such as A23187 or ionomycin have a similar effect (56). This consequence of PKC activation is believed to be mediated by so-called phorbol ester sensitive elements in the CFTR promoter, although the details have not yet been elucidated. Prolonged exposure of epithelial cells to phorbol ester has also been found to cause degradation of CFTR protein (58). Although this effect is diminished by kinase inhibitors, and hence appears to require activation of PKC, the mechanism has not been determined. In general, since treatment of cells with phorbol ester is also known to cause rapid down regulation of sensitive PKC isoforms, long-term effects of phorbol esters on CFTR have to be interpreted with caution.

cGMP-Dependent Protein Kinase

There has been considerable interest in the regulation of CFTR by cGMP in intestinal epithelia because heat-stable enterotoxins secreted by pathogenic strains of *Escherichia coli,* and the hormone guanylin, activate CFTR channels apparently via their action on guanylyl cyclase (59). But whether this regulation involves activation of cGMP-dependent protein kinase (PKG) has been difficult to establish. *In vitro,* both the major PKG isoforms, PKGI and PKGII, phosphorylate CFTR (30,46,60). Moreover, both PKGI and PKGII phosphorylate the R domain efficiently and to high stoichiometry (>5 mol/mol) (30,60), at sites that largely overlap those phosphorylated by PKA (30,60). It came as a surprise, then, to find that although PKGII could activate recombinant CFTR channels in patches from NIH 3T3 or IEC CF7 cells to about the same extent that PKA could (60), PKGI could not (46,60).

This conundrum was resolved by studies suggesting that myristoylation-dependent membrane targeting of PKGII imparts this specificity to PKGII (61,62). Thus, expressed membrane-associated PKGII, but not expressed soluble PKGI, was able to phosphorylate and activate CFTR channels in response to stimulation of the cells with atrial natriuretic peptide as efficiently as activation by PKA (62). These results imply that colocalization with CFTR in the plasma membrane confers a crucial kinetic advantage on PKGII in regulating CFTR, in comparison with cytosolic PKGI. In support of this proposed role for PKGII in the regulation of CFTR, knockout mice deficient in PKGII have been shown to be resistant to the effects of *E. coli* enterotoxin on intestinal secretion (63).

So a rise in cellular cGMP will likely activate CFTR channels via PKG-mediated phosphorylation, but only in cells expressing the PKGII isoform (62,64). cGMP-mediated activation of CFTR channels does not appear to occur, for example, in airway epithilium (46), which lacks PKGII (60). The new findings also add perspective to studies in which cGMP was suggested to exert its effects either by promiscuously activating PKA (59,65,66) or by inhibiting a phosphodiesterase that destroys cAMP (67,68). Most likely, in cells expressing it, PKGII will directly mediate the effects of cGMP unless cGMP concentrations reach sufficiently high levels to "cross-activate" PKA, or unless the type III cGMP-inhibited phosphodiesterase is highly expressed (62).

Interestingly, these studies of CFTR regulation in intestinal epithelia by heat-stable enterotoxins acting via cGMP (also by heat-labile toxins, like cholera toxin, that act via cAMP) have raised the possibility that CF heterozygote status might confer a genetic advantage against enterotoxin-induced diarrhea (59,69), thereby explaining the strikingly high frequency of the ΔF508 mutation (1 in 25 Caucasians is a carrier) in the CFTR gene (2).

Experiments on CFTR channels expressed in *Xenopus* oocytes have suggested that intracellular cGMP might also directly elicit CFTR Cl^- conductance via a PKG-independent pathway (70). Wild-type CFTR includes in its third cytoplasmic loop a domain that resembles the cyclic nucleotide binding site(s) in a class of proteins related to the catabolite–gene activator protein. Mutation of residues within this domain of CFTR impaired the ability of cGMP to activate CFTR, without altering the response to cAMP (70).

Ca^{2+}/Calmodulin-Dependent Protein Kinases

Purified R domain of CFTR can also be phosphorylated, *in vitro,* by CaM kinase I (30), but not by CaM kinase II (30,46), nor by CaM kinase III, nor by casein kinase II (30). Possible functional consequences of CFTR phosphorylation by CaM kinase I have not yet been investigated, either *in vitro* or *in vivo.* CaM kinase I is found in highest concentrations in neurons in the brain (71), cells that do not express CFTR. And, a recent study has found little evidence for coexpression of CaM kinase I and CFTR in nonneuronal tissues, including the intestine (72). However, CaM kinase I (73) and CFTR (74) are both expressed in the choroid plexus, sustaining the possibility that CFTR might be regulated by this enzyme in certain cells.

REGULATION OF CFTR BY DEPHOSPHORYLATION

Regulation by Protein Phosphatases

As noted earlier, the steady state level of phosphorylation of a protein *in vivo* depends on the relative rates of phosphorylation and dephosphorylation. In all cell types examined to date, CFTR Cl^- current activated by stimulation of PKA declines rapidly upon withdrawal of the PKA agonist, indicating that highly active endogenous phosphatases promptly dephosphorylate and inactivate CFTR (9,15–17). There are four main classes of serine/threonine protein phosphatases in eukaryotic cells, namely, PP1, PP2A, PP2B, and PP2C (75,76). However, at the biochemical level little is known about the specificity of these various enzymes toward phosphorylated CFTR. *In vitro,* CFTR phosphorylated by PKA is dephosphorylated by the catalytic subunit of PP2A, but not PP1 or PP2B (46,77). More recent studies have also shown that PP2C, as well as PP2A, can dephosphorylate phosphorylated R-domain peptide (38,78).

Further information has come from the use of protein phosphatase inhibitors in functional studies of CFTR. In patches excised from NIH 3T3 or CHO cells, purified PP2A (46,77), but not PP1 or PP2B (46), substantially deactivated PKA-phosphorylated epithelial CFTR channels. Also consistent with a deactivating role for PP2A, okadaic acid or microcystin (potent inhibitors of PP2A as well as of PP1; ref. 79) enhanced forskolin-activated CFTR Cl^- current in cardiac myocytes and slowed its deactivation on washout of forskolin (32). Okadaic acid also prevented deactivation of CFTR current in isolated sweat duct, though no enhancement of cAMP-activated Cl^- current was noted (80). In NIH 3T3 cells transfected with CFTR, calyculin A (another inhibitor of PP1 and PP2A) enhanced CFTR-dependent iodide efflux in the absence of any treatment to stimulate PKA activity, and this effect was parallelled by increased phosphorylation of CFTR measured biochemically (81). In insect cells transfected with CFTR, calyculin A had no effect by itself but increased dramatically

(>17-fold) forskolin-stimulated CFTR Cl^- current (82). A phosphorylated peptide inhibitor of PP1 was found to be ineffective in a preliminary test using cardiac myocytes, suggesting that, at least in those cells, PP2A was the target of microcystin and okadaic acid (83). However, the novel PP2A-like protein phosphatases PP4 and PP5 are also inhibited by microcystin and okadaic acid (see, e.g., Ref. 84). Any essential role for PP2B in deactivation of CFTR in cardiac myocytes can apparently be ruled out because of the nominal absence of Ca^{2+} ions (and presence of 10 mM EGTA) in the intracellular solutions in those studies (6,32). Also, FK506, a specific inhibitor of PP2B, did not augment the forskolin-induced activity of epithelial CFTR channels expressed in insect cells (82).

The use of selective protein phosphatase inhibitors in cardiac myocytes has also revealed further complexity in the regulation of CFTR (32). In that study, not only did maximally effective concentrations of okadaic acid or microcystin enhance CFTR current in the presence of forskolin and slow its deactivation following forskolin removal, but deactivation was incomplete, and up to 50% of the current persisted indefinitely in the continued presence of the phosphatase inhibitors. Analogous results have been obtained in studies of epithelial CFTR channels expressed in insect cells, where forskolin-induced CFTR current did not fully deactivate when forskolin was withdrawn in the maintained presence of calyculin A (82). The persistence of residual CFTR Cl^- conductance after full inhibition of PP2A implies that PP2A is essential for dephosphorylation of sites crucial to channel opening. But, the occurrence of partial deactivation even when PP2A is fully inhibited means that some phosphatase other than PP2A can dephosphorylate additional sites on CFTR, likely sites that modulate channel P_o (6,32). Based on its insensitivity to the aforementioned PP1/PP2A inhibitors, and its activity in the absence of Ca^{2+}, it seems probable that PP2C is the enzyme responsible for partial deactivation of CFTR (9,32,82,83), an idea now supported by the finding that PP2C dephosphorylates CFTR in vitro (38,78). Whether, under physiological conditions, the P_o modulatory sites can also be dephosphorylated by PP2A and/or PP1 is not yet clear (17,32).

Although these findings strongly argue that endogenous PP2A plays a major role in dephosphorylating and deactivating CFTR channels, some studies have failed to demonstrate any effect of PP2A inhibitors, and one has suggested an activating role for PP2A. Thus, okadaic acid did not affect channel currents in CHO and airway epithelial cells stably expressing CFTR, although in biochemical analyses it impaired CFTR dephosphorylation in membrane fractions prepared from the same cells (77). Tests with a range of phosphatase inhibitors on epithelial CFTR channels in NIH 3T3 cells have suggested (85) that whereas PP2B dephosphorylates a site essential for channel activation, and PP1 dephosphorylates a site that enhances channel P_o, PP2A dephosphorylates an inhibitory site (see Ref. 42). Remarkably, okadaic acid did not affect CFTR-dependent iodide efflux in another study using NIH 3T3 cells, despite the fact that the drug increased CFTR phosphorylation as measured by ^{32}P-labeling followed by immunoprecipitation (81). Yet, surprisingly, in that same study, forskolin, calyculin A, and genistein (a tyrosine kinase inhibitor, as discussed later) all also increased CFTR phosphorylation, but (unlike okadaic acid) they all increased iodide efflux. To explain these results, it was suggested that okadaic acid enhanced phosphorylation of a subset of sites on CFTR that were not adequate for channel activation, although preliminary phosphopeptide mapping analysis to address this point was inconclusive.

Exogenous alkaline phosphatase has been shown to deactivate CFTR channel currents in patches excised from CHO cells (5), pancreatic cells (86), and NIH 3T3 cells (46). In part, this deactivation was likely due to hydrolysis of bath ATP by the alkaline phosphatase (46), but alkaline phosphatase does also appear capable of dephosphorylating CFTR channels in patches (5). Support for a regulatory role for endogenous alkaline phosphatase comes from observations that a variety of inhibitors of alkaline phosphatase, including IBMX, theophylline, levamisole, and *p*-bromotetramisole, slowed deactivation of CFTR channels upon patch excision (77,86). Correspondingly, *in vitro* biochemical studies have shown that alkaline phosphatase can dephosphorylate CFTR protein, and that the various inhibitors can enhance PKA-dependent phosphorylation of CFTR in isolated CHO membranes (77). However, the specificity of the inhibitors is questionable, and alkaline phosphatase is known to nonspecifically dephosphorylate many protein and nonprotein substrates. Moreover, a physiological contribution of this enzyme to CFTR regulation seems rather unlikely because, although alkaline phosphatase is localized to the apical membrane of polarized pancreatic cells, it is oriented with its catalytic domain in the extracellular milieu (86), on the opposite side of the membrane from its proposed target, the R domain (1).

The reasons for the reported differences in the effects of inhibiting a particular phosphatase are not yet clear. They may reflect different levels of expression of specific protein phosphatases in the various cell types examined, or variable metabolism or cell membrane permeability of the different protein phosphatase inhibitors (see, eg., Ref. 87). Clearly, additional biochemical studies of CFTR dephosphorylation are warranted. It will be important to characterize the site(s) dephosphorylated by PP2A and PP2C, since it appears likely that these two enzymes are critical to the incremental activation and deactivation of CFTR now established both for native CFTR channels in mammalian cardiac myocytes (6,9,17,32,83) and for recombinant human epithelial CFTR channels expressed in insect cells (82). It will also be important to identify the phosphatases that dephosphorylate the site(s) regulated by PKC. Conceivably, sites phosphorylated by PKC might be dephosphorylated by PP1, PP2B, or even alkaline phosphatase, and such activity might account for some of the apparently contradictory results just presented.

Regulation by Tyrosine Kinase Inhibitors

No phosphorylation of threonyl or tyrosyl residues in CFTR has been detected so far, but a number of recent studies have implied that CFTR might be regulated either directly or indirectly by tyrosine phosphorylation. Application of the cellular oncogene p60$^{c\text{-}src}$, a tyrosine kinase, was found to restore fast flickery gating to CFTR channels in excised patches, an effect that could be reversed by addition of the nonspecific serine/threonine/tyrosine phosphatase, λ phosphatase, and that could be sustained by addition of orthovanadate to inhibit endogenous tyrosine phosphatases (88). Potential tyrosine phosphorylation sites in CFTR were noted (88), but no direct evidence was obtained that CFTR was, or can be, phosphorylated on tyrosine. Several studies have examined effects of the tyrosine kinase inhibitor genistein and its various analogs. In a variety of cell lines expressing recombinant (NIH-3T3, IEC-CF7) or native (HT-29/B6, T84) CFTR, genistein was found to activate iodide efflux, macroscopic CFTR Cl^- current, or single CFTR channel current (81,82,89,90), as well as to enhance phosphorylation of CFTR (81), in many cases in the absence of forskolin or other agent serving to elevate the concentration of cAMP; and genistein itself has no effect on cellular cAMP levels (89,91). Although initially interpreted as reflecting a novel cAMP-independent mechanism of CFTR channel activation (89), it is now clear that these effects of genistein are absolutely dependent on PKA activity. Thus, genistein had no effect in permeabilized HT-29/B6 monolayers in the absence of cAMP (90); genistein enhanced CFTR channel activity in Hi-5 insect and NIH 3T3 cells in which channels were basally active but had no effect in basally quiescent cells (82); and genistein's effects on CFTR phosphorylation and iodide efflux in 3T3 cells were inhibited by H-89, which inhibits PKA (81).

These results could all be explained if genistein were to inhibit a serine/threonine phosphatase that dephosphorylates PKA sites in CFTR (81,82,90), a conclusion suggested by two key results. First, the enhanced CFTR Cl^- current activated by cAMP agonists in intact or permeabilized monolayers, or in cell-attached patches, persisted following withdrawal of the agonists as long as genistein was present, but declined as soon as it was removed (90). Second, biochemical studies showed that genistein increased H-89-sensitive basal phosphorylation of CFTR in NIH 3T3 cells (81). Since, in the latter study (81) and in insect cells (82), the effect of genistein was additive with that of calyculin A (an inhibitor of PP1 and PP2A), a protein phosphatase distinct from PP1 or PP2A was considered a likely target of genistein, the obvious candidate being PP2C (82; cf. Ref. 81). The qualitative response to genistein was consistent among the different cell types, although quantitative aspects varied somewhat, presumably reflecting differences in the resting levels of cAMP, hence in the relative basal activities of PKA, PP2A, and PP2C in the different cells. Such differences presumably also account for the different quantitative responses to forskolin, calyculin A, and genistein, alone or in combination, in NIH 3T3, or insect, or other cells. Because no selective inhibitor of PP2C is presently available, the possibility that genistein might somehow regulate this enzyme could prove useful in identifying PKA sites involved in specific aspects of CFTR channel gating. A preliminary biochemical examination of CFTR phosphorylation has indeed suggested some differences in the sites affected by calyculin A and genistein (81), but much additional work will be needed to identify specific sites.

Although the foregoing results are consistent with genistein inhibiting a phosphatase, strictly speaking they demonstrate *only* that genistein in some way impairs dephosphorylation of certain PKA sites in CFTR, likely sites normally dephosphorylated by PP2C. Indeed, the most recent studies strongly suggest that genistein, in fact, interacts *directly* with CFTR (92,93). Clear evidence for this came from patch–clamp studies, using excised inside-out membrane patches from NIH 3T3 and IEC CF7 cells, in which ATP was replaced by ATPγS in the presence of PKA so that CFTR became thiophosphorylated, hence resistant to dephosphorylation (92). Results from those experiments, and from others on patches from insect cells (82,93), indicated that low concentrations of genistein increase CFTR current by stabilizing the open state of CFTR channels, presumably by interacting with NBD2. The response to genistein was found to be biphasic, however, with activation at low micromolar concentrations giving way to inhibition at higher (>50 μM) concentrations (see Ref. 90), suggesting that genistein binds to a high affinity site to slow channel closing and to a lower affinity site to reduce channel opening (93).

A parsimonious and unifying explanation for *all* the reported stimulatory effects of genistein on CFTR would be that genistein acts at NBD2 to stabilize the open conformation of CFTR and that the open channel is a poor substrate for dephosphorylation by PP2C (cf. Refs. 17, 82): such an action would enhance, in a PKA-dependent manner, single channel CFTR currents, macroscopic CFTR currents and fluxes, and steady state phosphorylation of a subset of serines. Interactions of this kind are not unexpected: since functional interactions between the phosphorylated R domain and the NBDs are believed to underlie the regulation of CFTR channel gating, simple thermodynamic constraints require that NBD function affect R-domain phosphorylation/dephosphorylation (94; cf. 95,96). Moreover, there is preliminary biochemical evidence for such an influence of NBD conformation on kinetics of R-domain phosphorylation by PKA (97).

An important unresolved question is whether any of these effects of genistein are mediated by tyrosine kinase inhibition. Two sets of observations appear, at first glance, consistent with a role for tyrosine phosphorylation. First, micromolar concentrations of orthovanadate inhibited the genistein-mediated stimulation of iodide efflux and of CFTR channel currents in cell-attached patches, effects interpreted as reflecting vanadate inhibition of the protein tyrosine phosphatase presumed to dephosphorylate the target of the tyrosine kinase (89). However, it is not easy to understand how phosphatase inhibition by addition of vanadate in the presence of a maximally effective concentration of genistein could enhance tyrosine phosphorylation, since genistein should have already fully inhibited the kinase. Second, Yang et al. (82) showed that in insect cells genistein's inactive analog, daidzein, failed to mimic the action of genistein (cf. Ref. 89), whereas a structurally unrelated tyrosine kinase inhibitor, tyrphostin A51, had a similar but smaller effect than genistein: tyrphostin B42 (89) in 3T3 cells, and tyrphostin 47 (91) in IEC-CF7 cells, were also found to be somewhat less effective than genistein.

On the other hand, Illek et al. (90) noted that in T84 and HT-29/B6 cells the tyrosine kinase inhibitors erbstatin, tyrphostin A23, tyrphostin A51, tyrphostin B42, and herbimycin A all failed to mimic the effects of genistein, which they therefore concluded were unlikely to be mediated via tyrosine kinase inhibition. De Jonge et al. (92) reached the same conclusion, citing the failure of tyrosine kinase inhibitors AG126, tyrphostin 25, catharidin, and herbimycin to activate CFTR in 3T3 cells, and the failure of the tyrosine phosphatase inhibitor pervanadate to inhibit genistein-stimulated iodide efflux in T84 or HT-29 colonocytes. Nor is there presently any evidence that PP2C can be regulated either directly or indirectly by tyrosine phosphorylation, although the catalytic subunit of PP2A (believed structurally unrelated to that of PP2C; cf. Refs. 98,99) can be phosphorylated on tyrosine by several kinases, including p60$^{v\text{-}src}$, thereby inhibiting PP2A's serine phosphatase activity (100). Dephosphorylation, and reactivation, of PP2A, however, apparently, occurs via an autodephosphorylation reaction, since this reactivation is sensitive to okadaic acid (100). Even if PP2C were to be similarly regulated, genistein would be expected to increase PP2C's phosphatase activity, hence to diminish serine phosphorylation in CFTR, an effect opposite (see Ref. 82) to the enhanced phosphorylation reported by Reenstra et al. (81). But a mechanism of this kind, acting on PP2A, could, at least qualitatively, account for the *inhibitory* effect on CFTR currents of high concentrations of genistein (93).

R-DOMAIN PHOSPHORYLATION BY PKA ALLOWS ATP HYDROLYSIS AT THE NBDs, HENCE GATING OF CFTR CHANNELS

CFTR Channel Opening Is Linked to ATP Hydrolysis

As already mentioned, PKA-phosphorylated CFTR channels require ATP or another hydrolyzable nucleoside triphosphate (e.g., GTP) to open (4,25). Sudden withdrawal of ATP from the cytoplasmic face of excised patches causes phosphorylated CFTR channels in the membrane to promptly close; the channels can be reopened by hydrolyzable nucleoside triphosphates, but not by poorly hydrolyzable analogs like AMP-PNP (4,25,101,102). Since AMP-PNP closely resembles ATP structurally but has a poorly hydrolyzable γ-phosphate bond (103), it seems that opening of CFTR channels depends on ATP hydrolysis. Further evidence of strict coupling between channel gating and ATP hydrolysis came from analysis of the extreme stabilization of the open conformation of CFTR channels caused by the inorganic phosphate (P_i) analogs orthovanadate and beryllium fluoride (28). These analogs are known to interrupt ATPase cycles, but only after the ATP has been hydrolyzed, by binding tightly in place of the released hydrolysis product, P_i (104). That channel opening requires ATP hydrolysis is confirmed by the finding (4,105) that even supersaturating concentrations of ATP (2 mM) fail to open phosphorylated CFTR channels in the absence of Mg^{2+} ions. The further observation that channels open more slowly at low (micromolar) than at high (millimolar) concentrations of free Mg^{2+} (105) demonstrates that ATP hydrolysis, not ATP binding (7), is the rate-limiting step in CFTR channel opening. This is consistent with the finding that increases in MgATP concentration enhance the open probability P_o of phosphorylated CFTR channels almost exclusively by increasing their opening rate (29,105–107).

Closing of CFTR Channels Is Also Mediated by ATP Hydrolysis

It is now clear that ATP hydrolysis controls not only channel opening, but also channel closing. Thus, although AMP-PNP by itself cannot open CFTR channels,

exposing them to a mixture of ATP and AMP-PNP allows channels to open via ATP hydrolysis, whereupon AMP-PNP acts to markedly stabilize their open conformation (6,8,29; but cf. Ref. 101). This implies that AMP-PNP can bind tightly at one NBD in CFTR, but only after ATP hydrolysis at the other NBD has opened the channel. It also suggests that AMP-PNP prolongs the channel open state because channel closing normally depends on hydrolysis of a second ATP. Prolongation of the channel open state by ATPγS (29), pyrophosphate (8,29) and tripolyphosphate (29), all of which require the simultaneous presence of ATP for their effect, further supports the conclusion that ATP hydrolysis at a second site permits closing of CFTR channels; presumably, when these poorly hydrolyzable analogs bind at that site they preclude the conformational change that attends nucleotide hydrolysis. And because channel closing is markedly slowed when ATP is applied in the presence of only micromolar levels of free Mg^{2+}, ATP hydrolysis must be the rate-limiting step for channel closure (105).

Which NBD Opens, and Which Closes, a CFTR Channel?

For several reasons, NBD1 was suggested (9,17) to be the site of the ATP hydrolysis event linked to CFTR channel opening, and NBD2 suggested to be where ATP is hydrolyzed to close CFTR channels. First, disease-associated NBD1 mutations are more numerous (2) and severe (108) than NBD2 mutations. Next, ATP could still open CFTR channels with a mutation at the conserved Walker A lysine in NBD2, K1250M, expected to impair ATP hydrolysis there (4). Finally, certain mutations anticipated to interfere with NBD function were found to favor the channel-closed state when introduced in NBD1 but to favor the channel-open state when introduced into NBD2 (109). That assignment of NBD1 to the crucial role in channel opening, and of NBD2 to channel closing, has now been largely confirmed by single channel recordings from CFTR mutants with alterations at the Walker lysines. Mutations at K464 in NBD1 reduced the rate of channel opening, while equivalent mutations at K1250 reduced the rate of channel closing (7,8). Essentially the same conclusions were drawn from detailed analyses of the macroscopic rates of activation and deactivation of CFTR channels with mutations at conserved NBD residues (110).

Phosphorylation of CFTR Controls ATP Hydrolysis and Channel Gating

Biochemical measurements have provided direct confirmation that purified and reconstituted epithelial CFTR indeed hydrolyzes ATP, and does so at approximately the same rate ($0.1–1\ s^{-1}$) as that of channel opening and closing (52). In keeping with the requirement of phosphorylation by PKA for channel gating, this ATPase activity of purified CFTR also depended on phosphorylation and, at 1 mM MgATP, was stimulated two- to threefold by PKA. Interestingly, the ATPase activity was markedly inhibited by pretreating the CFTR with PP2A, suggesting both that CFTR was basally phosphorylated and that it remained so throughout the purification procedure (52). This PKA-dependent stimulation of ATPase activity reflected an increased apparent affinity for ATP (with no alteration of maximal hydrolysis rate), and was accompanied by a shift of the Hill coefficient from 1 toward 2 for the fully phosphorylated protein, suggested to reflect recruitment of a second ATPase site (52). Unfortunately, whether CFTR channels dephosphorylated with PP2A were capable of substantial ATP hydrolysis or opening at very high ATP concentrations was not tested. In the same study, the cystic fibrosis associated CFTR mutant G551D was shown to hydrolyze ATP at only one-tenth the rate of wild-type CFTR and, correspondingly, G551D channels opened only infrequently (52). ATP hydrolysis at a low rate ($\sim 0.03\ s^{-1}$) has also been reported for CFTR's NBD1 fused to a maltose-binding protein (111).

Although these rates are relatively low for ATPases, they are well within the range of nucleotide hydrolysis by G proteins, and both structural and functional considerations suggest that the NBDs of ABC transporters might more closely resemble those of GTPases than of ion-motive ATPases (9,112–114). For instance, the apparent stabilization of the active channel-open conformation by ATP, and the extreme stabilization by AMP-PNP, binding at NBD2 shares striking analogy with the action of GTP (and of GMP-PNP) to stabilize activated G proteins. And amino acid sequence homologies (112), and consequences of mutating key catalytic residues (113), do indeed suggest that CFTR's NBDs might function like GTPases. This in turn suggests that it might be instructive to consider the inferred interactions (6) between the two NBDs, and between the phosphorylated R domain and the NBDs, in terms of the interactions between GTPases and activator proteins (GAPs), exchange factors (GEFs), and dissociation inhibitors (GDIs) (115).

Role of the R Domain in CFTR Channel Gating

CFTR channels with an intact R domain can neither hydrolyze ATP (52) nor be opened by ATP (4,5,25) until they are phosphorylated by PKA, whereas mutant channels (CFTRΔR) lacking much (residues 708–835) of the R domain can be opened by ATP without first being phosphorylated by PKA (4,43). Because 8-azido-ATP reportedly labels phosphorylated and dephosphorylated CFTR channels equally well (116), this would seem to suggest that the dephosphorylated R domain acts to prevent not ATP binding but ATP hydrolysis, and hence channel gating. There is, indeed, evidence that individual mutation of certain R-domain serines slows both the macroscopic activation rate of CFTR conductance (42) and the opening rate of single CFTR channels (117); in the latter experiments, the concomitant reduction in apparent affinity for ATP activation of channel P_o need not reflect an effect on ATP binding per se. Similarly, the substantially reduced P_o of 10SA mutant channels (44) appeared to be associated with a reduced opening rate (see discussion in Ref. 48), and possibly also a reduced apparent affinity for nucleotides (118).

The early suggestion that the dephosphorylated R domain might itself form the gate, plugging the channel and moving out of the way of permeant anions only upon phosphorylation (34,43,119,120), is easily ruled out. Thus, CFTRΔR channels, with most of the R domain missing, stay closed until they are provided with hydrolyzable nucleoside triphosphate (4,43), and even wild-type CFTR channels with phosphorylated R domains (i.e., in which block should be relieved) do not open unless exposed to ATP. So CFTR channels are gated open by ATP hydrolysis, and direct measurements have shown that phosphorylation enhances the capability of purified CFTR channels to hydrolyze ATP, whereas dephosphorylation markedly impairs ATP hydrolysis (52).

How Does Phosphorylation by PKA Modify R Domain Function?

Initial studies showed that strong stimulation by PKA could still activate CFTR channels with up to 10 Ser/Thr-to-Ala mutations, in various combinations, at the dibasic consensus sites (34,43,44). The evident decline of channel activity as the number of mutations was increased led to the notion of functional redundancy, in which only the number of phosphorylated residues, not their specific location, was suggested to govern activation of CFTR (34,43,44). A possible mechanism for such an effect was suggested to be an accumulation of negative charge on the R domain, due to phosphorylation, with a consequent gross electrostatic effect causing the R domain to move in a way that somehow facilitates ion flux through the channel pore (43,120). Some support for this mechanism came from findings that introduction of six to eight negative charges (6SD, 7SD, 8SD: ref. 43; 8SE: ref. 44) did allow

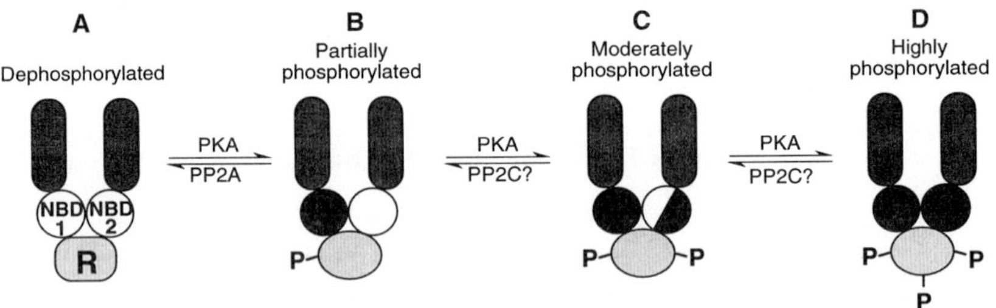

FIG. 3. Cartoon of proposed regulation of CFTR channel gating by incremental phosphorylation. Sequential, incremental phosphorylation of at least three sites (P), presumably in the R domain, progressively enhances the open probability (P_o) of a CFTR channel by regulating the likelihood of ATP binding tightly to, and being hydrolyzed by, NBD2. In the dephosphorylated channel **(A)** neither NBD hydrolyzes ATP and the channel remains closed. In a partially phosphorylated channel **(B)**—phosphorylated at a site(s) that requires PP2A for dephosphorylation—NBD1 can hydrolyze ATP, whereupon the channel opens; but ATP does not bind tightly to NBD2 and so the channel closes rapidly, yielding only brief open times and a low P_o. In a moderately phosphorylated channel **(C)**—additionally phosphorylated at a site(s) that can be dephosphorylated by an okadaic acid insensitive phosphatase, likely PP2C—ATP hydrolysis at NBD1 opens the channel, whereupon ATP may bind tightly to NBD2 and be hydrolyzed there, yielding some longer openings: but, during many openings, ATP is not bound and hydrolyzed at NBD2, yielding a proportion of brief openings, and an overall intermediate P_o. In highly phosphorylated channels **(D)**—further phosphorylated at a site(s) also likely dephosphorylated by PP2C—ATP hydrolysis at NBD1 opens the channel whereupon ATP invariably binds tightly at, and is hydrolyzed by, NBD2, yielding exclusively long openings and a high P_o. Although PP2C likely dephosphorylates the modulatory site(s) (as indicated) in the presence of inhibitors of PP1 and PP2A, it is entirely possible that those sites might be dephosphorylated by PP2A under normal physiological conditions (17,32).

channels to be opened by ATP alone, albeit with substantially diminished P_o, whereas insertion of only four or five negative charges (43,44) by substitution for the major sites (Serines 660, 737, 795, 813, plus or minus Ser-700) left channels still dependent on phosphorylation, but also with impaired activity.

But there are at least three difficulties with such a simple mechanism. First, phosphorylation of the R domain seems to be incremental, and to cause reproducible, discrete conformational changes, evident from mobility shifts of recombinant R domain on SDS–polyacrylamide gels (30,31,40). In addition, the CD spectrum of R-domain peptide is altered after phosphorylation by PKA, implying a change in its structure (31). Second, recent results demonstrate that phosphorylation of either of two of the Ser residues (737 and 768) in the R domain exerts an inhibitory effect on CFTR channel activation, whereas phosphorylation of others is generally stimulatory (42,109). Third, the CD spectrum of mutant 8SE CFTR R domain was found to differ markedly from those of either dephosphorylated or phosphorylated wild-type CFTR R domain (39). Those observations argue that activation of CFTR channels by phosphorylation cannot readily be explained by a simple buildup of negative charges (9). On the contrary, phosphorylation of stimulatory sites seems to counter the opposite influence of phosphorylation of inhibitory sites (42).

As a final cautionary note, the CD spectrum of mutant 9SA R-domain peptide was quite unlike that of either dephosphorylated or phosphorylated wild-type CFTR R domain, a reminder that seemingly benign mutations may themselves cause significant structural changes (39).

Modulation of CFTR Channel Gating by Incremental Phosphorylation

All the work to date suggests that during each gating cycle, there are strong interactions between the two NBDs, and between the R domain and each of the NBDs, and that phosphorylation acts by modifying these interactions (9). The simplified model in Fig. 3 is based on findings in myocytes (6,32,83; cf. Ref. 82) and provides a convenient framework for discussing these interactions as well as a proposed mechanism for the modulation of channel P_o by incremental phosphorylation by PKA (6,9). As already described, dephosphorylated channels (Fig. 3A) hydrolyze ATP only poorly, if at all and cannot be opened by ATP, whereas highly phosphorylated CFTR channels (Fig. 3D) hydrolyze ATP well (and show a relatively high apparent affinity for ATP), are readily opened by low ATP concentrations, and display a high P_o characterized by relatively long dwell times in the open state. Since the highly phosphorylated channels both open and close more slowly at low than at high concetrations of free Mg^{2+} (7,105,114), and since AMP-PNP alone cannot open them but can markedly slow their closing in the presence of the ATP that is required to open them, both NBDs must be capable of ATP hydrolysis in highly phosphorylated channels (6–8).

The relatively long durations in the open state have been attributed (9,17) to a stabilization caused by tight binding of ATP to NBD2. And, much like the analogous action of GTP in G proteins, that stabilization is terminated by hydrolysis of the ATP at NBD2. Partially phosphorylated channels (Fig. 3B), on the other hand, such as those underlying the residual CFTR current that persists after washing away PKA agonists in myocytes or insect cells treated with inhibitors of PP1 and PP2A (32,82), or those still functional in excised patches several minutes after withdrawal of PKA (6,101), have a low P_o characterized by only brief openings, and they do not (or do not readily) become "locked" open by AMP-PNP in the presence of ATP. We have proposed that when those partially phosphorylated channels are opened by ATP hydrolysis at NBD1, ATP does not bind tightly to NBD2 (nor does AMP-PNP), and so the channels can close quickly (presumably by release of the NBD1 hydrolysis products ADP and P_i; Refs. 6,28) because they do not have to wait for ATP hydrolysis at NBD2 (6,9). Because AMP-PNP can bind tightly to open, but not to closed, CFTR channels, and then only when they are highly phosphorylated (6), NBD1 must signal to NBD2 upon channel opening, and phosphorylation must alter either the signal or its consequence. Whereas the partially phosphorylated channels (e.g., in the presence of okadaic acid long after PKA withdrawal) show only brief openings, and highly phosphorylated channels (e.g., those in the presence of PKA) show only long openings, an intermediate phosphorylation state ("moderately" phosphorylated; Fig. 3C), in which a channel can show openings of both kinds, can sometimes be discerned as endogenous, okadaic acid resistant, and Ca^{2+}-independent phosphatases progressively dephosphorylate sensitive sites (33). In those additionally phosphorylated channels, signaling between the NBDs must be modified, since it appears that channel opening by ATP hydrolysis at NBD1 may or may not be followed by tight binding of ATP at NBD2, and hence may or may not result in a closing that is rate-limited by hydrolysis of that ATP.

It seems, then, that at least three functionally distinct phosphoforms of CFTR channels can be identified. The phosphorylation status of at least one site (or group of sites) controls ATP hydrolysis at NBD1, hence channel opening. The phosphorylation status of an additional site (or group of sites) seems to determine whether NBD2 will have the opportunity to tightly bind ATP; and the phosphorylation status of yet another site (or group of sites) appears to regulate whether that tight binding occurs upon every opening of the channel or follows only some of the openings. Preliminary observations of slowed opening of CFTR channels with individual R-domain Ser-to-Ala mutations (117), or even of 10SA mutant channels (44), imply that still other phosphorylation sites modulate the rate of ATP hydrolysis at NBD1. An important goal is to identify the site (or sites) associated with each of these kinds of channel behavior. It will be interesting to learn which, if any, of these functionally distinct phosphorylated states corresponds to each of the different bands seen on acrylamide gels of phosphorylated R domain (38,40). It will also be interesting to learn whether the incremental phosphorylation occurs strictly sequentially, as implied by the scheme in Fig. 3, or more randomly (see refs. 32,82). Preliminary mass spectrometric analysis (38) suggests that indeed some of the phosphorylation events may be sequential, although the precise nature and mechanism of the ordering remain to be determined.

How Does Incremental Phosphorylation of the R Domain Control NBD Function?

As long as MgATP is present, CFTRΔR or CFTRΔR-S660A channels (both lacking a substantial fraction of the R domain) open and close with no need for phosphorylation by PKA (4,43), but with a P_o far below that of phosphorylated wild-type CFTR channels. This suggests that, normally, phosphorylation by PKA not only relieves the inhibition of the NBD ATPase activity that is mediated by the dephosphorylated R domain, but also facilitates some aspect of NBD function in a way that requires an intact and phosphorylated R domain. Recent measurements have indeed demonstrated that expressed, purified R-domain peptide, when dephosphorylated, does inhibit intact wild-type CFTR channels, and the inhibition is removed upon phosphorylation of the R-domain peptide by PKA (121). Moreover, in the presence of PKA but not in its absence, R-domain peptide was able to increase the P_o of CFTRΔR-S660A channels (117). And, in keeping with the proposed role of the R domain in modulating the function of NBD2, it has recently been shown that neither AMP-PNP nor PP_i can lock open CFTRΔR channels (122).

Of course, if the rate of channel opening is controlled by ATP hydrolysis at NBD1 and the rate of channel closing is regulated by ATP hydrolysis at NBD2, the interpretation of observed changes in P_o, say resulting from mutation of one of the R-domain serines (or from exposure to some pharmacological agent, like genistein), becomes ambiguous. A phosphorylation site determined to have an inhibitory influence on CFTR conductance or channel P_o, like Ser-737 or Ser-768, could (among other possible mechanisms) slow channel opening by impairing ATP hydrolysis at NBD1 or speed channel closing by accelerating ATP hydrolysis at NBD2. And given the proposed interactions between the NBDs and the R domain, a mutation in one of the domains may be expected to result in altered function of another. Most likely, if we are to begin to unravel the complex mechanisms by which phosphorylation of the R domain at multiple sites regulates the gating of CFTR channels, we will need measurements of the rates of opening and closing of individual CFTR channels, each bearing just a single mutation.

ACKNOWLEDGMENTS

We are indebted to Kate Egnatz and Peter Hoff for invaluable assistance with the bibliography and illustrations. Preparation of this chapter, and our research reviewed here, were supported by grants HL-49907and DK-51767 from the National Institutes of Health.

REFERENCES

1. Riordan JR, Rommens JM, Kerem BS, Alon N, Rozmahel R, Grzelczak Z, Zielenski J, Lok S, Plavsic N, Chou J, Drumm ML, Iannuzzi MC, Collins FS, Tsui L-C: Identification of the cystic fibrosis gene: cloning and characterization of complementary DNA. *Science* 1989;245:1066–1073.

2. Tsui L-C: The spectrum of cystic fibrosis mutations. *Trends in Genet* 1992;8:392–398.
3. Higgins CF: ABC transporters: from microorganisms to man. *Annu Rev Cell Biol* 1992;8:67–113.
4. Anderson MP, Berger HA, Rich DP, Gregory RJ, Smith AE, Welsh MJ: Nucleoside triphosphates are required to open the CFTR chloride channel. *Cell* 1991;67:775–784.
5. Tabcharani JA, Chang X-B, Riordan JR Hanrahan, JW: Phosphorylation-regulated $C1^-$ channel in CHO cells stably expressing the cystic fibrosis gene. *Nature* 1991;352:628–631.
6. Hwang T-C, Nagel G, Nairn AC, Gadsby DC: Regulation of the gating of CFTR Cl channels by phosphorylation and ATP hydrolysis. *Proc Natl Acad Sci U S A* 1994;91:4698–4702.
7. Gunderson KL, Kopito RR: Conformational states of CFTR associated with channel gating: the role of ATP binding and hydrolysis. *Cell* 1995;82:231–239.
8. Carson MR, Travis SM, Welsh MJ: The two nucleotide-binding domains of cystic fibrosis transmembrane conductance regulator (CFTR) have distinct functions in controlling channel activity. *J Biol Chem* 1995;270:1711–1717.
9. Gadsby DC, Nairn AC: Regulation of CFTR channel gating. *Trends Biochem Sci* 1994;19:513–518.
10. Gill DR, Hyde SC, Higgins CF, Valverde MA, Mintenig GM, Sepulveda FV: Separation of drug transport and chloride channel functions of the human multidrug resistance P-glycoprotein. *Cell* 1992;71:23–32.
11. Mintenig GM, Valverde MA, Sepulveda FV, Gill DR, Hyde SC, Kirk J, Higgins CF: Specific inhibitors distinguish the chloride channel and drug transporter functions associated with the human multidrug resistance P-glycoprotein. *Recept Channels* 1993;1:305–313.
12. Inagaki N, Gonoi T, Clement JP IV et al: Reconstitution of I_{KATP}: An inward rectifier subunit plus the sulfonylurea receptor. *Science* 1995;270:1166–1169.
13. Stutts MJ: Regulation of other airway epithelial ion channels by CFTR. In: Dodge JA et al, eds. *Cystic Fibrosis—Current Topics.* Vol. 3. Chichester: Wiley, 1996;91–106.
14. Kunzelmann K, Kiser GL Schreiber R, Riordan JR: Inhibition of epithelial Na+ currents by intracellular domains of the cystic fibrosis transmembrane conductance regulator. *FEBS Lett* 1997;400:341–434.
15. Welsh MJ, Anderson MP, Rich DP, Berger HA, Denning GM, Ostedgaard LS, Sheppard DN, Cheng SH, Gregory RJ, Smith AE: Cystic fibrosis transmembrane conductance regulator: a chloride channel with novel regulation. *Neuron,* 1992;8:821–829 (1992).
16. Riordan JR: The cystic fibrosis transmembrane conductance regulator. *Ann Rev Physiol* 1993;55: 609–630.
17. Gadsby DC, Nagel GA, Hwang T-C: The CFTR chloride channel of mammalian heart. *Annu Rev Physiol* 1995;57:387–416.
18. Bear CE, Canhui L, Kartner N, Bridges RJ, Jensen TJ: Purification and functional reconstitution of the cystic fibrosis transmembrane conductance regulator (CFTR). *Cell* 1992;68:809–818.
19. Anderson MP, Gregory RJ, Thompson S, Souza DW, Paul I, Mulligan RC, Smith AE, Welsh MJ: Demonstration that CFTR is a chloride channel by alteration of its anion selectivity. *Science* 1991; 253:202–205.
20. McDonough S, Davidson N, Lester HA, McCarty NA: Novel pore-lining residues in CFTR that govern permeation and open channel block. *Neuron* 1994;13:623–634.
21. Akabas MH, Kaufmann C, Cook TA, Archdeacon P: Amino acid residues lining the chloride channel of the cystic fibrosis transmembrane conductance regulator. *J Biol Chem* 1994;269:14865–14868.
22. Cheung M, Akabas MH: Locating the anion-selectivity filter of the cystic fibrosis transmembrane conductance regulator (CFTR) chloride channel. *J Gen Phys* 1997;109:289–299.
23. Tabcharani JA, Rommens JM, Hou Y-X, Chang X-B, Tsui L-C, Riordan JR, Hanrahan JW: Multi-ion pore behaviour in the CFTR chloride channel. *Nature* 1993;366:79–82.
24. Cotten JF, Welsh MJ: Proton transfer at HIS-347 in the CFTR pore reveals two distinct conductance states. *Biophys J* (Abstr) 1997;72:A365.
25. Nagel G, Hwang T-C, Nastiuk KL, Nairn AC, Gadsby DC: The protein kinase A-regulated cardiac Cl^- channel resembles the cystic fibrosis transmembrane conductance regulator. *Nature* 1992;360: 81–84.
26. Cantiello HF, Prat AG, Reisin IL, Ercole LB, Abraham EH, Amara JF, Gregory RJ, Ausiello DA: External ATP and its analogs activate the cystic fibrosis transmembrane conductance regulator by a cyclic AMP-independent mechanism. *J Biol Chem* 1994;269:11224–11232.
27. Prat AG, Xiao Y-F, Ausiello DA, Cantiello HF: cAMP-independent regulation of CFTR by the actin cytoskeleton. *Am J Physiol* 1995;268:C1552–C1561.
28. Baukrowitz T, Hwang T-C, Nairn AC, Gadsby DC: Coupling of CFTR Cl^- channel gating to an ATP hydrolysis cycle. *Neuron* 1994;12:473–482.

29. Gunderson KL, Kopito RR: Effects of pyrophosphate and nucleotide analogs suggest a role for ATP hydrolysis in cystic fibrosis transmembrane regulator channel gating. *J Biol Chem* 1994;269:19349–19353.
30. Picciotto M, Cohn J, Bertuzzi G, Greengard P, Nairn A: Phosphorylation of the cystic fibrosis transmembrane conductance regulator. *J Biol Chem* 1992;267:12742–12752 .
31. Dulhanty AM, Riordan JR: Phosphorylation by cAMP-dependent protein kinase causes a conformational change in the R domain of the cystic fibrosis transmembrane conductance regulator. *Biochemistry* 1994;33:4072–4079.
32. Hwang T-C, Horie M, Gadsby DC: Functionally distinct phospho-forms underlie incremental activation of PKA-regulated Cl^- conductance in mammalian heart. *J Gen Physiol* 1993;101:629–650
33. Dousmanis AG, Nairn AC, Gadsby DC: Three functionally distinct phosphoforms of CFTR identified by patterns of single channel gating. *J Gen Physiol* 1996;108:11a.
34. Cheng SH, Rich DP, Marshall J, Gregory RJ, Welsh MJ, Smith AE: Phosphorylation of the R domain by cAMP-dependent protein kinase regulates the CFTR chloride channel. *Cell* 1991;66:1027–1036.
35. Seibert FS, Tabcharani JA, Chang X-B, Dulhanty AM, Mathews C, Hanrahan JW, Riordan JR: cAMP-Dependent protein kinase-mediated phosphorylation of cystic fibrosis transmembrane conductance regulator residue Ser-753 and its role in channel activation. *J Biol Chem* 1995;270:2158–2162.
36. Townsend, RR, Lipniunas, PH, Tulk, BM, Verkman, AS: Identification of protein kinase A phosphorylation sites on NBD1 and R domains of CFTR using electrospray mass spectrometry with selective phosphate ion monitoring. *Protein Sci* 1996;5:1865–1873.
37. Pearson RB, Kemp BE: Protein kinase phosphorylation site sequences and consensus specificity motifs: tabulations. *Methods Enzymol* 1991;200:62–81.
38. Nairn AC, Qin J, Chait BT, Gadsby DC: Identification of sites in the R domain of CFTR phosphorylated by cAMP-dependent protein kinase and dephosphorylated by protein phosphatases 2A and 2C. *Pediatr Pulmonol* 1996;Suppl 13:21.
39. Dulhanty AM, Chang X-B, Riordan JR: Mutation of potential phosphorylation sites in the recombinant R domain of the cystic fibrosis transmembrane conductance regulator has significant effects on domain conformation. *Biochem Biophys Res Commun* 1995;206:207–214.
40. Borchardt R, Kole J, Cohn J: Phosphorylation of CFTR Ser-737 by protein kinase A. *Pediatr Pulmonol* 1996;Suppl 13:212.
41. Seibert FS, Loo TW, Clarke DM, Riordan JR: Phosphorylation by cAMP-dependent protein kinase (PKA) of non-dibasic consensus sequences in CFTR. *Pediatr Pulmonol* 1995;Suppl 12:189.
42. Wilkinson DJ, Strong TV, Mansoura MK, Wood DL, Smith SS, Collins FS, Dawson DC: CFTR activation: additive effects of stimulatory and inhibitory phosphorylation sites in the R domain. *Am J Physiol* 1997;273:L127–L133.
43. Rich DR, Berger HA, Cheng SH, Travis SM, Saxene M, Smith AE, Welsh MJ: Regulation of the cystic fibrosis transmembrane conductance regulator C1 channel by negative charge in the R domain. *J Biol Chem* 1993;268:20259–20267.
44. Chang X-B, Tabcharani JA, Hou YX, Jensen TJ, Kartner N, Alon N, Hanrahan JW, Riordan JR: Protein kinase A (PKA) still activates CFTR chloride channels after mutagenesis of all 10 PKA consensus phosphorylation sistes. *J Biol Chem* 1993;268:11304–11311.
45. Drumm ML, Wilkinson DJ, Smit LS, Worrell RT, Strong TV, Frizell RA, Dawson DC, Collins FS: Chloride conductance expressed by ΔF508 and other mutant CFTRs in *Xenopus* oocytes. *Science* 1991;254:1797–1799.
46. Berger HA, Travis SM, Welsh MJ: Regulation of the cystic fibrosis transmembrane conductance regulator Cl channels by specific kinases and protein phosphatases. *J Biol Chem* 1993;268:2037–2047.
47. Graff JM, Rajan RR, Randall RR, Nairn AC, Blackshear PJ: Protein kinase C substrate and inhibitor characteristics of peptides derived from the myristoylated alanine-rich C kinase substrate (MARCKS) protein phosphorylation site domain. *J Biol Chem* 1991;266:14390–14398.
48. Jia YL, Mathews CJ, Hanrahan JW: Phosphorylation by protein kinase C is required for acute activation of cystic fibrosis transmembrane conductance regulator by protein kinase A. *J Biol Chem* 1997;272:4978–4984.
49. Dechecchi MC, Tamanini A, Berton G, Cabrini G: Protein kinase C activates chloride conductance in C127 cells stably expressing the cystic fibrosis gene. *J Biol Chem* 1993;268:11321–11325.
50. Bajnath RB, Groot JA, De Jonge HR, Kansen M, Bijman J: Synergistic activation of non-rectifying small conductance chloride channels by forskolin and phorbol esters in cell-attached patches of the human colon carcinoma cell line HT-29cl.19A. *Pfluegers Arch* 1993;425:100–108.

51. Collier ML, Hume JR: Unitary chloride channels activated by protein kinase C in guinea pig ventricular myocytes. *Circ Res* 1995;76:317–324.
52. Li C, Ramjeesingh M, Wang W, Garami E, Hewryk M, Lee D, Rommens JM, Galley K, Bear CE: ATPase activity of the cystic fibrosis transmembrane conductance regulator. *J Biol Chem* 1996;271: 28463–28468.
53. Winpenny JP, McAlroy HL, Gray MA, Argent BE: Protein kinase C regulates the magnitude and stability of CFTR currents in pancreatic duct cells. *Am J Physiol* 1995;268:C823–C828.
54. Newton AC: Protein kinase C: structure, function, and regulation. *J Biol Chem* 1995;270:28495–28498.
55. Trapnell BC, Zeitlin PL, Chu C-S et al: Down-regulation of cystic fibrosis gene mRNA transcript levels and induction of the cystic fibrosis chloride secretory phenotype in epithelial cells by phorbol ester. *J Biol Chem* 1991;266:10319–10323.
56. Bargon J, Trapnell BC, Yoshimura K et al: Expression of the cystic fibrosis transmembrane conductance regulator gene can be regulated by protein kinase C. *J Biol Chem* 1992;267:16056–16060.
57. Dechecchi MC, Rolfini R, Tamanini A, Gamberi C, Berton G, Cabrini G: Effect of modulation of protein kinase C on the cAMP-dependent chloride conductance in T84 cells. *FEBS Lett* 1992;311: 25–28.
58. Breuer W, Glickstein H, Kartner N, Riordan JR, Ausiello DA, Cabantchik IZ: Protein kinase C mediates down-regulation of cystic fibrosis transmembrane conductance regulator levels in epithelial cells. *J Biol Chem* 1993;268:13935–13939.
59. Chao AC, de Sauvage FJ, Dong Y-J, Wagner JA, Goeddel DV, Gardner P: Activation of intestinal CFTR Cl^- channel by heat-stable enterotoxin and guanylin via cAMP-dependent protein kinase. *EMBO J* 1994;13:1065–1072.
60. French PJ, Bijman J, Edixhoven M, Vaandrager AB, Scholte BJ, Lohmann SM, Nairn AC, De Jonge HR: Isotype-specific activation of CFTR–chloride channels by cGMP-dependent protein kinase II. *J Biol Chem* 1995;270:26626–26631.
61. Vaandrager AB, Ehlert EME, Jarchau T, Lohmann SM, De Jonge HR: N- terminal myristoylation is required for membrane localization of cGMP-dependent protein kinase type II. *J Biol Chem* 1996; 271:7025–7029.
62. Vaandrager AB, Tilly BC, Smolenski A et al: cGMP stimulation of cystic fibrosis transmembrane conductance regulator Cl^- channels co-expressed with cGMP-dependent protein kinase type II but not type Iβ. *J Biol Chem* 1997;272:4195–4200.
63. Pfeifer A, Aszodi A, Seidler U, Ruth P, Hofmann F, Fassler R: Intestinal secretory defects and dwarfism in mice lacking cGMP-dependent protein kinase II. *Science* 1996;274:2082–2086.
64. Markert T, Vaandrager AB, Gambaryan S, Pöhler D, Häusler C, Walter U, De Jonge HR, Jarchau TH, Lohman SM: Endogenous expression of type II cGMP-dependent protein kinase mRNA and protein in rat intestine: Implication for cystic fibrosis transmembrane conductance regulator. *J Clin Invest* 1995;96:822–830.
65. Forte LR, Thorne PK, Eber SL et al: Stimulation of intestinal C1- transport by heat-stable enterotoxin: activation of cAMP-dependent protein kinase by cGMP. *Am J Physiol* 1992;263:C607–C615.
66. Tien XY, Brasitus TA, Kaetzel MA, Dedman JR, Nelson DJ: Activation of the cystic fibrosis transmembrane conductance regulator by cGMP in the human colonic cancer cell line, Caco-2. *J Biol Chem* 1994;269:51–54.
67. Ono K, Tareen FM, Yoshida A, Noma A: Synergistic action of cyclic GMP on catecholamine-induced chloride current in guinea-pig ventricular cells. *J Physiol (Lond)* 1992;453:647–661.
68. Kelley TJ, Al-Nakkash L, Drumm ML: C-type natriuretic peptide increases chloride permeability in normal and cystic fibrosis airway cells. *Am J Respi Cell Mol Biol* 1997;16(4):464–470.
69. Quinton PM: What is good about cystic fibrosis? *Curr Biol* 1994;4:742–743.
70. Sullivan SK, Agellon LB, Schick R: Identification and partial characterization of a domain in CFTR that may bind cyclic nucleotides directly. *Curr Biol* 1995;5:1159–1167.
71. Picciotto MR, Czernik AJ, Nairn AC: Calcium/calmodulin-dependent protein kinase I: cDNA cloning and identification of autophosphorylation site. *J Biol Chem* 1993;268:26512–26521.
72. Matovcik LM, Nairn AC, Gorelick FS: Localization of Ca^{2+}/calmodulin-dependent protein kinase I to α-cells and Ca^{2+}/calmodulin-dependent protein kinase II to δ-cells of the rat endocrine pancreas. *J Histochem Cytochem* 1998;46:S19–S26.
73. Picciotto MR, Zoli M, Bertuzzi G, Nairn AC: Immunochemical localization of calcium/calmodulin-dependent protein kinase I. *Synapse* 1995;20:75–84.

74. Hincke MT, Nairn AC, Staines WA: Cystic fibrosis transmembrane conductance regulator is found within brain ventricular epithelium and choroid plexus. *J Neurochem* 1995;64:1662–1668.
75. Cohen P: The structure and regulation of protein phosphatases. *Annu Rev Biochem* 1989;58:453–508.
76. Shenolikar S, Nairn AC: Protein phosphatases: recent progress. In: Greengard P, Robison GA, eds *Advances in Second Messenger Phosphoprotein Research.* Vol. 23. New York: Raven, 1991;1–121.
77. Becq F, Jensen TJ, Chang X-B, Savoia A, Rommens JM, Tsui L-C, Buchwald M, Riordan JR, Hanrahan JW: Phosphatase inhibitors activate normal and defective CFTR chloride channels. *Proc Natl Acad Sci U S A* 1994;91:9160–9164.
78. Berger HA, Baldursson O, Welsh MJ: Phosphorylation site mutations alter cystic fibrosis transmembrane conductance regulator activity in an intact epithelium. *Pediatr Pulmonol* 1996;Suppl 13:211–212.
79. MacKintosh C, MacKintosh RW: Inhibitors of protein kinases and phosphatases. *Trends Biochem Sci* 1994;19:444–447.
80. Reddy MM, Quinton PM: Deactivation of CFTR-C1 conductance by endogenous phosphatases in the native sweat duct. *Am J Physiol* 1996;270:C474–C480.
81. Reenstra WW, Yurko-Mauro K, Dam A, Raman S, Shorten S: CFTR chloride channel activation by genistein: the role of serine/theonine protein phosphatases. *Am J Physiol* 1996;271:C650–C657.
82. Yang ICH, Cheng TH, Wang F, Price EM, Hwang TC: Modulation of CFTR chloride channels by calyculin A and genistein. *Am J Physiol* 1997;272:C142–C155.
83. Hwang T-C, Nagel G, Nairn AC, and Gadsby DC: Dephosphorylation of cardiac Cl^- channels requires multiple protein phosphatases. *Biophy J* 1993;64:A343
84. Chen MX, McPartlin AE, Brown L, Chen YH, Barker HM, Cohen PTW: A novel human protein serine/threonine phosphatase, which possesses four tetratricopeptide repeat motifs and localizes to the nucleus. *EMBO J* 1994;13:4278–4290.
85. Fischer H, Illek B, Machen TE: CFTR's activation, steady-state activity, and inactivation are controlled by distinct phosphatases. *Pediatr Pulmonol* 1995;Suppl 12:186–187.
86. Becq F, Fanjul M, Merten M, Figarella C, Hollande E, Gola M: Possible regulation of CFTR-chloride channels by membrane-bound phosphatases in pancreatic duct cells. *FEBS Lett* 1993;327:337–342.
87. Favre B, Turowski P, Hemmings BA: Differential inhibition and posttranslational modification of protein phosphatase 1 and 2A in MCF7 cells treated with calyculin-A, okadaic acid, and tautomycin. *J Biol Chem* 1997;272:13856–13863.
88. Fischer H, Machen TE: The tyrosine kinase $p60^{c\text{-}src}$ regulates the fast gate of the cystic fibrosis transmembrane conductance regulator chloride channel. *Biophys J* 1996;71:3073–3082.
89. Illek B, Fischer H, Santos GF, Widdicombe JH, Machen TE, Reenstra WW: cAMP-independent activation of CFTR Cl channels by the tyrosine kinase inhibitor genistein. *Am J Physiol* 1995;268:C886–C893.
90. Illek B, Fischer H, Machen TE: Alternate stimulation of apical CFTR by genistein in epithelia. *Am J Physiol* 1996∉65–C275.
91. Sears CL, Firoozmand F, Mellander A et al: Genistein and tyrphostin 47 stimulate CFTR-mediated $C1^-$ secretion in T84 cell monolayers. *Am J Physiol* 1995;269:G874–G882.
92. De Jonge, HR, French, PJ, Bot, AGM, Bijman, J: CFTR-C1 channel opening by genistein is not mediated by protein tyrosine kinase- or protein phosphatase-inhibition but may involve a direct interaction with CFTR. *Pediatr Pulmonol* 1996;Suppl 13:218.
93. Wang F, Hwang T-C: Effects of genistein on CFTR chloride channel gating. *Biophys J* (Abstr) 1997;72:A113.
94. Smit LS, Wilkinson DJ, Mansoura MK, Collins FS, Dawson DC: Functional roles of the nucleotide-binding folds in the activation of the cystic fibrosis transmembrane conductance regulator. *Proc Natl Acad Sci U S A* 1993;90:9963–9967.
95. Hill, TL: *Free energy transduction in biology.* New York: Academic Press, 1977.
96. Jackson, MB: Perfection of a synaptic receptor: kinetics and energetics of the acetylcholine receptor. *Proc Natl Acad Sci U S A* 1989;86:2199–2203.
97. Rozanas CR, Neville DCA, Townsend RR, Verkman AS: Evidence for interactions between the NBD1 and R domains of CFTR. *Pediatr Pulmonol* 1996;Suppl 13:223.
98. Goldberg J, Huang HB, Kwon YG, Greengard P, Nairn AC, Kuriyan J: Three-dimensional structure of the catalytic subunit of protein serine/threonine phosphatase-1. *Nature* 1995;376:745–753.

99. Das AK, Helps NR, Cohen PTW, Barford D: Crystal structure of the protein serine/threonine phosphatase 2C at 2.0 Å resolution. *EMBO J* 1996;15:6798–6809.
100. Chen J, Martin BL, Brautigan DL: Regulation of protein serine-threonine phosphatase type 2A by tyrosine phosphorylation. *Science* 1992;257:1261–1264.
101. Carson MR, Welsh MJ: 5'-Adenylylimidodiphosphate does not activate CFTR chloride channels in cell-free patches of membrane. *Am J Physiol* 1993;265:L27–L32.
102. Schultz BD, Venglarik CJ, Bridges RJ, Frizzell RA: Regulation of CFTR Cl^- channel gating by ADP and ATP analogues. *J Gen Physiol* 1995;105:329–361.
103. Yount RG: ATP analogs. *Adv Enzymol* 1975;43:1–56.
104. Chabre M: Aluminofluoride and beryllofluoride complexes: new phosphate analogs in enzymology. *TIBS* 1990;15:6–10.
105. Dousmanis AG, Nairn AC, Gadsby DC: [Mg^{2+}] governs CFTR $C1^-$ channel opening and closing rates, confirming hydrolysis of two ATP molecules per gating cycle. *Biophys J* 1996;70;127a.
106. Venglarik CJ, Schultz BD, Frizzell RA, Bridges RJ: ATP alters current fluctuations of cystic fibrosis transmembrane conductance regulator: Evidence for a three-state activation mechanism. *J Gen Physiol* 1994;104:123–146.
107. Winter MC, Sheppard DN, Carson MR, Welsh MJ: Effect of ATP concentration on CFTR Cl-channels: A kinetic analysis of channel regulation. *Biophys J* 1994;66:1398–1403.
108. Gregory RJ, Rich DP, Cheng SH, Souza DW, Paul S. Manavalan P, Anderson MP, Welsh MJ, Smith AE: Maturation and function of cystic fibrosis transmembrane conductance regulator variants bearing mutations in putative nucleotide-binding domains 1 and 2. *Mol Cell Biol* 1991;11:3886–3893.
109. Wilkinson DJ, Mansoura MK, Watson PY, Smit LS, Colllins FS, Dawson DC: CFTR activation: distinct roles for the nucleotide binding folds. *J Gen Physiol* 1994;104:34a.
110. Wilkinson DJ, Mansoura MK, Watson PY, Smit LS, Collins FS, Dawson DC: CFTR: The nucleotide binding folds regulate the accessibility and stability of the active state. *J Gen Physiol* 1996;107: 103–119.
111. Ko YH, Pedersen PL: The first nucleotide binding fold of the cystic fibrosis transmembrane conductance regulator can function as an active ATPase. *J Biol Chem* 1995;270:22093–22096.
112. Manavalan P, Dearborn DG, McPherson JM, Smith AE: Sequence homologies between nucleotide binding regions of CFTR and G-proteins suggest structural and functional similarities. *FEBS Lett* 1995;366:87–91.
113. Carson MR, Welsh MJ: Structural and functional similarities between the nucleotide-binding domains of CFTR and GTP-binding proteins. *Biophys J* 1995;69:2443–2448.
114. Schultz BD, Bridges RJ, Frizzell, RA: Lack of conventional ATPase properties in CFTR chloride channel gating. *J Membr Biol* 1996;151:63–75.
115. Boguski MS, McCormick F: Proteins regulating Ras and its relatives. *Nature* 1993;366:643–654.
116. Travis SM, Carson MR, Ries DR, Welsh MJ: Interaction of nucleotides with membrane-associated cystic fibrosis transmembrane conductance regulator. *J Biol Chem* 1993;268:15336–15339.
117. Winter MC, Welsh MJ: ATP and phosphorylation dependence of the R domain mediated regulation of CFTR. *Pediatr Pulmonol* 1996;Suppl 13:212.
118. Mathews CJ, Tabcharani JA, Chang X-B, Riordan JR, Hanrahan JW: Characterization of nucleotide interactions with CFTR channels. *Pediatr Pulmonol* 1996;Suppl 13:221.
119. Kartner N, Hanrahan JW, Jensen TJ, Naismith AL, Sun S, Ackerley CA, Reyes EF, Tsui L-C, Rommens JM, Bear CE, Riordan JR: Expression of the cystic fibrosis gene in non-epithelial invertebrate cells produces a regulated union conductance. *Cell* 1991;64:681–691.
120. Welsh MJ, Smith AE: Molecular mechanisms of CFTR chloride channel dysfunction in cystic fibrosis. *Cell* 1993;73:1251–1254.
121. Ma J, Tasch JE, Tao T et al: Phosphorylation-dependent block of cystic fibrosis transmembrane conductance regulator chloride channel by exogenous R domain protein. *J Biol Chem* 1996;271: 7351–7356.
122. Zhao J, Drumm ML, Xie J, Davis PB, Ma J: Essential role of the R domain in mediating the interaction between NBD1 and NBD2 for the gating of the CFTR chloride channel. *Pediatr Pulmonol* 1996;Suppl 13:220.
123. Horowitz B, Tsung SS, Hart P, Levesque PC, Hume JR: Alternative splicing of CFTR Cl^- channels in heart. *Am J Physiol* 1993;264:H2214–H2220.
124. Pai EF, Krengel U, Petsko GA, Goody RS, Kabsch W, Wittinghofer A: Refined crystal structure of the triphosphate conformation of H-ras p21at 1.35 Å resolution: implications for the mechanism of GTP hydrolysis. *EMBO J* 1990;9(8):2351–2359.

Advances in Second Messenger and Phosphoprotein Research, Vol. 33

1040-7952/99 $30.00

5

Ion Channels as Physiological Effectors for Growth Factor Receptor and Ras/ERK Signaling Pathways

Stanley G. Rane

Department of Biological Sciences, Purdue University, West Lafayette, Indiana 47907

INTRODUCTION

The study of ion channel regulation was initiated with the classic investigations into the mechanisms of voltage- and ligand-dependent channel gating. Recognition of the fundamental biological importance of these two modes of channel activation resulted in an informal, but easily discernible functional classification scheme: voltage-dependent channels (e.g., squid axon sodium and delayed rectifier potassium channels), and ligand-gated channels (e.g., acetylcholine and glutamate receptor channels in vertebrate and crustacean skeletal muscle, respectively). A mechanistic demonstration that a single channel type could be responsive to both regulatory mechanisms was first provided for β-adrenergic receptor potentiation of dihydropyridine-sensitive, voltage-gated calcium channels in mammalian heart. Although β-adrenergic ligands bind to a receptor distinct from the channel protein, and channel activity is not ligand dependent, it was clear for the first time that a receptor-activated chemical signal (PKA-dependent phosphorylation) could modulate the voltage-dependent gating function. Ultimately, similar biochemical modulation has also been shown to be an important feature of classically defined voltage-dependent channels in neurons. These studies were the leading edge of what is now a considerable body of work from many labs focusing on the biochemical modulation of channel activity, a concept that dominates thinking about how channel function is modified to suit different biological requirements. Indeed it is probably reasonable to state that most contemporary investigations of channel function are in some way concerned with the biochemical signals that alter the channel's primary gating determinant, and the concept of an ion channel being either purely voltage or ligand dependent now seems remote.

What has emerged then is a field of study concerned mainly with understanding the rapid modulation of channel activity as a result of biochemical signaling from members of the seven transmembrane spanning α-helix receptor superfamily (e.g., α- and β-adrenergic, opioid, and muscarinic receptors). These receptors generate relatively transient signals, such as α or βγ GTP-binding protein subunits, cyclic nucleotides, and calcium, which provide for fast and readily reversible channel modification suitable for altering episodic electrical signaling (see Chapters 1–3 of this volume). A somewhat slower and more sustained form of signaling is provided by another receptor superfamily, the receptor tyrosine kinases (RTKs). These receptors and their signaling pathways are critical in controlling cell proliferation, fate determination, and differentiation, yet a challenge to a full mechanistic understanding of these processes remains the identification of the final physiologic mediators of RTK-induced signaling events. The studies to be reviewed here, showing that control of ion channel function is an important aspect of RTK-mediated signaling, have been important in beginning to identify how ion channels may contribute to the long-term aspects of cellular regulation that depend on RTK activation.

RTKs are characterized by an intracellular tyrosine kinase catalytic region, a single transmembrane-spanning domain, and an external ligand-binding domain for peptides such as nerve, epidermal, and platelet-derived growth factors (NGF, EGF, PDGF) (Fig. 1). Peptide ligand binding produces receptor dimerization, activation of tyrosine kinase catalytic activity, and resultant tyrosine autophosphorylation. These phosphotyrosines then act as docking sites for a variety of signaling proteins, all of which utilize 100 amino acid (approximately) Src homology 2 (SH2) sequences as phosphotyrosine-binding domains. Upon docking to the activated RTK, some of these signaling proteins in turn become catalytically active. This group of proteins includes the Src family of non–receptor tyrosine kinases (Src, Yes, Fyn), the phosphoinositide anabolic enzyme phosphoinositide 3-kinase (PI3-K), the phosphoinositide-specific phospholipase C-γ (PLC-γ), Syp tyrosine phosphatase, and the Ras GTPase activating protein (Ras-GAP). Other SH2 containing proteins that dock to activated RTKs have no obvious catalytic function. Instead, these proteins appear to function as adaptors, which tether together other signaling molecules and in this way allow their activation as part of a larger signaling complex.

A prime example of adaptor-mediated signal activation involves the Grb2 and Shc proteins, which coordinate the interaction of the guanine nucleotide exchange factor SOS with its target, the membrane-associated homomeric GTP-binding protein, Ras. This initiates a signaling pathway common to all RTKs in which SOS promotes exchange of GTP for GDP on Ras, followed by Ras-induced sequential activation of the serine/threonine Raf kinase (aka MAPKKK), the dual-specificity kinase MEK-1 (aka MAPKK), and the serine/threonine extracellular receptor kinases, ERK1 and ERK2 (aka MAPK1 and 2). Ultimately, ERK activation results in changes in gene transcription, and in this way growth factor signals at the cell surface are transmitted to the nucleus to be translated into the appropriate changes in cellular physiology. The Ras/ERK signaling backbone is a key element in both RTK-induced proliferation and differentiation, and these profoundly different outcomes suggest that selective activation of auxiliary signal paths, along with cellular context, will be significant contributors to Ras/ERK-driven cellular end points (1–4).

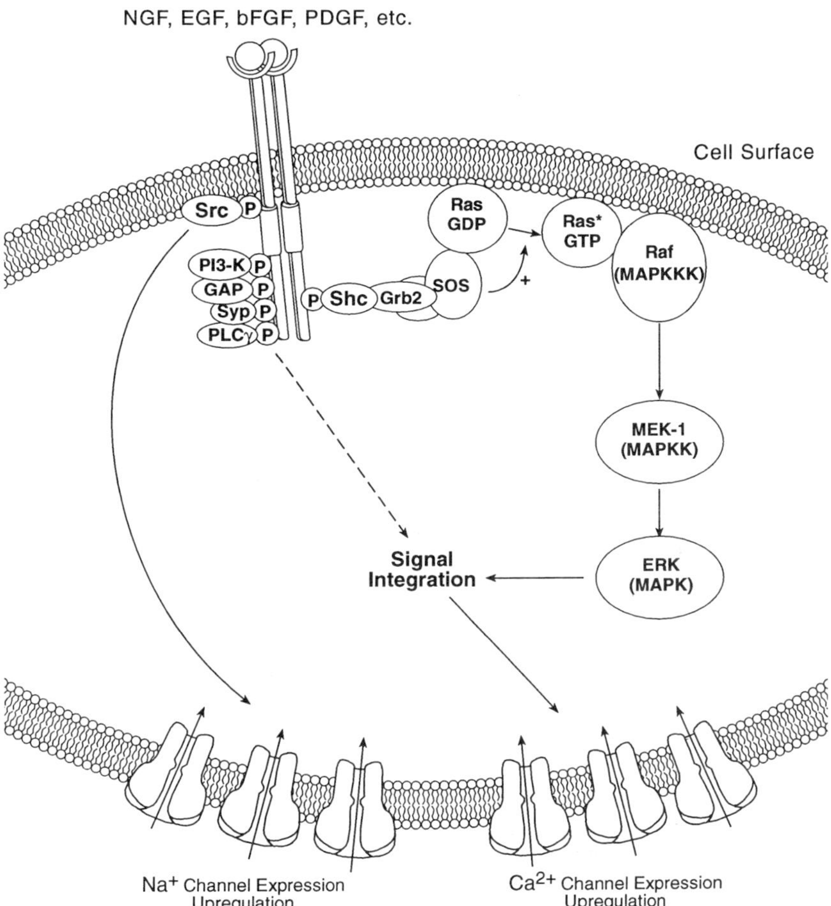

FIG. 1. A schematic representation of some of the key growth factor receptor tyrosine kinase activated signaling pathways likely to be important in ion channel regulation. Binding of peptide growth factors to their respective receptors results in receptor dimerization and activation of intracellular tyrosine kinase catalytic domains (thickened portions of the receptor). The resultant autophosphorylation at multiple tyrosine residues provides docking sites for a number of signaling molecules, which then become active. The evidence for which pathways are thought to be necessary for long-term upregulation of voltage-gated sodium and calcium channels was obtained in the PC12 model neuronal cell system. Calcium channel upregulation requires Ras/ERK activation as well as activation of a second as yet unidentified signal (dashed line). See text for details. Abbreviations for signaling molecules are as follows: Src, Src-family of non–receptor tyrosine kinases (includes Src, Yes, Fyn); PI3-K, phosphoinositide 3-kinase; PLC-γ, phospholipase C-γ; Syp, SH2-domain-containing tyrosine phosphatase; GAP, Ras GTPase-activating protein; Grb2, growth factor receptor bound protein- 2; Shc, SH2-domain-containing protein; SOS, son-of-sevenless guanine nucleotide exchange factor; MAPK, mitogen-activated protein kinase; MEK-1, mitogen and extracellular signal regulated protein kinase; ERK, extracellular signal regulated protein kinase. (Figure graphics by Michael Hilborn.)

One challenge to a full mechanistic understanding of RTK action remains the identification of the final physiologic mediators of RTK–Ras/ERK-induced signaling events. A number of studies now indicate that regulation of ion channel function is an important physiological target of these essential signal transduction systems. In terms of biochemical mechanisms, time course, and cellular function, RTK–Ras/ERK-mediated ion channel regulation clearly differs from the regulation provided by the seven-transmembrane-spanning α-helix receptor superfamily. The material to follow concentrates particularly on some of the unique biological functions subserved by RTK–Ras/ERK-actuated channel regulation, as well as what is known about the signals essential to each specific type of regulation.

POTASSIUM CHANNEL REGULATION IN THE CONTROL OF CELL PROLIFERATION AND DIFFERENTIATION

Stimulation or enhancement of ion fluxes in response to peptide growth factors is a well-established property of a number of cell types including murine fibroblast cell lines, in which many of the signaling aspects of mitogenic stimulation have been investigated (5). However, the initial work establishing a role for potassium channels in mitogenically activated cation flux has come from studies on mitogenic regulation of channel function in T lymphocytes. A comprehensive review of this work is available (6), but this discussion highlights mainly the aspects that bear on the role of potassium channels in RTK–Ras/ERK-mediated proliferative control in fibroblast cells. Together, the fibroblast and lymphocyte systems have provided much of the evidence for the importance of potassium channels in mitogenic signaling events.

T lymphocytes express at least three types of voltage-gated potassium channels (K_v) as well as two to four voltage-independent, calcium-activated potassium channel types (K_{Ca}). The channels that are most responsive to mitogenic regulation are the $K_{v1.3}$ *Shaker* family, voltage-gated *n* channel (prevalent in *normal* human T cells), and small conductance K_{Ca} channels. Mitogenic stimulation of T lymphocytes (with phytohemaglutinin, concanavalin A, or interleukin 2) results in increases in channel densities of approximately 10 to 20-fold for *n* (7,8), and 20- to 25-fold for K_{Ca} (9) channels, as a precedent to increased DNA synthesis over the course of 1–2 days. Acute regulation of the *n* K_v channel has also been reported, although results vary among labs, and it is unclear what significance this regulation may have for cell proliferative control (10–12). Mitogenic stimulation of B lymphocytes also results in K_v and K_{Ca} upregulation, the latter being prevented by inhibitors of transcription (13).

Pharmacological blockers have been used to test whether the temporal sequence of mitogenic stimulation, potassium channel upregulation, and increased DNA synthesis is indicative of a causal relation between the latter two events. Relatively broad spectrum channel blockers (e.g., 4-aminopyridine, quinine, TEA) were first shown to have an inhibitory effect on measures of lymphocyte proliferation (8,10,12). More specific potassium channel blockers derived from scorpion venoms have been used to further this analysis. Charybdotoxin (ChTX) blocks both K_v and K_{Ca} channels in lymphocytes, and inhibits mitogen-stimulated lymphocyte proliferation and interleukin production (14–16; but see ref. 17). The structurally related peptides margatoxin (MTX) and noxiustoxin (NTX), which specifically block K_v, also inhibit T-cell activation (the stimulated increase in interleukin production and intracellular calcium), suggesting a central role of K_v in this process (although effects on cell proliferation were not directly assessed) (15). However, the unrelated venom peptide kaliotoxin (KTX), which also preferentially blocks K_v, has been shown to reduce proliferation by only about one-half the amount produced by ChTX, arguing against exclusive control of proliferation by K_v (16). Thus, the issue of whether one or more classes of potassium channel are critical for controlling lymphocyte proliferation remains to be resolved.

Several intriguing questions are raised by the work just described. Mitogen or antigen binding to the T-cell receptor (TCR) initiates a number of signaling events including activation of Src-related tyrosine kinases and PLC-γ mediated metabolism of membrane phospholipids. What are the signaling mechanisms that transduce mitogen-stimulated potassium channel upregulation? Defining these mechanisms will be important to understanding how channel regulation integrates into the complex of biochemical and genetic events that occur during T-cell activation.

How does potassium channel regulation contribute to proliferative control, and does the lymphocyte system have applicability to understanding proliferation in other cell types? Lymphocyte activation is known to require receptor-activated calcium influx, and although activation of this influx is voltage independent, its magnitude is thought to be a rather direct function of the electrochemical driving force for calcium (6,18). There is general agreement that TCR-mediated potassium channel upregulation will serve to maintain a relatively hyperpolarized cell membrane potential, which in turn will indirectly contribute to TCR-activated, voltage-independent calcium influx via calcium release activated calcium channels or (CRAC) (6,15,16,18; but see ref. 19 for another view). Furthermore, the antimitogenic peptide potassium channel blockers depolarize lymphocytes (16,20), which should blunt calcium entry. Does a similar dynamic exist in other cell types, and does it necessarily involve CRAC, K_v, and K_{Ca}? Some of these issues have begun to be addressed in other cell types, including murine fibroblast cell lines.

In fibroblasts and other nonneuronal cells, peptide mitogens drive multiple signaling events including RTK–Ras/ERK activation and mobilization of cation fluxes. In normal cells, RTK activation produces relatively transient activation of Ras/Raf/MEK–1/ERK and their final effectors, and these signals coordinate normal cell proliferation. One very useful model for understanding which signals and effectors are crucial in mitogenic signaling has involved analysis of the actions of constitutively active mutant forms (oncoproteins) of the RTK–Ras/ERK signaling proto-oncoproteins. For example, constitutively GTP-bound and active oncogenic Ras persistently stimulates the Raf/MEK–1/ERK cascade, producing deregulated cell growth presumably as a result of constitutive expression of the ultimate Ras/Raf/MEK–1/ERK effectors. Therefore, to ask whether ion channels are downstream physiological effectors for Ras/ERK, we looked for constitutive alterations in electrophysiological phenotype due to ectopic expression of genes coding for either Harvey or Kirsten Ras oncoproteins. In each of four fibroblast lines examined, stable transfection with Ras oncogenes causes a persistent 3- to 10-fold increase in the membrane area normalized density of a small conductance, K_{Ca} current; and over-expression of normal cellular Ras increases K_{Ca} current density two-fold (21,22). Transfection of cells with constitutively active forms of Raf, the kinase immediately downstream of Ras, produces a 20-fold upregulation of the same current. In separate experiments, expression of two dominant negative Raf proteins (23) reversed oncogenic Ras-induced K_{Ca} upregulation (22). This last result (and other conventional patch–clamp experiments; see ref. 22) suggest that increased K_{Ca} current density is not a result of direct activation of K_{Ca} channels by Ras G protein. More importantly, these studies clearly implicate the fibroblast K_{Ca} channel as a physiological endpoint for Ras/ERK signaling, although formal proof of the precise signaling sequence beyond Raf kinase must still be accomplished.

Analysis of the properties of the Ras/Raf upregulated channel in fibroblasts indicate that it, or a very similar channel, is expressed in other mitogenically active cells. The channel in fibroblasts is potassium-selective ($P_{Na}:P_K < 0.02$), and shows weak inward rectification in symmetric 150 mM potassium chloride solutions (single channel conductances at −60 and 60 mV of 33 and 17 pS). Based on its small conductance relative to the BK large-conductance potassium channel family, the fibroblast K_{Ca} channel has been designated as SK. The calcium activation threshold for the SK channel in excised patches is less than 0.2 μM, and it has weak voltage sensitivity exhibited as an increase in opening frequency between −60 and 60 mV (24). Extracellular application of 100 nM ChTX blocks more than 90% of whole-cell SK current, while 10 mM TEA is somewhat less effective, and apamin at 0.5 μM is ineffective (21). The fibroblast SK channel is essentially identical in terms of conductance, rectification, calcium sensitivity, and pharmacology to the mitogen-upregulated K_{Ca} channel observed in B and T lymphocytes (9,25,26), and partially characterized channels reported for mitogenically active, tumor-derived cells (27,28). Thus, SK upregulation is a consistent feature of mitogenic activation in a number of cell types.

Oncogenic transformation as a model for normal mitogenic signaling events has been used to predict the physiological parallel of oncogenically induced SK upregulation. In nontransformed cells, stimulation of the endogenous Ras/ERK proto-oncoprotein signal cascade by the mitogenic peptides EGF and PDGF was also found to upregulate SK channel density. This effect is blocked by inhibitors of RTK activity and by protein synthesis inhibition, suggestive of *de novo* SK channel synthesis in response to mitogenic RTK-driven Ras/ERK activation. EGF and PDGF upregulation of SK is rapid, with significant increases in current density being observed within an hour of mitogen application (22). Similarly, SK is upregulated in primary chick embryo fibroblasts infected with a temperature-sensitive mutant of $pp60^{v\text{-}src}$, a non-RTK capable of stimulating Ras/ERK, within 1–2 hours of a shift to the permissive temperature (29,30). SK channel upregulation is sustained for at least 24 hours in the presence of PDGF, a full mitogen, but SK channel density reverses to nonstimulated levels after 6 hours of treatment with EGF, a partial mitogen (i.e., requires presence of a co-mitogen for efficient proliferative stimulation). Furthermore, PDGF-stimulated cell proliferation is inhibited by the SK channel blocker ChTX (22). Thus growth factor activation of the central Ras/ERK mitogenic signaling path produces rapid and persistent SK channel upregulation, and functional blockade of SK arrests proliferation. For fibroblasts, therefore, there is good evidence that the SK channel is a physiologic effector for mitogenic signaling via Ras/ERK.

Together, the fibroblast and lymphocyte studies make a strong case for potassium channels, particularly K_{Ca} types, as controlling elements in mitogenic cell growth. A possible corollary to this mitogenic role is that potassium channel activity will also govern, perhaps negatively, cell differentiation. In support of this idea, overexpression of ectopic *Shaker*-like potassium channels in early stage *Xenopus* embryos has been shown to increase delayed rectifier potassium current densities, as well as to reduce the number of morphologically differentiated neurons (31). In both immature and mature neurons, the increase in delayed rectifier current shortens action potential duration, which would attenuate the primary source of calcium influx via voltage-gated calcium channels. However, it is not known whether changes in calcium entry are responsible for the observed reduction in neuronal morphological differentiation. Endogenous, delayed-rectifier potassium channel currents are enhanced in mitogen stimulated Schwann cells and oligodendrocytes, and mitogen-stimulated proliferation is inhibited by block of this current (32,33).

A rather different interpretation regarding potassium channel function and neuronal development is suggested by the effects of the *weaver* mutation on cerebellar granule cell precursors. The *weaver* mutation corresponds to a single amino acid substitution in a G protein gated, inwardly rectifying K^+ channel (GIRK2), which causes both constitutive activation of the channel and a loss of selectivity for potassium (i.e., the channel becomes a nonselective cation channel) (34–36). Proliferation of granule cell precursors in the *weaver* mutant is unaffected by this "loss" of potassium channel function. However, these cells fail to differentiate, although differentiation can be rescued by blockers of the mutant GIRK2 conductance (36). Thus it appears that GIRK2 potassium conductance is permissive for differentiation and is not required for proliferation in this neuronal lineage.

The control of both cell proliferation and differentiation requires signaling via RTK–Ras/ERK, and upregulation of the SK channel is integral to proliferation. Therefore, we asked whether the SK channel could have a corollary role to play during differentiation in the C3H10T1/2 (10T1/2) cell model system (37). A multipotent, fibroblast-like cell line, 10T1/2 proliferates in response to RTK–Ras/ERK stimulation by basic fibroblast growth factor (bFGF). Ectopic expression of the MRF4 myogenic regulatory transcription factor in these cells (10T1/2-MRF4) allows selective expression of a muscle phenotype (i.e., α-actin and myosin heavy chain expression, and ultimately formation of multinucleate myotubes), but only after withdrawal of the bFGF mitogenic stimulus, which negatively regulates MRF4 (38).

Withdrawal of bFGF from 10T1/2-MRF4 cells, and thus removal of MRF4 negative regulation, produces upregulation of acetylcholine (ACh) receptor channels within 24 hours, a classic index of myogenesis.

In bFGF-stimulated 10T1/2-MRF4 cells, MRF4 is negatively regulated, myogenesis is suppressed, and no ACh receptor expression is observed. These cells concomitantly proliferate and express the SK channel known to be associated with mitogenesis in other fibroblast lines (21,22,24,29,30). Like PDGF in NIH 3T3 cells, bFGF produces persistent SK upregulation in 10T1/2-MRF4 cells, consistent with maintenance of a proliferative versus a differentiated cell state. However, chronic application to bFGF-stimulated 10T1/2-MRF4 cells of the SK channel blocker ChTX induces ACh receptor channel expression within 24 hours. Furthermore, ACh receptor currents could be recorded in cell cultures maintained in bFGF and ChTX for as long as 5 days. In terms of ACh receptor expression, therefore, SK channel functional blockade precisely mimics the removal of MRF4 negative regulation seen with bFGF withdrawal, suggesting that the SK channel contributes to RTK–Ras/ERK negative regulation of MRF4-dependent transcriptional control. A charybdotoxin- and TEA-sensitive, apamin-insensitive K_{Ca} channel is also upregulated in mitogenically stimulated human muscle satellite cells, coincident with inhibition of myogenic progression (39). Thus it appears that among mesoderm-derived cells (e.g., muscle progenitors, fibroblasts, T lymphocytes), there is conservation of a central role for K_{Ca} channels in the signaling events controlling proliferation as well as differentiation. These signals must ultimately drive changes in gene expression, and analysis of the 10T1/2-MRF4 system should tell us how K_{Ca} channel activity is coupled to these nuclear events.

Among key issues yet to be resolved is what occurs downstream of changes in potassium channel expression levels (and presumably activity) to produce changes in transcriptional activity governing cell growth. Coupling between K_v or K_{Ca} and transcription could rely on the channel's ability to set membrane potential. This would result in an indirect electrochemical effect on either voltage-independent calcium influx, an important element in mitogenic stimulation, or intracellular levels of other ions (e.g., sodium) that are coupled to transport of metabolically critical molecules (see Ref. 40 for an in-depth analysis of this hypothesis). For lymphocytes there is good evidence that potassium channels control cell growth via an indirect influence on CRAC or CRAC-like calcium influx pathways, but the applicability of this mechanism to other cell systems has not been addressed. In fibroblasts and myogenic cells, as discussed later, growth factor activated, voltage-independent calcium channels may take the place of CRAC.

Also to be resolved are the signaling elements upstream of the SK K_{Ca} channel that mediate its upregulation. In fibroblasts, the results with Ras and Raf oncoproteins and Raf-dominant negatives are consistent with the Ras/ERK pathway being necessary and sufficient for SK upregulation. However, this issue must be formally demonstrated for upregulation in response to growth factor RTK stimulation of the endogenous signaling systems, which may invoke both redundant and divergent pathways (1–4). For T lymphocytes, TCR stimulation both upregulates K_{Ca} (and

K_v) and activates the Ras/ERK pathway; however, a number of other important signaling pathways are also simultaneously activated by the TCR, most notably those driven by Src family non-RTKs, particularly Lck (see Ref. 41 for review). Lck can activate Ras signaling, and for fibroblasts the presence of activated Src emulates Ras/ERK activation in upregulating SK. It seems reasonable then to postulate that upregulation of SK by the TCR will involve an activation sequence of Src family non-RTKs and Ras/ERK. Active Ras mutations are the most frequently observed aberration in human hematologic malignancies, and Ras signaling is required for chronic myelogenous leukemia associated with the constitutively active Abl non-RTK oncogenes (42,43). The fibroblast work again predicts that active Ras will cause chronic upregulation of SK K_{Ca} channels in these abnormally proliferative hematologic cells, or, in the case of Abl oncogenes, their undifferentiated, blast precursors. Fibroblasts do not appear to express K_v-type channels, and thus for this channel type they are not predictive for hematologic cells.

The presence of SK may be diagnostic for oncogenic signaling that results in constitutive activation of the central Ras/ERK pathway. Further, SK and other potassium channels may offer new sites for therapeutic developments (19,44), particularly in the areas of immune system function and cancer, where RTK–Ras/ERK signaling is of central importance. For example, blockers for ATP-sensitive potassium channels arrest proliferation in the human mammary carcinoma cell line MCF-7, at the G_0/G_1 phase of the cell cycle (40,45,46), a point associated with potassium channel modulation in embryonic cells as well (47). Thus regulated potassium channel expression may be viewed as having general importance to cell proliferation, although the specific channel types could vary through development stages or with disruptions in the central mitogenic and proliferative signaling pathways.

RECEPTOR TYROSINE KINASE ACTIVATION OF CALCIUM-PERMEABLE ION CHANNELS

Rapid mobilization of calcium occurs as an integral event in growth factor activated signaling. Influx of extracellular calcium is involved in the mitogenic activity of growth factor RTKs (5,48,49), as well as the proliferative events driven by non-RTKs coupled to the TCR (18). In neuronal cells, calcium influx has been suggested as providing a stimulatory input to the RTK–Ras/ERK pathway, by virtue of its ability to directly activate either Src, Ras, or a neuronal Ras guanine nucleotide exchange factor (50–52). However, it is not clear from these studies whether calcium activation of Ras/ERK is a significant event within the context of RTK activation of the central Ras/ERK pathway. It also remains to be seen whether calcium activation of Ras and Src occurs in nonneuronal cells, although there is evidence that extracellular calcium stimulates fibroblast growth via activation of ERKs (53).

For neurons (and other excitable cells), membrane depolarization and activation of voltage-dependent calcium channels constitutes the primary calcium influx pathway. Although calcium influx is clearly important to normal growth of nonneuronal cells and contributes significantly to their oncogenic growth processes (48), electrophysiological studies indicate that voltage-dependent calcium channels are conspicuously absent, or expressed at very low levels, particularly in fibroblasts (21,37,54–56). Furthermore, one of the earliest studies to look at the effect of oncogenes on ion channel expression or activity showed that the small T-type voltage-dependent calcium currents recorded in 3T3 fibroblasts were suppressed in response to transformation with *ras*, v-*fms*, or polyoma middle T oncogenes (54). Mitogenic calcium influx in the fibroblast model systems appears instead to involve activation of voltage-independent calcium influx pathways. Fluorescence-based intracellular calcium imaging has been used to show PDGF and EGF stimulated, voltage-independent calcium influx in quiescent cells, and this action has been correlated with stimulation of DNA synthesis and ultimately cell proliferation (49,57–59). In NIH 3T3 cells overexpressing EGF receptors, EGF-stimulated calcium influx also evokes a ChTX-sensitive membrane hyperpolarization, consistent with activation of the SK channel in these cells (see preceding section). In addition, EGF-stimulated proliferation is partially inhibited by ChTX (49). This result is consistent with the electrogenic model for proliferative control, in which SK clamps the membrane to a relatively hyperpolarized potential, thus maintaining a strong electrochemical driving force for calcium entry. Therefore, it is postulated that mitogenic calcium influx is a function of membrane potential but is voltage independent in terms of its activation.

To completely understand proliferative cell signaling, it will be essential to identify and characterize the voltage-independent pathways responsible for RTK-activated calcium influx. Because growth factor RTKs activate PLC-γ and therefore mobilize calcium stores that are sensitive to inositol 1,4,5-triphosphate (IP_3) some of their sustained calcium-mobilizing activity may depend on activation of the calcium-refilling pathway responsible for replenishing the IP_3-sensitive store (60). The terms "capacitative calcium entry," "storage dependent calcium influx," and "calcium release activated calcium current" (I_{CRAC}) all refer to calcium-refilling pathways, although they have different mechanistic implications. The light-sensitive, calcium-permeable channel *trp,* initially cloned from *Drosophila,* with homologs later identified in mammalian cells, has been proposed as a putative plasma membrane localized refilling channel. However, the relationship between *trp* and I_{CRAC} remains to be clarified (61). For the purpose of discussing mitogenic calcium entry, "calcium-refilling pathway" is used here to refer solely to the compensatory calcium influx that occurs in response to mobilization of internal calcium stores.

What is the relationship between calcium refilling as just defined and mitogen-activated calcium influx in the fibroblast model systems? The calcium-refilling path present in a number of cell types, including neutrophils, platelets, thymocytes, macrophages, and astrocytoma cells, is typified by its permeability to Mn^{2+} (62–66). An inwardly rectifying, voltage-independent, calcium channel in Jurkat T cells is a likely candidate for the underlying conductance in hematologic cells (67). Although fluorescence quenching by Mn^{2+} is often used as a surrogate indicator for the stores-refilling calcium influx, electrophysiological study of calcium-refilling events has identified both Mn^{2+} permeable and Mn^{2+}-blockable channels, suggesting the existence of multiple refilling paths (68–70). The Mn^{2+} sensitivity of calcium-refilling pathways in fibroblasts has not been conclusively addressed (66,71,72).

In any case, there is evidence for RTK-induced calcium mobilization in fibroblasts that is independent of the calcium-refilling path. In C3H10T1/2 cells, PDGF activates a calcium influx that is pharmacologically distinct from the calcium-refilling flux activated by thapsigargin, the stores-emptying, microsomal calcium-ATPase inhibitor (57). In this system La^{3+}, which at 0.3 mM blocks calcium-refilling currents in Jurkat cells (67), blocks RTK- but not thapsigargin-activated calcium influx when applied at a low concentration (10 μM). Interestingly, La^{3+} was also shown to block PDGF-stimulated cell cycle progression (57). These results suggest the existence of multiple pharmacologically distinct calcium influx paths in fibroblasts, and show that La^{3+} may prove useful in distinguishing the mitogenic importance of these pathways. Further, both the PDGF-stimulated, La^{3+}-sensitive calcium influx path and the influx path activated by thapsigargin are Mn^{2+} permeable, indicating that Mn^{2+} permeability alone cannot be used to discriminate among multiple calcium influx paths in fibroblast cells.

In mesangial cells a voltage-independent, calcium-permeable channel (conductance approximately 1 pS) is also activated by PDGF, with channel activity being observed when PDGF is applied via the patch pipette in cell-attached patch mode. However, PDGF fails to evoke channel activity in cell-attached patches when the peptide is applied to the extrapatch membrane (73). Similar results have been reported for activation of a 19-pS, calcium-permeable cation channel in response to insulin-like growth factor II application to BALB/c 3T3 fibroblasts (74). The lack of channel activation in the patch in response to extrapatch ligand application is typically taken as evidence that the channel and ligand receptor are tightly coupled (i.e., that channel activation is not dependent on generation of a diffusible signaling molecule).

The results described earlier, therefore, seem inconsistent with the idea that channel activation by these mitogens constitutes a calcium-refilling pathway, since activation of any refilling pathway almost certainly requires a diffusible messenger (60). Munaron and colleagues (75) have identified a bFGF-activated, 8-pS, calcium-permeable channel in BALB-c 3T3 fibroblasts, and this channel can be activated in cell-attached patches when the peptide is applied to the extrapatch membrane. However, this bFGF-evoked current was clearly shown to be independent of thapsigargin-induced calcium mobilization, which is considered diagnostic for activation of the calcium-refilling path. Therefore, there seems to be ample evidence in fibroblast cells for the existence of mitogen-activated, voltage independent calcium channels, which are not associated with the prototypical calcium-refilling pathway.

Obviously there is still much to be determined about how RTKs activate ion channels. Indeed, the basic issue of whether the tyrosine kinase activity of these receptors is required for channel activation remains to be addressed for most of the heretofore mentioned calcium-permeable channels, as well as the other nonselective channels shown to be mitogen sensitive (76,77). Other channel types have been shown to be directly regulated by TKs (78–80), and in human fibroblasts TK inhibitors suppress calcium influx stimulated by both thapsigargin and bradykinin, a mitogenic agonist whose receptor has no intrinsic TK activity (71). However, a comprehensive understanding of the role of calcium influx in RTK-driven mitogenesis, particularly in fibroblasts, still requires identification of the actuating pathways specific to RTKs. In addition, characterization of the channels themselves is at a very early stage, and thus there has not been extensive development of specific channel blockers as either investigative or therapeutic tools. One possible candidate, SK&F 96365, has been shown both to inhibit serum-stimulated calcium influx and to arrest fibroblast proliferation in $G_{2/M}$ phase (81). There are a number of potential loci for calcium regulation of cell proliferation, including control of kinase activities, transcription factor translocation and activation (82,83), and transit of molecules across the nuclear envelope (84,85). Establishing the RTK-activated calcium influx signaling mechanisms and channel effectors will be an essential part of understanding calcium as a mitogenic signal.

RECEPTOR TYROSINE KINASE REGULATION OF NEURONAL ION CHANNELS

Like the peptide growth factors discussed earlier, the neurotrophins NGF, brain-derived neurotrophic factor (BDNF) and neurotrophin-3, bind to their cognate neuronal RTKs and activate Ras/ERK, as well as other signaling systems. In doing so they produce in neuroblast cells a profound shift from mitogenic growth to cell cycle withdrawal and expression of a terminally differentiated phenotype (86). A significant aspect of this relatively slowly occurring phenotypic change is the development of electrical excitability via the upregulation of voltage-gated ion channels. In terms of mechanistically understanding the development of neuronal cell physiology, therefore, it is important to identify the neurotrophin RTK-activated signaling functions that are critical for ion channel regulation. Also, it is becoming increasingly clear that the electrophysiology of mature neurons can be acutely modulated by neurotrophin RTK activation, in a manner analogous to the short-term modulation produced by the seven-transmembrane-spanning α-helix receptor superfamily. Thus there is a substantial impetus for analyzing RTK–Ras/ERK signaling in the context of its action on neuronal electrophysiology.

A role for Ras/ERK signaling in ion channel regulation in neuronal cells was initially assessed via both direct injection of activated Ras and stable transfection of cell lines with genes coding for activated Ras. In mature *Hermissenda* neurons, somatal injection of v-Ras has been observed to enhance voltage-gated calcium currents within minutes, suggestive of a fairly direct action of Ras or an immediate downstream signaling molecule (87). Ras has also been suggested to have a direct channel modulatory action in guinea pig atrial myocytes which, although not neurons, are certainly fully differentiated, excitable cells. Exogenous application of either Ras or its GTPase-activating protein (Ras-GAP) was shown to inhibit muscarinic receptor activation of its inwardly rectifying potassium channel target (88,89). These results were interpreted to indicate that the exogenously applied proteins interact with their endogenous, membrane-associated counterparts to form a complex that then blocks muscarinic receptor activation of the endogenous G_K coupling G protein. Since both Ras and Ras-GAP are already present in the native membrane, it remains unclear why no tonic inhibitory action was observed to result from complexation of the endogenous proteins. Because the studies on *Hermissenda* neurons and cardiac myocytes involved terminally differentiated cells and acutely applied Ras, their results suggest actions of Ras separate from those that may occur during neurotrophin-induced signaling and neuronal development.

Long-term regulation of channel function in the pituitary-derived AtT20 cell line has been demonstrated in response to transfection with the H-*ras* oncogene. Normal AtT20 cells and cells transfected with antibiotic resistance marker only express both tetrodotoxin (TTX)-resistant and TTX-sensitive sodium channels. However, *ras*-transfected cells fail to express the TTX-resistant current (90). Expression of activated H-Ras has also been suggested to suppress sodium channel functional expression in human bladder carcinoma–fibroblast hybrid cells (91). In addition, *ras* transfection of AtT20 cells produces a two- to threefold increase in the density of voltage-gated potassium currents and a reduction in the rate of potassium current inactivation, concomitant with the appearance of two neuron-specific, potassium channel mRNA species absent from nontransfected AtT20 cells (92). The relationship between the appearance of these mRNAs and the electrophysiological changes remains to be determined; however, the resultant shortening of action potential duration and the extension of neurites in *ras*-transfected AtT20 cells could be interpreted as a evidence for involvement of the Ras G protein in the development of a neuronal phenotype.

The PC12 sympathetic neuron–like cell line has been used extensively as a model to understand neuronal differentiation in response to activation of neurotrophin RTKs. It is now also a prototypical model system for understanding how the different neurotrophin RTK signaling functions, including the Ras/ERK pathway, contribute to development of an electrophysiologically mature neuron. Chronic NGF activation of the *trkA* receptor results in a stereotypic differentiation response from PC12 cells, including cell cycle withdrawal, somal flattening, extension of neurites, and the development of electrical excitability. The initial evidence for Ras/ERK signaling being involved in *trkA*-induced neuronal differentiation was the demonstration that PC12 cells develop a morphology similar to that evoked by NGF in response to either injection of activated Ras (93), infection with sarcoma viruses incorporating *ras* oncogenes (94), or transfection with activated forms of *src* or *raf* (95,96). Transfection with a dominant negative form of Ras, N17Ras (Ser^{17}–Asn^{17}), was used to show that *trkA*-stimulated PC12 cell morphological differentiation is dependent on the activation of endogenous Ras. N17Ras interacts with the endogenous Ras GTP exchange protein, but the interaction fails to cause off-loading of GDP in favor of GTP as for normal Ras. This futile interaction is thought to be the mechanism by which N17Ras abrogates the activation of normal Ras by growth factor RTKs, including *trkA* and the fibroblast growth factor receptor (FGF-R). For PC12 cells transfected with N17Ras, application of NGF or bFGF fails to significantly stimulate neurite outgrowth, although the cells do morphologically differentiate in response to db-cAMP and TPA (97).

Additional evidence for the necessity of Ras/ERK signaling to neuronal differentiation was provided by Cowley et al. (98), who showed that a constitutively active MEK-1 kinase mutant causes neurite outgrowth in PC12 cells. Furthermore,

NGF-induced neurite outgrowth was inhibited by expression of MEK-1 mutants with phosphorylation site substitutions that inhibit their activation by Raf kinase, the immediate downstream effector of Ras (98). But Vaillancourt et al. (99), using ectopic expression of wild-type and mutant PDGF-Rs in PC12 cells, have shown that Ras/ERK signaling alone is not sufficient to drive morphological differentiation. Application of PDGF to cells expressing wild-type PDGF-R causes morphological differentiation very similar to that caused by activation of endogenous *trkA* or bFGF-R. However, neurite extension was blocked in cells expressing selective PDGF-R phosphotyrosine mutants, which retain Ras/ERK signaling capability but have phenylalanine substituted for tyrosine at binding sites for either Src family TKs or PLC-γ (99). Thus RTK-induced differentiation in the PC12 model system, at least as defined by morphological changes, requires activation of Ras/ERK as well as other signaling molecules including at least Src family TKs or PLC-γ. An important question remains regarding whether this analysis is relevant to physiological differentiation, another major determinant of neuronal function both in the developing and terminally differentiated cell.

Chronic stimulation of either endogenous or ectopically expressed growth factor RTKs in PC12 cells causes increased mRNA levels for the α subunit of brain (II/IIA) and peripheral (PN1) neuron sodium channel types, and causes corresponding increases in their functional membrane current densities (100–102) (Fig. 1). Unlike the case with respect to morphological differentiation, it is clear that Ras signaling is not required for type II/IIA sodium channel upregulation, since NGF- and bFGF-induced increases in either mRNA or current levels are readily observed in N17Ras-expressing cells (103,104). Also, activated forms of Ras and Raf fail to significantly upregulate sodium channel mRNA levels (105), arguing further against the necessity of the Ras/ERK pathway.

As with morphological differentiation, however, use of PDGF-R phosphotyrosine mutants shows that this aspect of physiological differentiation likely involves two distinct signaling pathways, one being the Src family TKs. A PDGF-R with tyrosine substitutions that abrogate simultaneously coupling to PI3-K, PLC-γ, Syp tyrosine phosphatase, and Ras-GAP produces normal upregulation of both type II/IIA sodium channel mRNA and functional expression (106). Receptors with this background, along with substitutions in either one of two tyrosines that bind Src TKs, produce significantly less mRNA and functional channels, implicating Src binding as necessary for sodium channel regulation. Interestingly however, a receptor with substitutions in both Src family binding sites, but with the PI3-K, PLC-γ, Syp, and Ras-GAP sites restored, shows normal upregulation of sodium channel mRNA, and reduced but relatively robust channel functional expression (106). One possible interpretation of these results is that signaling from either a fully functional Src TK binding site, or from one or more of PLC-γ, Syp, and Ras-GAP, acts as redundant mechanisms for sodium channel upregulation by growth factor RTKs. Alternatively, perhaps there is a common, as-yet unidentified, downstream signaling effector for both Src and PLC-γ, Syp, and Ras-GAP, so that all these signaling moieties must be removed if sodium channel upregulation is to be blocked.

Neurotrophin upregulation of voltage-gated calcium channels in PC12 cells has signaling requirements that differ from those for sodium channel upregulation. PC12 cells express three pharmacologically distinct calcium channel types: N (ω-conotoxin sensitive), L (dihydropyridine sensitive), and a third known as resistant, which is insensitive to either ω-conotoxin or dihydropyridine. Chronic application of either NGF or bFGF causes a pronounced increase in both N-type current and the corresponding α_{1B} pore-forming subunit mRNA. Increases in current levels for the L and resistant channel types are more modest, although the resistant current appears to be more responsive than the L (107–111). Ectopic expression of a temperature-sensitive form of v-*src* in PC12 cells, and subsequent growth at the permissive temperature, causes upregulation of N and resistant currents similar to that seen in response to chronic neurotrophin application (108,109). It is not known whether oncogenic Src may simply act as a surrogate for a more complex set of signals underlying neurotrophin induced calcium channel upregulation, in much the same way as constitutively active MEK-1 mimics neurotrophin-induced morphological differentiation (98).

Unlike sodium channel upregulation, calcium channel upregulation requires Ras signaling, as evidenced by the complete blockage of NGF- or bFGF-induced increases in calcium channel densities in cells expressing N17Ras (104). However, it appears unlikely that Ras signaling alone can account for neurotrophin-induced calcium channel upregulation, since calcium current levels are not increased either by sustained activation of endogenous Ras proteins or by induced expression of oncogenic Ras (104,113). Signaling requirements for neurotrophin induction of calcium channels are further distinguished from those for sodium channels, in that the involvement of cAMP signaling in long-term sodium channel regulation (100) is not paralleled for calcium channels (113). Taken together, the studies on calcium channel upregulation by neurotrophin receptors argue convincingly for a necessary role for Ras/ERK signaling, with a complementary signal perhaps being provided via Src (or Src family TKs). The results with oncogenic Src induction of calcium channels are consistent with this idea, in that Src is known to increase gene expression in PC12 cells via both Ras/ERK-dependent and Ras/ERK-independent pathways. Ras/ERK is also recognized as a downstream effector for Src in mitogenic signaling. Therefore, in terms of calcium channel upregulation, the presence of oncogenic Src could drive the necessary Ras/ERK pathway, as well as complementary Src-dependent (but Ras/ERK-independent) paths. Further analysis using the ectopically expressed mutant PDGF-R approach will likely resolve these issues for calcium channels.

The analyses in the PC12 system of signaling requirements for the combination of physiological (i.e., ion channel regulation) and morphological differentiation clearly point to a complex of partially overlapping RTK-activated signaling components. The commonality of signals for neurite extension and calcium channel upregulation, including Ras/ERK activation and very likely activation of pathways dependent on Src family TKs, suggest that these phenomena may be coupled during neurotrophin-driven neuronal development. Indeed, calcium influx via voltage-gated calcium channels has been shown to stimulate Ras/ERK signaling, leading to neurite extension (50,52,114,115). These findings suggest that an in-depth understanding of neuronal development must come from an analysis of the intimate interplay among neurotrophin signaling systems, ion channel activities, and morphological development.

ACUTE MODULATION OF ION CHANNEL FUNCTION

The most intensively studied examples of ion channel regulation by growth factor RTKs and their prototype signaling pathways have involved long-term actions typically involving changes in ion channel expression. The exception to this has been acute RTK activation of calcium-permeable channels in nonneuronal cells, although mechanistic information from these systems is still quite limited (see preceding section, "Receptor Tyrosine Kinase Activation of Calcium-Permeable Ion Channels"). It seems entirely likely that acute modulation of channel physiology by growth factor RTKs will become an increasingly important area of investigation. There is substantial evidence that neurotrophins have short-term actions on the physiology and morphology of both central (116–118) and peripheral neurons (119,120), and it is probable that some of these actions will be attributable to direct modulation of ion channel activity. NGF has been shown to potentiate voltage-gated calcium current in motor neurons of the freshwater snail *Lymnaea* (121), and NGF and neurotrophin-3 activate calcium-dependent potassium channels in neurons isolated from embryonic mouse brain (122). Physiological and mechanistic aspects of these neurotrophin actions have yet to be assessed.

Determination of the signaling mechanisms for acute growth factor RTK modulation of ion channel function is an area of great potential significance. Src has been implicated as a key signaling molecule that closely associates with and acutely regulates potassium (123) and NMDA receptor (124) channels (although it is not known whether these associations mediate RTK actions in these systems). It is important to also recognize that ligand binding to the classical seven-transmembrane-spanning α-helix receptor superfamily, and resultant GTP-binding protein activation, can act as a stimulator of RTK-associated signaling pathways (125,126). Thus, the possibilities for modulation of neuronal physiology become even more complex when the many levels of cross talk possible between these two signal transduction networks are considered. As the importance of growth factor RTKs and Ras/ERK signaling to cell growth and proliferation emerged, an intense effort arose to use this information for the development of therapeutics, particularly anticancer agents. In the future, perhaps a similar effort will be warranted in exploiting these signaling systems as neuropharmacological targets for affecting the behavior of both developing and mature neurons.

ACKNOWLEDGMENTS

The author thanks Michael D. Hilborn and Teresa Peña for helpful comments on this manuscript. The Whitehall Foundation and the National Science Foundation are gratefully acknowledged for their support.

REFERENCES

1. Johnson GL, Vaillancourt RR: Sequential protein kinase reactions controlling cell growth and differentiation. *Curr Opin Cell Biol* 1994;6:230–238.
2. Schlessinger J, Bar-Sagi D: Activation of Ras and other signaling pathways by receptor tyrosine kinases. *Cold Spring Harbor Symp Quant Biol* 1994;59:173–179.
3. Blumer KJ, Johnson GL: Diversity in function and regulation of MAP kinase pathways. *Trends Biochem Sci* 1994;19:236–240.
4. Marshall CJ: Specificity of receptor tyrosine kinase signaling: transient versus sustained extracellular signal-regulated kinase activation. *Cell* 1995;80:179–185.
5. Rozengurt E: Early signals in the mitogenic response. *Science* 1986;234:161–166.
6. Lewis RS, Cahalan MD: Potassium and calcium channels in lymphocytes. *Annu Rev Immunol* 1995; 13:623–653.
7. DeCoursey TE, Chandy KG, Gupta S, Cahalan MD: Mitogen induction of ion channels in murine T lymphocytes. *J Gen Physiol* 1987;89:405–420.
8. Lee SC, Sabath DE, Deutsch C, Prystowsky MB: Increased voltage-gated potassium conductance during interleukin 2-stimulated proliferation of a mouse helper T lymphocyte clone. *J Cell Biol* 1986;102:1200–1208.
9. Grissmer S, Nguyen AN, Cahalan MD: Calcium-activated potassium channels in resting and activated human T lymphocytes. Expression levels, calcium dependence, ion selectivity, and pharmacology. *J Gen Physiol* 1993;102:601–630.
10. DeCoursey TE, Chandy KG, Gupta S, Cahalan MD: Voltage-gated K^+ channels in human T lymphocytes: a role in mitogenesis? *Nature* 1984;307:465–468.
11. Schlichter L, Sidell N, Hagiwara S: K channels are expressed early in human T-cell development. *Proc Natl Acad Sci U S A* 1986;83:5625–5629.
12. Deutsch C, Krause D, Lee SC: Voltage-gated potassium conductance in human T lymphocytes stimulated with phorbol ester. *J Physiol (Lond)* 1986;372:405–423.
13. Partiseti M, Korn H, Choquet D: Pattern of potassium channel expression in proliferating B-lymphocytes depends upon the mode of activation. *J Immunol* 1993;151:2462–2470.
14. Price M, Lee SC, Deutsch C: Charybdotoxin inhibits proliferation and interleukin 2 production in human peripheral blood lymphocytes. *Proc Natl Acad Sci U S A* 1989;86:10171–10175.
15. Lin CS, Boltz RC, Blake JT, Nguyen M, Talento A, Fischer PA, Springer MS, Sigal NH, Slaughter RS, Garcia ML, Kaczorowski GJ, Koo GC: Voltage-gated potassium channels regulate calcium-dependent pathways involved in human T lymphocyte activation. *J Exp Med* 1993;177: 637–645.
16. Rader RK, Kahn LE, Anderson GD, Martin CL, Chinn KS, Gregory SA: T cell activation is regulated by voltage-dependent and calcium-activated potassium channels. *J Immunol* 1996;156: 1425–1430.
17. Gelfand EW: Or R. Charybdotoxin-sensitive, Ca^{2+}-dependent membrane potential changes are not involved in human T or B cell activation and proliferation. *J Immunol* 1991;147:3452–3458.
18. Premack BA, Gardner P: Role of ion channels in lymphocytes. *J Clin Immunol* 1994;11: 225–238.
19. Dubois JM, Rouzaire-Dubois B: Role of potassium channels in mitogenesis. *Prog Biophys Mol Biol* 1993;59:1–21.
20. Leonard RJ, Garcia ML, Slaughter RS, Reuben JP: Selective blockers of voltage-gated K^+ channels depolarize human T-lymphocytes: mechanism of the antiproliferative effect of charybdotoxin. *Proc Natl Acad Sci U S A* 1992;89:10094–10098.
21. Rane SG: A Ca^{2+}-activated K^+ current in *ras*-transformed fibroblasts is absent from nontransformed cells. *Am J Physiol* 1991;260:C104–C112.
22. Huang Y, Rane SG: Potassium channel induction by the Ras/Raf signal transduction cascade. *J Biol Chem* 1994;269:31183–31189.
23. Bruder JT, Heidecker G, Rapp UR: Serum-, TPA-, and Ras-induced expression from Ap-1/Ets-driven promoters requires Raf-1 kinase. *Genes Dev* 1992;6:545–556.
24. Huang Y, Rane SG: Single channel study of a Ca^{2+}-activated K^+ current associated with *ras*-induced cell transformation. *J Physiol (Lond)* 1993;461:601–618.
25. Mahaut-Smith MP, Schlichter LC: Ca^{2+}-activated K^+ channels in human B lymphocytes and rat thymocytes. *J Physiol* 1989;415:69–83.
26. Partiseti M, Choquet D, Diu A, Korn H: Differential regulation of voltage- and calcium-activated potassium channels in human B lymphocytes. *J Immunol* 1992;148: 3361–3368.
27. Sauvé R, Simoneau C, Monette R, Roy G: Single-channel analysis of the potassium permeability in

HeLa cancer cells: evidence for a calcium-activated potassium channel of small unitary conductance. *J Membr Biol* 1986;92: 269–282.
28. Enomoto K, Furuya K, Maeno T, Edwards C, Oka T: Oscillating activity of a calcium-activated K^+ channel in normal and cancerous mammary cells in culture. *J Membr Biol* 1991;119: 133–139.
29. Repp H, Draheim H, Ruland J, Seidel G, Beise J, Presek P, Dreyer F: Profound differences in potassium current properties of normal and Rous sarcoma virus-transformed chicken embryo fibroblasts. *Proc Natl Acad Sci U S A* 1993;90:3403-3407.
30. Draheim HJ, Repp H, Dreyer F: *Src*-transformation of mouse fibroblasts induces a Ca^{2+}-activated K^+ current without changing the T-type Ca^{2+} current. *Biochim Biophys Acta* 1995;1269:57–63.
31. Jones SM, Ribera AB: Overexpression of a potassium channel gene perturbs neural differentiation. *J Neurosci* 1994;14:2789–2799.
32. Wilson GF, Chiu SY: Mitogenic factors regulate ion channels in Schwann cells cultured from newborn rat sciatic nerve. *J Physiol (Lond)* 1993;470:501–520.
33. Gallo V, Zhou JM, Mcbain CJ, Wright P, Knutson PL, Armstrong RC: Oligodendrocyte progenitor cell proliferation and lineage progression are regulated by glutamate receptor-mediated K^+ channel block. *J Neurosci* 1996;16:2659–2670.
34. Patil N, Cox DR, Bhat D, Faham M, Myers RM, Peterson AS: A potassium channel mutation in *weaver* mice implicates membrane excitability in granule cell differentiation. *Nat Genet* 1995;11: 126–129.
35. Slesinger PA, Patil N, Liao YJ, Jan YN, Jan LY, Cox DR: Functional effects of the mouse *weaver* mutation on G protein-gated inwardly rectifying K^+ channels. *Neuron* 1996;16:321–331.
36. Kofuji P, Hofer M, Millen KJ, Millonig JH, Davidson N, Lester HA, Hatten ME: Functional analysis of the *weaver* mutant GIRK2 K^+ channel and rescue of *weaver* granule cells. *Neuron* 1996;16: 941–952.
37. Peña TL, Rane SG: The small conductance calcium-activated potassium channel regulates ion channel expression in C3H10T½ cells ectopically expressing the muscle regulatory factor MRF4. *J. Biol Chem* 1997;272:21909–21916.
38. Rhodes SJ, Konieczny SF: Identification of MRF4: a new member of the muscle regulatory factor gene family. *Genes Dev* 1989;3:2050–2061.
39. Hamann M, Widmer H, Baroffio A, Aubry J-P, Krause RM, Kaelin A, Bader CR: Sodium and potassium currents in freshly isolated and in proliferating human muscle satellite cells. *J Physiol (Lond)* 1994;475:305–317.
40. Wonderlin WF, Strobl JS: Potassium channels, proliferation and G1 progression. *J Memb Biol* 1996; 154:91–107.
41. Weiss A, Littman DR: Signal transduction by lymphocyte antigen receptors. *Cell* 1994;76:263–274.
42. Hawley RG, Fong AZ, Ngan BY, Hawley TS: Hematopoietic transforming potential of activated *ras* in chimeric mice. *Oncogene* 1995;11:1113–1123.
43. Sawyers CL, McLaughlin J, Witte ON: Genetic requirement for Ras in the transformation of fibroblasts and hematopoietic cells by the Bcr-Abl oncogene. *J Exp Med* 1995;181:307–313.
44. Kaczorowski GJ, Koo GC: Lymphocyte ion channels as a target for immunosuppression. *Perspect Drug Discovery Des* 1994;2:233–248.
45. Wonderlin WF, Woodfork KA, Strobl JS: Changes in membrane potential during the progression of MCF-7 human mammary tumor cells through the cell cycle. *J Cell Physiol* 1995;165:177–185.
46. Woodfork KA, Wonderlin WF, Peterson VA, Strobl JS: Inhibition of ATP-sensitive potassium channels causes reversible cell-cycle arrest of human breast cancer cells in tissue culture. *J Cell Physiol* 1995;162:163–171.
47. Day ML, Pickering SJ, Johnson MH, Cook DI: Cell-cycle control of a large-conductance K^+-channel in mouse early embryos. *Nature* 1993;365:560–562.
48. Cole K, Kohn E: Calcium-mediated signal transduction: biology, biochemistry, and therapy. *Cancer Metastasis Rev* 1994;13:31–44.
49. Magni M, Meldolesi J, Pandiella A: Ionic events induced by epidermal growth factor. *J Biol Chem* 1991;266:6329–6335.
50. Rosen LB, Ginty DD, Weber MJ, Greenberg ME: Membrane depolarization and calcium influx stimulate MEK and MAP kinase via activation of Ras. *Neuron* 1994;12:1207–1221.
51. Farnsworth CL, Freshney NW, Rosen LB, Ghosh A, Greenberg ME, Feig LA: Calcium activation of Ras mediated by neuronal exchange factor Ras-GRF. *Nature* 1995;376:524–527.
52. Rusanescu G, Qi HQ, Thomas SM, Brugge JS, Halegoua S: Calcium influx induces neurite growth through a *src–ras* signaling cassette. *Neuron* 1995;15:1415–1425.

53. Huang S, Maher VM, McCormick JJ: Extracellular Ca^{2+} stimulates the activation of mitogen-activated protein kinase and cell growth in human fibroblasts. *Biochem J* 1995;310:881–885.
54. Chen CF, Corbley MJ, Roberts TM, Hess P: Voltage-sensitive calcium channels in normal and transformed 3T3 fibroblasts. *Science* 1988;239:1024–1026.
55. Estacion M. Characterization of ion channels seen in subconfluent human dermal fibroblasts. *J Physiol (Lond)* 1991;436:579–601.
56. Wang Z, Estacion M, Mordan LJ: Ca^{2+} influx via T-type channels modulates PDGF-induced replication of mouse fibroblasts. *Am J Physiol* 1993;265:C1239–C1246.
57. Estacion M, Mordan LJ: Competence induction by PDGF requires sustained calcium influx by a mechanism distinct from storage-dependent calcium influx. *Cell Calcium* 1993;14:439–454.
58. Mogami H, Kojima I: Stimulation of calcium entry is prerequisite for DNA synthesis induced by platelet-derived growth factor in vascular smooth muscle cells. *Biochem Biophys Res Commun* 1993;196:650–658.
59. Kojima I, Mogami H, Shibata H, Ogata E: Role of calcium entry and protein kinase C in the progression activity of insulin-like growth factor-I in Balb/c 3T3 cells. *J Biol Chem* 1993;268: 10003–10006.
60. Putney JW, Bird GStJ: The signal for capacitative calcium entry. *Cell* 1993;75:199–201.
61. Clapham DE: TRP is cracked but is CRAC TRP? *Neuron* 1996;16:1069–1072.
62. Merritt JE, Jacob R, Hallam TJ: Use of manganese to discriminate between calcium influx and mobilization from internal stores in stimulated human neutrophils. *J Biol Chem* 1989;264:1522–1527.
63. Sage SO, Merritt JE, Hallam TJ, Rink TJ: Receptor-mediated calcium entry in fura-2-loaded human platelets stimulated with ADP and thrombin. Dual-wavelengths studies with Mn^{2+}. *Biochem J* 1989; 258:923–926.
64. Mertz LM, Baum BJ, Ambudkar IS: Refill status of the agonist-sensitive Ca^{2+} pool regulates Mn^{2+} influx into parotid acini. *J Biol Chem* 1990;265:15010–15014.
65. Alvarez J, Montero M, Garcia-Sancho J: Cytochrome P-450 may link intracellular Ca^{2+} stores with plasma membrane Ca^{2+} influx. *Biochem J* 1991;274:193–197.
66. Randriamampita C, Tsien RY: Emptying of intracellular Ca^{2+} stores releases a novel small messenger that stimulates Ca^{2+} influx. *Nature* 1993;364:809–814.
67. Premack BA, McDonald TV, Gardner P: Activation of Ca^{2+} current in Jurkat T cells following the depletion of Ca^{2+} stores by microsomal Ca^{2+}-ATPase inhibitors. *J Immunol* 1994;152:5226–5240.
68. Hoth M, Penner R: Depletion of intracellular calcium stores activates a calcium current in mast cells. *Nature* 1993;355:353–356.
69. Hoth M, Penner R: Calcium release-activated calcium current in rat mast cells. *J Physiol (Lond)* 1993;465:359–386.
70. Parekh AB, Terlau H, Stuhmer W: Depletion of $InsP_3$ stores activates a Ca^{2+} and K^+ current by means of a phosphatase and a diffusible messenger. *Nature* 1993;364:814–818.
71. Lee K-M, Toscas K, Villereal ML: Inhibition of bradykinin- and thapsigargin-induced Ca^{2+}-entry by tyrosine kinase inhibitors. *J Biol Chem* 1993;268:9945–9948.
72. Razani-Boroujerdi S, Partridge LD, Sopori ML: Intracellular calcium signaling induced by thapsigargin in excitable and inexcitable cells. *Cell Calcium* 1994;16:467–474.
73. Matsunaga H, Ling BN, Eaton DC: Ca^{2+}-permeable channel associated with platelet-derived growth factor receptor in mesangial cells. *Am J Physiol* 1994;267:C456–C465.
74. Matsunaga H, Nishimoto I, Kojima I, Yamashita N, Kurokawa K, Ogata E: Activation of a calcium-permeable cation channel by insulin-like growth factor II in BALB/c 3T3 cells. *Am J Physiol* 1988; 255:C442–C446:
75. Munaron L, Distasi C, Carabelli V, Baccino FM, Bonelli G, Lovisolo D: Sustained calcium influx activated by basic fibroblast growth factor in Balb-c 3T3 fibroblasts. *J Physiol (Lond)* 1995;484: 557–566.
76. Frace AM, Gargus JJ: Activation of single-channel currents in mouse fibroblasts by platelet-derived growth factor. *Proc Natl Acad Sci U S A* 1989;86:2511–2515.
77. Jung F, Selvaraj S, Gargus JJ: Blockers of platelet-derived growth factor-activated nonselective cation channel inhibit cell proliferation. *Am J Physiol* 1992;262:C1464–C1470.
78. Huang X, Morielli AD, Peralta, EG: Tyrosine kinase-dependent suppression of a potassium channel by the G-protein-coupled m1 muscarinic acetylcholine receptor. *Cell* 1993;75:1145–1156.
79. Catarsi S, Ching S, Merz DC, Drapeau P: Tyrosine phosphorylation during synapse formation between identified leech neurons. *J Physiol (Lond)* 1995;485:775–786.

80. Lev S, Moreno H, Martinez R, Canoll P, Peles E, Musacchio JM, Plowman GD, Rudy B, Schlessinger J: Protein tyrosine kinase PYK2 involved in Ca^{2+}-induced regulation of ion channel and MAP kinase functions. *Nature* 1995;376:737–745.
81. Barbiero G, Munaron L, Antoniotti S, Baccino FM, Bonelli G, Lovisolo D: Role of mitogen-induced calcium influx in the control of the cell cycle in Balb-c 3T3 fibroblasts. *Cell Calcium* 1995;18: 542–556.
82. Timmerman LA, Clipstone NA, Ho SN, Northrop JP, Crabtree GR: Rapid shuttling of NF-AT in discrimination of Ca^{2+} signals and immunosuppression. *Nature* 1996;383:837–840.
83. Dolmetsch RE, Lewis RS, Goodnow CC, Healy JI: Differential activation of transcription factors induced by Ca^{2+} response amplitude and duration. *Nature* 1997;386:855–858.
84. Stehno-Bittel L, Perez-Terzic C, Clapham DE: Diffusion across the nuclear envelope inhibited by depletion of the nuclear Ca^{2+} store. *Science* 1995;270:1835–1838.
85. Perez-Terzic C, Pyle J, Jaconi M, Stehno-Bittel L, Clapham DE: Conformational states of the nuclear pore complex induced by depletion of nuclear Ca^{2+} stores. *Science* 1996;273:1875–1877.
86. Davies AM: The role of neurotrophins in the developing nervous system. *J Neurobiol* 1994;25: 1334–1348.
87. Collin C, Papageorge AG, Lowy DR, Alkon DL: Early enhancement of calcium currents by H-ras oncoproteins injected into *Hermissenda* neurons. *Science* 1990;250:1743–1745.
88. Yatani A, Okabe K, Polakis P, Halenbeck R, McCormick F, Brown AM: *ras* p21 and GAP inhibit coupling of muscarinic receptors to atrial K^+ channels. *Cell* 1990;61:769–776.
89. Martin GA, Yatani A, Clark R, Conroy L, Polakis P, Brown AM, McCormick F: GAP domains responsible for *ras* p21-dependent inhibition of muscarinic atrial K^+ channel currents *Science* 1992; 255:192–195.
90. Flamm RE, Birnberg NC, Kaczmarek LK: Transfection of activated *ras* into an excitable cell line (AtT-20) alters tetrodotoxin sensitivity of voltage-dependent sodium current. *Pfluegers Arch* 1990; 416:120–125.
91. Estacion M: Inhibition of voltage-dependent Na^+ current in cell-fusion hybrids containing activated c-Ha-*ras*. *J Memb Biol* 1990;113:169–175.
92. Hemmick LM, Perney TM, Flamm RE, Kaczmarek LK, Birnberg NC: Expresson of the H-*ras* oncogene induces potassium conductance and neuron-specific potassium channel mRNAs in the AtT20 cell line. *J Neurosci* 1992;12:2007–2014.
93. Bar-Sagi D, Feramisco JR: Microinjection of the *ras* oncogene protein into PC12 cells induces morphological differentiation. *Cell* 1985;42:841–848.
94. Noda M, Ko M, Ogura A, Liu DG, Amano T, Takano T, Ikawa Y: Sarcoma viruses carrying *ras* oncogenes induce differentiation-associated properties in a neuronal cell line. *Nature* 1985;318: 73–75.
95. Wood KW, Qi HQ, Darcangelo G, Armstrong RC, Roberts TM, Halegoua S: The cytoplasmic *raf* oncogene induces a neuronal phenotype in PC12 cells: a potential role for cellular *raf* kinases in neuronal growth factor signal transduction. *Proc Natl Acad Sci U S A* 1993;90:5016–5020.
96. Thomas SM, Hayes M, D'Arcangelo G, Armstrong RC, Meyer BE, Zilberstein A, Brugge JS, Halegoua S: Induction of neurite outgrowth by v-*src* mimics critical aspects of nerve growth factor-induced differentiation. *Mol Cell Biol* 1991;11:4739–4750.
97. Szeberényi J, Cai H, Cooper GM: Effect of a dominant inhibitory Ha-*ras* mutation on neuronal differentiation of PC12 cells. *Mol Cell Biol* 1990;10:5324–5332.
98. Cowley S, Paterson H, Kemp P, Marshall CJ: Activation of MAP kinase kinase is necessary and sufficient for PC12 differentiation and for transformation of NIH 3T3 cells. *Cell* 1994;77:841–852.
99. Vaillancourt RR, Heasley LE, Zamarripa J, Storey B, Valius M, Kazlauskas A, Johnson GL: Mitogen-activated protein kinase activation is insufficient for growth factor receptor-mediated PC12 cell differentiation. *Mol Cell Biol* 1995;15:3644–3653.
100. D'Arcangelo G, Paradiso K, Shepherd D, Brehm P, Halegoua S, Mandel, G: Neuronal growth factor regulation of two different sodium channel types through distinct signal transduction pathways. *J Cell Biol* 1993;122:915–921.
101. Fanger GR, Jones JR, Maue RA: Differential regulation of neuronal sodium channel expression by endogenous and exogenous tyrosine kinase receptors expressed in rat pheochromocytoma cells. *J Neurosci* 1995;15:202–213.
102. Toledoaral JJ, Brehm P, Halegoua S, Mandel G: A single pulse of nerve growth factor triggers long-term neuronal excitability through sodium channel gene induction. *Neuron* 1995;14:607–611.

103. Fanger GR, Erhardt P, Cooper GM, Maue R: *ras*-Independent induction of rat brain type II sodium channel expression in nerve growth factor-treated PC12 cells. *J Neurochem* 1993;61:1977–1980.
104. Pollock JD, Rane SG: p21^{ras} Signaling is necessary but not sufficient to mediate neurotrophin induction of calcium channels in PC12 cells. *J Biol Chem* 1996;271:8008–8014.
105. D'Arcangelo G, Halegoua S: A branched signaling pathway for nerve growth factor is revealed by src-mediated, *ras*-mediated, and *raf*-mediated gene inductions. *Mol Cell Biol* 993;13:3146–3155.
106. Fanger GR, Vaillancourt RR, Heasley LE, Montmayeur JPR, Johnson GL, Maue RA: Analysis of mutant platelet-derived growth factor receptors expressed in PC12 cells identifies signals governing sodium channel induction during neuronal differentiation. *Mol Cell Biol* 1997;17:89–99.
107. Usowicz MM, Porzig H, Becker C, Reuter H: Differential expression by nerve growth factor of two types of Ca^{2+} channels in rat phaeochromocytoma cell lines. *J Physiol (Lond)* 1990;426:95–116.
108. Rausch DM, Lewis DL, Barker JL, Eiden LE: Functional expression of dihydropyridine-insensitive calcium channels during PC12 cell differentiation by nerve growth factor (NGF), oncogenic *ras*, or *src* tyrosine kinase. *Cell Mol Neurobiol* 1990;10:237–255.
109. Lewis DL, Deaizpurua HJ, Rausch DM: Enhanced expression of Ca^{2+} channels by nerve growth factor and the v-src oncogene in rat phaeochromocytoma cells. *J Physiol (Lond)* 1993;465: 325–342.
110. Rane SG, Pollock JD: Fibroblast growth factor-induced increases in calcium currents in the PC12 pheochromocytoma cell line are tyrosine phosphorylation dependent. *J Neurosci Res* 1994;38: 590–598.
111. Liévano A, Bolden A, Horn R: Calcium channels in excitable cells: divergent genotypic and phenotypic expression of α_1-subunits. *Am J Physiol* 1994;267:C411–C424.
112. Liu HY, Felix R, Gurnett CA, Dewaard M, Witcher DR, Campbell KP: Expression and subunit interaction of voltage-dependent Ca^{2+} channels in PC12 cells. *J Neurosci* 1996;16:7557–7565.
113. Hilborn MD, Rane SG, Pollock JD: EGF in combination with depolarization or cAMP produces morphological but not physiological differentiation in PC12 cells. *J Neurosci Res* 1996;47:16–26.
114. Rosen LB, Greenberg ME: Stimulation of growth factor receptor signal transduction by activation of voltage-sensitive calcium channels. *Proc Natl Acad Sci U S A* 1996;93:1113–1118.
115. Williams EJ, Furness J, Walsh FS, Doherty P: Characterisation of the second messenger pathway underlying neurite outgrowth stimulated by FGF. *Development* 1994;120:1685–1693.
116. Kang H, Schuman EM: Long-lasting neurotrophin-induced enhancement of synaptic transmission in the adult hippocampus. *Science* 1995;267:1658–1662.
117. Figurov A, Pozzo-Miller LD, Olafsson P, Wang T, Lu B: Regulation of synaptic responses to high-frequency stimulation and LTP by neurotrophins in the hippocampus. *Nature* 1996;381: 706–709.
118. Prakash N, Cohen-Cory S, Frostig RD: Rapid and opposite effects of BDNF and NGF on the functional organization of the adult cortex *in vivo*. *Nature* 1996;381:702–706.
119. Lohof AM, Ip NY, Poo MM: Potentiation of developing neuromuscular synapses by the neurotrophins NT-3 and BDNF. *Nature* 1993;363:350–353.
120. Stoop R, Poo MM: Synaptic modulation by neurotrophic factors: differential and synergistic effects of brain-derived neurotrophic factor and ciliary neurotrophic factor. *J Neurosci* 1996;16:3256–3264.
121. Wildering WC, Lodder JC, Kits KS, Bulloch AG: Nerve growth factor (NGF) acutely enhances high-voltage-activated calcium currents in molluscan neurons. *J Neurophysiol* 1995;74:2778–2781.
122. Holm NR, Christophersen P, Olesen SP, Gammeltoft S: Activation of calcium-dependent potassium channels in rat brain neurons by neurotrophin-3 and nerve growth factor. *Proc Natl Acad Sci U S A* 1997;94:1002–1006.
123. Holmes TC, Fadool DA, Ren RB, Levitan IB: Association of Src tyrosine kinase with a human potassium channel mediated by SH3 domain. *Science* 1996;274:2089–2091.
124. Yu XM, Askalan R, Keil GJ, Salter MW: NMDA channel regulation by channel-associated protein tyrosine kinase Src. *Science* 1997;275:674–678.
125. Crespo P, Xu NZ, Simonds WF, Gutkind JS: Ras-dependent activation of MAP kinase pathway mediated by G-protein beta gamma subunits. *Nature* 1994;369:418–420.
126. Luttrell LM, Hawes BE, Vanbiesen T, Luttrell DK, Lansing TJ, Lefkowitz RJ: Role of c-Src tyrosine kinase in G protein-coupled receptor and G $\beta\gamma$ subunit-mediated activation of mitogen-activated protein kinases. *J Biol Chem* 1996; 271:19443–19450.

Part II

Closely Associated Proteins

Advances in Second Messenger and Phosphoprotein Research, Vol. 33

1040-7952/99 $30.00

6

Voltage-Dependent Modulation of N-Type Calcium Channels: Role of G Protein Subunits

Stephen R. Ikeda* and Kathleen Dunlap†

**Laboratory of Molecular Physiology, Guthrie Research Institute, Sayre, Pennsylvania 18840, and †Departments of Physiology and Neuroscience, Tufts University School of Medicine, Boston, Massachusetts 02111*

INTRODUCTION

Voltage-gated Ca^{2+} channels (VGCC) are highly modulated by a wide variety of substances acting through a number of diverse signaling pathways. Here we examine ways in which N-type Ca^{2+} channels are modulated by heterotrimeric G proteins. A number of reviews cover heterotrimeric G proteins (1–4), VGCCs (5–8), and modulation (9–14), and no attempt is made to recapitulate this information (except as necessary for background). Instead, we focus on recent advances in our understanding of how heterotrimeric G proteins interact with N-type Ca^{2+} channels to produce a specific form of modulation that is voltage dependent, membrane delimited, and likely mediated by direct interaction of G protein and Ca^{2+} channel subunits. Such a mechanism raises new questions and suggests new avenues for investigating the physiological roles of this modulatory pathway.

N-TYPE CHANNELS: IDENTIFICATION, STRUCTURE, AND PHYSIOLOGICAL ROLE

N-type Ca^{2+} channels were initially identified in cultures of primary sensory neurons on the basis of electrophysiological properties, most notably the membrane potential range over which inactivation occurs (15–17). However, this criterion, used in numerous early studies, did not always adequately isolate the current component arising from the activation of N-type channels from other Ca^{2+} channels that activated and inactivated over a similar voltage range. The discovery of ω-CTx GVIA, a peptide isolated from the venom of the cone snail *Conus geographicus,* overcame this inadequacy by providing a pharmacological means of identifying N-type Ca^{2+}

channels in a variety of preparations, including ones in which membrane potential was not under experimental control (e.g., broken cell preparations). Potently, selectively, and irreversibly, ω-CTx GVIA blocks a single subtype of voltage-gated Ca^{2+} channel with properties identical to those of N-type channels originally described (18,19). Thus, a Ca^{2+} current component that is irreversibly blocked by ω-CTx GVIA is operationally defined as having originated from N-type channels. Using this criterion, it has been shown that the majority (65–95%) of Ca^{2+} current in peripheral sympathetic and sensory neurons arises from N-type channels (20–22). Conversely, neurons of the CNS tend to possess a more heterogeneous population of Ca^{2+} channel subtypes where N-type channels typically account for only 20–30% of the total Ca^{2+} current (23,24). N-type channels are not found in skeletal, smooth, or cardiac muscles (25).

Biochemical purification and molecular cloning techniques have led to advances in our understanding of the structure and function of Ca^{2+} channels. They comprise at least three subunits: α_1, β, and α_2/δ (7). The skeletal muscle Ca^{2+} channel has a γ subunit as well. The α_1 subunit, denoted α_{1B} for the N-type channel, is the primary determinant of the channel phenotype and consists of four nonidentical repeats—each repeat containing six putative transmembrane segments (TMS) homologous to those found in other voltage-gated channels (25,26). By analogy with K^+ channels, the voltage sensor is thought to reside in a group of positively charged residues on TMS 4, whereas the pore, which controls ion permeation and selectivity, is thought to comprise a region (S5) between TM5 and TM6. The binding site for ω-CTx GVIA is also located on the α_{1B} subunit (27). Thus, the characteristic properties of the N-type channel (i.e., gating, permeation, and pharmacology) are specified by the α_1 subunit. Several Ca^{2+} channel β subunits have been cloned (28); which of these associates with α_{1B} *in situ* has not been defined, although $Ca_{\beta 3}$ has been implicated (29). Ca^{2+} channel β subunits are cytosolic proteins that interact with the α_{1B} subunit on the intracellular linker between repeats I and II (30). The α_{1B}–β interaction influences the level of functional expression (31), the voltage over which the channel activates (32), and G protein mediated modulation (33–35). The α_2/δ subunit is encoded by a single gene and then post-translationally processed into the α_2 and δ subunits. The influence of these subunits has been less well studied, but coexpression of α_2/δ along with α_{1B} and β increases functional Ca^{2+} channel expression (31).

In regard to physiological roles, Ca^{2+} channels are considered unique among voltage-gated ion channels because one of their primary functions appears to be transducing changes in membrane potential into cellular signals rather than producing membrane potential changes per se. In neurons, for example, the magnitude of current produced by the flux of Ca^{2+} through VGCCs is small compared with that produced by Na^+ flowing through Na^+ channels. Hence, the contribution of VGCCs to action potential trajectory is likely to be minimal. Their contribution to Ca^{2+}-dependent effector responses, however, is significant. Ca^{2+} influx through VGCCs raises intracellular Ca^{2+} concentration $[Ca^{2+}]_i$ into the micromolar range, where it initiates a variety of cellular responses by way of Ca^{2+}-binding proteins with micromolar affinities for Ca^{2+}. The consequences of Ca^{2+} entry include triggering the exocytotic release of neurotransmitters from presynaptic nerve terminals activating a variety of enzymes [such as protein kinase C (PKC), adenylyl cyclase (AC), and Ca^{2+}/calmodulin-dependent protein kinase], opening Ca^{2+}-activated K^+ and Cl^- channels, and initiating gene transcription (36).

A particularly important physiological role of N-type channels involves supplying intracellular Ca^{2+} to the presynaptic apparatus responsible for exocytotic release of classical and peptide neurotransmitters (8). In the peripheral nervous system, N-type channels play a dominant role in the release of neurotransmitters. This has been demonstrated by using ω-CTx GVIA to inhibit release of norepinephrine (NE) (37) and substance P (38) from neurons in culture, synaptic transmission between cocultured peripheral neurons and their targets (39), and nerve stimulation mediated smooth muscle contraction in various sympathetically innervated end organs (40,41). In the CNS, the situation is more complex with both N-type and P/Q-type channels mediating synaptic transmission (42–44); evidence suggests that the two types are colocalized at synaptic release sites (43,45,46), since they coprecipitate with proteins comprising the secretion apparatus (47,48).

MODULATION: VOLTAGE DEPENDENT VERSUS VOLTAGE INDEPENDENT

N-type Ca^{2+} channels are modulated by most known and putative neurotransmitters (9). In almost all cases, the modulation is inhibitory and the response mediated by G protein coupled receptors of the seven-transmembrane family. Despite these similarities, discrete modulatory pathways have been identified based on electrophysiological and biochemical criteria (10,11). Modulation can be functionally divided into two main groups based on whether the Ca^{2+} current inhibition is voltage dependent (VD) or voltage independent (VI). Voltage dependence refers to the relationship between the membrane potential used to evoke the Ca^{2+} current (often referred to as test potential) and the fractional inhibition of Ca^{2+} current amplitude produced by a substance at that potential. Substances that produce VD inhibition generally produce large effects at moderate depolarizations (−10 to 10 mV), which decrease at more positive membrane potentials (49–52), whereas with VI inhibition, the magnitude of inhibition is not greatly influenced by membrane potential. Although the classification is seemingly arbitrary, it is becoming clear that the large number of substances that produce VD inhibition share a number of characteristics, and thus probably operate through a single molecular mechanism. In contrast, the characteristics of VI inhibition are more heterogeneous, involving a number of discrete signaling pathways (53–57).

VD inhibition is by the far the most common (and thus, the most intensely studied) form of N-type Ca^{2+} channel modulation. Examples of neurotransmitters that produce VD inhibition include norepinphrine, somatostatin, prostaglandins, adenosine, acetylcholine, γ-aminobutyric acid (GABA), opioids, serotonin, glutamate, and vasoactive intestinal polypeptide (VIP)/secretin (9,11). With the exception of VIP/secretin (52), VD inhibition is mediated by a G protein that is sensitive to pertussis toxin (PTX). Efforts to implicate additional signaling elements subsequent to G protein activation (e.g., cyclic nucleotides, protein kinases, Ca^{2+}, arachidonic acid, or phospholipases) have been largely unsuccessful, leading to the hypothesis that modulation occurs through a direct interaction of an activated G protein subunit with the N-type channel. This idea is corroborated by the finding that VD inhibition does not seem to involve small diffusible molecules (i.e., the pathway is membrane delimited, as discussed later).

VI inhibition is less common than VD inhibition and consequently has been studied in detail for only a few substances. However, the methodology used in the majority of studies, conventional whole-cell patch–clamp recording, favors preservation of signal transduction pathways with the minimum number of components and fewest metabolically labile steps. Thus, the contribution of pathways that result in VI inhibition may be underestimated. Unlike the case for VD inhibition, few unifying themes have emerged, and there appear to be a number of diverse pathways involved. In rat sympathetic neurons, substance P, oxotremorine-M, and angiotensin II produce VI inhibition through a PTX-insensitive pathway. However, the effect of substance P is membrane delimited (55), while the angiotensin II (57) and oxotremorine-M (acting through M_1 AChR) mediated inhibition is produced by a diffusible messenger (53). In chick sensory neurons, NE and GABA both produce VI inhibition (as well as VD inhibition) via a PTX-sensitive pathway (56,58). The effect of NE is mediated by a PKC-dependent pathway (54,59) whereas the GABA effect may be mediated by a tyrosine kinase dependent pathway (60). From these examples, it is clear that VI inhibition is produced by a number of converging pathways that share little commonality. Thus, the classification of a pathway as being VI has few mechanistic implications other than to distinguish it from the VD pathway.

VOLTAGE-DEPENDENT MODULATION

The remainder of the chapter is restricted to VD-dependent inhibition and is loosely organized into two sections. The first section covers established observations that form the basis for our current ideas about the mechanism of VD inhibition. Most of this work has focused on N-type channels natively expressed in neurons derived from the peripheral nervous system (e.g., autonomic and sensory ganglia) and confirmed in a number of laboratories. The second section describes more recent work aimed at extending our understanding of the mechanism of VD inhibition to the molecular level. Most of these studies capitalize on the knowledge gained from the cloning of the molecular elements that comprise the signaling pathway and the ability to heterologously express these elements, both wild type and mutant, in a variety of systems. This area of research is very dynamic, and many of the findings await confirmation by other investigators before the generality of the results can be determined.

Voltage Dependence Arises from a Shift in the Voltage Range over Which the N-Type Channel Activates

In 1989 Bean (61) proposed a theory according to which N-type channels could exist in either a "willing" or a "reluctant" mode, with unmodulated N-type channels existing primarily in a "willing" mode and, thus, able to open following modest depolarizations from the resting potential. During application of neurotransmitters (NE, dynorphin, somatostatin), a fraction of the channels adopt a "reluctant" mode in which channel opening is still possible but a much stronger depolarization is required. Consequently, the net effect of a neurotransmitter at modest depolarizations would be inhibition of current, since the fraction of channels in the "reluctant" mode would not open. During a strong depolarization, however, inhibition would be minimal inasmuch as channels in both the willing and reluctant modes would open. Thus, VD inhibition occurs because of a shift in the voltage range over which N-type channels open rather than a block of the channel pore or a decrease in the number of functional channels. Experimental evidence for the theory was provided by comparing activation curves (obtained from tail currents during deactivation of the N-type channel) in the absence or presence of neurotransmitters. During modulation, the activation curve contains a second component with a half-activation voltage (V_H) that is shifted to depolarized potentials.

A second aspect of the Bean theory is that the "willing" and "reluctant" modes of the channel can interconvert. The action of neurotransmitters, then, is to shift the equilibrium from the willing mode to the reluctant mode. The original model connected the two modes through a single closed state of the channel. Further refinement of the model allowed interconversion through the open state and subsequently, through multiple closed states of the channel (62–66). The significance of this modification is that it provides an explanation for two extremely distinctive characteristics of VD inhibition, namely, "kinetic slowing" and "facilitation." When a Ca^{2+} current, evoked at a single potential, is measured before and after application of a neurotransmitter (which produces VD inhibition), it is apparent that the rising phase of the current is significantly altered (Fig. 1A). Instead of the relatively rapid monophasic rise in current amplitude typical of control currents, the activation phase in the presence of neurotransmitter becomes biphasic, with a pronounced slow component. This phenomenon was noted in early studies (67,68) of Ca^{2+} channel modulation and is sometimes referred to as kinetic slowing (56). The net result is a time-dependent inhibition—that is, current inhibition is greater when measured earlier in the voltage pulse than later. The relief of inhibition can be rationalized as a voltage-dependent reequilibration of channels from the reluctant to willing mode during the voltage pulse (61).

Another way to examine the reequilibration of channels from the reluctant to willing mode is by using a double-pulse voltage protocol (Fig. 1B). The protocol consists of two identical voltage pulses (typically between 0 and +20 mV), termed test pulses, separated by a large depolarizing conditioning pulse (typically between +80 and +120 mV). In the absence of modulation, the conditioning pulse has only

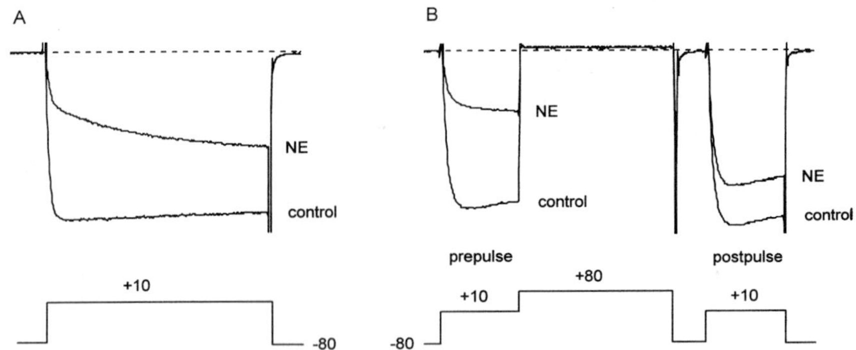

FIG. 1. Characteristics of voltage-dependent N-type Ca^{2+} channel modulation. **(A)** Kinetic slowing. Voltage–clamp recordings of whole-cell Ca^{2+} currents from adult rat sympathetic neurons. Superimposed current traces were produced in the absence (lower trace) or presence (upper trace) of 10 μM NE with the voltage protocol shown. Note that in the presence of NE, the current activation displays kinetic slowing—a biphasic activation phase with a prominent slow component. The relief of block is believed to result from channels converting from the reluctant to the willing mode during the 70-ms depolarizing pulse. **(B)** Facilitation. Superimposed Ca^{2+} current traces evoked with the "double-pulse" voltage protocol illustrated below in the absence (lower traces) or presence (upper traces) of NE (10 μM). Facilitation is defined as the ratio of Ca^{2+} current amplitude determined from the test pulse (+10 mV) occurring after (postpulse) and before (prepulse) the conditioning pulse to +80 mV. Note that both the current inhibition and kinetic slowing produced by NE are largely (although not completely) relieved by the conditioning pulse.

small effects on the current evoked during a subsequent test pulse (Fig. 1B, lower trace). However, when the N-type channels are inhibited, the conditioning pulse produces two prominent effects. First, the amplitude of the postpulse current is greatly increased, and second, the kinetics of activation are restored (Fig. 1B, upper trace). Thus, the effect of the conditioning pulse is to cause reequilibration of channels from the reluctant to the willing mode, thereby reversing, at least partially, the modulation. Facilitation was noted in early studies (69,70) and became a standard protocol for characterizing VD inhibition following the detailed analysis of Elmslie et al. (62). It should be noted that other forms of Ca^{2+} channel facilitation have been observed (71).

The significance of kinetic slowing and facilitation in terms of studying VD N-type Ca^{2+} channel inhibition is fourfold. First, these phenomenan provide a rapid and convenient signature of the VD pathway at both the whole-cell (50–52) and single channel levels (65,66). This is particularly useful when only a small component of the total Ca^{2+} current is modulated or when multiple forms of inhibition (i.e., VD and VI) are produced by the same agonist. Second, analysis of the detailed kinetics underlying these processes helps provide insight into mechanisms of VD modulation. The reblocking kinetics (i.e., the rate at which facilitation disappears following the conditioning pulse) have been shown to be dependent on agonist concentration, consistent with a mechanism in which reassociation of the modulating species and the channel occurs following the conditioning pulse (64,72,73). Third, tonic modulation through the VD pathway can be readily identified by means of basal facilitation (as a measure of VD current inhibition). Based on this criterion, modulation in some cells occurs in the absence of overt agonists or G protein activators (74,75). Fourth, kinetic slowing and facilitation may have functional implications. During a single action potential, modulated N-type channels have little time to convert from the reluctant to the willing mode. Conversely, a train of action potentials may produce some relief of Ca^{2+} channel inhibition, much as does a depolarizing conditioning pulse under voltage–clamp conditions (76–78). Such a mechanism might play a role in frequency-dependent alterations in neurotransmitter release.

Heterotrimeric G Proteins Couple Receptors to the VD Response

The coupling of neurotransmitter receptors to N-type channels via heterotrimeric G proteins has been investigated using (1) guanine nucleotide analogs that target G proteins in general, (2) bacterial toxins and chemical agents that target a specific class or subset of Gα subunits, (3) reconstitution with purified Gα subunits, (4) antibodies that functionally inhibit specific Gα subunits, and (5) antisense targeting of specific Gα subunits.

A general means of implicating G proteins in a signaling pathway is to introduce nonhydrolyzable or hydrolysis-resistant guanine nucleotide analogs into the cell via the patch electrode and to determine the effect on neurotransmitter-mediated modulation. Guanosine 5'-*O*-(2-thiodiphosphate) (GDPβS), a hydrolysis-resistant analog of GDP, acts as a competitive inhibitor of GTP binding (79) to Gα subunits and attenuates effects of neurotransmitters that utilize G protein dependent pathways (80). GDPβS is particularly useful for establishing G protein involvement when more specific agents are not available. Guanosine 5'-*O*-(3-thiotriphosphate) (GTPγS) and guanylylimidodiphosphate (GppNHp) are hydrolysis-resistant analogs of GTP. Since G protein actions are normally terminated by GTP hydrolysis, these analogs induce a persistently activated state.

Two effects of internally perfusing neurons with GTPγS or GppNHp have been observed. First, introduction of GTPγS or GppNHp into the cell may have minimal effects on basal N-type channel behavior but produce irreversible VD inhibition following treatment with an agonist (81). Second, spontaneous modulation may occur shortly after the whole-cell configuration has been attained, as evidenced by the progressive onset of kinetic slowing and facilitation (49,74,82). Presumably, whether spontaneous modulation occurs is dependent on the basal GDP-GTP exchange rate of the cell under study. An advantage of using GTPγS or GppNHp to activate G proteins is that receptor coupling is bypassed. For example, the observation that modulation produced by GTPγS or GppNHp produces kinetic slowing and facilitation rules out VD binding of ligands to the receptor as a mechanism underlying these effects.

Pertussis toxin provides a means of targeting specific subsets of G proteins. PTX ADP-ribosylates a C-terminal cysteine residue on G protein α subunits of the α_i (α_{i1}, α_{i2}, and α_{i3}) and α_o (α_{oA} and α_{oB}) subclasses, thereby preventing coupling of the heterotrimer (Gαβγ) to receptors and disrupting receptor-mediated modulation (1). Holz et al. (80) were the first to use PTX to demonstrate that a G protein was involved in neurotransmitter-mediated Ca^{2+} channel modulation. Later studies demonstrated that nearly all neurotransmitters that produce VD inhibition of N-type channels do so via a PTX-sensitive mechanism (11). A potential obstacle to using PTX with intact cells is the preincubation period (usually >3 h) required for the holotoxin to bind to the cell surface and the catalytic subunit to enter the cytoplasm (80). The long preincubation time (typically overnight) precludes the use of PTX on cells of limited longevity, such as acutely isolated CNS neurons (83), and does not allow for pre- and post-treatment response to be determined in the same cell. In 1994 Shapiro et al. (84) showed that short applications (120 s) of *N*-ethylmaleimide (NEM), a sulfhydryl alkylating agent, selectively blocks PTX-sensitive G proteins in rat sympathetic neurons and can be used in situations where PTX treatment is impractical.

A number of techniques have been used to identify further the G protein α subunit responsible for coupling to receptors that produce VD inhibition. One strategy involves reconstituting receptor coupling after PTX pretreatment by introducing purified α subunits into the cell from a patch pipette. Opioid inhibition of Ca^{2+} currents in neuroblastoma x glioma cells was reconstituted by introduction of either α_o or α_i (85). However, α_o was approximately 10 times more potent than α_i. A similar result was obtained in rat dorsal root ganglion (DRG) neurons for inhibition induced by neuropeptide Y (NPY) (86). In contrast, G protein subunits α_o and α_i were equally effective in reconstituting bradykinin inhibition of Ca^{2+} currents in DRG neurons. (It should be noted that VD inhibition was not established in either of these studies.)

A second strategy to identify Gα subunits is to employ function-blocking antibodies (made to the C-terminus of Gα) that disrupt receptor–Gαβγ coupling (87). Antibodies to α_o, but not those to α_i, when introduced into the cell (via the patch pipette or by microinjection), attenuated the effect of NE and baclofen (a $GABA_B$ agonist) in rat sympathetic (88) and DRG neurons (89), respectively. A third strategy utilizes antisense oligonucleotides to prevent the synthesis of specific G protein subunits (90–92). In rat DRG neurons, baclofen-induced N-type Ca^{2+} channel inhibition was attenuated 24–32 hours following injection of antisense oligonucleotides complementary to mRNA from G protein subunit α_o but not α_i (93). Taken together, these studies all support a role for α_o in the coupling of neurotransmitter receptors to N-type channels.

Although the vast majority of receptors producing VD inhibition couple to N-type channels via a PTX-sensitive G protein (presumably $\alpha_o\beta\gamma$), a receptor in sympathetic neurons activated by VIP and secretin utilizes G protein subunit α_s to produce an identical form of inhibition (52). Application of VIP results in classical VD inhibition, as indicated by kinetic slowing and enhanced facilitation. The response is PTX insensitive but blocked by pretreatment with cholera toxin and anti-α_s antibodies. Furthermore, the kinetic properties of this form of inhibition are indistinguishable from those mediated by the PTX-sensitive pathway in these cells (94). Thus, α_o is the predominant, but not exclusive, Gα subunit involved in coupling neurotransmitter receptors to the VD pathway.

Lack of Evidence for Additional Signaling Pathways

VD inhibition is believed to be a direct effect of G proteins on the N-type Ca^{2+} channel. Although the idea remains to be rigorously tested, two lines of evidence support it. First, VD inhibition, where tested, has been shown to be "membrane delimited" based on the experimental paradigm developed by Forscher et al. (95). In this experiment, Ca^{2+} currents are recorded using the cell-attached rather than the whole-cell configuration of the patch–clamp method. Usually, a large patch pipette is employed and the average current flowing through many Ca^{2+} channels is recorded. With the cell-attached configuration, the seal between the pipette glass and the cell membrane effectively isolates a patch of membrane both electrically and chemically (96). Only Ca^{2+} channels within the confines of the pipette tip contribute to the recorded current, and (non-lipid-soluble) agents applied outside the pipette do not gain access to the active area of membrane. Thus, for an agent applied outside the patch pipette to produce modulation of channels within the patch, a molecular messenger that diffuses through the cytosol or laterally through the plasma membrane is thought to be necessary. For pathways that produce VD inhibition, application of agonist outside the patch pipette produces no discernible current inhibition (52,53). Conversely, in rat sympathetic neurons, activation of M_1-muscarinic (53) and angiotensin II (57) receptors produces a VI inhibition that produces modulation using this paradigm.

The second line of evidence consists of experiments designed to test for the participation of common signaling elements that occur downstream from G protein activation. In general, substances that activate or inhibit a pathway are either introduced into the interior of the cell via the patch pipette or applied to the exterior of the cell in the bath solution, whereupon the effect on VD inhibition is determined. This strategy has not produced convincing evidence implicating the requirement for cAMP/PKA, cGMP/PKG, PKC, arachidonic acid, or IP_3 in the signaling pathway responsible for VD inhibition (49,53,97–100). Both PKC (101–103) and cGMP (104,105) have been reported to secondarily modify VD inhibition, as noted later, but no evidence suggests that cGMP or PKC is essential for VD inhibition. The inability to demonstrate the participation of common second messenger pathways indirectly supports the idea of direct G protein Ca^{2+} channel interaction.

Activation of PKC Prevents VD Inhibition

Although PKC does not directly participate in VD inhibition, activation of PKC with phorbol esters appears to prevent (and even reverse) VD inhibition. Swartz (106) has shown that application of phorbol-12-myristate-13-acetate (PMA), an activator of PKC, produces an increase in basal Ca^{2+} current (consistent with the relief of tonic G protein mediated modulation), prevents transmitter-induced VD inhibition, and reverses GTPγS-induced inhibition. Thus, activation of PKC appears to regulate the inhibitory effects of G proteins. These observations have been confirmed by others as (101,103), although G protein independent modulation has been noted as well (107). Shapiro et al. (108) demonstrated that PKC-induced regulation was specific for VD inhibition by showing that PMA pretreatment did not suppress the VI inhibition produced by M_1 muscarinic receptors in rat sympathetic neurons. Interestingly, PKC activation has been reported to interfere with presynaptic inhibition of neurotransmitter release (a process linked in some cases to N-type channel modulation), suggesting that the phenomenon may be of great physiological significance (102). To date, however, receptor-mediated initiation of such effects has not been reported.

G Protein βγ Subunit Mediates VD Inhibition

Because tools (e.g., GDPβS, GTPγS, PTX, antibodies) used to investigate G protein *coupling* of receptors to N-type channels interact with the G_α subunit, it became widely assumed that Gα subunits *mediate* VD inhibition (12,109). The fact that subunits $\alpha_o\beta\gamma$ and $\alpha_s\beta\gamma$ evoke identical VD inhibition of N-type current in sympathetic neurons, however, suggested either that binding characteristics of the two α subunits were indistinguishable or that Gβγ mediated the inhibition. Experiments by Ikeda (110) and Herlitze et al. (111) demonstrated that Gβγ is the active entity. Both studies bypassed receptor coupling by directly perturbing the normal Gα-to-Gβγ stoichiometry and monitoring tonic VD modulation. Ikeda (110) overexpressed G protein α and βγ subunits in rat sympathetic neurons by intranuclear cDNA injection. Expression of several different G protein βγ subunits produced profound kinetic slowing, facilitation, and occlusion of NE-induced inhibition consistent with a tonic effect of Gβγ. Expression of Gα, Gβ, or Gγ alone produced minimal effects, and $G\alpha_o$ (wild-type and a constitutively active forms) attenuated NE-mediated modulation in a manner consistent with its buffering of endogenous Gβγ. Herlitze et al. (111) performed similar experiments on both heterologously expressed and native Ca^{2+} channels. In the former, P/Q-type channels (Ca^{2+} channel α_{1A}, β_{1b}, $\alpha_2\delta$) were transiently expressed in tSA-201 cells, and the effects of expressing Gα and Gβγ were examined by determining shifts in tail current activation curves produced by a depolarizing conditioning pulse. The latter experiments were performed in rat sympathetic neurons in which either purified Gβγ protein was injected into the cytoplasm or G protein subunits were overexpressed by injection of cRNA. With both approaches, Gβγ produced tonic modulation, whereas Gα did not. In contrast with the work of Ikeda, Herlitze et al. (111) observed tonic modulation when Gβ but not Gγ was expressed alone. These initial accounts of Gβγ-induced inhibition have been corroborated in experiments on α_{1A} (P/Q) channels in *Xenopus* oocytes (112) and α_{1B} (N) channels in HEK 293 cells (113).

Heterogeneity both in VD and in Gβγ-mediated inhibition appears to exist. In chick sensory neurons, introduction of recombinant Gβγ (at concentrations $\leq$100 nM) via the patch pipette produces a VI inhibition that is mediated by PKC; VD inhibition was not observed with Gβγ (58). Furthermore, preliminary observations indicate that introduction of $G\alpha_o$–GTPγS (but not $G\alpha_i$–GTPγS) mimics VD inhibition in this system (60). Uncovering the basis for differences between chick sensory neurons and the other preparations that have been studied to date might provide mechanistic clues to VD inhibition. It will be interesting to test whether the $G\alpha_o$-mediated VD inhibition in the chick cells results from a difference in: (1) the N-type channel expressed, (2) the concentration of G protein subunits, or (3) the presence of unidentified modulating factors.

Ca^{2+} Channel β Subunits Interfere with G Protein Modulation

Ca^{2+} channel β subunits (Caβ) comprise a family of cytosolic proteins (β_1–β_4) that bind to Ca^{2+} channel α_1 subunits on the intracellular linker between repeats I and II. Coexpression of Caβ with α_{1B} or α_{1A} subunits in *Xenopus* oocytes produces an increase in the whole-cell Ca^{2+} current, accelerates its activation phase, and shifts inactivation to more hyperpolarized potentials (31,32,114). Since these actions of Caβ are opposite to those of Gβγ (described earlier), investigators decided to test the effects of Caβ on G protein mediated inhibition. Several recent studies indicate that Caβ interferes with VD inhibition of N- and P/Q-type Ca^{2+} channels. Campbell et al. (33) produced depletion of natively expressed Caβ in rat DRG neurons by injecting antisense oligonucleotides. Following injection (108–116 h), Caβ immunoreactivity and Ca^{2+} current amplitude were greatly reduced. The fraction of Ca^{2+} current inhibition produced by the $GABA_B$ agonist baclofen was increased about twofold. Effects consistent with this finding are seen with heterologously expressed N- and P/Q-type channels in *Xenopus* oocytes.

Roche et al. (34) showed that expression of Ca^{2+} channel α_{1B} or α_{1A} (but not α_{1C}) subunit alone (or with α_2/δ) results in channels that are tonically modulated by endogenous G proteins, as indicated by the increase in current amplitude and a decrease in facilitation produced by injection of GDPβS or treatment with NEM. Coexpression of Ca^{2+} channel α_{1B} or α_{1A} with the β_3 (and α_2/δ) subunits abolished these effects, leading to the conclusion that Caβ interferes with G protein dependent modulation. Similarly, Bourinet et al. (35) showed that opioid-induced inhibition was about threefold larger for currents arising from expression of α_{1A} alone compared with coexpressed with β_4 and α_2/δ subunits. Thus, Caβ attenuates the efficacy of G proteins in modulating N- and P/Q-type Ca^{2+} channels.

The Domain I–II Linker Is Involved in VD Modulation

Several lines of evidence have focused attention on the intracellular linker connecting repeats I and II of the Ca^{2+} channel α_1 subunit as a structure involved in G protein mediated VD inhibition. First, Caβ both interacts with the repeat I–II linker and interferes with VD modulation of N-type channels (34,35,115). Second, PKC-dependent up-modulation of α_{1B} currents appears to involve a site on the repeat I–II linker (116). Third, the repeat I–II linker contains a motif (QXXER) essential for Gβγ binding to adenylyl cyclase type 2 (117). Several strategies have been used to address the role of the repeat I–II linker in VD inhibition.

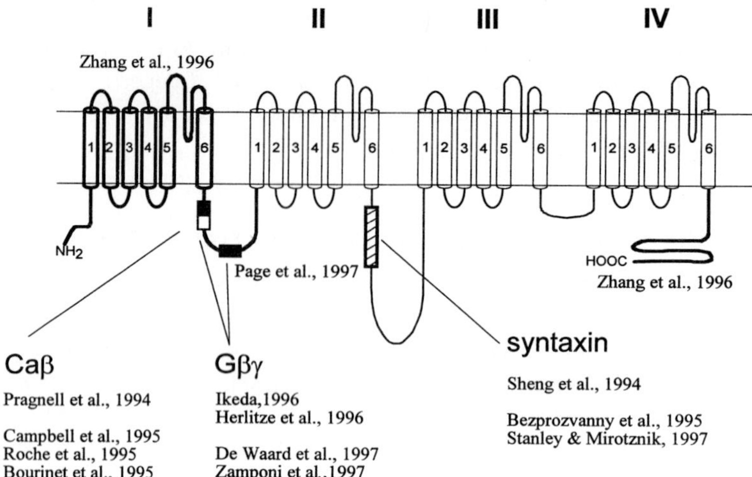

FIG. 2. Calcium channel α_{1B} subunit: proposed topology of the Ca^{2+} channel α_{1B} subunit, indicating areas believed to be involved in voltage-dependent modulation.

One strategy involves making chimeric constructs of α_{1B} (see Fig. 2) and other Ca^{2+} channel α_1 subunits, taking advantage of the observation that VD G protein inhibition is more robust for α_{1B} than for α_{1A} (P/Q type) or α_{1E} (R type) and is negligible for α_{1C} (L type). Zhang et al. (118) examined the somatostatin receptor induced VD inhibition in an extensive series of chimeras of $\alpha_{1B/A}$ and $\alpha_{1B/C}$ expressed in *Xenopus* oocytes. Chimeric constructs of α_{1B} in which the domain I–II linker was replaced with that from α_{1A} or α_{1C} responded to somatostatin (SST) with current inhibition and facilitation comparable to that obtained with wild-type α_{1B}. The latter result was particularly striking because SST did not produce measurable modulation of wild-type α_{1C}. Examination of the role played by each of the four repeats and the C-terminus showed that exchanging repeat I (including the N-terminus) and the C-terminus appeared to be sufficient to confer modulatory phenotype. That is, expression of a chimeric α_{1B} construct in which repeat I and the C-terminus were replaced with the cognate sequences from α_{1A} or α_{1C}, produced channels that displayed, respectively, little or no VD modulation. Thus, these data suggest that repeat I and the C-terminus of the Ca^{2+} channel α_1 subunit are important structural determinants of G protein modulation but provide little evidence for the participation of the repeat I–II linker.

In contrast to Zhang et al. (118), Page et al. (119) and Zamponi et al. (113) showed that when the I–II linker of α_{1B} was used to replace the cognate sequence in either α_{1A} or α_{1E}, the modulatory phenotype of α_{1B} was at least partially transferred. Thus, these two studies provide evidence supporting the involvement of the I–II linker in G protein modulation. Although the reasons for the discrepant conclusions are unclear, it should be noted that experimental conditions and approaches were not identical.

A second strategy utilized synthetic peptides derived from the sequence of the α_{1B} repeat I–II linker to probe for sites involved with G protein modulation. Zamponi et al. (113) used a protein overlay experiment to demonstrate that G protein subunit $\beta_1\gamma_2$, but not α_o, bound to the I–II linker region but not to other regions (II–III linker, domain II S5-6, carboxy terminus) of α_{1B}. The site of G$\beta\gamma$ interaction was further refined by determining the effect of 18- to 20-mer peptides (derived from the α_{1B} I–II linker sequence) on Ca^{2+} current inhibition induced by intracellular application of purified G$\beta\gamma$ (10 nM). The rationale underlying the experiment was that peptides derived from interacting regions on α_{1B} would bind G$\beta\gamma$ and reduce modulation. Two regions of interaction were found. One was at the N-terminus of the α_{1B} I–II linker and contained the QXXER G$\beta\gamma$-binding site just mentioned. The second site was about 20 residues C-terminal to the first site. Interestingly, the second region includes consensus phosphorylation sequences for PKC. Furthermore, phosphorylation *in vitro* of a peptide from this second site reduced the peptide's ability to attenuate G$\beta\gamma$-dependent modulation.

A third strategy measured binding of G protein subunits to Ca^{2+} channel α_1 subunit fusion proteins (mostly α_{1A} in these studies). De Waard et al. (112) showed that G protein subunits $\beta_1\gamma_2$ bound to the I–II linker but not to the II–III linker from α_{1A}. Conversely, subunits α_o and γ did not bind to either protein. When $\beta_1\gamma_2$ binding to the I–II linkers from different α_1 subunits was compared, binding was detected to α_{1A}, α_{1B}, and α_{1E} but not to α_{1S} or α_{1C}. Binding of $\beta_1\gamma_2$ to subdomains of the α_{1A} I–II linker identified two discrete sites. One was in the α_1 interaction domain (AID), previously identified as the Caβ binding site. The second binding site was downstream from the AID, and was termed D2. The affinity of G protein subunit $\beta_1\gamma_2$ for AID_A (i.e., AID from α_{1A}) and $D2_A$, as determined from binding studies to fusion proteins, revealed K_D values of 63 and 24 nM, respectively. Point mutagenesis studies of the AID_A protein identified several residues that were essential for $\beta_1\gamma_2$ binding, including Q343 and R387 of the QXXER G$\beta\gamma$ binding motif found within this sequence. An R387E mutant of α_{1A}, when expressed in *Xenopus* oocytes, was not susceptible to the GTPγS-induced kinetic slowing observed in wild-type α_{1A}, providing a functional correlate to the binding data. It should be noted that for reasons that are unclear, the cognate mutation in α_{1B} (R383E) did not alter G protein modulation in the study of Zhang et al. (118).

Taken together, these studies largely support the idea that G$\beta\gamma$ binds directly to sites on the domain I–II linker of α_{1B} (as well as α_{1A} and α_{1E}), to produce VD inhibition. There may be additional sites on the α_1 subunit that are involved in VD modulation as indicated by Zhang et al. (118). This would not be surprising given the size of the α_1 subunit and the identification of multiple sites of G protein interaction on G protein modulated, inwardly rectifying K^+ channels (120–123).

Other Molecules Essential to Ca^{2+} Channel Inhibition?

Syntaxin is a 35-kDa synaptic protein thought to be important in exocytosis of synaptic vesicles. Together with SNAP-25 and synaptobrevin/VAMP, syntaxin is believed to form a core complex that primes the vesicle for the Ca^{2+}-triggered step (124). Syntaxin is also found complexed with the N-type Ca^{2+} channel α_{1B} subunit and synaptotagmin, the putative Ca^{2+} sensor that triggers vesicle fusion. Approximately 80% of ω-CTx GVIA binding sites can be immunoprecipitated with anti-syntaxin antibodies, indicating a high degree of association (47). Using fusion proteins from α_{1B}, Sheng et al. (125) narrowed the syntaxin-binding site to an 87 amino acid sequence on the C-terminus of the large domain II–III linker. Further work by this group (126) showed that syntaxin binding displayed a biphasic Ca^{2+} dependence, with maximum binding occurring near 25 μM Ca^{2+}. Finally, fusion proteins of α_{1B} that contain the syntaxin binding site have been shown to block binding of GST-syntaxin to N-type channels purified from rat brain. When introduced into presynaptic neurons, these fusion proteins inhibit synaptic transmission between pairs of sympathetic ganglion neurons in culture, suggesting that Ca^{2+} channel/syntaxin association is essential to exocytosis (127).

The effect of syntaxin on Ca^{2+} channel function is only beginning to be explored. Coexpression of syntaxin 1A with N-type Ca^{2+} channels ($\alpha_{1B}\beta_3\alpha_2/\delta$) in *Xenopus* oocytes caused a −20 mV shift in the steady state inactivation curve, drastically reducing Ca^{2+} channel availability at moderate holding potentials (128). A similar effect of syntaxin was seen on P/Q-type channels (α_{1A}) but not L-type channels (α_{1C}). In contrast, a shift in the inactivation of natively expressed N-type Ca^{2+} channel was not observed in sympathetic neurons injected with a fusion peptide that disrupts syntaxin binding or in the calyx-type nerve terminal of the chick ciliary ganglion in which syntaxin was cleaved with botulinum toxin C_1 (129).

Although botulinum toxin C_1 proved ineffective in shifting the inactivation of N-type channels, it attenuates GTPγS-induced N-type channel inhibition in calyx nerve terminals (129). These results, together with the heterologous expression studies, demonstrate that the function and modulation of N-type channels can be influenced by proteins other than Caβ and α_2/δ which interact with intracellular domains of the α_1 subunit. Such results imply that heterogeneity in VD inhibition might be expected and, may be explained at least in part by the differential expression of regulatory molecules like syntaxin.

CONCLUSIONS

Information not available until the late 1990s provides the basis for our current understanding of a mechanism underlying VD inhibition. The proposed pathway is remarkably compact and consists of G protein coupled receptor, heterotrimeric G protein, and N-type Ca^{2+} channel. The sequence of events is believed to be: (1) ligand binding to receptor, (2) activation of the heterotrimeric G protein via GDP–GTP exchange, (3) dissociation of Gαβγ into Gα–GTP and Gβγ, (4) binding of Gβγ to sites on the Ca^{2+} channel α_1 subunit, and (5) Gα-GTP hydrolysis and reassociation of Gα–GDP with Gβγ to terminate the reaction.

The recent elucidation of events occurring subsequent to Gαβγ dissociation provides plausible (but as yet unproven) explanations for several past observations. The finding that $G\beta\gamma$, rather than $G\alpha$, mediates VD inhibition explains why activation of receptors that couple to different G proteins produces nearly identical forms of modulation. For example, in sympathetic neurons, the forms of VD inhibition produced by NE and VIP/secretin, which couple to N-type Ca^{2+} channels via G_o and G_s, respectively, are kinetically indistinguishable (52,94). This is consistent with the finding that many (but not all) $G\beta\gamma$ combinations interact with effectors in a similar fashion (130,131). The demonstration that $G\beta\gamma$ binds directly to the Ca^{2+} channel α_1 subunit (112,113) is consistent with a membrane-delimited action, rapid signaling, and failed attempts to invoke additional elements such as protein kinases in the transduction pathway.

Although a direct action of $G\beta\gamma$ on the Ca^{2+} channel remains to be proven, the binding data make the idea more plausible. The identification of a specific binding site for $G\beta\gamma$ on the Ca^{2+} channel α_1 I–II linker (30) provides insight into why $Ca\beta$, which binds to an overlapping region (112,113), interferes with G protein modulation (34,35,115). In addition, conservation of sequences in the I–II linker region (e.g., the QXXER motif) may explain why the DHP-insensitive α_1 subunits (i.e., α_{1A}, α_{1B}, α_{1E}) are all susceptible to G protein mediated VD inhibition, albeit to different extents. Finally, the demonstration that PKC-mediated phosphorylation (*in vitro*) of sites within one of the putative $G\beta\gamma$ interaction domains attenuates $G\beta\gamma$ binding (113) provides a possible mechanism for the prevention or reversal of G protein mediated VD inhibition following phorbol ester application (102,106).

Despite recent progress, several fundamental aspects regarding VD modulation remain unresolved. First, the molecular explanation for VD modulation requires further refinement, since it is not obvious how binding of $G\beta\gamma$ to the Ca^{2+} channel α_1 subunit alters gating. What is the role of the $Ca\beta$ in regard to $G\beta\gamma$ binding and gating shifts (132)? What is the physical basis underlying kinetic slowing and facilitation? For example, does $G\beta\gamma$ dissociate from the α_1 subunit at depolarized potentials and rebind upon returning to hyperpolarized potentials, as suggested by kinetic models (64,72,73,118)? Does $G\alpha$ or $G\alpha\beta\gamma$ play a direct role in VD modulation (133)? Does binding of $G\beta\gamma$ produce alterations in the permeability of the pore in addition to gating shifts (134)?

Second, the origins of receptor coupling specificity are unclear. Since activation of all G protein coupled receptors liberates $G\beta\gamma$, does specificity result because specific combinations of $G\alpha\beta\gamma$ couple to particular receptors (92) or because only certain $G\beta\gamma$ combinations are capable of binding with the Ca^{2+} channel? Alternatively, is the minimal transduction unit (i.e., receptor, G protein, and channel) assembled in a "microdomain" conferring specificity from spatial constraints?

Third, the physiological consequences of VD inhibition require further investigation. Does the relief of G protein inhibition, as occurs during large depolarizing pulses under voltage–clamp conditions, also occur during high frequency trains of action potentials, accounting for frequency-dependent changes in synaptic transmission (77,78)? Does PKC-mediated attenuation of VD inhibition occur during physiological or pathological processes? In light of the reported actions of syntaxin, are N-type channels in neuronal cell bodies identical to those found in nerve terminals with respect to their biophysical properties and their susceptibility to G protein dependent modulation? Similarly, how comparable are native N-type channels and those heterologously expressed in *Xenopus* oocytes or clonal cell lines?

VD inhibition of N-type Ca^{2+} channels appears to represent a minimalist approach to signal transduction. In a teleological sense, perhaps this is reasonable. When compared to other G protein pathways in which several enzymatic steps (e.g., adenylyl cyclase, protein kinase A) are interposed between ligand binding and effector modulation, the direct modulation of N-type channels by G protein subunits is a compact and elegant mechanism. Such simplicity sacrifices gain for speed. At presynaptic nerve terminals, ligand is in abundance and timing is essential. Thus, the mechanism underlying VD modulation seems well suited for producing autoinhibitory feedback at nerve terminals—a proposed physiological role for N-type Ca^{2+} channel modulation (135).

ACKNOWLEDGMENTS

Supported in part by National Institutes of Health grants GM 56180-01 (S. R. I.), DA 10350-01 (S. R. I), NS 16483 (K. D.), and a Jacob Javits Award (KD). Dedicated to the memory of Xiangyang "Chris" Wei, whose insights and generosity contributed greatly to the field of Ca^{2+} channel modulation.

REFERENCES

1. Hepler JR, Gilman AG: G proteins. *Trends Biochem Sci* 1992;17:383–387.
2. Clapham DE, Neer EJ: New roles for G-protein $\beta\gamma$-dimers in transmembrane signalling. *Nature* 1993;365:403–406.
3. Neer EJ: Heterotrimeric G proteins: organizers of transmembrane signals. *Cell* 1995;80:249–257.
4. Offermanns S, Simon MI: Organization of transmembrane signalling by heterotrimeric G proteins. *Cancer Surv* 1996;27:177–198.
5. Tsien RW, Ellinor PT, Horne WA: Molecular diversity of voltage-dependent Ca^{2+} channels. *Trends Pharmacol Sci* 1991;12:349–354.
6. Hofmann F, Biel M, Flockerzi V: Molecular basis for Ca^{2+} channel diversity. *Annu Rev Neurosci* 1994;17:399–418.
7. Perez-Reyes E, Schneider T: Calcium channels: structure, function, and classification. *Drug Dev* 1994;33:295–318.
8. Dunlap K, Luebke JI, Turner TJ: Exocytotic Ca^{2+} channels in mammalian central neurons. *Trends Neurosci* 1995;18:89–98.
9. Anwyl R: Modulation of vertebrate neuronal calcium channels by transmitters. *Brain Re Rev* 1991; 16:265–281.
10. Hille B: G protein-coupled mechanisms and nervous signaling. *Neuron* 1992;9:187–195.
11. Hille B: Modulation of ion-channel function by G-protein-coupled receptors. *Trends Neurosci* 1994; 17:531–536.
12. Dolphin AC: The G. L. Brown Prize Lecture. Voltage-dependent calcium channels and their modulation by neurotransmitters and G proteins. *Exp Physiol* 1995;80:1–36.
13. Wickman K, Clapham DE: Ion channel regulation by G proteins. *Physiol Rev* 1995;75:865–885.
14. Wickman KD, Clapham DE: G-protein regulation of ion channels. *Curr Opin Neurobiol* 1995;5: 278–285.
15. Nowycky MC, Fox AP, Tsien RW: Three types of neuronal calcium channel with different calcium agonist sensitivity. *Nature* 1985;316:440–443.

16. Fox AP, Nowycky MC, Tsien RW: Single-channel recordings of three types of calcium channels in chick sensory neurones. *J Physiol (Lond)* 1987;394:173–200.
17. Fox AP, Nowycky MC, Tsien RW: Kinetic and pharmacological properties distinguishing three types of calcium currents in chick sensory neurones. *J Physiol (Lond)* 1987;394:149–172.
18. Kasai H, Aosaki T, Fukuda J: Presynaptic Ca-antagonist ω-conotoxin irreversibly blocks N-type Ca-channels in chick sensory neurons. *Neurosci Res* 1987;4:228–235.
19. McCleskey EW, Fox AP, Feldman DH, Cruz LJ, Olivera BM, Tsien RW, Yoshikami D: ω-Conotoxin: direct and persistent blockade of specific types of calcium channels in neurons but not muscle. *Proc Natl Acad Sci U S A* 1987;84:4327–4331.
20. Plummer MR, Logothetis DE, Hess P: Elementary properties and pharmacological sensitivities of calcium channels in mammalian peripheral neurons. *Neuron* 1989;2:1453–1463.
21. Cox DH, Dunlap K: Pharmacological discrimination of N-type from L-type calcium current and its selective modulation by transmitters. *J Neurosci* 1992;12:906–914.
22. Boland LM, Morrill JA, Bean BP: ω-Conotoxin block of N-type calcium channels in frog and rat sympathetic neurons. *J Neurosci* 1994;14:5011–5027.
23. Regan LJ, Sah DW, Bean BP: Ca^{2+} channels in rat central and peripheral neurons: high-threshold current resistant to dihydropyridine blockers and omega-conotoxin. *Neuron* 1991;6:269–280.
24. Mintz IM, Adams ME, Bean BP: P-type calcium channels in rat central and peripheral neurons. *Neuron* 1992;9:85–95.
25. Dubel SJ, Starr TV, Hell J, Ahlijanian MK, Enyeart JJ, Catterall WA, Snutch TP: Molecular cloning of the α-1 subunit of an ω-conotoxin-sensitive calcium channel. *Proc Natl Acad Sci U S A* 1992;89: 5058–5062.
26. Williams ME, Brust PF, Feldman DH, Patthi S, Simerson S, Maroufi A, McCue AF, Velicelebi G, Ellis SB, Harpold MM: Structure and functional expression of an φ-conotoxin-sensitive human N-type calcium channel. *Science* 1992;257:389–395.
27. Ellinor PT, Zhang JF, Horne WA, Tsien RW: Structural determinants of the blockade of N-type calcium channels by a peptide neurotoxin. *Nature* 1994;372:272–275.
28. Isom LL, De Jongh KS, Catterall WA: Auxiliary subunits of voltage-gated ion channels. *Neuron* 1994;12:1183–1194.
29. Witcher DR, De Waard M, Sakamoto J, Franzini-Armstrong C, Pragnell M, Kahl SD, Campbell KP: Subunit identification and reconstitution of the N-type Ca^{2+} channel complex purified from brain. *Science* 1993;261:486–489.
30. Pragnell M, De Waard M, Mori Y, Tanabe T, Snutch TP, Campbell KP: Calcium channel β-subunit binds to a conserved motif in the I–II cytoplasmic linker of the α_1-subunit. *Nature* 1994;368: 67–70.
31. Brust PF, Simerson S, McCue AF, Deal CR, Schoonmaker S, Williams ME, Velicelebi G, Johnson EC, Harpold MM, Ellis SB: Human neuronal voltage-dependent calcium channels: studies on subunit structure and role in channel assembly. *Neuropharmacology* 1993;32:1089–1102.
32. Stea A, Dubel SJ, Pragnell M, Leonard JP, Campbell KP, Snutch TP: A β-subunit normalizes the electrophysiological properties of a cloned N-type Ca^{2+} channel α_1-subunit. *Neuropharmacology* 1993;32:1103–1116.
33. Campbell V, Berrow NS, Fitzgerald EM, Brickley K, Dolphin AC: Inhibition of the interaction of G protein G_o with calcium channels by the calcium channel β-subunit in rat neurones. *J Physiol (Lond)* 1995;485:365–372.
34. Roche JP, Anantharam V, Treistman SN: Abolition of G protein inhibition of α_{1A} and α_{1B} calcium channels by co-expression of β_3 subunit. *FEBS Lett* 1995;371:43–46.
35. Bourinet E, Soong TW, Stea A, Snutch TP: Determinants of the G protein-dependent opioid modulation of neuronal calcium channels. *Proc Natl Acad Sci U S A* 1996;93:1486–1491.
36. Clapham DE: Calcium signaling. *Cell* 1995;80:259–268.
37. Hirning LD, Fox AP, McCleskey EW, Olivera BM, Thayer SA, Miller RJ, Tsien RW: Dominant role of N-type Ca^{2+} channels in evoked release of norepinephrine from sympathetic neurons. *Science* 1988;239:57–61.
38. Holz GG, Dunlap K, Kream RM: Characterization of the electrically evoked release of substance P from dorsal root ganglion neurons: methods and dihydropyridine sensitivity. *J Neurosci* 1988;8: 463–471.
39. Toth PT, Bindokas VP, Bleakman D, Colmers WF, Miller RJ: Mechanism of presynaptic inhibition by neuropeptide Y at sympathetic nerve terminals. *Nature* 1993;364:635–639.

40. De Luca A, Li CG, Rand MJ, Reid JJ, Thaina P, Wong-Dusting HK: Effects of ω-conotoxin GVIA on autonomic neuroeffector transmission in various tissues. *Br J Pharmacol* 1990;101:437–447.
41. Lundy PM, Frew R: Pharmacological characterization of voltage-sensitive Ca^{2+} channels in autonomic nerves. *Eur J Pharmacol* 1993;231:197–202.
42. Turner TJ, Adams ME, Dunlap K: Calcium channels coupled to glutamate release identified by ω-Aga-IVA. *Science* 1992;258:310–313.
43. Turner TJ, Adams ME, Dunlap K: Multiple Ca^{2+} channel types coexist to regulate synaptosomal neurotransmitter release. *Proc Natl Acad Sci U S A* 1993;90:9518–9522.
44. Wheeler DB, Randall A, Tsien RW: Roles of N-type and Q-type Ca^{2+} channels in supporting hippocampal synaptic transmission. *Science* 1994;264:107–111.
45. Westenbroek RE, Hell JW, Warner C, Dubel SJ, Snutch TP, Catterall WA: Biochemical properties and subcellular distribution of an N-type calcium channel α_1 subunit. *Neuron* 1992;9:1099–1115.
46. Westenbroek RE, Sakurai T, Elliott EM, Hell JW, Starr TV, Snutch TP, Catterall WA: Immunochemical identification and subcellular distribution of the α_{1A} subunits of brain calcium channels. *J Neurosci* 1995;15:6403–6418.
47. Leveque C, el Far O, Martin-Moutot N, Sato K, Kato R, Takahashi M, Seagar MJ: Purification of the N-type calcium channel associated with syntaxin and synaptotagmin. A complex implicated in synaptic vesicle exocytosis. *J Biol Chem* 1994;269:6306–6312.
48. Rettig J, Sheng ZH, Kim DK, Hodson CD, Snutch TP, Catterall WA: Isoform-specific interaction of the α_{1A} subunits of brain Ca^{2+} channels with the presynaptic proteins syntaxin and SNAP-25. *Proc Natl Acad Sci U S A* 1996;93:7363–7368.
49. Ikeda SR, Schofield GG: Somatostatin blocks a calcium current in rat sympathetic ganglion neurones. *J Physiol (Lond)* 1989;409:221–240.
50. Ikeda SR: Prostaglandin modulation of Ca^{2+} channels in rat sympathetic neurones is mediated by guanine nucleotide binding proteins. *J Physiol (Lond)* 1992;458:339–359.
51. Zhu Y, Ikeda SR: Adenosine modulates voltage-gated Ca^{2+} channels in adult rat sympathetic neurons. *J Neurophysiol* 1993;70:610–620.
52. Zhu Y, Ikeda SR: VIP inhibits N-type Ca^{2+} channels of sympathetic neurons via a pertussis toxin-insensitive but cholera toxin-sensitive pathway. *Neuron* 1994;13:657–669.
53. Bernheim L, Beech DJ, Hille B: A diffusible second messenger mediates one of the pathways coupling receptors to calcium channels in rat sympathetic neurons. *Neuron* 1991;6:859–867.
54. Diversé-Pierluissi M, Dunlap K: Distinct, convergent second messenger pathways modulate neuronal calcium currents. *Neuron* 1993;10:753–760.
55. Shapiro MS, Hille B: Substance P and somatostatin inhibit calcium channels in rat sympathetic neurons via different G protein pathways. *Neuron* 1993;10:11–20.
56. Luebke JI, Dunlap K: Sensory neuron N-type calcium currents are inhibited by both voltage-dependent and -independent mechanisms. *Pfluegers Arch* 1994;428:499–507.
57. Shapiro MS, Wollmuth LP, Hille B: Angiotensin II inhibits calcium and M current channels in rat sympathetic neurons via G proteins. *Neuron* 1994;12:1319–1329.
58. Diversé-Pierluissi M, Goldsmith PK, Dunlap K: Transmitter-mediated inhibition of N-type calcium channels in sensory neurons involves multiple GTP-binding proteins and subunits. *Neuron* 1995;14: 191–200.
59. Rane SG, Walsh MP, McDonald JR, Dunlap K: Specific inhibitors of protein kinase C block transmitter-induced modulation of sensory neuron calcium current. *Neuron* 1989;3:239–245.
60. Diversé-Pierluissi M, Remmers AE, Neubig RR, Dunlap K: Novel form of crosstalk between G protein and tyrosine kinase pathways. *Proc Natl Acad Sci U S A* 1997; 94:5417–5421.
61. Bean BP: Neurotransmitter inhibition of neuronal calcium currents by changes in channel voltage dependence. *Nature* 1989;340:153–156.
62. Elmslie KS, Zhou W, Jones SW: LHRH and GTP-γ-S modify calcium current activation in bullfrog sympathetic neurons. *Neuron* 1990;5:75–80.
63. Boland LM, Bean BP: Modulation of N-type calcium channels in bullfrog sympathetic neurons by luteinizing hormone-releasing hormone: kinetics and voltage dependence. *J Neurosci* 1993;13: 516–533.
64. Golard A, Siegelbaum SA: Kinetic basis for the voltage-dependent inhibition of N-type calcium current by somatostatin and norepinephrine in chick sympathetic neurons. *J Neurosci* 1993;13: 3884–3894.
65. Carabelli V, Lovallo M, Magnelli V, Zucker H, Carbone E: Voltage-dependent modulation of single N-type Ca^{2+} channel kinetics by receptor agonists in IMR32 cells. *Biophys J* 1996;70:2144–2154.
66. Patil PG, de Leon M, Reed RR, Dubel S, Snutch TP, Yue DT: Elementary events underlying voltage-dependent G-protein inhibition of N-type calcium channels. *Biophys J* 1996;71:2509–2521.

67. Marchetti C, Carbone E, Lux HD: Effects of dopamine and noradrenaline on Ca channels of cultured sensory and sympathetic neurons of chick. *Pfluegers Arch* 1986;406:104–111.
68. Tsunoo A, Yoshii M, Narahashi T: Block of calcium channels by enkephalin and somatostatin in neuroblastoma–glioma hybrid NG108-15 cells. *Proc Natl Acad Sci U S A* 1986;83:9832–9836.
69. Grassi F, Lux HD: Voltage-dependent GABA-induced modulation of calcium currents in chick sensory neurons. *Neurosci Lett* 1989;105:113–119.
70. Scott RH, Dolphin AC: Voltage-dependent modulation of rat sensory neurone calcium channel currents by G protein activation: effect of a dihydropyridine antagonist. *Br J Pharmacol* 1990;99: 629–630.
71. Dolphin AC: Facilitation of Ca^{2+} current in excitable cells. *Trends Neurosci* 1996;19:35–43.
72. Lopez HS, Brown AM: Correlation between G protein activation and reblocking kinetics of Ca^{2+} channel currents in rat sensory neurons. *Neuron* 1991;7:1061–1068.
73. Elmslie KS, Jones SW: Concentration dependence of neurotransmitter effects on calcium current kinetics in frog sympathetic neurones. *J Physiol (Lond)* 1994;481:35–46.
74. Ikeda SR: Double-pulse calcium channel current facilitation in adult rat sympathetic neurones. *J Physiol (Lond)* 1991;439:181–214.
75. Kasai H: Tonic inhibition and rebound facilitation of a neuronal calcium channel by a GTP-binding protein. *Proc Natl Acad Sci U S A* 1991;88:8855–8859.
76. Toth PT, Miller RJ: Calcium and sodium currents evoked by action potential waveforms in rat sympathetic neurones. *J Physiol (Lond)* 1995;485:43–57.
77. Song WJ, Surmeier DJ: Voltage-dependent facilitation of calcium channels in rat neostriatal neurons. *J Neurophysiol* 1996;76:2290–2306.
78. Williams S, Serafin M, Mühlethaler M, Bernheim L: Facilitation of N-type calcium current is dependent on the frequency of action potential–like depolarizations in dissociated cholinergic basal forebrain neurons of the guinea pig. *J Neurosci* 1997;17:1625–1632.
79. Eckstein F, Cassel D, Levkovitz H, Lowe M, Selinger Z: Guanosine 5'-*O*-(thiodiphosphate): An inhibitor of adenylate cyclase stimulation by guanine nucleotides and fluoride ions. *J Biol Chem* 1979;254:9829–9834.
80. Holz GG, Rane SG, Dunlap K: GTP-binding proteins mediate transmitter inhibition of voltage-dependent calcium channels. *Nature* 1986;319:670–672.
81. Lewis DL, Weight FF, Luini A: A guanine nucleotide-binding protein mediates the inhibition of voltage-dependent calcium current by somatostatin in a pituitary cell line. *Proc Natl Acad Sci U S A* 1986;83:9035–9039.
82. Dolphin AC, Scott RH: Calcium channel currents and their inhibition by (–)-baclofen in rat sensory neurones: modulation by guanine nucleotides. *J Physiol (Lond)* 1987;386:1–17.
83. Swartz KJ, Bean BP: Inhibition of calcium channels in rat CA3 pyramidal neurons by a metabotropic glutamate receptor. *J Neurosci* 1992;12:4358–4371.
84. Shapiro MS, Wollmuth LP, Hille B: Modulation of Ca^{2+} channels by PTX-sensitive G-proteins is blocked by *N*-ethylmaleimide in rat sympathetic neurons. *J Neurosci* 1994;14:7109–7116.
85. Hescheler J, Rosenthal W, Trautwein W, Schultz G: The GTP-binding protein G_o, regulates neuronal calcium channels. *Nature* 1987;325:445–447.
86. Ewald DA, Pang IH, Sternweis PC, Miller RJ: Differential G protein-mediated coupling of neurotransmitter receptors to Ca^{2+} channels in rat dorsal root ganglion neurons in vitro. *Neuron* 1989;2: 1185–1193.
87. McFadzean I, Mullaney I, Brown DA, Milligan G: Antibodies to the GTP binding protein, G_o, antagonize noradrenaline-induced calcium current inhibition in NG108-15 hybrid cells. *Neuron* 1989;3:177–182.
88. Caulfield MP, Jones S, Vallis Y, Buckley NJ, Kim GD, Milligan G, Brown DA: Muscarinic M-current inhibition via $G\alpha_{q/11}$ and α-adrenoceptor inhibition of Ca^{2+} current via $G\alpha_o$ in rat sympathetic neurones. *J Physiol (Lond)* 1994;477:415–422.
89. Menon-Johansson AS, Berrow N, Dolphin AC: G_o transduces $GABA_B$-receptor modulation of N-type calcium channels in cultured dorsal root ganglion neurons. *Pfluegers Arch* 1993;425:335–343.
90. Kleuss C, Hescheler J, Ewel C, Rosenthal W, Schultz G, Wittig B: Assignment of G-protein subtypes to specific receptors inducing inhibition of calcium currents. *Nature* 1991;353:43–48.
91. Kleuss C, Scherubl H, Hescheler J, Schultz G, Wittig B: Different β-subunits determine G-protein interaction with transmembrane receptors. *Nature* 1992;358:424–426.
92. Kleuss C, Scherubl H, Hescheler J, Schultz G, Wittig B: Selectivity in signal transduction determined by γ subunits of heterotrimeric G proteins. *Science* 1993;259:832–834.
93. Campbell V, Berrow N, Dolphin AC: $GABA_B$ receptor modulation of Ca^{2+} currents in rat sensory neurones by the G protein G_o: antisense oligonucleotide studies. *J Physiol (Lond)* 1993;470:1–11.

94. Ehrlich I, Elmslie KS: Neurotransmitters acting via different G proteins inhibit N-type calcium current by an identical mechanism in rat sympathetic neurons. *J Neurophysiol* 1995;74:2251–2257.
95. Forscher P, Oxford GS, Schulz D: Noradrenaline modulates calcium channels in avian dorsal root ganglion cells through tight receptor-channel coupling. *J Physiol (Lond)* 1986;379:131–144.
96. Hamill OP, Marty A, Neher E, Sakmann B, Sigworth FJ: Improved patch–clamp techniques for high-resolution current recording from cells and cell-free membrane patches. *Pfluegers Arch* 1981; 391:85–100.
97. Wanke E, Ferroni A, Malgaroli A, Ambrosini A, Pozzan T, Meldolesi J: Activation of a muscarinic receptor selectively inhibits a rapidly inactivated Ca^{2+} current in rat sympathetic neurons. *Proc Natl Acad Sci U S A* 1987;84:4313–4317.
98. Bley KR, Tsien RW: Inhibition of Ca^{2+} and K^{+} channels in sympathetic neurons by neuropeptides and other ganglionic transmitters. *Neuron* 1990;4:379–391.
99. Plummer MR, Rittenhouse A, Kanevsky M, Hess P: Neurotransmitter modulation of calcium channels in rat sympathetic neurons. *J Neurosci* 1991;11:2339–2348.
100. Abrahams TP, Schofield GG: Norepinephrine-induced Ca^{2+} current inhibition in adult rat sympathetic neurons does not require protein kinase C activation. *Eur J Pharmacol* 1992;227:189–197.
101. Golard A, Role LW, Siegelbaum SA: Protein kinase C blocks somatostatin-induced modulation of calcium current in chick sympathetic neurons. *J Neurophysiol* 1993;70:1639–1643.
102. Swartz KJ, Merritt A, Bean BP, Lovinger DM: Protein kinase C modulates glutamate receptor inhibition of Ca^{2+} channels and synaptic transmission. *Nature* 1993;361:165–168.
103. Zhu Y, Ikeda SR: Modulation of Ca^{2+}-channel currents by protein kinase C in adult rat sympathetic neurons. *J Neurophysiol* 1994;72:1549–1560.
104. Meriney SD, Gray DB, Pilar GR: Somatostatin-induced inhibition of neuronal Ca^{2+} current modulated by cGMP-dependent protein kinase. *Nature* 1994;369:336–339.
105. Chen C, Schofield GG: Nitric oxide donors enhanced Ca^{2+} currents and blocked noradrenaline-induced Ca^{2+} current inhibition in rat sympathetic neurons. *J Physiol (Lond)* 1995;482:521–531.
106. Swartz KJ: Modulation of Ca^{2+} channels by protein kinase C in rat central and peripheral neurons: disruption of G protein-mediated inhibition. *Neuron* 1993;11:305–320.
107. Yang J, Tsien RW: Enhancement of N- and L-type calcium channel currents by protein kinase C in frog sympathetic neurons. *Neuron* 1993;10:127–136.
108. Shapiro MS, Zhou J, Hille B: Selective disruption by protein kinases of G-protein-mediated Ca^{2+} channel modulation. *J Neurophysiol* 1996;76:311–320.
109. Kurachi Y: G protein regulation of cardiac muscarinic potassium channel. *Am J Physiol* 1995;269: C821–C830.
110. Ikeda SR: Voltage-dependent modulation of N-type calcium channels by G-protein $\beta\gamma$ subunits. *Nature* 1996;380:255–258.
111. Herlitze S, Garcia DE, Mackie K, Hille B, Scheuer T, Catterall WA: Modulation of Ca^{2+} channels by G-protein $\beta\gamma$ subunits. *Nature* 1996;380:258–262.
112. De Waard M, Liu H, Walker D, Scott VES, Gurnett CA, Campbell KP: Direct binding of G-protein $\beta\gamma$ complex to voltage-dependent calcium channels. *Nature* 1997;385:446–450.
113. Zamponi GW, Bourinet E, Nelson D, Nargeot J, Snutch TP: Crosstalk between G proteins and protein kinase C mediated by the calcium channel α_1 subunit. *Nature* 1997;385:442–446.
114. De Waard M, Campbell KP: Subunit regulation of the neuronal α_{1A} Ca^{2+} channel expressed in *Xenopus* oocytes. *J Physiol (Lond)* 1995;485:619–634.
115. Berrow NS, Campbell V, Fitzgerald EM, Brickley K, Dolphin AC: Antisense depletion of β-subunits modulates the biophysical and pharmacological properties of neuronal calcium channels. *J Physiol (Lond)* 1995;482:481–491.
116. Stea A, Soong TW, Snutch TP: Determinants of PKC-dependent modulation of a family of neuronal calcium channels. *Neuron* 1995;15:929–940.
117. Chen J, DeVivo M, Dingus J, Harry A, Li J, Sui J, Carty DJ, Blank JL, Exton JH, Stoffel RH, Inglese J, Lefkowitz RJ, Logothetis DE, Hildebrandt JD, Iyengar R: A region of adenylyl cyclase 2 critical for regulation by G protein $\beta\gamma$ subunits. *Science* 1995;268:1166–1169.
118. Zhang JF, Ellinor PT, Aldrich RW, Tsien RW: Multiple structural elements in voltage-dependent Ca^{2+} channels support their inhibition by G proteins. *Neuron* 1996;17:991–1003.
119. Page KM, Stephens GJ, Berrow NS, Dolphin AC: The intracellular loop between domains I and II of the B-type calcium channel confers aspects of G-protein sensitivity to the E-type calcium channel. *J Neurosci* 1997;14:1330–1338.
120. Huang CL, Slesinger PA, Casey PJ, Jan YN, Jan LY: Evidence that direct binding of the $G\beta\gamma$ to the GIRK1 G protein-gated inwardly rectifying K^{+} channel is important for channel activation. *Neuron* 1995;15:1133–1143.

121. Krapivinsky G, Krapivinsky L, Wickman K, Clapham DE: Gβγ binds directly to the G protein-gated K^+ channel, IK_{ACh}. *J Biol Chem* 1995;270:29059–29062.
122. Kunkel MT, Peralta EG: Identification of domains conferring G protein regulation of inward rectifier potassium channels. *Cell* 1995;83:443–449.
123. Slesinger PA, Reuveny E, Jan YN, Jan LY: Identification of structural elements involved in G protein gating of the GIRK1 potassium channel. *Neuron* 1995;15:1145–1156.
124. Südhof TC: The synaptic vesicle cycle: a cascade of protein–protein interactions. *Nature* 1995;375: 645–653.
125. Sheng ZH, Rettig J, Takahashi M, Catterall WA: Identification of a syntaxin-binding site on N-type calcium channels. *Neuron* 1994;13:1303–1313.
126. Sheng ZH, Rettig J, Cook T, Catterall WA: Calcium-dependent interaction of N-type calcium channels with the synaptic core complex. *Nature* 1996;379:451–454.
127. Mochida S, Sheng ZH, Baker C, Kobayashi H, Catterall WA: Inhibition of neurotransmission by peptides containing the synaptic protein interaction site of N-type Ca^{2+} channels. *Neuron* 1996;17: 781–788.
128. Bezprozvanny I, Scheller RH, Tsien RW: Functional impact of syntaxin on gating of N-type and Q-type calcium channels. *Nature* 1995;378:623–626.
129. Stanley EF, Mirotznik RR: Cleavage of syntaxin prevents G-protein regulation of presynaptic calcium channels. *Nature* 1997;385:340–343.
130. Ueda N, Iniguez-Lluhi JA, Lee E, Smrcka AV, Robishaw JD, Gilman AG: G protein βγ subunits. Simplified purification and properties of novel isoforms. *J Biol Chem* 1994;269:4388–4395.
131. Wickman KD, Iniguez-Lluhl JA, Davenport PA, Taussig R, Krapivinsky GB, Linder ME, Gilman AG, Clapham DE: Recombinant G-protein βγ-subunits activate the muscarinic-gated atrial potassium channel. *Nature* 1994;368:255–257.
132. Clapham DE: Intracellular signalling: more jobs for Gβγ. *Curr Biol* 1996;6:814–816.
133. McEnery MW, Snowman AM, Snyder SH: The association of endogenous $G_o\alpha$ with the purified ω-conotoxin GVIA receptor. *J Biol Chem* 1994;269:5–8.
134. Kuo CC, Bean BP: G-protein modulation of ion permeation through N-type calcium channels. *Nature* 1993;365:258–262.
135. Dunlap K, Fischbach GD: Neurotransmitters decrease the calcium conductance activated by depolarization of embryonic chick sensory neurones. *J Physiol (Lond)* 1981;317:519–535.

Advances in Second Messenger and Phosphoprotein Research, Vol. 33

1040-7952/99 $30.00

7

L-Type Calcium Channel Modulation

Annette C. Dolphin

Department of Pharmacology, University College of London, London WC1E 6BT, England

INTRODUCTION

L-type calcium channels were initially defined functionally as those that are sensitive to the calcium channel antagonist drugs, including 1,4-dihydropyridines (DHPs). These drugs were first identified as being particularly effective ligands for calcium channels in the heart and blood vessels. The cardiac calcium channel therefore formed the prototypical L-type calcium channel and as such has been extensively studied, in terms of both its intrinsic biophysical properties, and its modulation by second messengers, principally cyclic AMP (1,2). However, it has been found that α1D, the major L-type calcium channel mRNA in neurons and neurosecretory cells, is the product of a different gene and this channel would appear to have somewhat different properties in terms of its biophysics and pharmacology (3,4). In this chapter I shall concentrate on modulation of L-type calcium currents in neurons and neurosecretory cells, but will also draw on studies from cardiac and skeletal muscle L-type channels.

PROPERTIES OF L-TYPE CHANNELS

Subtypes of L Channel

Cardiac calcium channels are activated by large depolarizations, carry Ba^{2+} as a charge carrier better than Ca^{2+}, and inactivate rapidly in the presence of Ca^{2+} but not Ba^{2+}. Block by 1,4-DHP antagonists is strikingly voltage dependent. For example, the IC_{50} for nimodipine is 4 and 450 nM at −30 and −80 mV respectively (5). A similar voltage dependence was observed for nifedipine (6). The single channel currents show brief openings and have a single channel conductance of 20–24 pS. A hallmark of L-type calcium currents is their enhancement by 1,4-DHP agonists, particularly *S*-(−)-BayK8644. The efficacy, but not the potency of calcium channel activators, is affected by voltage, efficacy being markedly increased at hyperpolarized holding potentials (5). Several DHP antagonists have also been found to have an agonist phase at hyperpolarized holding potentials (7), suggesting that many of these drugs have partial agonist activity. At the single channel level this enhancement is characterized by an increase in frequency of openings, a reduction of sweeps with no openings, and an increase in the duration of openings (7). This behavior has been referred to as a mode switch (8), with the null sweeps being termed mode 0, the brief opening mode being mode 1, and the long opening mode called mode 2. The presence of modes has been called into question by some groups (7), but what is without question is the ability of these agonist drugs to enhance L-type calcium channel currents, most distinctively by prolonging the channel open time. This is an important characteristic, since the specificity of the antagonist drugs is not absolute. Although the picture is clear in cardiac myocytes, where calcium current is completely blocked by submicromolar concentrations of DHP antagonists, the use of these drugs to specify components of current in cell types, particularly neurons, containing multiple types of channel is less reliable, as will be discussed.

Calcium channels were first identified in skeletal muscle (9); these channels have properties very different from cardiac L-type channels, being very slowly activating. However, the sensitivity of skeletal muscle calcium channels to DHP agonists and antagonists results in their classification as L-type channels. The skeletal muscle DHP receptor is thought to function as a voltage sensor rather than as a calcium-conducting pore. By a mechanism not clearly understood, membrane depolarization is thus coupled to release of calcium from the sarcoplasmic reticulum. Thus the rapid intramembrane charge movement that is the initial response of the channel resulting from depolarization of the membrane is the trigger for opening of the ryanodine receptor. This charge movement can be observed as an outward gating current at the start of depolarization. The open times of single skeletal muscle L-type calcium channels are enhanced by DHP agonists, and the probability of opening is reduced by DHP antagonists, which also affect charge movement (10). Several groups have suggested that there is a discrepancy between the number of DHP receptors in skeletal muscle and the calcium channel amplitude, indicating that a large proportion of these receptors do not represent functional channels (11). However, skeletal muscle L channels also show an anomalous gating behavior, not seen in cardiac muscle cells, in that there is a time-dependent enhancement of the calcium tail current, upon repolarization, which may represent normally silent channels (12,13).

In neurons and neurosecretory cells, L-type calcium channels are present, usually in conjunction with other types of channel. The distinction between the L-type and channels of other subtypes is usually made using DHP antagonists, at concentrations that completely inhibit cardiac calcium channels (usually 1 μM). However, the use of DHP antagonists does not always result in a clean definition of L current in neurons, particularly, as already mentioned, since block by DHP antagonists is voltage dependent (7). For example, in cortical neurons some authors have observed an almost complete inhibition of calcium current by DHP antagonists (14), whereas a proportion of the current can also be inhibited by antagonsts of other channel types (14,15). We have recently shown that the DHP antagonist nicardipine inhibits expressed α1E calcium currents, although the agonist BayK8644 produces no enhancement (16). Therefore the effect of DHP agonists represents a much more secure definition of the presence of L-type current. The prolongation of tail current by a DHP agonist, resulting from delayed closing of mode 2 channels, is clear evidence of the presence of L-type channels. However, this does not allow determination of the proportion of current that is L-type. Furthermore examination of the ability of neurotransmitters or modulators to inhibit L current by examination of the effect on the DHP agonist prolonged tail current, is fraught with problems of interpretation (17,18), since the modulator may be ineffective to inhibit channels in their mode 2 (DHP-modified) state.

L-Type Calcium Channel Structure

The skeletal muscle DHP receptor was the first to be purified, a task aided by its high concentration in the T tubular system (19). It was found to consist of a number of subunits, including the DHP receptor itself, termed the α_1 subunit, which is now known to be the main pore-forming subunit of the channel. In addition, there was a glycosylated α_2 subunit, attached by a disulfide bond to a δ subunit, and a β subunit; γ subunits were also found associated with skeletal muscle DHP receptors (for reviews see Refs. 20,21). The relatively large yield of purified skeletal muscle calcium channel subunits allowed peptide sequences to be obtained, and cDNAs encoding these proteins to be isolated (22–24). Further clones were then obtained by homology screening (4,25–32).

There are now four cloned L-type calcium channels, corresponding to skeletal muscle (α1S) (22), cardiac (α1C) (25), neuronal/neurosecretory (α1D) (3,33,34), and retinal (α1F) (34a). The cardiac calcium channel has been studied most extensively, following expression in *Xenopus* oocytes and mammalian cell lines (25,35,36). Several splice variants have been identified, and it is present not only in cardiac myocytes but also in smooth muscle (37), in neurons, and elsewhere. Some splice variants confer differences in pharmacological properties (38). The skeletal muscle calcium channel clone has been more difficult to study because it does not express well in *Xenopus* oocytes, but expresses well in dysgenic mouse skeletal myocytes, which do not themselves show any functional endogenous calcium channels (39). One of the reasons for this problem of expression may be that the II–III loop of α1S interacts directly with the endogenous ryanodine receptors, which aid functional expression (40). More recently α1S has been heterologously expressed in tsA-201 cells and shown to exhibit properties similar to those of the calcium currents in a skeletal muscle cell line (13). There are little functional data on α1D, which has been cloned both from human and from rat tissue (3,24,41).

All α1 subunits show increased functional expression in the presence of a β subunit (42–44), and several studies also show an effect of the α2δ subunit (36,45,46). It is thought that all native calcium channels contain these subunits (47,48) and possibly others as well, and that the skeletal muscle γ subunit has a counterpart (γ2 in other tissues (48a) (for review see Ref. 20).

MODULATION OF CARDIAC AND SKELETAL MUSCLE L-TYPE CHANNELS

Role of G_s in Stimulation of Cardiac Calcium Channels

Cardiac calcium channels are the substrates for a number of different modulatory systems. In cardiac tissue it has long been known that calcium current is enhanced by norepinephrine acting at β-adrenoceptors, and inhibited by activation of muscarinic M2 receptors (1). The effect of β-adrenoceptor stimulation is to activate the G protein G_s, which by activation of adenylyl cyclase causes an increase in cyclic AMP levels, stimulation of cAMP-dependent protein kinase (PKA), and phosphorylation (49,50), possibly of one of the calcium channel subunits (51). Both calcium channel β subunits and the α1C subunit are substrates *in vitro* for PKA (52,53) and have been shown to be phosphorylated by PKA under physiological conditions (54). However, it is not yet clear which sites are responsible for enhanced activity (55), and systematic mutation of α1C PKA sites from serine to alanine has little effect on calcium channel properties (56). Cloned and expressed α1S is modulated by PKA, and this requires an A kinase anchoring protein, which may be situated in close association with the calcium channel (13). Clearly, a similar situation is likely to prevail for α1C, and the other L-type α1 subunits.

Direct Stimulation by G_s versus Elevation of cAMP

There is also a body of work suggesting that there is a direct pathway for activation of cardiac calcium channels by G_s, which of course would be faster than a pathway involving cAMP. In experiments in which elevation of cyclic AMP or activation of PKA was blocked, it was observed that a small rapid stimulation of calcium currents was still present (57–59). This was attributed to direct effect of G_s on the calcium channel, which could be mimicked by GTPγS-activated $G\alpha_s$. Further, under normal conditions it was suggested that a direct rapid effect of G_s activation could be observed following activation of β-adrenoceptors, followed by the larger slower effect, which could be attributed to a rise in cyclic AMP (58). This was suggested to have physiological significance as a rapid response to sympathetic stimulation (58). A similarly fast response to G_s has been observed when skeletal muscle and cardiac calcium channels are reconstituted in lipid bilayers (57,60). However, the finding of a biphasic response to β-adrenoceptor stimulation has been disputed and could not be replicated in several different species (61,62).

An alternative and entirely credible explanation is that elevation of cAMP is not uniform in a cardiac cell, and application of β-adrenoceptor agonists may produce an initial and very rapid localized increase in cAMP whose ability to spread throughout the cell is delayed by phosphodiesterase action. Thus the response would be more complex than that due to forskolin (63). Nevertheless, biochemical experiments also showed the association of $G\alpha_s$ with purified skeletal L-type calcium channels (64). However, no experiments have been performed with cloned cardiac α1C subunits to determine whether $G\alpha_s$ has any affinity to bind with intracellular domains of the channel. Such evidence is essential if the direct $G\alpha_s$–calcium channel hypothesis is to retain credibility. What is not disputed is that by far the major effect on cardiac calcium channels of G_s activation is cAMP elevation and resultant phosphorylation.

Inhibitory Modulation of Cardiac Calcium Channels

The effect of stimulation of muscarinic receptors is to reduce the current flow through cardiac calcium channels. This occurs by activation of pertussis toxin sensitive G proteins, largely the G_i species in cardiac tissue. It is thought that inhibition results from inhibition of adenylyl cyclase, resulting in a fall in cAMP levels and a reduction in phosphorylation (65,66). Other mechanisms also probably contribute to muscarinic inhibition of cardiac calcium current, as explored in 1997 (67).

In neuronal tissue, one of the hallmarks of direct inhibition of calcium currents by pertussis toxin sensitive G proteins of the G_i/G_o family is slowed current activation, which is interpreted as a voltage-dependent inhibition of calcium currents (20). It is suggested that this phenomenon represents voltage-dependent relief of block of the channels, possibly a dissociation between activated G protein ($G\beta\gamma$ subunits) and a component of the channel itself at depolarized voltages. This process can also be reversed, producing a relief of inhibition, by applying a large depolarizing prepulse shortly before the test pulse (for a recent review see Ref. 20,20a). There is no clear evidence for such a phenomenon associated with muscarinic inhibition of cardiac calcium currents, suggesting that there is no direct interaction of these G proteins with α1C subunits. However, facilitation of calcium current by depolarizing prepulses does occur in native and cloned cardiac calcium channels (68,69), although the effect is not enhanced by G protein activation and is believed to be a property of the channels themselves (166).

Modulation of Skeletal Muscle Excitation–Contraction Coupling by G Proteins

The L-type channels in skeletal muscle T tubules are involved in excitation–contraction coupling, and several pieces of evidence suggest that G proteins are also present in the T-tubular membrane, in close association with calcium channels. $G\alpha_o$ is present in great abundance, as well as $G\beta$ and $G\gamma$ subunits (70,71). $G\alpha_s$ is also present (72). It has been shown by several groups that G protein activation by nonhydrolyzable GTP analogs causes contraction in skinned muscle fibers, an effect blocked by pertussis toxin (73,74). GTP analogs also enhanced caffeine-induced calcium release from sarcoplasmic reticulum (71), and increased calcium current and charge movement in skinned muscle fibers (75). Although not tested in every case, it would appear that the G protein involved is probably of the G_o or G_i subtype, because of its pertussis toxin sensitivity. In contrast, Hamilton et al. (64) showed that purified activated $G\alpha_s$ enhanced the open probability of DHP receptors from skeletal muscle T tubules incorporated into lipid bilayers.

MODULATION OF NEURONAL/NEUROSECRETORY L-TYPE CALCIUM CHANNELS

Identification of L-Type Calcium Currents in Neurons and Secretory Cells

In most neurons the major high voltage activated calcium channel subtypes appear to be non-L, that is, either N-type, blocked by ω-conotoxin GVIA, or P/Q- type, blocked by ω-agatoxin IVA. However, in most neuronal cell types there is a proportion of current that is blocked by DHP antagonists, and enhanced by the DHP agonist BayK8644. In some types of neuron the proportion of L current is significant: for example, in cerebellar granule cells and several cell types in the retina (76–80). It should also be noted that there are many DHP binding sites in the brain, even associated with cell types such as Purkinje neurons, which display little L-type calcium current (81). This may be explained by the very low probability of opening of a subset of L-type channels, unless enhanced ("facilitated") by DHP agonists (12). Indeed some neuronal L-type channels may be functionally silent in terms of gating Ca^{2+} (81). Therefore it has been proposed that in some neurons, L-type calcium channels may function in a similar way to their role as voltage sensors in skeletal muscle (81). This exciting possibility certainly merits further investigation.

In neuroendocrine tissue, by contrast, a large proportion of the current is generally L-type (82,83). This is true not only for cells prepared directly from animal tissue, or placed in primary culture, but also for the numerous cell lines derived from various neurosecretory cell types (84). Activation of the L-type current is essential for entry of calcium for the secretory process (85), which can often be almost completely blocked by L-type calcium channel antagonists.

Modulation of Neuronal and Neurosecretory Calcium Currents by G Proteins

Hille (85a) identified a number of different pathways for inhibitory calcium channel modulation; some were membrane delimited and thought to be directly G protein mediated, and others involved soluble second messengers including Ca^{2+}. It was suggested that the fast membrane-delimited pathways were confined to an action on non-L-type channels, and generally utilized a pertussis toxin sensitive pathway, although examples of pertussis toxin insensitive pathways are present in the literature, particularly inhibition by LHRH and substance P (86,87). Further, this schema indicates that modulation of L-type channels generally occurs by activation of a pertussis toxin insensitive pathway utilizing a G protein of the G_q/G_{11} family, and involves a soluble second messenger that is sensitive to intracellular Ca^{2+} chelation. This messenger may be the rise in intracellular Ca^{2+} itself, since L-type Ca^{2+} channels are subject to Ca^{2+}-dependent inhibition (88). Ca^{2+} may additionally activate downstream Ca^{2+}-dependent pathways. As demonstrated next, however, there are in fact several examples in the literature of L channels being modulated by what appears to be a pertussis toxin sensitive direct G protein pathway.

Membrane-Delimited Inhibition of Neuronal and Neurosecretory L-Type Currents

Many neurotransmitters and neuromodulators activate G proteins (usually pertussis toxin sensitive) to produce a characteristic inhibition of neuronal calcium currents that is independent of elevation of internal Ca^{2+}. Some workers have subdivided this type of inhibition into two subsets (89,90). One, the voltage-dependent aspect, is characterized by slowed activation kinetics, with inhibition diminishing the greater the depolarization, and reversal of inhibition by large depolarizing prepulses (91–93). The second type of inhibition, which shows simply a scaled reduction of the current with no slowed activation phase, is often termed steady state inhibition. Whether these represent inhibition of two structurally different types of calcium channels or two different mechanisms of inhibition remains to be settled categorically. However, by the use of selective blockers, a number of groups have identified scaled inhibition with L-type channels, and voltage-dependent inhibition with non-L (N or P/Q) channels (90,94). In conflict with these results, G protein modulation of cloned α1B channels involves both a kinetic slowing component and a steady state inhibition that cannot be reversed by a depolarizing prepulse (95,96), so that the categorical distinction between these two modes of inhibition remains unclear.

Although in many systems G protein mediated inhibition involves predominantly non-L channels, modulation of L-type channels has been observed in many cell types, particularly in secretory systems and in certain neurons. When the criteria of block by DHP antagonists are used to define an L-type current component, there are a number of reports of L current inhibition resulting from the activation of several classes of receptor. In a study using selective calcium channel blockers, serotonin acting at 5HT1A and 5HT2C receptors inhibited an L-type as well as a Q-type current in rat pituitary melanotrophs (90). Dopamine and selective agonists acting at D2 receptors on pituitary melanotrophs inhibited L-type current in a voltage-dependent manner (97).

In goldfish retinal ganglion cells, where the calcium current appears to be entirely L-type, activation of a $GABA\text{-}_B$-like receptor and a substance P receptor produced a marked inhibition (76). In these cells, transmitter release results from calcium influx through L-type channels and is also inhibited by $GABA\text{-}_B$-like receptors (76,98). This was also true in retinal bipolar cells (99). Activation of $GABA\text{-}_B$ receptors in hippocampal neurons inhibited both N- and L-type currents (100).

In cerebellar granule neurons, L-type current makes up an unusually high proportion of the total macroscopic current (77,80,101). (–)-Baclofen acting at $GABA\text{-}_B$ receptors inhibits maximally over 50% of the current in these cells (102,103), and therefore must be inhibiting L current, at least in part. It has further been shown that following inhibition of cerebellar granule cell calcium current with nimodipine, the amount of inhibition produced by (–)-baclofen was reduced (18). However, (–)-baclofen had no effect on BayK8644 prolonged tail currents, suggesting that mode 2 calcium currents are not available for inhibitory modulation by G proteins (18). However, Haws and colleagues (104) showed that whereas a DHP agonist still stimulated residual current in these cells following infusion of GTPγS, this enhancement gradually decayed, suggesting that the L current also is inhibited by activated G protein. It has further been shown that metabotropic glutamate receptors and dopamine D4 receptors inhibit L-type channels in the same neurons, by a mechanism that appeared to be direct (105,106). Similarly, activation of mGluR receptors in acutely isolated cortical neurons caused an inhibition of L-type calcium currents that was apparently direct, inasmuch as it was not attenuated by intracellular Ca^{2+} buffering (107).

In rat dorsal root ganglion (DRG) neurons, the GABA-$_B$ agonist (–)-baclofen inhibits about 30% of the voltage-dependent calcium current, measured using Ba^{2+} as the charge carrier (108). This inhibition can be completely reversed in a voltage-dependent manner by a depolarizing prepulse (Fig. 1A). The current in rat DRGs is made up of approximately 40–50% N-type current (inhibited by ω-conotoxin GVIA) (109), and a slightly smaller proportion of L-type current (inhibited by 1–5 μM nifedipine) (110). A further 15–20% is inhibited by 100–200 nM ω-agatoxin-IVA (111), indicating that in these cells there is only a small proportion of P/Q current. In cultured DRGs the residual current following irreversible block of N channels with ω-CTX GVIA was still substantially inhibited by (–)-baclofen, indicating that L and P/Q channels also are likely to be modulated (Fig.1 B,C).

The molecular nature of the L-type current in DRGs remains unclear, since there are no selective pharmacological tools to isolate α1C and α1D. However, we have identified α1D transcripts in cultured DRGs by reverse transcriptase PCR (112). Surprisingly, in DRGs that have been cultured and then acutely replated nonenzymatically, to isolate cell bodies, no L-type current was observed, suggesting that it may have been associated with neurites (113). In these cells inhibition of the calcium current by (–)- baclofen was completely abolished by ω-CTX GVIA, indicating that only N channels were modulated (113). In DRGs, as in many cell types, the poorly reversible activator of G proteins, GTPγS, produced a marked inhibition and slowed activation of calcium current (114–116). This inhibition may be up to 90%, and may be potentiated by activation of receptors such as those for GABA-$_B$ (114), which enhance the exchange of GTPγS for GDP on the G protein (117). When the effect of GTPγS is maximal, the entire high voltage activated current is slowly activating (115), without an unaffected fast component, indicating that all subtypes of high voltage activated current may be modulated by interaction with activated G protein. This has been shown most clearly by the use of "caged" photoactivated GTPγS (115).

Mechanism of Inhibition of Neuronal and Neurosecretory L Channels by G Proteins

There is no clear body of evidence that inhibition of L channels *always* utilizes a radically different G protein or mechanism from inhibition of other high voltage activated channels, although as Hille has discussed (85a), L channels of the α1C class are inhibited by mechanisms that elevate intracellular Ca^{2+}, probably because they are

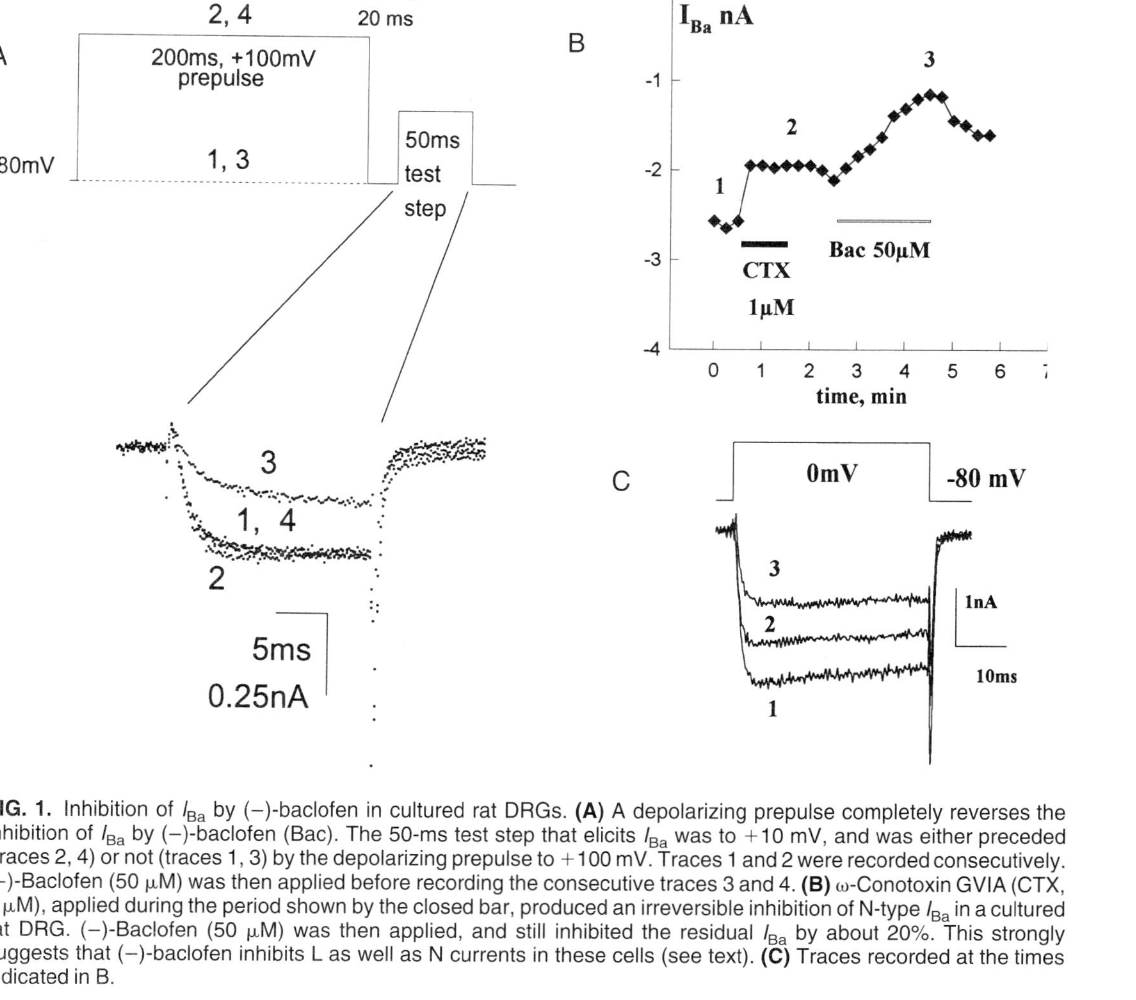

FIG. 1. Inhibition of I_{Ba} by (−)-baclofen in cultured rat DRGs. **(A)** A depolarizing prepulse completely reverses the inhibition of I_{Ba} by (−)-baclofen (Bac). The 50-ms test step that elicits I_{Ba} was to +10 mV, and was either preceded (traces 2, 4) or not (traces 1, 3) by the depolarizing prepulse to +100 mV. Traces 1 and 2 were recorded consecutively. (−)-Baclofen (50 µM) was then applied before recording the consecutive traces 3 and 4. **(B)** ω-Conotoxin GVIA (CTX, 1 µM), applied during the period shown by the closed bar, produced an irreversible inhibition of N-type I_{Ba} in a cultured rat DRG. (−)-Baclofen (50 µM) was then applied, and still inhibited the residual I_{Ba} by about 20%. This strongly suggests that (−)-baclofen inhibits L as well as N currents in these cells (see text). **(C)** Traces recorded at the times indicated in B.

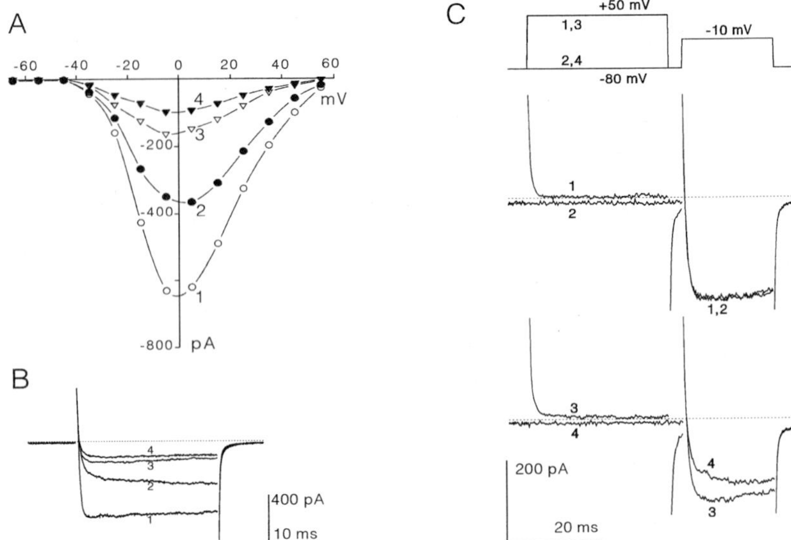

FIG. 2. Inhibition of the Ca^{2+}-channel current (I_{Ca}) by galanin in RINm5F cells. **(A)** The I–V relations of I_{Ca} in RINm5F cells: 1, I_{Ca} under control conditions; 2, I_{Ca} after application of galanin (0.5 μM); 3, I_{Ca} after addition of (+)-isradipine (10 μM); 4, I_{Ca} after subsequent application of a mixture of galanin and (+)-isradipine. **(B)** Representative traces of I_{Ca} at a test potential of −5 mV; numbers refer to the experimental conditions outlined in A. **(C)** Consecutive traces of I_{Ca} recorded in RINm5F cell with a standard (2, 4) and a double pulse (traces 1, 3) protocol, shown in the top panel above the current traces. Traces 1 and 2 are in the absence, and traces 3 and 4 in the presence, of galanin, which induces a slowed current activation, reversed by the depolarizing prepulse (120).

extremely sensitive to direct Ca^{2+}-dependent inhibition (67,118). In most cases the G proteins involved in L-type calcium channel inhibition appear to be of the G_i/G_o type (119). The G protein subunits involved in the signal transduction pathway leading to calcium current inhibition have been studied in an extensive series of experiments by Schultz and colleagues, largely examining inhibition of L-type channels in neurosecretory cells (82). An example from their work on a pancreatic β cell line (120) is shown in Fig. 2. It is clear from Fig. 2A that the great majority of the current is inhibited by the DHP antagonist used. Using both antibodies and antisense oligonucleotide injection, the subtypes of Gα, Gβ, and Gγ subunit implicated in inhibition of calcium channels resulting from activation of several different receptors in a number of cell types have been determined (121). For the same receptor in different cell types, the combination appears to be the same: it always involves $G_o\alpha 1$ or $G_o\alpha 2$ and a combination of β and γ subunits; for example, muscarinic M_4 receptors tend to use $G_o\alpha 1$, β3, γ4, and somatostatin receptors $G_o\alpha 2$, β1, γ3 (84), with some overlap. The combination reflects the receptor type across different tissues, suggesting that certain G protein combinations are preferred to recognize particular receptors. It is a common finding that a $G_o\alpha$ rather than $G_i\alpha$ species is involved

(84,113,122–124). This may reflect the abundance of these Gα subunit species in the cell types involved, or may indicate a specificity of $G_o\alpha$ in the downstream processes, or compartmentalization of specific receptors, G proteins, and calcium channels. Certainly the receptors involved are also able to utilize $G_i\alpha$ and do so to inhibit adenylyl cyclase, which occurs via both $G_i\alpha$ and Gβγ (125).

G Protein Subunits Involved in Calcium Channel Modulation

The mechanism of G protein mediated inhibition therefore involves agonist activation of a particular receptor, which recognizes and activates a particular G protein–heterotrimer combination. This results in exchange of GTP for GDP on the Gα subunit and possibly the separation of the activated Gα subunit from Gβγ. Recent evidence suggests that Gβγ subunits are responsible for direct inhibition of N-type channels in sympathetic neurons and of cloned α1A and α1B channels (126–129). The direct role (if any) of $G_o\alpha$ in this schema is therefore unclear, although it acts as a molecular switch and a timer, having an intrinsic GTPase activity, such that when it returns to its GDP-bound ground state, it rebinds βγ, thus terminating the signal transduction process (130). This process is now known to be speeded up by regulators of G protein signaling (RGS) proteins (130a). However, it is of interest that activated $G_o\alpha$ has been found to be associated with purified N channels, and it may be that it has a binding site on the calcium channel, for proximity to Gβγ and termination of the inhibitory process. It is also possible that the calcium channel or an associated protein acts, in concert with RGS proteins, as a GTPase-activating protein (131,132) and may provide a binding site for specific RGS proteins. The intrinsic GTPase activity of $G_o\alpha$ is lower than for $G_i\alpha$ (133), and the affinity of $G_o\alpha$ subunits for Gβγ is generally lower than that of $G_i\alpha$ (134). Furthermore, there is normally a large proportion of $G_o\alpha$ in neurons and neurosecretory cells relative to $G_i\alpha$ (135). These three pieces of evidence may explain why $G_o\alpha$ is found to be most involved in calcium channel inhibition (108,113,122,136), because once activated, it releases Gβγ subunits for a sufficient length of time to have an inhibitory effect. It is also generally thought to be characteristic of signaling via Gβγ compared to α subunits that higher concentrations are required (137). If it is true for calcium channels that a high concentration of Gβγ is needed for modulation, there may normally be insufficient G_q or G_s to provide this in most cell types. Thus the major pool of Gβγ in neurons would be found associated with G_o.

Voltage-Dependent Inhibition of Neuronal and Neurosecretory L Currents

Voltage dependence of calcium current inhibition normally refers to inhibition that is greater at submaximal depolarizations and can be overcome, either by large depolarizations or by long steps, such that the current at the end of a long depolarizing step is less inhibited by agonist than is the current near the start of the step (see Fig. 2B, C for an example of the effect of galanin on L-type calcium currents in RINm5F cells) (120). Such data suggest a model that in a simple form would have an activated

G protein binding to the calcium channel with highest affinity at hyperpolarized potentials, when the channels are closed. Then, following depolarization of the membrane, a conformational change in the calcium channel resulting from movement of the voltage sensors would possibly result in opening of the modulated or "reluctant" channel (92), but more importantly either gradual voltage-dependent dissociation of the activated G protein or removal of its ability to inhibit the channel, such that increasing concentrations of the unmodulated "willing" channels would be available to open normally. An even simpler form of the model suggests that G protein bound channels do not open at all, and the slowed activation seen represents dissociation of activated G protein from the calcium channel (138), such that the time constant of current activation in the presence, for example, of GTPγS is dominated by the dissociation rate of activated G protein from the channel.

The simple model just described was rejected by Elmslie and Jones (139), who found no G protein concentration dependence of the slowed activation. However, more recently a clear concentration dependence of reinhibition following a depolarizing prepulse has been observed (140,140a), and single channel recordings of α1B channels support the hypothesis that slowed activation stems from the requirement for unbinding of activated G protein (141). Originally, models suggested that up to four G proteins might bind per calcium channel, possibly one per voltage sensor (92). However, other work has shown, at least in the case of skeletal muscle calcium channels, that in practice one or two of the voltage sensors provide the rate-limiting step. From studies of chimeras, one group has suggested that domain I is most important for sensing voltage (142) whereas another group has found that the voltage sensors in domains III and IV are responsible for the slow activation of skeletal muscle calcium channels (143). Thus it is possible that only one activated G protein is involved, and its influence is associated only with the rate-limiting voltage sensor domain. This would be in agreement with the recent finding that the effects of G$\beta\gamma$ subunits are mediated via the VDCC α1 I–II loop (96,128,129,144), domain I and the N terminus (144a,b).

Although most work on the voltage dependence of calcium current inhibition has examined modulation of N-type current, particularly in sympathetic ganglion neurons, there are several examples of similar phenomena occurring for L channel modulation (cf. the example given in Fig. 2). In particular, slowed activation and reversal of inhibition by depolarizing prepulses has been reported for neurosecretory L currents (97,103,145). Thus the distinction mentioned earlier concerning scaled inhibition of L currents, contrasting with voltage-dependent inhibition of N currents (90,146), cannot be classed as absolute. To take another example, in pancreatic β cells, GTPγS inhibited L-type current in a voltage-dependent manner, and the inhibition was relieved by a depolarizing prepulse (145). A similar result was observed in a pancreatic β-cell line (Fig. 2C). However, one caveat related to these results is that we observed α1E to be partially blocked by DHP antagonists (16), and the G protein modulation of "L current" in these cells may actually represent modulation of α1E (144b).

Site of Interaction between Calcium Channel and Activated G Protein

The calcium channel β subunit has been observed by several groups to affect the biophysical properties of expressed α1 subunits of calcium channels (for review see Ref. 147). Several effects may be involved, both on expression of functional channels at the plasma membrane (148–151) and on activation and inactivation kinetics and voltage dependence (45,152,153). The effects of coexpression of the calcium channel β subunit are to some extent opposite of those resulting from G protein modulation, involving a hyperpolarizing shift in activation and an enhancement of activation kinetics. We have confirmed this in native neurons by the depletion, using antisense oligonucleotides, of endogenous calcium channel β subunits in DRGs (152). We observed that the current decreased in amplitude, its activation was slowed, and its voltage dependence shifted to more depolarized potentials (152). However, a novel and intriguing finding was that there was greater inhibition of the residual current by (−)-baclofen (108). This result suggested that the calcium channel β subunit may occlude the interaction of activated G protein subunits with the channel (108). It is known that calcium channel β subunits bind to a region on the intracellular loop between transmembrane domains I and II on the α1 subunits (154). The binding domain on the α1 subunit has been called the alpha interaction domain (AID), and it interacts with a domain on the calcium channel β subunit called beta interaction domain (BID) (155,156). It has recently been reported that the α1 I–II loop contains binding sites for Gβγ (128,129). If the calcium channel β subunit does physically occlude the binding of activated G protein, this suggests, first that there is only one binding site for activated G protein per calcium channel oligomer, and second that the binding of calcium channel β subunit should be reversible. However, the affinity of interaction of AID and BID has been measured *in vitro* to be in the low nanomolar range for a number of different combinations of calcium channel α1 and β subunits (155). Therefore it is possible, if competition for interaction cannot be demonstrated, that there is an allosteric interaction with another site or sites on the calcium channel, with which activated G protein βγ subunits bind. However, the evidence from biophysical studies that the inhibitory modulation by G proteins is voltage dependent suggests the possibility that the calcium channel α1β subunit interaction may be reciprocally voltage dependent, showing greater affinity at depolarized potentials. Furthermore, the essential role of the N terminus of the α1B and α1E subunits implicates this region in a complex binding site (144b).

Modulation of DHP Binding and Function by Calcium Channel β Subunits and G Proteins

The binding sites for DHP agonists and antagonists on skeletal muscle and cardiac calcium channels have been mapped by several groups and have been shown to overlap, with an additional epitope being involved in DHP agonist action (157,158). The principal regions appear to be on domains III and IV of the channels. The calcium channel β subunit appears to increase the affinity of DHP ligand binding to expressed channels (159), as well as increasing the apparent number of channels expressed (149,160). We also observed a reduction in the agonist response to BayK8644 following antisense depletion of the calcium channel β subunit in DRGs (152). Therefore it appears that there is a remote effect of calcium channel β binding to the intracellular I–II loop on the conformation of the DHP binding site. If there are also G protein interactions with this loop on the neurosecretory α1D channel (but see later), G protein activation may also affect DHP binding or function, and vice versa. Indeed in the literature there are several examples of such phenomena.

In cultured DRGs we used DHP antagonists to determine whether a particular type of calcium current remained following modulation of the current by GTPγS. Surprisingly, we observed that application of several DHP antagonists resulted in a marked agonist phase when cells were dialyzed with intracellular GTPγS (110,161,162). This effect was blocked by pertussis toxin, and was much more prolonged at hyperpolarized than at depolarized holding potentials. Calcium channel antagonists often show an initial agonist phase to their response (7), especially at hyperpolarized potentials, and the effect of activation of a pertussis toxin sensitive G protein appeared to promote this phase (110,161). The mechanism of this enhancement of the ability of DHP antagonists to show an initial transient agonist response remains unclear, although it may involve a change in phosphorylation state (163,164). It may be that altered G protein binding or second messenger–induced phosphorylation affects the ability of DHPs to bind to the agonist and antagonist sites on the calcium channel.

There is other evidence that a pertussis toxin sensitive G protein is involved in the behavior of neuronal L-type calcium channels, for example, pertussis toxin treatment of cerebellar granule cells prolongs the lifetime of excised single L-type channels, suggesting that a pertussis toxin sensitive G protein is normally involved in their tonic inhibitory modulation (165). In agreement with this, we observed that pertussis toxin increased the number of high affinity DHP binding sites in cerebellar granule neurons (103).

Facilitation of Neuronal and Neurosecretory L Channels

Neuronal and neurosecretory L-type calcium channels may be facilitated or enhanced by several G protein dependent mechanisms. This area has been reviewed recently (166). One mechanism, studied particularly in chromaffin cells (167,168), appears to involve G_s and cAMP-dependent phosphorylation (167,169), whereas other mechanisms involve voltage-dependent reversal of G protein mediated inhibition (170,171). The latter mechanism has been studied particularly in relation to non-L channels (171,172), whereas the former mechanism appears to involve largely L-type channels. It remains unclear whether facilitation of calcium currents and reversal of G protein mediated inhibition of calcium currents are clearly distinct events, since tonic inhibition of calcium current, due to tonic autocrine release of substances such as ATP and opiates, may be a common phenomenon (171,172) (for review see Ref. 166). It is likely that this process will affect all calcium channels that are G protein modulated, possibly including neurosecretory L channels. Yet another mechanism involves activation of receptors linked to elevation of intracellular Ca^{2+} concentration and has been shown to occur in cerebellar granule cells as a result of activation of mGluR1 or 5 (173). This novel mechanism of facilitation has recently been shown to involve a direct link to neuronal ryanodine receptors (174). As yet there is no evidence in neurons for any direct G_s-linked facilitatory mechanism of L-type calcium channels, such as has been proposed for cardiac and skeletal muscle L-type channels (57,58,64).

BIOCHEMICAL STUDIES OF G PROTEIN INTERACTIONS WITH CALCIUM CHANNELS

Although it has been shown only for native N-type calcium channels (126) and for expressed α1A, α1B, and α1E channels (127) that Gβγ mediates calcium channel inhibition, it is possible that this is also the mechanism of inhibition of neuronal/neurosecretory (α1D) L channels. However, to date there have been no biochemical studies of Gβγ binding to α1D calcium channel domains, and the I–II loop of α1D does not contain a QQIER motif, unlike α1A and α1B. Instead this is QQLEE, the same sequence as in α1C. As suggested earlier, it is possible that there is also a binding site for $G\alpha_o$ that ensures proximity to, and rapid rebinding of Gβγ. Such a site has been observed on the inwardly rectifying class of K^+ channels that are regulated by Gβγ (175,176). There are several observations that suggest a biochemical interaction between voltage-dependent calcium channels and G proteins. First, $G\alpha_s$ was found to copurify with skeletal muscle L-type calcium channels (64). Second, activated $G\alpha_o$ was associated with purified N channels (177). Less directly, GTPγS increased the affinity of neuronal membrane DHP receptors for DHP ligands (178) and reduced the amount of ω-conotoxin GVIA binding (179). Both these effects were blocked by pertussis toxin. Both these results indicate that calcium channel interaction with an activated G protein produces a conformational change that affects binding of the ligand. Since this occurred in cell membranes, it is likely to involve a direct interaction rather than generation of a second messenger. However, such evidence of direct interaction of G protein α subunits with endogenous calcium channels requires direct experimental examination with cloned calcium channels, and none has been reported to date.

In a series of functional experiments to determine whether the calcium channel as an effector had any influence on the enzymatic activity of the G protein, we examined G protein GTPase activity. We have observed that GTPase activity in cortical membranes can be stimulated by DHPs, particularly agonists (131), (–)-BayK8644 having an EC_{50} of about 0.5 nM (132). For potentiation of cardiac L-type channels, its EC_{50} has been reported to be 3 nM for a high affinity site (7). Since the enhancement of GTPase can be blocked by anti-G_o but not G_i antibodies, it is due to stimulation of the intrinsic GTPase activity of G_o (131). The effect is additive or synergistic with receptor agonists, which increase GTPase by enhancing GTP exchange for GDP (131,180). Furthermore, DHP agonists, unlike receptor agonists, do not increase the binding of nonhydrolyzable GTP analogs (131). This suggests that the GTPase enhancement is due to an increase in intrinsic GTPase activity consistent with an increase in V_{max} of the enzyme with no change in K_m (131). We put forward the hypothesis that the L-type calcium channel is acting as a GTPase activating protein (GAP) (131). Furthermore, we have shown that the DHP agonist stimulated increase in GTPase is prevented by an antibody against the calcium channel β subunit, suggesting that the β subunit is involved, either directly or indirectly, in the enhanced GTPase (132). Following the recent identification of RGS proteins (130a), it is possible that calcium channel α1 subunits act to bind and localize these proteins.

The concept of a GAP was first mooted for monomeric G proteins because they have very low rates of intrinsic GTPase activity and are associated with specific proteins that enhance this GTPase (for review see Ref. 181). For example p21-*ras* is associated with *ras*-GAP (182). For heterotrimeric G proteins the intrinsic GTPase rate is faster, but in several instances where it has been measured, this rate is insufficiently fast to account for the off-rate of the G protein mediated response (183). Several effectors have been found or suggested to have GAP activity, the process involving activated G protein interacting with the effector, and producing a response, while at the same time, the effector reciprocally terminates the process by enhancing the intrinsic GTPase activity of the G protein α subunit (184,185). In principle this would be effective as a direct mechanism only if the activated Gα subunit interacted with the effector. However, if Gβγ is involved it may activate the effector, which in its activated state then has less affinity for βγ. The effector–Gβγ complex would then dissociate and Gβγ would indirectly increase the GTPase activity of activated Gα by reassociating with it (186,187).

Alternatively, there may be a secondary binding site on the effector for $G\alpha_o$–GTP that has GAP activity, stimulating $G\alpha_o$–GDP production and rebinding of Gβγ. This site would be in addition to the Gβγ site on the calcium channel α1 subunit. It will be of interest to examine the role of the recently discovered group of RGS proteins in the process of terminating calcium channel modulation by G proteins (174,188,189). These RGS proteins stimulate the GTPase activity of specific heterotrimeric G proteins, in an analogous way to the GAP proteins for the Ras superfamily of monomeric G proteins (190).

CONCLUSIONS

Much work has concentrated on modulation by inhibitory G proteins of N- and P/Q-type channels, but there has been less examination of whether neurosecretory L channels also undergo modulation. I have tried to summarize in this short review the work that has been done implicating interactions between both stimulatory and inhibitory G proteins and L-type calcium channels. Some of it is controversial, particularly whether there is a direct G_s coupling to cardiac L-type calcium channels, which remains uncertain. In the case of inhibitory modulation, it will be of great interest in the forthcoming years to determine whether the cloned neurosecretory L-type channel (α1D) is modulated by G proteins by a similar mechanism to α1A and

α1B. This would certainly be predicted from much of the work outlined here on native neurosecretory L channels, but it may not involve an identical mechanism.

ACKNOWLEDGMENTS

I would like to thank Dr. B. Nuernberg, Frei Universität Berlin, for supplying Fig. 2. The author's work is supported by the Wellcome Trust and the Medical Research Council (U.K.).

REFERENCES

1. Reuter H: Calcium channel modulation by neurotransmitters, enzymes and drugs. *Nature* 1983;301: 569–574.
2. Di Virgilio F, Salviati G, Pozzan T, Volpe P: Is a guanine nucleotide-binding protein involved in excitation–contraction coupling in skeletal muscle? *EMBO J* 1986;5:259–262.
3. Seino S, Chen L, Seino M, Blondel O, Takeda J, Johnson JH, Bell GI: Cloning of the α_1 subunit of a voltage-dependent calcium channel expressed in pancreatic β cells. *Proc Natl Acad Sci U S A* 1992;89:584–588.
4. Williams ME, Feldman DH, McCue AF, Brenner R, Velicelebi G, Ellis SB, Harpold MM: Structure and functional expression of α_1, α_2, and β subunits of a novel human neuronal calcium channel subtype. *Neuron* 1992;8:71–84.
5. Hamilton SL, Yatani A, Brush K, Schwartz A, Brown AM: A comparison between the binding and electrophysiological effects of dihydropyridines on cardiac membranes. *Mol Pharmacol* 1986;31: 221–231.
6. Doring F, Degtiar VE, Grabner M, Striessnig J, Hering S, Glossman H: Transfer of L-type calcium channel IVS6 segment increases phenylalkylamine sensitivity of α_{1A}. *J Biol Chem* 1996;271: 11745–11749.
7. Brown AM, Kunze DL, Yatani A: Dual effects of dihydropyridines on whole cell and unitary calcium currents in single ventricular cells of guinea-pig. *J Physiol (Lond)* 1986;379:495–514.
8. Nowycky MC, Fox AP, Tsien RW: Long-opening mode of gating of neuronal calcium channels and its promotion by the dihydropyridine calcium agonist Bay K 8644. *Proc Natl Acad Sci U S A* 1985; 82:2178–2182.
9. Fatt P, Katz B: The electrical properties of crustacean muscle fibres. *J Physiol (Lond)* 1953;120: 171–204.
10. Rios E, Pizarro G, Stefani E: Charge movement and the nature of signal transduction in skeletal muscle excitation–contraction coupling. *Annu Rev Physiol* 1991;54:109–133.
11. Schwartz LM, McCleskey EW, Almers W: Dihydropyridine receptors in muscle are voltage-dependent but most are not functional calcium channels. *Nature* 1985;314:747–751.
12. Fleig A, Penner R: Excessive repolarization-dependent calcium currents induced by strong depolarizations in rat skeletal myoballs. *J Physiol (Lond)* 1995;489:41–53.
13. Johnson BD, Brousal JP, Peterson BZ, Gallombardo PA, Hockerman GH, Lai Y, Scheuer T, Catterall WA: Modulation of the cloned skeletal muscle L-type Ca^{2+} channel by anchored cAMP-dependent protein kinase. *J Neurosci* 1997;17:1243–1255.
14. Allen TGJ, Sim JA, Brown DA: The whole-cell calcium current in acutely dissociated magnocellular cholinergic basal forebrain neurones of the rat. *J Physiol (Lond)* 1993;460:91–116.
15. Brown AM, Schwindt PC, Crill WE: Voltage dependence and activation kinetics of pharmacologically defined components of the high-threshold calcium current in rat neocortical neurons. *J Neurophysiol* 1993;70:1530–1543.
16. Stephens GJ, Page KM, Burley JR, Berrow NS, Dolphin AC: Functional expression of rat brain cloned α1E calcium channels in COS-7 cells. *Pflugers Arch* 1997;433:523–532.
17. Plummer MR, Rittenhouse A, Kanevsky M, Hess P: Neurotransmitter modulation of calcium channels in rat sympathetic neurons. *J Neurosci* 1991;11:2339–2348.
18. Amico C, Marchetti C, Nobile M, Usai C: Pharmacological types of calcium channels and their modulation by baclofen in cerebellar granules. *J Neurosci* 1995;15:2839–2848.

19. Takahashi M, Seager MJ, Jones JF, Reber BFX, Catterall WA: Subunit structure of dihydropyridine-sensitive calcium channels from skeletal muscle. *Proc Natl Acad Sci U S A* 1987;84:5478–5482.
20. Dolphin AC: Voltage-dependent calcium channels and their modulation by neurotransmitters and G proteins: G. L. Brown Prize Lecture. *Exp Physiol* 1995;80:1–36.
20a. Dolphin AC: Mechanisms of modulation of voltage-dependent calcium channels by G proteins. *J. Physiol. (Lond)* 1998;506:3–11.
21. Tsien RW, Lipscombe D, Madison D, Bley K, Fox A: Reflections on Ca^{2+}-channel diversity, 1988–1994. *Trends Neurosci* 1995;18:52–54.
22. Tanabe T, Takeshima H, Mikami A, Flockerzi V, Takahashi H, Kangawa K, Kojima M, Matsuo H, Hirose T, Numa S: Primary structure of the receptor for calcium channel blockers from skeletal muscle. *Nature* 1987;328:313–318.
23. Lacerda AE, Kim HS, Ruth P, Perez-Reyes E, Flockerzi V, Hofmann F, Birnbaumer L, Brown AM: Normalization of current kinetics by interaction between the α_1 and β subunits of the skeletal muscle dihydropyridine-sensitive Ca^{2+} channel. *Nature* 1991;352:527–530.
24. Ellis SB, Williams ME, Ways NR, Brenner R, Sharp AH, Leung AT, Campbell KP, McKenna E, Koch WJ, Hui A, Schwartz A, Harpold MM: Sequence and expression of mRNAs encoding the $\alpha 1$ and $\alpha 2$ subunits of a DHP-sensitive calcium channel. *Science* 1988;241:1661–1664.
25. Mikami A, Imoto K, Tanabe T, Niidome T, Mori Y, Takeshima H, Narumiya S, Numa S: Primary structure and functional expression of the cardiac dihydropyridine-sensitive calcium channel. *Nature* 1989;340:230–233.
26. Williams ME, Brust PF, Feldman DH, Patthi S, Simerson S, Maroufi A, McCue AF, Veliçelebi G, Ellis SB, Harpold MM: Structure and functional expression of an ω-conotoxin-sensitive human N-type calcium channel. *Science* 1992;257:389–395.
27. Kim H-L, Kim H, Lee P, King RG, Chin H: Rat brain expresses an alternatively spliced form of the dihydropyridine-sensitive L-type calcium channel $\alpha 2$ subunit. *Proc Natl Acad Sci U S A* 1992;89: 3251–3255.
28. Mori Y, Friedrich T, Kim M-S, Mikami A, Nakai J, Ruth P, Bosse E, Hofmann F, Flockerzi V, Furuichi T, Mikoshiba K, Imoto K, Tanabe T, Numa S: Primary structure and functional expression from complementary DNA of a brain calcium channel. *Nature* 1991;350:398–402.
29. Hui A, Ellinor PT, Krizanova O, Wang J-J, Diebold RJ, Schwartz A: Molecular cloning of multiple subtypes of a novel brain isoform of the $\alpha 1$ subunit of the voltage-dependent calcium channel. *Neuron* 1991;7:35–44.
30. Kameyama M, Hofmann F, Trautwein W: On the mechanism of β-adrenergic regulation of the Ca channel in the guinea-pig heart. *Pfluegers Arch* 1985;405:285–293.
31. Castellano A, Wei X, Birnbaumer L, Perez-Reyes E: Cloning and expression of a neuronal calcium channel β subunit. *J Biol Chem* 1993;268:12359–12366.
32. Yaney GC, Wheeler MB, Wei X, Perez-Reyes E, Birnbaumer L, Boyd AE, III, Moss LG: Cloning of a novel α_1-subunit of the voltage-dependent calcium channel from the β-cell. *Mol Endocrinol* 1992;6:2143–2152.
33. Thompson SM, Gahwiler BH: Comparison of the actions of baclofen at pre- and postsynaptic receptors in the rat hippocampus *in vitro*. *J Physiol (Lond)* 1992;451:329–345.
34. Birnbaumer L, Campbell KP, Catterall WA, Harpold MM, Hofmann F, Horne WA, Mori Y, Schwartz A, Snutch TP, Tanabe T, Tsien RW: The naming of voltage-gated calcium channels. *Neuron* 1994; 13:505–506.
34a. Bech-Hansen NT, Naylor MJ, Maybaum TA, Pearce WG, Koop B, Fishman GA, Mets M, Mosarella MA, Boycott KM: Loss of function matations in a calcium channel α_1 subunit gene in Xp11.23 cause incomplete X-linked congenital stationary night blindness. *Nature Genetics* 1998;19: 264–267.
35. Tanabe T, Mikami A, Numa S, Beam KG: Cardiac-type excitation–contraction coupling in dysgenic skeletal muscle injected with cardiac dihydropyridine receptor cDNA. *Nature* 1990;344:451–453.
36. Itagaki K, Koch WJ, Bodi I, Klöckner U, Slish DF, Schwartz A: Native-type DHP-sensitive calcium channel currents are produced by cloned rat aortic smooth muscle and cardiac α_1 subunits expressed in *Xenopus laevis* oocytes and are regulated by α_2- and β-subunits. *FEBS Lett* 1992;297:221–225.
37. Feron O, Octave J-N, Christen M-O, Godfraind T: Quantification of two splicing events in the L-type calcium channel α-1 subunit of intestinal smooth muscle and other tissues. *Eur J Biochem* 1994;222:195–202.
38. Soldatov NM, Bouron A, Reuter H: Different voltage-dependent inhibition by dihydropyridines of human Ca^{2+} channel splice variants. *J Biol Chem* 1995;270:10540–10543.

39. Tanabe T, Beam KG, Powell JA, Numa S: Restoration of excitation–contraction coupling and slow calcium current in dysgenic muscle by dihydropyridine receptor complementary DNA. *Nature* 1988; 336:134–139.
40. Nakai J, Dirksen RT, Nguysen HT, Pessah IN, Beam KG, Allen PD: Enhanced dihydropyridine receptor channel activity in the presence of ryanodine receptor. *Nature* 1996;380:72–75.
41. Matthews E, Snutch TP: Molecular cloning of a class D L-type calcium channel with an elongated carboxyl terminus. *Soc Neurosci Abstr* 1995;21:618.8.
42. Varadi G, Lory P, Schultz D, Varadi M, Schwartz A: Acceleration of activation and inactivation by the β subunit of the skeletal muscle calcium channel. *Nature* 1991;352:159–162.
43. Stea A, Dubel SJ, Pragnell M, Leonard JP, Campbell KP, Snutch TP: A β-subunit normalizes the electrophysiological properties of a cloned N-type Ca^{2+} channel α_1-subunit. *Neuropharmacology* 1993;32:1103–1116.
44. Lory P, Varadi G, Slish DF, Varadi M, Schwartz A: Characterization of β subunit modulation of a rabbit cardiac L-type Ca^{2+} channel α1 subunit as expressed in mouse L cells. *FEBS Lett* 1993;315: 167–172.
45. Singer D, Biel M, Lotan I, Flockerzi V, Hofmann F, Dascal N: The roles of the subunits in the function of the calcium channel. *Science* 1991;253:1553–1557.
46. Gurnett CA, De Waard M, Campbell KP: Dual function of the voltage-dependent Ca^{2+} channel $\alpha_2\delta$ subunit in current stimulation and subunit interaction. *Neuron* 1996;16:431–440.
47. McEnery MW, Snowman AM, Seagar MJ, Copeland TD, Takahashi M: Immunological characterization of proteins associated with the purified omega-conotoxin GVIA receptor. *Ann N Y Acad Sci* 1993;707:386–391.
48. Liu H, De Waard M, Scott VES, Gurnett CA, Lennon VA, Campbell KP: Identification of three subunits of the high affinity ω-conotoxin MVIIC-sensitive Ca^{2+} channel. *J Biol Chem* 1996;271: 13804–13810.
48a. Letts VA, Felix R, Biddlecome GH, Arikkath J, Mahaffey CL, Valenzuela A, Bartlett IS, Mori Y, Campbell KP, Frankel WN: The mouse stargazer gene encodes a neuronal Ca^{2+} channel gamma subunit. *Nature Genetics* 1998;19:340–347.
49. Cachelin AB, De Peyer JE, Kokubun S, Reuter H: Ca^{2+} channel modulation by 8-bromocyclic AMP in cultured heart cells. *Nature* 1983;304:462–464.
50. Bean BP, Nowycky MC, Tsien RW: β-Adrenergic modulation of calcium channels in frog ventricular heart cells. *Nature* 1984;307:371–375.
51. Haase H, Karczewski P, Beckert R, Krause EG: Phosphorylation of the L-type calcium channel β subunit is involved in β-adrenergic signal transduction in canine myocardium. *FEBS Lett* 1993;335: 217–222.
52. Perez-Reyes E, Yuan W, Wei X, Bers DM: Regulation of the cloned L-type cardiac calcium channel by cyclic-AMP-dependent protein kinase. *FEBS Lett* 1994;342:119–123.
53. Charnet P, Lory P, Bourinet E, Collin T, Nargeot J: cAMP-dependent phosphorylation of the cardiac L-type Ca channel: a missing link. *Biochimie* 1995;77:957–962.
54. Sculptoreanu A, Rotman E, Takahashi M, Scheuer T, Catterall WA: Voltage-dependent potentiation of the activity of cardiac L-type calcium channel α1 subunits due to phosphorylation by cAMP-dependent protein kinase. *Proc Natl Acad Sci U S A* 1993;90:10135–10139.
55. Hosey MM, Chien AJ, Puri TS: Structure and regulation of L-type calcium channels—a current assessment of the properties and roles of channel subunits. *Trends Cardiovasc Med* 1996;6: 265–273.
56. Allen TJA, Mikala G, Wu X, Dolphin AC: Effects of 2,3-butanedione-monoxime (BDM) on calcium channels expressed in *Xenopus* oocytes. *J. Physiol (Lond)* 1998;508:1–14.
57. Yatani A, Imoto Y, Codina J, Hamilton SL, Brown AM, Birnbaumer L: The stimulatory G protein of adenylyl cyclase, Gs, also stimulates dihydropyridine-sensitive Ca^{2+} channels. *J Biol Chem* 1988; 263:9887–9895.
58. Yatani A, Brown AM: Rapid β-adrenergic modulation of cardiac calcium channel currents by a fast G protein pathway. *Science* 1989;245:71–74.
59. Cavalié A, Allen TJA, Trautwein W: Role of the GTP binding protein G_s in the β-adrenergic modulation of cardiac Ca channels. *Pfluegers Arch* 1991;419:433–443.
60. Imoto Y, Yatani A, Reeves JP, Codina J, Birnbaumer L, Brown AM: Alpha-subunit of Gs directly activates cardiac calcium channels in lipid bilayers. *Am J Physiol* 1988;255:H722–H728.
61. Hartzell HC, Méry P-F, Fischmeister R, Szabo G: Sympathetic regulation of cardiac calcium current is due exclusively to cAMP-dependent phosphorylation. *Nature* 1991;351:573–576.

62. Hartzell HC, Fischmeister R: Direct regulation of cardiac Ca^{2+} channels by G proteins: neither proven nor necessary. *Trends Pharmacol Sci* 1992;13:380–385.
63. Jurevicius J, Fischmeister R: cAMP compartmentation is responsible for a local activation of cardiac Ca^{2+} channels by β-adrenergic agonists. *Proc Natl Acad Sci U S A* 1996;93:295–299.
64. Hamilton SL, Codina J, Hawkes MJ, Yatani A, Sawada T, Strickland FM, Froehner SC, Spiegel AM, Toro L, Stefani E, Birnbaumer L, Brown AM: Evidence for direct interaction of $G_s\alpha$ with the Ca^{2+} channel of skeletal muscle. *J Biol Chem* 1991;266:19528–19535.
65. Fischmeister R, Hartzell HC: Mechanism of action of acetylcholine on calcium current in single cells from frog ventricle. *J Physiol (Lond)* 1986;376:183–202.
66. Jurevicius J, Fischmeister R: Acetylcholine inhibits Ca^{2+} current by acting exclusively at a site proximal to adenylyl cyclase in frog cardiac myocytes. *J Physiol (Lond)* 1996;491:669–675.
67. Pemberton KE, Jones SVP: Inhibition of the L-type calcium channel by the five muscarinic receptors (m1–m5) expressed in NIH 3T3 cells. *Pfluegers Arch* 1997;433:505–514.
68. Bourinet E, Charnet P, Tomlinson WJ, Stea A, Snutch TP, Nargeot J: Voltage-dependent facilitation of a neuronal α_{1C} L-type calcium channel. *EMBO J* 1994;13:5032–5039.
69. Tiaho F, Piot C, Nargeot J, Richard S: Regulation of the frequency-dependent facilitation of L-type Ca^{2+} currents in rat ventricular myocytes. *J Physiol (Lond)* 1994;477:237–252.
70. Toutant M, Gabrion J, Vandaele S, Peraldi-Roux S, Barhanin J, Bockaert J, Rouot B: Cellular distribution and biochemical characterization of G-proteins in skeletal muscle: comparative location with voltage-dependent calcium channels. *EMBO J* 1990;9(2):363–369.
71. Villaz M, Robert M, Carrier L, Beeler T, Rouot B, Toutant M, Dupont Y: G-protein dependent potentiation of calcium release from sarcoplasmic reticulum of skeletal muscle. *Cell Signal* 1989;1: 493–506.
72. Scherer NM, Toro M-J, Entman ML, Birnbaumer L: G-protein distributon in canine cardiac sarcoplasmic reticulum and sarcolemma: comparison to rabbit skeletal muscle membranes and to brain and erythrocyte G-proteins. *Arch Biochem Biophys* 1987;259(2):431–440.
73. Somasundaram B, Tregear RT, Trentham DR: GTPyS causes contraction of skinned frog skeletal muscle via the DHP-sensitive Ca^{2+} channels of sealed T-tubules. *Pfluegers Arch* 1991;418: 137–143.
74. Codina J, Grenet D, Yatani A, Birnbaumer L, Brown AM: Hormonal regulation of pituitary GH3 cell K^+ channels by G_K is mediated by its alpha-subunit. *FEBS Lett* 1987;216:104–106.
75. Garcia J, Gamboa-Aldeco R, Stefani E: Charge movement and calcium currents are regulated by G-proteins in skeletal muscle fibres. *Pfluegers Arch* 1990;417:114–116.
76. Heidelberger R, Matthews G: Inhibition of calcium influx and calcium current by gamma-aminobutyric acid in single synaptic terminals. *Proc Natl Acad Sci U S A* 1991;88:7135–7139.
77. De Waard M, Feltz A, Bossu JL: Properties of a high-threshold voltage-activated calcium current in rat cerebellar granule cells. *Eur J Neurosci* 1991;3:771–777.
78. Slesinger PA, Lansman JB: Inactivating and non-inactivating dihydropyridine-sensitive Ca^{2+} channels in mouse cerebellar granule cells. *J Physiol (Lond)* 1991;439:301–323.
79. Tachibana M, Okada T, Arimura T, Kobayashi K, Piccolino M: Dihydropyridine-sensitive calcium current mediates neurotransmitter release from bipolar cells of the goldfish retina. *J Neurosci* 1993; 13:2898–2909.
80. Pearson HA, Sutton KG, Scott RH, Dolphin AC: Characterization of Ca^{2+} channel currents in cultured rat cerebellar granule neurones. *J Physiol (Lond)* 1995;482:493–509.
81. Melliti K, Bournaud R, Bastide B, Shimahara T: Nifedipine-sensitive intramembrane charge movement in Purkinje cells from mouse cerebellum. *J Physiol (Lond)* 1996;490:363–372.
82. Schmidt A, Hescheler J, Offermanns S, Spicher K, Hinsch K-D, Klinz F-J, Codina J, Birnbaumer L, Gausephol H, Frank R, Schultz G, Rosenthal W: Involvement of pertussis toxin-sensitive G-proteins in the hormonal inhibition of dihydropyridine-sensitive Ca^{2+} currents in an insulin-secreting cell line (RINm5F). *J Biol Chem* 1991;266:18025–18033.
83. Kleuss C, Hescheler J, Ewel C, Rosenthal W, Schultz G, Wittig B: Assignment of G-protein subtypes to specific receptors inducing inhibition of calcium currents. *Nature* 1991;353:43–48.
84. Hescheler J, Schultz G: Heterotrimeric G proteins involved in the modulation of voltage-dependent calcium channels of neuroendocrine cells. *Ann N Y Acad Sci* 1994;733:306–312.
85. Bokvist K, Eliasson L, Ämmälä C, Renström E, Rorsman P: Co-localization of L-type Ca^{2+} channels and insulin-containing secretory granules and its significance for the initiation of exocytosis in mouse pancreatic B-cells. *EMBO J* 1995;14:50–57.
85a. Hille, B: G protein coupled mechanisms and nervous signalling. *Neuron* 1992;9:187–195.

86. Shapiro MS, Hille B: Substance P and somatostatin inhibit calcium channels in rat sympathetic neurons via different G protein pathways. *Neuron* 1993;10:11–20.
87. Elmslie KS: Calcium current modulation in frog sympathetic neurones: multiple neurotransmitters and G proteins. *J Physiol (Lond)* 1992;451:229–246.
88. Imredy JP, Yue DT: Mechanism of Ca^{2+}-sensitive inactivation of L-type Ca^{2+} channels. *Neuron* 1994;12:1301–1318.
89. Diversé-Pierluissi M, Inglese J, Stoffel RH, Lefkowitz RJ, Dunlap K: G protein-coupled receptor kinase mediates desensitization of norepinephrine-induced Ca^{2+} channel inhibition. *Neuron* 1996; 16:579–585.
90. Ciranna L, Feltz P, Schlichter R: Selective inhibition of high voltage-activated L-type and Q-type Ca^{2+} currents by serotonin in rat melanotrophs. *J Physiol (Lond)* 1996;490:595–609.
91. Grassi F, Lux HD: Voltage-dependent GABA-induced modulation of calcium currents in chick sensory neurons. *Neurosci Lett* 1989;105:113–119.
92. Boland LM, Bean BP: Modulation of N-type calcium channels in bullfrog sympathetic neurons by luteinizing hormone-releasing hormone: kinetics and voltage dependence. *J Neurosci* 1993;13: 516–533.
93. Scott RH, Dolphin AC: Voltage-dependent modulation of rat sensory neurone calcium channel currents by G protein activation: effect of a dihydropyridine antagonist. *Br J Pharmacol* 1990;99: 629–630.
94. Pollo A, Lovallo M, Sher E, Carbone E: Voltage-dependent noradrenergic modulation of omega-conotoxin-sensitive Ca^{2+} channels in human neuroblastoma IMR32 cells. *Pfluegers Arch* 1992;422: 75–83.
95. Toth PT, Shekter LR, Ma GH, Philipson LH, Miller RJ: Selective G-protein regulation of neuronal calcium channels. *J Neurosci* 1996;16:4617–4624.
96. Page KM, Stephens GJ, Berrow NS, Dolphin AC: The intracellular loop between domains I and II of the B type calcium channel confers aspects of G protein sensitivity to the E type calcium channel. *J Neurosci* 1997;17:1330–1338.
97. Keja JA, Kits KS: Voltage dependence of G-protein-mediated inhibition of high-voltage-activated calcium channels in rat pituitary melanotropes. *Neuroscience* 1994;62:281–289.
98. Matthews G, Ayoub GS, Heidelberger R: Presynaptic inhibition by GABA is mediated via two distinct GABA receptors with novel pharmacology. *J Neurosci* 1994;14:1079–1090.
99. Maguire G, Maple B, Lukasiewicz P, Werblin F: γ-Aminobutyrate type B receptor modulation of L-type calcium channel current at bipolar cell terminals in the retina of the tiger salamander. *Proc Natl Acad Sci U S A* 1989;86:10144–10147.
100. Scholz KP, Miller RJ: $GABA_B$ receptor-mediated inhibition of Ca^{2+} currents and synaptic transmission in cultured rat hippocampal neurones. *J Physiol (Lond)* 1991;444:669–686.
101. Bossu JL, De Waard M, Fagni L, Tanzi F, Feltz A: Characteristics of calcium channels responsible for voltage-activated calcium entry in rat cerebellar granule cells. *Eur J Neurosci* 1994;6:335–344.
102. Wojcik WJ, Travagli RA, Costa E, Bertolino M: Baclofen inhibits with high affinity an L-type-like voltage-dependent calcium channel in cerebellar granule cell cultures. *Neuropharmacology* 1990; 29:969–972.
103. Huston E, Cullen G, Sweeney MI, Pearson H, Fazeli MS, Dolphin AC: Pertussis toxin treatment increases glutamate release and dihydropyridine binding sites in cultured rat cerebellar granule neurons. *Neuroscience* 1993;52:787–798.
104. Haws CM, Slesinger PA, Lansman JB: Dihydropyridine- and ω-conotoxin-sensitive Ca^{2+} currents in cerebellar neurons: persistent block of L-type channels by a pertussis toxin-sensitive G-protein. *J Neurosci* 1993;13:1148–1156.
105. Chavis P, Shinozaki H, Bockaert J, Fagni L: The metabotropic glutamate receptor types 2/3 inhibit L-type calcium channels via a pertussis toxin-sensitive G-protein in cultured cerebellar granule cells. *J Neurosci* 1994;14:7067–7076.
106. Mei YA, Griffon N, Buquet C, Martres MP, Vaudry H, Schwartz J-C, Sokoloff P, Cazin L: Activation of dopamine D_4 receptor inhibits an L-type calcium current in cerebellar granule cells. *Neuroscience* 1995;68:107–116.
107. Sayer RJ, Schwindt PC, Crill WE: Metabotropic glutamate receptor-mediated suppression of L-type calcium current in acutely isolated neocortical neurons. *J Neurophysiol* 1992;68:833–842.
108. Campbell V, Berrow NS, Fitzgerald EM, Brickley K, Dolphin AC: Inhibition of the interaction of G protein G_o with calcium channels by the calcium channel β-subunit in rat neurones. *J Physiol (Lond)* 1995;485:365–372.

109. Scott RH, Dolphin AC, Bindokas VP, Adams ME: Inhibition of neuronal Ca^{2+} channel currents by the funnel web spider toxin ω-Aga-1A. *Mol Pharmacol* 1990;38:711–718.
110. Scott RH, Dolphin AC: Activation of a G protein promotes agonist responses to calcium channel ligands. *Nature* 1987;330:760–762.
111. Mintz IM, Adams ME, Bean BP: P-type calcium channels in rat central and peripheral neurons. *Neuron* 1992;9:85–95.
112. Wyatt CN, Campbell V, Brodbeck P, Brice NL, Page KM, Berrow NS, Brickley K, Terracciano R, Naqvi RV, Macleod KT, Dolphin AC: Voltage-dependent binding and calcium current inhibition by an anti-α1D subunit antibody in rat dorsal root ganglion neurones and guinea pig myocytes. *J Physiol (Lond)* 1997;502:307–319.
113. Menon-Johansson AS, Berrow N, Dolphin AC: G_o transduces GABAB-receptor modulation of N-type calcium channels in cultured dorsal root ganglion neurons. *Pfluegers Arch* 1993;425:335–343.
114. Scott RH, Dolphin AC: Regulation of calcium currents by a GTP analogue: potentiation of (−)-baclofen-mediated inhibition. *Neurosci Lett* 1986;69:59–64.
115. Dolphin AC, Wootton JF, Scott RH, Trentham DR: Photoactivation of intracellular guanosine triphosphate analogues reduces the amplitude and slows the kinetics of voltage-activated calcium channel currents in sensory neurones. *Pfluegers Arch* 1988;411:628–636.
116. Dolphin AC, Scott RH: Calcium channel currents and their inhibition by (−)-baclofen in rat sensory neurones: modulation by guanine nucleotides. *J Physiol (Lond)* 1987;386:1–17.
117. Gilman AG: G proteins: transducers of receptor-generated signals. *Annu Rev Biochem* 1987;56: 615–649.
118. DE Leon M, Wang Y, Jones L, Perez-Reyes E, Wei XY, Soong TW, Snutch TP, Yue DT: Essential Ca^{2+}-binding motif for Ca^{2+}-sensitive inactivation of L-type Ca^{2+} channels. *Science* 1995;270: 1502–1506.
119. Gudermann T, Kalkbrenner F, Schultz G: Diversity and selectivity of receptor–G protein interaction. *Annu Rev Pharmacol Toxicol* 1996;36:429–459.
120. Degtiar V, Harhammer R, Nuernberg B: Receptors couple to L-type channels via distinct G_o proteins in rat neuroendocrine cell lines. *J Physiol (Lond)* 1997;502:321–333.
121. Gudermann T, Nürnberg B, Schultz G: Receptors and G proteins as primary components of transmembrane signal transduction. Part 1. G-protein-coupled receptors: structure and function. *Clin Invest* 1995;73:51–63.
122. McFadzean I, Mullaney I, Brown DA, Milligan G: Antibodies to the GTP binding protein, G_o, antagonize noradrenaline-induced calcium current inhibition in NG108-15 hybrid cells. *Neuron* 1989;3:177–182.
123. Campbell V, Berrow N, Dolphin AC: $GABA_B$ receptor modulation of Ca^{2+} currents in rat sensory neurones by the G protein G_o: antisense oligonucleotide studies. *J Physiol (Lond)* 1993;470:1–11.
124. Baertschi AJ, Audigier Y, Lledo P-M, Israel J-M, Bockaert J, Vincent J-D: Dialysis of lactotropes with antisense oligonucleotides assigns guanine nucleotide binding protein subtypes to their channel effectors. *Mol Endocrinol* 1992;6:2257–2265.
125. Hartl FU: Molecular chaperones in cellular protein folding. *Nature* 1996;381:571–580.
126. Ikeda SR: Voltage-dependent modulation of N-type calcium channels by G protein βy subunits. *Nature* 1996;380:255–258.
127. Herlitze S, Garcia DE, Mackie K, Hille B, Scheuer T, Catterall WA: Modulation of Ca^{2+} channels by G-protein βγ subunits. *Nature* 1996;380:258–262.
128. De Waard M, Liu HY, Walker D, Scott VES, Gurnett CA, Campbell KP: Direct binding of G-protein βγ complex to voltage-dependent calcium channels. *Nature* 1997;385:446–450.
129. Zamponi GW, Bourinet E, Nelson D, Nargeot J, Snutch TP: Crosstalk between G proteins and protein kinase C mediated by the calcium channel α_1 subunit. *Nature* 1997;385:442–446.
130. Bourne HR, Sanders DA, McCormick F: The GTPase superfamily: a conserved switch for diverse cell functions. *Nature* 1990;348:125–132.
130a. Arshavsky VY, Pugh EN: Lifetime regulation of G protein effector complex: Emerging importance of RGS proteins. *Neuron* 1998;20:11–14.
131. Sweeney MI, Dolphin AC: 1,4-Dihydropyridines modulate GTP hydrolysis by Go in neuronal membranes. *FEBS Lett* 1992;310:66–70.
132. Campbell V, Berrow N, Brickley K, Page K, Wade R, Dolphin AC: Voltage-dependent calcium channel β-subunits in combination with $alpha_1$ subunits have a GTPase activating effect to promote hydrolysis of GTP by G $alpha_o$ in rat frontal cortex. *FEBS Lett* 1995;370:135–140.

133. Higashijima T, Ferguson KM, Smigel MD, Gilman AG: The effect of GTP and Mg^{2+} on the GTPase activity and the fluorescent properties of G_o. *J Biol Chem* 1987;262:757–761.
134. Katada T, Oinuma M, Ui M: Two guanine nucleotide-binding proteins in rat brain serving as the specific substrate of islet-activating protein, pertussis toxin. Interaction of the alpha-subunits with beta gamma-subunits in development of their biological activities. *J Biol Chem* 1986;261:8182–8191.
135. Brabet P, Dumis A, Sebben M, Pantaluni C, Bockaert J, Humburger V: Immunocytochemical localization of the guanine nucleotide-binding protein G_o in primary cultures of neuronal and glial cells. *J Neurosci* 1988;8:701–708.
136. Ewald DA, Sternweis PC, Miller RJ: Guanine nucleotide-binding protein G_o-induced coupling of neuropeptide Y receptors to Ca^{2+} channels in sensory neurons. *Proc Natl Acad Sci U S A* 1988;85: 3633–3637.
137. Camps M, Carozzi A, Schnabel P, Scheer A, Parker PJ, Gierschik P: Isozyme-selective stimulation of phospholipase C-β2 by G protein βγ-subunits. *Nature* 1992;360:684–686.
138. Dolphin AC: Regulation of calcium channel activity by GTP binding proteins and second messengers. *Biochim Biophys Acta* 1991;1091:68–80.
139. Elmslie KS, Jones SW: Concentration dependence of neurotransmitter effects on calcium current kinetics in frog sympathetic neurones. *J Physiol (Lond)* 1994;481:35–46.
140. Zhou J, Shapiro M, Hille B: Speed of calcium channel modulation by neurotransmitters in rat sympathetic neurons. *J Neurophysiol* 1997;77:2040–2048.
140a. Stephens GJ, Brice NL, Berrow NS, Dolphin AC: Facilitation of rabbit α1B calcium channels: involvement of endogenous Gβgamma subunits. *J. Physiol (Lond)* 1998;509:15–27.
141. Patil PG, De Leon M, Reed RR, Dubel S, Snutch TP, Yue DT: Elementary events underlying voltage-dependent G-protein inhibition of N-type calcium channels. *Biophys J* 1996;71:2509–2521.
142. Nakai J, Adams BA, Imoto K, Beam KG: Critical roles of the S3 segment and S3-S4 linker of repeat I in activation of L-type calcium channels. *Proc Natl Acad Sci U S A* 1994;91:1014–1018.
143. Wang Z, Grabner M, Berjukow S, Savchenko A, Glossmann H, Hering S: Chimeric L-type calcium channels expressed in *xenopus laevis* oocytes reveal role of repeats III and IV in activation gating. *J Physiol (Lond)* 1995;486:131–137.
144. Herlitze S, Hockerman GH, Scheuer T, Catterall WA: Molecular determinants of inactivation and G protein modulation in the intracellular loop connecting domains I and II of the calcium channel α_{1A} subunit. *Proc Natl Acad Sci U S A* 1997;94:1512–1516.
144a. Stephens GJ, Canti C, Page KM, Dolphin AC: Role of domain I of neuronal CA^{2+} channel α1 subunits in G protein modulation. *J. Physiol.(Lond.)* 1998: 509: 163–169.
144b. Page KM, Canti C, Stephens GJ, Berrow NS, Dolphin AC: Identification of the amino terminus of neuronal Ca^{2+} channel α1 subunits α1B and α1E as an essential determinant of G protein modulation. *J. Neuroscience* 1998;18:4815–4824.
145. Ämmälä C, Berggren P-O, Bokvist K, Rorsman P: Inhibition of L-type calcium channels by internal GTP [gammaS] in mouse pancreatic β cells. *Pfluegers Arch* 1992;420:72–77.
146. Albillos A, Carbone E, Gandia L, Garcia AG, Pollo A: Opioid inhibition of Ca^{2+} channel subtypes in bovine chromaffin cells: selectivity of action and voltage-dependence. *Eur J Neurosci* 1996;8:1561–1570.
147. Perez-Reyes E, Schneider T: Calcium channels: structure, function, and classification. *Drug Dev Res* 1994;33:295–318.
148. Shistik E, Ivanina T, Puri T, Hosey M, Dascal N: Ca^{2+} current enhancement by αa2/δ and β subunits in *Xenopus* oocytes: contribution of changes in channel gating and α1 protein level. *J Physiol (Lond)* 1995;489:55–62.
149. Chien AJ, Zhao XL, Shirokov RE, Puri TS, Chang CF, Sun D, Rios E, Hosey MM: Roles of a membrane-localized β subunit in the formation and targeting of functional L-type Ca^{2+} channels. *J Biol Chem* 1995;270:30036–30044.
150. Berrow NS, Brice NL, Tedder I, Page K, Dolphin AC: Properties of cloned rat α1A calcium channels transiently expressed in the COS-7 cell line. *Eur J Neurosci* 1997;9:739–748.
151. Brice NL, Berrow NS, Campbell V, Page KM, Brickley K, Tedder I, Dolphin AC: Importance of the different β subunits in the membrane expression of the α1A and α1 calcium channel subunits: studies using a depolarisation-sensitive α1A antibody. *Eur J Neurosci* 1997;9:749–759.
152. Berrow NS, Campbell V, Fitzgerald EG, Brickley K, Dolphin AC: Antisense depletion of β-subunits modulates the biophysical and pharmacological properties of neuronal calcium channels. *J Physiol (Lond)* 1995;482:481–491.

153. Wakamori M, Niidome T, Furutama D, Furuichi T, Mikoshiba K, Fujita Y, Tanaka I, Katayama K, Yatani A, Schwartz A et al: Distinctive functional properties of the neuronal BII (class E) calcium channel. *Recept Channels* 1994;2:303–314.
154. Witcher DR, De Waard M, Liu H, Pragnell M, Campbell KP: Association of native Ca^{2+} channel β subunits with the α_1 subunit interaction domain. *J Biol Chem* 1995;270:18088–18093.
155. De Waard M, Witcher DR, Pragnell M, Liu H, Campbell KP: Properties of the α_1-β anchoring site in voltage-dependent Ca^{2+} channels. *J Biol Chem* 1995;270:12056–12064.
156. De Waard M, Scott VES, Pragnell M, Campbell KP: Identification of critical amino acids involved in α_1-β interaction in voltage-dependent Ca^{2+} channels. *FEBS Lett* 1996;380:272–276.
157. Grabner M, Wang Z, Hering S, Striessnig J, Glossmann H: Transfer of 1,4-dihydropyridine sensitivity from L-type to class A (BI) calcium channels. *Neuron* 1996;16:207–218.
158. Kalasz H, Watanabe T, Yabana H, Itagaki K, Natto K, Nakayama H, Schwartz A, Vaghy PL: Identification of 1,4-dihydropyridine binding domains within the primary structure of the α1 subunit of the skeletal muscle L-type calcium channel. *FEBS Lett* 1993;331:177–181.
159. Mitterdorfer J, Froschmayr M, Grabner M, Striessnig J, Glossmann H: Calcium channels: The β-subunit increases the affinity of dihydropyridine and Ca^{2+} binding sites of the α_1-subunit. *FEBS Lett* 1994;352:141–145.
160. Neely A, Wei X, Olcese R, Birnbaumer L, Stefani E: Potentiation by the β subunit of the ratio of the ionic current to the charge movement in the cardiac calcium channel. *Science* 1993;262: 575–578.
161. Dolphin AC, Scott RH: Interaction between calcium channel ligands and guanine nucleotides in cultured rat sensory and sympathetic neurones. *J Physiol (Lond)* 1989;413:271–288.
162. Scott RH, Dolphin AC: The agonist effect of Bay K8644 on neuronal calcium channel currents is promoted by G-protein activation. *Neurosci Lett* 1988;89:170–175.
163. Dolphin AC: Ca^{2+} channel currents in rat sensory neurones: interaction between guanine nucleotides, cyclic AMP and Ca^{2+} channel ligands. *J Physiol (Lond)* 1991;432:23–43.
164. Dolphin AC: The effect of phosphatase inhibitors and agents increasing cyclic-AMP-dependent phosphorylation on calcium channel currents in cultured rat dorsal root ganglion neurones: interaction with the effect of G protein activation. *Pfluegers Arch* 1992;421:138–145.
165. Lambert RC, Feltz A: Maintained L-type Ca^{2+} channel activity in excised patches of PTX-treated granule cells of the cerebellum. *J Neurosci* 1995;15:6014–6022.
166. Dolphin AC: Facilitation of Ca^{2+} current in excitable cells. *Trends Neurosci* 1996;19:35–43.
167. Artalejo CR, Ariano MA, Perlman RL, Fox AP: Activation of facilitation calcium channels by D_1 dopamine receptors through a cyclic AMP/protein kinase A-dependent mechanism. *Nature* 1990; 348:239–242.
168. Sculptoreanu A, Figourov A, De Groat WC: Voltage-dependent potentiation of neuronal L-type calcium channels due to state-dependent phosphorylation. *Am J Physiol Cell Physiol* 1995;269: C725–C732.
169. Arialejo CR, Rossie S, Perlman RL, Fox AP: Voltage-dependent phosphorylation may recruit Ca^{2+} current facilitation in chromaffin cells. *Nature* 1992;358:63–66.
170. Doupnik CA, Pun RYK: G-protein activation mediates prepulse facilitation of Ca^{2+} channel currents in bovine chromaffin cells. *J Membr Biol* 1994;140:47–56.
171. Currie KMP, Fox AP: ATP serves as a negative feedback inhibitor of voltage-gated Ca^{2+} channel currents in cultured bovine adrenal chromaffin cells. *Neuron* 1996;16:1–20.
172. Albillos A, Gandia L, Michelena P, Gilabert J-A, Del Valle M, Carbone E, Garcia AG: The mechanism of calcium channel facilitation in bovine chromaffin cells. *J Physiol (Lond)* 1996;494:687–695.
173. Chavis P, Nooney JM, Bockaert J, Fagni L, Feltz A, Bossu JL: Facilitatory coupling between a glutamate metabotropic receptor and dihydropyridine-sensitive calcium channels in cultured cerebellar granule cells. *J Neurosci* 1995;15:135–143.
174. Chavis P, Fagni L, Lansman JB, Bockaert J: Functional coupling between ryanodine receeptors and L-type calcium channels in neurons. *Nature* 1996;382:719–722.
175. Schreibmayer W, Dessauer CW, Vorobiov D, Gilman AG, Lester HA, Davidson N, Dascal N: Inhibition of an inwardly rectifying K^+ channel by G-protein α-subunits. *Nature* 1996;380: 624–627.
176. Huang CL, Slesinger PA, Casey PJ, Jan YN, Jan LY: Evidence that direct binding of $G_{\beta\gamma}$ to the GIRK1 G protein-gated inwardly rectifying K^+ channel is important for channel activation. *Neuron* 1995;15:1133–1143.

177. McEnery MW, Snowman AM, Snyder SH: The association of endogenous $G_o\alpha$ with the purified ω-conotoxin GVIA receptor. *J Biol Chem* 1994;269:5–8.
178. Heidelberger R, Matthews G: Calcium influx and calcium current in single synaptic terminals of goldfish retinal bipolar neurons. *J Physiol (Lond)* 1992;447:235–256.
179. Bergamaschi S, Govoni S, Battaini F, Trabucchi M, Del Monaco S, Parenti M: G protein modulation of omega-conotoxin binding sites in neuroblastoma X glioma NG 108-15 hybrid cells. *J Neurochem* 1992;59:536–543.
180. Sweeney MI, Dolphin AC: Adenosine A_1 agonists and the Ca^{2+} channel agonist Bay K8644 produce a synergistic stimulation of the GTPase activity of G_o in rat frontal cortical membranes. *J Neurochem* 1995;64:2034–2042.
181. Takai Y, Kaibuchi K, Kikuchi A, Kawata M: Small GTP-binding proteins. *Int Rev Cytol* 1992;133: 187–222.
182. Huang DCS, Marshall CJ, Hancock JF: Plasma membrane-targeted *ras* GTPase-activating protein is a potent suppressor of $p21^{ras}$ function. *Mol Cell Biol* 1993;13:2420–2431.
183. Bourne HR, Stryer L: G proteins. The target sets the tempo [news]. *Nature* 1992;358:541–543.
184. Arshavsky VY, Bownds MD: Regulation of deactivation of photoreceptor G protein by its target enzyme and cGMP. *Nature* 1992;357:416–417.
185. Berstein G, Blank JL, Jhon DY, Exton JH, Rhee SG, Ross EM: Phospholipase C-beta 1 is a GTPase-activating protein for Gq/11, its physiologic regulator. *Cell* 1992;70:411–418.
186. Enomoto K, Asakawa T: Inhibition of catalytic unit of adenylate cyclase and activation of GTPase of N_i protein by βγ-subunits of GTP-binding proteins. *FEBS Lett* 1986;202:63–68.
187. Fung BK-K: Characterisation of transducin from bovine retinal rod outer segments. I. Separation and reconstitution of the subunits. *J Biol Chem* 1983;258:10495–10502.
188. Hunt TW, Fields TA, Casey PJ, Peralta EG: RGS10 is a selective activator of $G\alpha_i$ GTPase activity. *Nature* 1996;383:175–177.
189. Berman DM, Wilkie TM, Gilman AG: GAIP and RGS4 are GTPase-activating proteins for the G_i subfamily of G protein α subunits. *Cell* 1996;86:445–452.
190. Boguski MS, McCormick F: Proteins regulating Ras and its relatives. *Nature* 1993;366:643–654.

Advances in Second Messenger and Phosphoprotein Research, Vol. 33

1040-7952/99 $30.00

8

G Protein Gated Potassium Channels

Jin-Liang Sui,* Kim Chan,[†] Marie-Noëlle Langan, Michel Vivaudou,[§] and Diomedes E. Logothetis

Department of Physiology and Biophysics, Mount Sinai School of Medicine, City University of New York, New York, New York 10029

INTRODUCTION

Extracellular signals of a physical or chemical nature such as light, hormones, neurotransmitters, tastants, and odorants can be sensed by surface receptors and transmitted through membrane-associated molecules to effector proteins. The link between receptors and effectors is in many cases mediated by the heterotrimetic GTP-binding (G) proteins. When the end effector molecule is a membrane-associated protein and signaling does not require cytosolic-soluble second messengers, the process is rapid and is referred to as "membrane delimited."

The majority of the G protein linked receptors are composed of a single polypeptide chain with seven relatively hydrophobic domains presumed to form membrane-spanning helices interspersed by three sets of intra- and extracellular loops. The ligand-binding site is thought to be formed by the bundling of the seven α helices, based on the cryoelectron microscopic studies of two-dimensional membrane crystals of the proteins bacteriorhodopsin and bovine rhodopsin (1,2).

The heterotrimetic G proteins consist of three different subunits: alpha (Gα, 39–52 kDa), beta (Gβ, 35–36 kDa) and gamma (Gγ, 6–8 kDa). The Gα and Gγ subunits undergo post-translational modification, resulting in the covalent attachment of specific lipophilic groups essential for membrane association. Physiologically, the Gβ and Gγ subunits exist always as a complex. Depending on the guanine nucleotide bound to the Gα subunit, G proteins can switch between an active and an inactive form. In the inactive form the Gα subunit is bound to GDP and is associated with the Gβγ subunits. It is the heterotrimeric G protein form that interacts with the G protein

*Present address: Cambridge Neuroscience, One Kendall Square, Building 700, Cambridge, Massachusetts 02139.

[†]Present address: Laboratory of Cardiac/Membrane Physiology, The Rockefeller University, New York, New York 10021.

[§]Permanent address: CEA, DBMS, Biophysique Moléculaire et Cellulaire (URA CNRS 520), 17 Rue des Martyrs, 38054 Grenoble, France.

coupled receptor. Specifically, it is thought that the second and third intracellular loops and the first part of the C-terminal tail of the receptor are important for interactions with the Gα-GDP subunit. Upon agonist binding to the receptor, the Gα subunit is allosterically stimulated to bind efficiently intracellular GTP and to dissociate from the receptor and the complex of Gβγ subunits. This state is considered to be the activated form of the G protein, where the dissociated Gα-GTP or the Gβγ complex is free to interact with effector proteins. The process is turned off by the intrinsic GTPase activity of the Gα subunit (its ability to hydrolyze GTP to GDP), which can be accelerated by target enzymes or independently associated proteins (3). The inactivated Gα-GDP subunit is able to bind the Gβγ complex and associate with the G protein linked receptor, and the cycle of events can be repeated. Since 1993, crystal structures of the heterotrimeric G protein complex, of the Gα in the inactive state (i.e., bound to GDP) or in the active state (i.e., bound to GTPγS, a nonhydrolyzable GTP analog), as well as of the Gβγ complex have all been elucidated (4–8). This wealth of structural information has offered great insights toward the key structural changes induced by the (γ-phosphate of GTP causing subunit dissociation of the heterotrimer into its active components.

Effector molecules, such as adenylate cyclases, phospholipases, potassium or calcium ion channels, and kinases interact in the sea of membrane lipids with one active G protein component, the Gα subunit or the Gβγ complex, or even with both of them simultaneously to regulate a host of cellular functions. Many basic questions remain unanswered. For example, the physiological importance of signaling through Gα versus Gβγ or through both G protein subunits remains poorly understood, as does the role of lipids in membrane-delimited signaling.

The family of G protein gated potassium (K_G) channels is a prime example of effectors regulated in a membrane-delimited manner. The exquisite resolution of the patch–clamp technique (9), allowing assay of single molecular functions in real time, presents a powerful system for the study of signaling mediated through protein–protein and lipid–protein interactions. Since 1993, five recombinant K_G channels have been isolated. In this chapter we review current understanding of the regulation of K_G channel activity.

G PROTEIN GATED POTASSIUM CHANNELS: THE PRE-CLONING ERA

Direct activation of potassium (K^+) channels by G proteins is involved in the rapid inhibition of membrane excitability, such as in the slowing of heart rate by the vagus nerve or perhaps the autoinhibitory release of dopamine by midbrain neurons. The atrial (or nodal) K^+ channel activated by acetylcholine (ACh: i.e., K_{ACh}) (10) through muscarinic m_2-type receptors has served as the prototypical K_G channel. Pertussis toxin sensitive, neurotransmitter-activated, inwardly rectifying K^+ currents have also been reported in other peripheral tissues such as the pancreas (11–13) and the pituitary (14–16). In brain, activation through G protein coupled receptors has been described in hippocampus (17–22), dorsal raphe (23), substantia nigra (24), locus coeruleus (25–28), nucleus basalis (29–31), and submucous plexus (25,32). In these tissues, activation of G protein coupled K^+ channels is thought to underlie suppression of firing (18). Since 1984, when the membrane-delimited nature of K_{ACh} activation was first appreciated (33), numerous studies have attempted to elucidate its precise gating mechanism. In the ensuing years of intense studies, we continue to be surprised by the complexity of the molecular picture depicting the regulation of K_{ACh}. As a prelude to our presentation of the current understanding of the regulation of activity of these channels, we highlight some of the major discoveries that did not require the use of recombinant K_G channels, mainly prior to 1993, the time at which the first member (K_{G1}) of this subfamily of K^+ channels was cloned (34,35). Details on advances during this precloning era can be found in earlier reviews (36–39).

Membrane-Delimited, Direct G Protein Gating of K_{ACh}

Using the cell-attached mode of the patch–clamp technique Soejima and Noma (33) showed that extracellular application of ACh was effective in stimulating channel activity in an atrial patch when perfused through the pipette, but not through the bath. In their isolated patch, external signaling was limited to ACh in the patch pipette (since the membrane patch is physically isolated at the sites of the gigaseal between the glass electrode and the plasma membrane from substances in the bath and from membrane molecules other than those within the patch), whereas internally, soluble second messengers (e.g., GTP) did have access to the cytoplasmic surface of the patch. The investigators interpreted this result as evidence for the membrane-delimited nature of the action of ACh. A year later, using the whole-cell mode of the patch–clamp technique, other workers showed that the ACh effect proceeds via a pertussis toxin (PTX) sensitive G protein (40) and that nonhydrolyzable GTP analogs can bypass signaling through the receptor and cause persistent stimulation of channel activity (41).

Experiments with inside-out patches provided further evidence for the membrane-delimited nature of the ACh signaling and the involvement of G proteins (42). Perfusion of inside-out patches with purified Gβγ subunits caused persistent stimulation of K_{ACh} activity in a Mg^{2+}-independent manner (43). The Gβγ activation of K_{ACh} provided the first example of the effector function of the Gβγ complex in any system (for a review of Gβγ effector functions see ref. 44). Purified Gα-GTPγS subunits were also shown to stimulate K_{ACh} activity (45). However, the Gα-GTPγS stimulation of activity occurred in approximately one-third of the patches tested (36,46), and the maximal activation was only 20% of the maximal activity induced by GTPγS or Gβγ when all reagents were applied at saturating concentrations (36). The mechanism of the Gα-GTPγS stimulation of K_{ACh} activity remains unclear. Six recombinant combinations of Gβγ subunits activated K_{ACh} with similar potency, suggesting promiscuity in the Gβγ channel interaction (47). These results further underscored the gap in our understanding of how Gβγ specificity is conferred. Recently, a peptide mimicking a carboxyl terminal region of the adenylate cyclase type II isoform (called QEHA after the first four amino acids in its sequence), prevented or reversed Gβγ effects on five distinct effectors, including K_{ACh} (48). The ability of the QEHA peptide to block general Gβγ effectiveness was interpreted as evidence that the peptide bound to a common Gβγ site, preventing an important functional interaction of this site with different effectors. In addition, the QEHA peptide blocked almost completely the GTPγS-induced stimulation of K_{ACh} activity (Fig. 1), suggesting that it is indeed the Gβγ complex that is the predominant transducer conveying the agonist signal. Gβγ subunits have been shown to be responsible also for the activation of neuronal potassium channels, such as those in cultured noradrenergic neurons from the rat locus coeruleus (49).

K_{ACh} Current Characteristics

Single K_{ACh} channel activity exhibits several identifying features. In the absence of agonist, (basal) K_{ACh} channel activity is very low. Channel activity induced by ACh (in the pipette of cell-attached or in the bath of outside-out patches), GTP, or GTPγS and Gβγ (in the intracellular side of the membrane) shows inward rectification due to the presence of internal Mg^{2+}. In symmetrical 145 mM K^+ solutions, K_{ACh} has a unit conductance of ~ 40 pS and a mean open time of ~ 1 ms (in the absence of hydrolyzable forms of ATP in the cytosolic side). Figure 2A shows the inwardly rectifying character in a cell-attached patch with single ACh-activated K_{ACh} currents at different membrane potentials. Single channel openings during a ramp protocol in a cell-attached patch reveal the inwardly rectifying characteristics and a conductance of ~ 40 pS under symmetrical K^+ conditions (Fig. 2B). Stimulation of K_{ACh} whole-cell currents by application of external ACh reveals another distinguishing feature. Figure 2C shows whole-cell currents from a chick embryonic atrial cell. In high potassium (HK) external solutions (symmetrical to the internal K^+ concentration), an agonist-independent or basal inwardly rectifying current develops, which passes through distinct but related channels described in these cells. At 10 μM, ACh elicits a partially desensitizing ionic current, in which two distinct phases can be discerned: a rapid (< 1 min) and a slow phase (4–5 min) (Refs. 36,50; Fig. 2C). The underlying molecular basis for K_{ACh} current desensitization remains unclear.

Homologous receptor desensitization by G protein receptor kinase 2 (GRK2) has been proposed as a possible mechanism responsible for the slow phase of current desensitization (51). Whole-cell currents induced by GTPγS diffusing into the cell through the pipette reached desensitized levels of activity rather than the peak ACh-induced current levels (36,50). This observation led Kurachi and colleagues to postulate that some unknown factor may be necessary for the initial rapid activation of K_{ACh} (36). Further work will be required to definitively show whether the magnitude and kinetics of the K_{ACh} current desensitization can be accounted for by specific intracellular factors.

MgATP-Dependent Effects on K_{ACh} Channel Activity

MgATP-dependent effects on K_{ACh} activity or channel characteristics have been described by several groups (52–55). The relative activation of K_{ACh} by millimolar

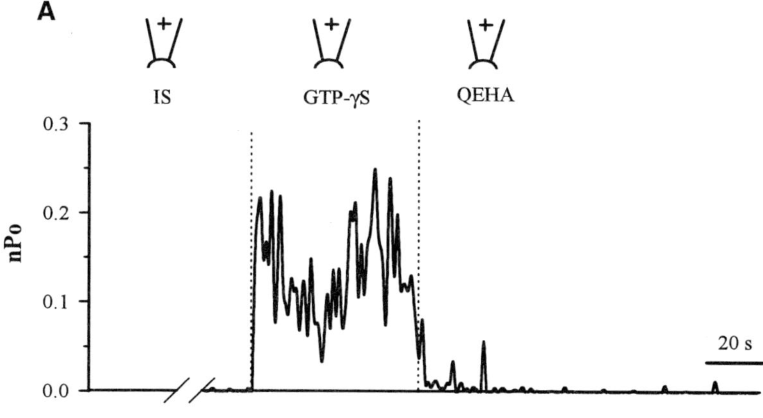

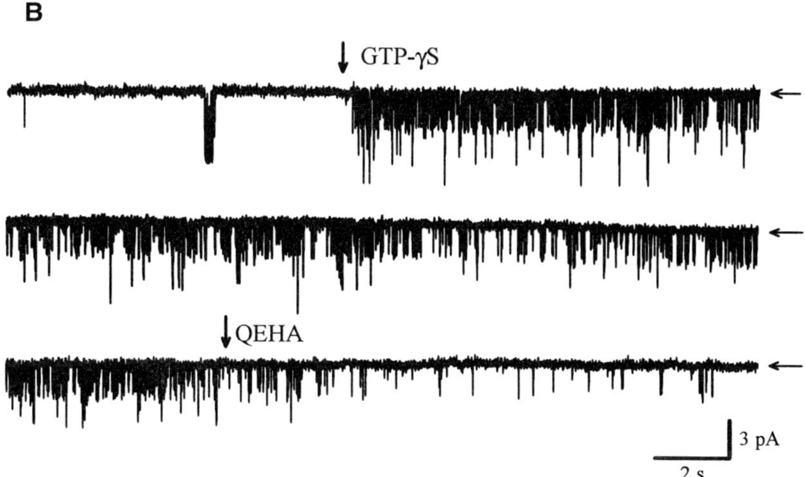

FIG. 1. Effects of QEHA on the atrial K_{ACh} channel activity induced by GTPγS. **(A)** nP_o plot of K_{ACh} channel activity from an embryonic chick atrial inside-out patch as a function of time (5 μM ACh in the pipette). Bath perfusion of the inside surface of the patch with intracellular saline (IS), caused channels to close due to the lack of GTP ($nP_o < 0.001$); GTP γS added in the bath (1 μM) increased channel activity by activating G proteins present in the membrane patch ($nP_o = 0.135$). Application of 50 μM QEHA specifically blocked the effects of GTPγS-activated K_{ACh} channels. Since the QEHA peptide has been shown to block the ability of G protein βγ subunits to stimulate the channel, the QEHA block of the GTPγS-induced activity implies that it is the Gβγ complex that mediates the GTPγS channel activation. **(B)** Single channel currents (raw data) before and after adding GTPγS and QEHA peptide. Data were recorded from 14-day embryonic chick atrial membranes, in symmetrical potassium solutions (140/140 mM), sampled at 5 kHz and filtered at 1 kHz at room temperature.

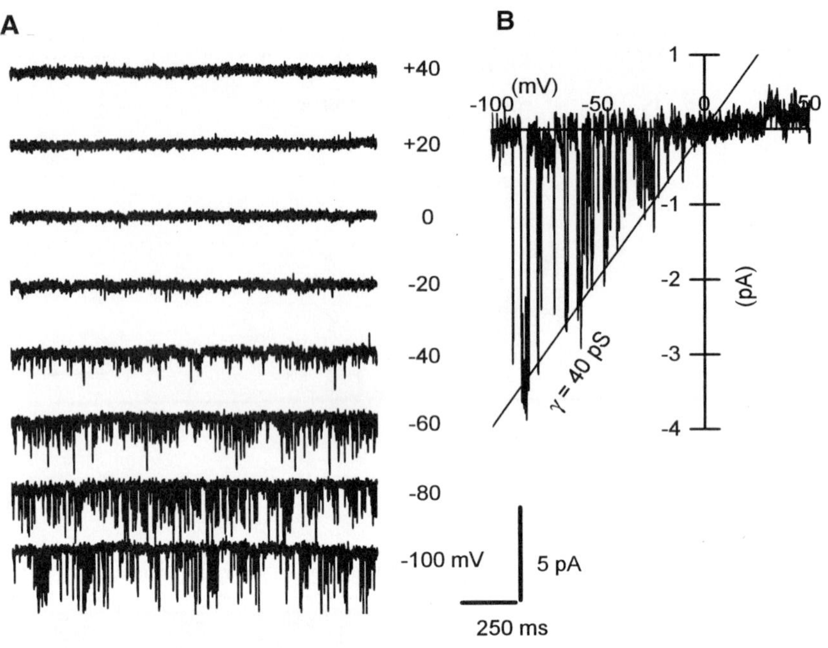

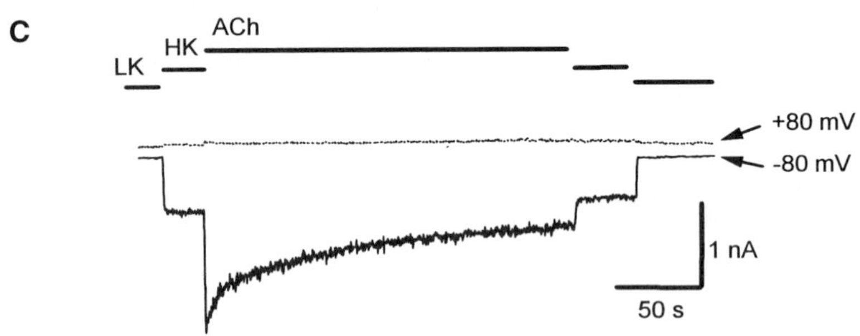

FIG. 2. Atrial K_{ACh} single channel currents and whole cell current desensitization. **(A)** K_{ACh} channels recorded in the cell-attached configuration from chick embryonic atrial myocytes at different holding membrane potentials; 5 μM ACh was present in the pipette solution. Membrane potentials are indicated on each current trace. **(B)** Single K_{ACh} channel currents recorded on a voltage ramp protocol, also recorded in the cell-attached configuration from an embryonic chick atrial myocyte. The straight line indicates a linear conductance slope of 40 pS. **(C)** Whole-cell current time course recorded from an embryonic chick atrial myocyte in the whole-cell clamp mode. Both current time courses are shown by arrows at +80 mV (top line) and at −80 mV (bottom line) As indicated by the bars, HK (high K^+) solution induced a large inward current, which was further increased by application of ACh. The two desensitizing phases, rapid (<1 min) and slow (4–5 min) of the ACh-induced current can be appreciated. Data were from a 17-day embryonic chick atrial cell. For single channel data in A and B, symmetrical potassium solution was used (140/140 mM), with 5 μM ACh in the pipette. Data were sampled at 5kHz, filtered at 1kHz. For whole-cell current in C, intracellular solution had 100 GTP, 5 mM ATP, 140 K^+, and 20 mM Na^+; extracellular solution included 5mM (LK) or 140 (HK) K^+, 135 (LK) or 0 (HK) Na^+.

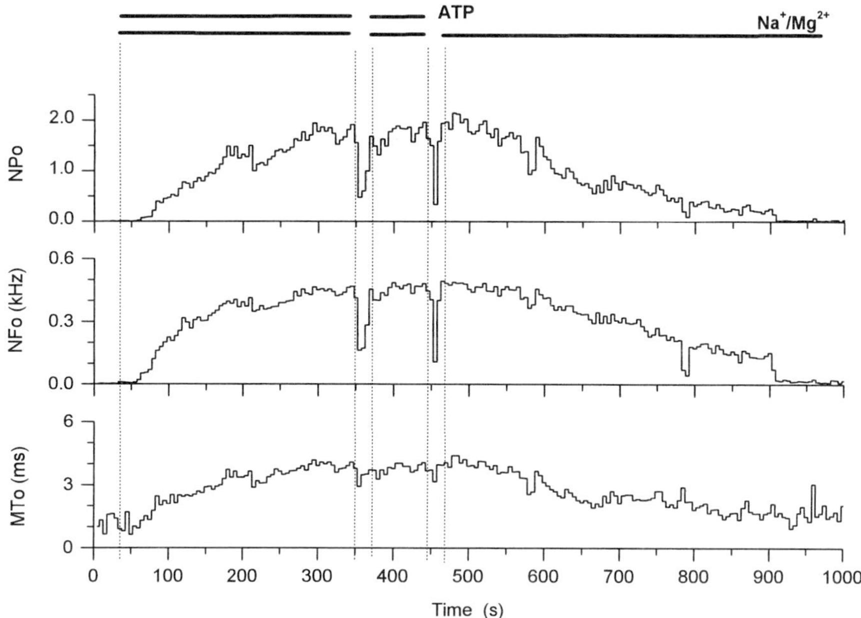

FIG. 3. Kinetics of MgATP/Na activation and deactivation of recombinant K_G channels: NP_o (activity), NF_o (frequency of channel opening), and MT_o (mean open time) plots of single channel activity recorded from an oocyte injected with K_{G4} and K_{G1} cRNAs. The inside-out patch was voltage-clamped at −80 mV under symmetrical K^+ solutions (96/96 mM). Following extensive wash of the patch in the absence of ATP there was very little basal activity left. Perfusion of the patch with solutions, containing a combination of ATP (5 mM), Mg^{2+} (1.2 mM free), and Na^+ (20 mM), caused a slow increase in the activity of the K_G channels, where both MT_o and NF_o increased. Normally 2–3 minutes is needed to reach maximal activation. Interruptions of MgATP/Na^+ applications caused brief and rapid decline in activity, which immediately returned to maximal activity levels when MgATP/Na^+ was reapplied. Once ATP perfusion had stopped, a slow declining of activity took place and Mg^{2+} and Na^+ could no longer maintain the activity. Oocytes were injected with K_{G4} (4 ng), K_{G1} (4 ng) RNAs. Pipette solution had (in mM): KCl 96, $CaCl_2$ 2, HEPES 10, $MgCl_2$ 1, gadolinium chloride 0.1, pH 7.35; bath solution had (in mM): KCl 96, EGTA 5, HEPES 10, $MgCl_2$ 0.6, pH 7.35. The membrane was voltage-clamped at −80 mV.

concentrations of MgATP in the absence of agonist constitutes a small fraction of the agonist-induced activity. Sui et al. (55) showed that hydrolysis of ATP sensitized K_{ACh} channels to gating by intracellular Na^+, which at physiological concentrations caused appreciable levels of activity. Earlier work had shown channel activity also to be dependent on the concentration of free Mg^{2+} (53). Other ions however, such as K^+ or Li^+, did not seem to affect channel gating (55). Thus, specific ions at physiologic concentrations (such as Na^+ and Mg^{2+}) are capable of gating K_{ACh} channels. The ATP effect on K_{ACh} activity could be reliably tracked by continuously monitoring kinetic changes in single channel parameters. The results of such analysis were consistent with conclusions drawn previously by Kim (54). Figure 3 shows continuous monitoring of activity (NP_o), frequency of channel opening (NF_o), and mean

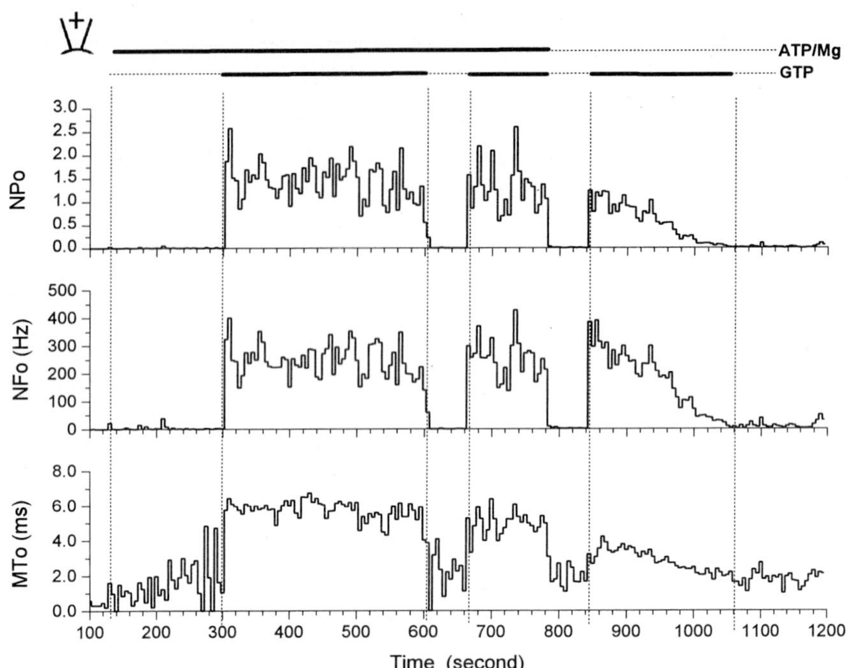

FIG. 4. Rundown of GTP activation of atrial K_{ACh} channels depends on ATP: NP_o, NF_o, and MT_o plots of K_{ACh} channel activity from a chick embryonic atrial myocyte. ATP (2 mM, Na^+ salt, in the presence of 0.6 mM Mg^{2+} in the buffer) maintains GTP (100 μM) activation from running down. The bars on top of the plots indicate application of each nucleotide. Plots are means of each bin (5 s). In the absence of MgATP, channel activity runs down, and open times populate short-lived states. Similar effects were seen in several other records ($N > 5$). Data were from 17-day embryonic chick atrial membranes, in symmetrical potassium solution (140/140 mM) and 10 μM ACh in the pipette. Cell membrane was clamped at −80 mV.

open time (MT_o) as a function of time. The slow kinetics of the current activation and deactivation correlate well with the changes in the mean open time duration due to the MgATP-induced channel modification. Brief interruptions in MgATP/Na^+ caused rapid decline in the activity but not in the channel-open time, while removal of ATP perfusion resulted in the slow reversal of the MgATP effect.

The molecular mechanism of the MgATP effect has been under active investigation. It appears to be independent of direct Gβγ gating, since it occurs in the absence of agonist (55) and when Gβγ gating is blocked by QEHA peptide (120). Nucleoside diphosphate kinase mediating phosphotransfer to the G protein from ATP or a membrane-bound kinase phosphorylating the channel protein has been proposed (52,54). However, conclusive evidence in favor of such a specific mechanism has not been presented yet. Not only is hydrolysis of ATP important in G protein subunit independent gating, but it seems to affect G protein dependent gating as well. Figure 4 demonstrates a MgATP-dependent rundown of agonist-induced activation, which we described previously (56). In the presence of MgATP, repeated applications of internal GTP are capable of stimulating K_{ACh} activity (in the continuous presence of agonist in the pipette) for a much longer period than in the absence of internal MgATP.

Since MgATP seems to affect agonist-independent as well as agonist-dependent K_{ACh} activity, it is possible that the same molecular mechanism is utilized for both processes.

Modulation of K_{ACh} Activity by Lipid Derivatives

The essential fatty acid arachidonate was shown to stimulate K_{ACh} channel activity (57,58). Arachidonic acid (AA) effectiveness seemed to be limited exclusively (57) or predominantly (58) to bath application during a cell-attached experiment (with no agonist in the pipette). It could be blocked by blockers of the 5-lipooxygenase pathway. Moreover leukotriene C_4 (LTC_4) could mimic the effects of AA. These fatty acid effects on K_{ACh} were present even in PTX-treated cells. Yet, upon excision into inside-out patches, AA- and LTC4-induced activity could be restored with cytosolic GTP alone, suggesting the involvement of a PTX-insensitive G protein. Block of phospholipase A2 (PLA_2) by an antibody seemed to prevent $G\beta\gamma$ activation (58), suggesting that the $G\beta\gamma$ effect on channel activity may have been indirect, mediated via activation of PLA_2. However, this view was no longer supported when other blockers of the lipooxygenase pathway failed to block $G\beta\gamma$ activation, which was 6–16 times greater than AA or LTC_4 stimulation of activity (57). The precise mechanism of K_{ACh} activation by AA and its metabolites remains to be clarified.

More recently, other lipid derivatives, such as lysosphingomyelin and sphingosine 1-phosphate, have been shown to stimulate K_{ACh} currents through atrial receptors coupled to PTX-sensitive G proteins (59,60). Finally, anionic phospholipids have been shown to stimulate and to be involved in the rundown of the ATP-sensitive K^+ channel activity, another close relative to K_{ACh} belonging to the inwardly rectifying channel family (61,62). Recent findings in our laboratory and others indicate that phosphatidylinositol 4,5-bisphosphate [$PI(4,5)P_2$] underlies the ATP-dependence of K_{ACh} channel activity and even exerts a permissive role on both $G\beta\gamma$ and Na^+ gating (56,121). Lipid regulation of ion channel activity is just beginning to receive long overdue attention, promising exciting new insights on the role of lipids in the regulation of activity of these integral membrane proteins.

THE CLONING ERA

Members of the K_G Subfamily and Tissue Distribution

Five unique cDNAs coding for G protein gated K^+ channels have been isolated. We refer to these cloned channels as K_{G1-5} instead of the common nomenclature of GIRK1–5 (119), to avoid confusion with GRK1–6, which refer to the family of G protein receptor kinases. The K_G channels belong to the larger inwardly rectifying family (K_{ir}) and have been classified as $K_{ir3.0}$ channels (63), where the number after the decimal corresponds to that used in other nomenclatures. K_{G1}, the first member of the K_G subfamily, was cloned from cardiac atrial tissues (34,35), where it was found to be predominantly expressed among peripheral tissues (cloned also from pancreas; Refs. 64,65). In the brain, K_{G1} was expressed in areas such as the hippocampus, the cerebral cortex, the cerebellar granular layer, the olfactory bulb, the anterior pituitary, the thalamic nuclei, the paraventricular nucleus of the hypothalamus, the pontine nuclei and amygdala, and several distinct nuclei of the lower brain stem (66–69).

K_{G2} and K_{G3} were first isolated from brain (70). K_{G2} was found to be expressed in the hippocampus, substantia nigra, pontine nuclei, cerebellar granular layer, olfactory bulb, cerebral cortex, septum, and amygdala (67). K_{G3} was shown to be expressed in the cerebral cortex, thalamus, and cerebellar granular and molecular layers, as well as throughout the olfactory system, hippocampus, hypothalamus, amygdala, caudate putamen, and most midbrain and brain stem nuclei (67,71). K_{G4} was first cloned from heart (72,73) where it is found exclusively in atria.

K_{G4} was found to be predominantly expressed in pancreas but also in kidney, testis, lung, and brain (65,74). In brain, expression was much lower than K_{G1-3} and was mainly localized to the hippocampus, superior colliculus, cerebellar Purkinje cells, inferior olive, medial habenular, paraventricular thalamic, and ventromedial hypothalamic nucleus (75). K_{G5} was cloned from *Xenopus* and is expressed in *Xenopus* oocytes, a system widely used for heterologous overexpression of K_G subunits (76).

Figure 5A compares the deduced amino acid sequences of the K_{G1-5} recombinant channels. All members of the K_G subfamily show high homology for a stretch of ~300 residues, with the greatest divergence in the end N- and C-terminal segments. K_{G1} is the largest K_G protein, comprising 501 amino acid residues. In Fig. 5B we have sketched the presumed structure of K_{G1} and highlighted similarities with K_{G4} but also those residues found in both K_{G1} and K_{G4} but not in the two prototypical G protein insensitive channels, IRK1($K_{ir2.1}$) and ROMK1($K_{ir1.1}$). This representation reveals that, within the K_{G1}/K_{G4} conserved intracellular regions (i.e., the proximal C- and N-termini), there is only one stretch of K_G-specific residues: a segment of four residues, CMFI, beginning at position 179. This segment is within a region of 22 conserved residues—among which are three other scattered K_G-specific residues—which extends from the end of M2 into the proximal C-terminus. Furthermore, comparison with other known inward rectifier sequences reveals that the CMFI sequence is absolutely conserved in all K_G subunits and absolutely different in all other inwardly rectifying K^+ channel subunits. However, published reports have shown that portions of the C-terminal of K_G proteins that did not incorporate this stretch of residues can either bind Gβγ (77,78) or confer G protein sensitivity to chimeras between IRK1 and K_{G1} (78–81) or K_{G4} (82) (see later discussion). Thus, this sequence is probably not part of a region critical for G protein interactions. Since as noted shortly, the proximal C-terminal has been implicated in selective assembly of inward rectifiers (83), the CMFI stretch could determine the other specific property of K_G subunits, preferential intrafamilial assembly.

Outside this proximal C-terminal region, there are a few K_G-specific residues equally scattered in the conserved N-terminal and C-terminal regions. This suggests that interaction sites depend on the folding pattern of large regions of both C- and N-termini. This would explain the difficulty in pinpointing a specific sequence region as responsible for G protein gating by simple linear "cutting and pasting" (77,80) (as discussed later).

Heteromeric Nature of K_G Channels and Stoichiometry

Krapivinsky and colleagues (73) provided strong evidence for the heteromeric nature of the atrial K_{ACh} channel. Antibodies raised against K_{G1} coimmunoprecipitated K_{G1} together with K_{G4} from bovine atrial membranes, indicating the tight association of these proteins. Moreover, heterologous coexpression of K_{G1} and K_{G4} resulted in "K_{ACh}-like," inwardly rectifying currents, 8–18 times greater than currents resulting from similar expression of either channel alone. Multiple studies have confirmed these results and have shown that association of K_{G1} with any of the other subfamily members resulted in great enhancement of channel activity (84–87). Coexpression of various concatenated trimeric constructs with different monomers gave results consistent with a tetrameric channel comprising two of each of the heteromeric subunits $(K_{G1})_2/(K_{G4})_2$ (88), showing no positional preference for a given subunit, unlike other inwardly rectifying heterotetramers (89).

Immunolocalization studies of the K_{G1} and K_{G4} subunits expressed in mammalian cells (90) or *Xenopus* oocytes (91) suggested that K_{G1} subunits are not targeted efficiently to the cell surface unless they are coexpressed with K_{G4} subunits. Figure 6 shows similar results from experiments in which each of the two subunits has been tagged at its carboxy terminal end with the cDNA for the green fluorescence protein (GFP). Confocal fluorescence images of 50-μm oocyte sections revealed that K_{G1}-GFP was predominantly localized intracellularly. Coexpression of K_{G1}-GFP with nontagged K_{G4} subunits translocated K_{G1} subunits to the cell surface. In contrast, K_{G4}-GFP subunits could be localized at the cell surface regardless of the presence of nontagged K_{G1} subunits. The basis for the targeting defect of K_{G1} subunits to the cell surface is currently unknown.

Biochemical and electrophysiological studies have attempted to localize regions in the primary amino acid sequence that are involved in homomeric and heteromeric channel assembly. Unlike voltage-gated K^+ channels, where the N-terminus and the first transmembrane domain seem to be critical for assembly, K_{ir} channels rely on their second transmembrane segment and the proximal C-terminus for assembly (83). In particular, K_G channels, seem to use regions throughout their length for intersubunit interactions (91). Woodward and colleagues showed that C-terminal polypeptides from the same (K_G) or different (K_{G1}/K_{G2}) subunits recognized each other but did not seem to recognize the N-terminal or transmembrane regions of K_{G1}, suggesting that the N- and C-termini may not cross-interact. Coexpression of truncated K_{G1} channels with full-length K_{G1}/K_{G2} heteromers inhibited current magnitudes when the second transmembrane or the second transmembrane and C-terminal segments were present, in close agreement with the results by Tinker and colleagues on K_{ir} channels (83). When the K_{G1} C-terminus was divided into proximal (Met^{180}–Gly^{307}) and distal

A

```
KG1  MSALRRK            FGDDYQVVTTSSSGSGLQPQGPGQDPQQ     QLV  PKKKRQRFVDKNGRCNVQHGNLGSETSRYLSD  70
KG2  -TMAKLTESMTNVLEGDSMDQDVESPVAIHQPK LPKQARDDLPRHISRDR    TKRKI--YVR-D-K---HH--V R--Y--LT-  81
KG3  -AQEN                        AAFSP       GSEEPPR       RRGR--YVE-D-R---QQ--V R--Y--LT-  47
KG4  -AGDSRN          AMNQDMEIGVTPWDPKKIPKQARDYVPIATDRTRLLAEGKKPR--YME-S-K---HH--V Q--Y--LS-  76
KG5  -ARDLRV          SMNQDLKVDITP    KKLPKQAREDPKRSIDRTHLLFEHKKPR--YME-D-K---HH--V K--Y--FS-  73
      *                         *         ***                **    **  *  *   **  *      **
                           M1                                                  H5
KG1  LFTTLVDLKWRWNLFIFILTYTVAWLFMASMWWVIAYTRGDLNKAHVGNYTPCVANVYNFPSAFLFFIETEATIGYGYRYITDKCPE  157
KG2  I-------K-RFN-LI-VMV-TVT---FGMI--L---I---MDHIEDPSWT---T-LNG-VS----S----T-----Y-V--DK---  168
KG3  L-------Q-RLR-LF-VLA-ALT---FGVI--L---G---LEHLEDTAWT---N-LNG-VA----S----T-----H-V--DQ---  134
KG4  L-------K-RFN-LV-TMV-TVT---FGFI--L---I---LDHVGDQEWI---E-LSG-VS----S----T-----F-V--EK---  163
KG5  L-------K-HFS-FI-TLV-TVT---FGLI--L---I---LDHLGDKNWV---E-LNG-VS----S----T-----Y-V--EK---  160
     *        * * * *  *** ***   ** *  *   *   ***    ***   * * *  *         *         **
               M2
KG1  GIILFLFQSILGSIVDAFLIGCMFIKMSQPKKRAETLMFSEHAVISMRDGKLTLMFRVGNLRNSHMVSAQIRCKLLKSRQTPEGEFL  244
KG2  --I-L-I-SV---I-N-FMV----V-I---K---E--V--TH--I-M--GK-C------D--N--I-E-S--A--IK-K--S----I  255
KG3  --V-L-L-AI---M-N-FMV----V-I---N---A--V--SH--V-L--GR-C------D--S--I-E-S--A--IR-R--L----I  221
KG4  --I-L-V-AI---I-N-FMV----V-I---K---E--M--NN--I-M--EK-C------D--N--I-E-S--A--IK-R--K----I  250
KG5  --V-L-V-AI---I-N-LMV----V-I---K---E--M--NN--I-L--GK-C------D--S--I-E-S--A--IK-K--K----I  247
       * *   **   * * ***    * *    *    *   *    *  *  *  **         *  *  * *   *  ** *      *
KG1  PLDQLELDVGFSTGADQLFLVSPLTICHVIDAKSPFYDLSQRSMQTEQFEIVVILEGIVETTGMTCQARTSYTEDEVLWGHRFFPVI  331
KG2  --N-SDIN--YY--D-R-------I-S-E-NQQ---WEI-KAQLPKEEL-I------I--A--------S--ITS-I---Y--TP-L  342
KG3  --H-TDLS--FD--D-R-------V-S-E-DAA---WEA-RRALERDDF-I------M--A--------S--LVD-V---H--TS-L  308
KG4  --N-TDIN--FD--D-R-------I-S-E-NEK---WEM-QAQLHQEEF-V------M--A--------S--MDT-V---H--TP-L  337
KG5  --N-TDIN--FD--D-R-------I-S-E-NEK---WEM-RTQLEIEEF-I------M--A--------S--VDS-V---H--IP-L  334
       *  ***  *   * *        * *   **     **  *  *  *** *      *  *        *    * *   *   * *
KG1  SLEEGFFKVDYSQFHATFEVPTPPYSVKEQEEMLLMSSPLIAPAITNSKERHNSVECLDGLDDITTKLPSKLQKITGREDFPKKLLR  418
KG2  TM-D--YE---NS--ETY-TS--SLSAK-LA-L                                                       375
KG3  TL-D--YE---AS--ETF-VP--SCSAR-LA-A                                                       341
KG4  TL-K--YE---NT--DTY-TN--SCCAK-LA-M                                                       370
KG5  TL-K--YE---TT--DNY-TP--SCSAK-LK-I                                                       367
     ** *   **    *    *** *    * ***
KG1  MSSTTSEKAYSLGDLPMKLQRISSVPGNSEEKLVSKTTKMLSDPMSQSVADLPPKLQKMAGGAARMEGNLPAKLRKMNSDRFT    501
KG2               ANRAEVP          LSWSVSSKLNQHAELETEEEEKNPEELT                     ERNG     414
KG3               AARLDAHLYWSIPS  RL             DEKVEEEGAGEGG                      RCGRWS   376
KG4               KREGRLLQYLPSPP  LLGGCAEAGLDAEAEQNEEDEPKGLGGSR                     EARGSV   419
KG5               QHGEQVLH              HTYTGQRKNSAQNGAPLGSEDLN                     RRTQPD   404
```

FIG. 5. K_{G1-5} deduced amino acid comparisons and the sequence of K_{G1} subunit represented within the presumed topology of inward rectifier K^+ channels. **(A)** Deduced amino acid sequence comparisons of the five members of the K_{G1-5} subfamily: dashes indicate residues identical to the K_{G1} residue shown; asterisks indicate that residues at the particular position are well

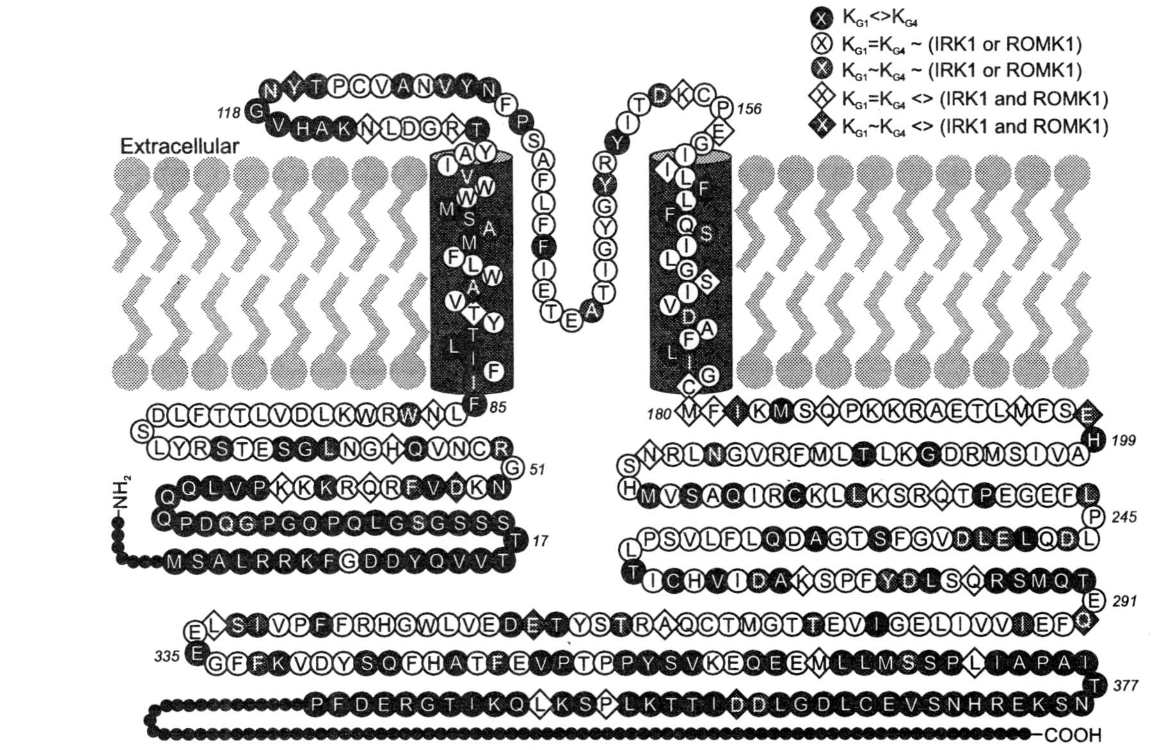

FIG. 5 *(Continued).* conserved. Gaps between amino acid were introduced by the alignment algorithm. Sequence alignments were performed using ClustalW, version 1.6. The putative transmembrane (M1 and M2) and pore-forming (H5) regions are indicated. **(B)** Symbol shading indicates the degree of similarity between a given residue and the aligned residue in K_{G4} (white = identical; gray = similar; black = different). Symbol shape indicates whether a similar or identical residue is found also in IRK1 and ROMK1: a white (or gray) diamond indicates K_G-specific residues [i.e., that K_{G1} and K_{G4} have identical (or similar) residues at this position but that these residues have no similarity with the aligned residues in IRK1 and ROMK1]. A white or gray circle indicates that a similar residue is found in ROMK1 and IRK1. Dissimilarities between K_{G1} and K_{G4} are marked by black circles whatever the residue is in IRK1 or ROMK1. Smaller, unlabeled circles at either end of the sequence stand for the 7 excess residues of K_{G4} at the N-terminal and the 88 excess residues of K_{G1} at the C-terminal. Sequence alignments were performed using ClustalW version 1.6 (43) and optimized manually.

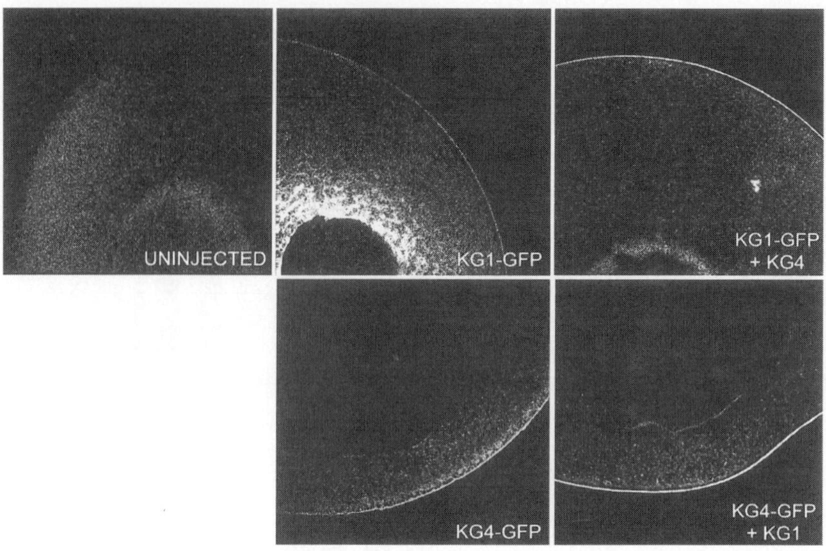

FIG. 6. Localization of K_{G1} or K_{G4} subunits tagged with the green fluorescence protein (GFP): sections of an uninjected oocyte illustrate the low fluorescence background; $K_{G1\text{-}GFP}$ localizes predominantly cytoplasmically. A faint surface localization may reflect the inefficient targeting of K_{G1} to the plasma membrane or its association with the low level endogenous K_{G5} protein. Coinjection of $K_{G1\text{-}GFP}$ with nontagged K_{G4} corrects the targeting problem of the $K_{G1\text{-}GFP}$ subunit. $K_{G4\text{-}GFP}$ localizes predominantly to the surface. $K_{G1\text{-}GFP}$ coinjected with nontagged K_{G1} also localizes predominantly to the oocyte surface. Oocytes were injected with ~4 ng cRNA. Two to three days following injections, the oocytes were fixed in 4% paraformaldehyde, and 50 μm sections were made with a vibratome. The sections were mounted on slides and viewed under fluorescence with a confocal microscope.

(Met^{308}–Thr^{501}) segments, both polypeptides coprecipitated with the full-length K_{G1} subunit. However, the distal segment (encompassing part of the C-terminus unique to the longest K_{G1} subfamily member) coprecipitated much more efficiently with K_{G1} than K_{G2}. These studies leave us with the strong sense of the complexity involved in K_G intersubunit interactions and point to the need for additional work to further localize and characterize the nature of such interactions.

Chimeric studies between K_{G4} and IRK1 exchanging the N-termini, the C-termini, or the transmembrane regions showed that upon association of chimeras, enhanced basal currents resulted only when the transmembrane region of K_{G4} was present (82). Other structure–function studies have identified structural elements of K_{G1} that contribute to the current enhancement of other subfamily members. Chan, Sui, and colleagues (92) identified a P-region position (137 in K_{G1}) which, depending on the amino acid present, greatly affected channel activity and open-time kinetics. Phe^{137} in K_{G1} is unique among inwardly rectifying channels, which possess a Ser residue at the corresponding position. Phe^{137} keeps K_{G1} homomers inactive, while when present in heteromers, which possess a Ser at the corresponding position, it contributes to changes in gating and the enhancement of their activity. In addition, Chan and colleagues (93) identified the unique C-terminal end of K_{G1} (Lys^{359}–Thr^{501}) as an important contributor to the current enhancement. The enhancement in current due to this unique K_{G1} 142 amino acid C-terminal segment did not involve changes in surface expression compared to the wild-type K_{G4} subunit; rather, it increased the frequency of single channel openings. The molecular details underlying this effect of the K_{G1} 142 amino acid C-terminal segment remain to be elucidated. In contrast, smaller peptides (17–18 amino acids) derived from the C-terminal end region of K_{G1} had inhibitory effects on K_G activity, when used to perfuse inside-out patches (94). Specifically, the 17-mere Glu^{485}–Thr^{501} prolonged silent interburst intervals without affecting intraburst kinetics (95). Another proximal peptide (Ser^{447}–Ser^{464}) was also shown by Schreibmayer and colleagues to have inhibitory effects. It will be interesting to determine possible conformational changes in the 142 amino acid C-terminal segment (Lys^{359}–Thr^{501}) that not only seem to ameliorate the inhibitory effects of the smaller peptides contained within its sequence (e.g., Glu^{485}–Thr^{501} and Ser^{447}–Ser^{464}) but also to enhance activity as part of the larger peptide.

Direct Gβγ Binding to K_G Subunits and Channel Gating

Reuveny and colleagues showed first (96) that coexpression of K_{G1} and Gβγ subunits in oocytes produced constitutively active channels that showed no dependence on intracellular GTP. G protein subunit activation of recombinant K_G activity was limited to Gβγ and not to activated Gα-GTPγS subunits. This study suggested a major role for Gβγ subunits in the stimulation of K_G channel activity, in agreement with earlier work on atrial cells (43,47,97). A number of studies have demonstrated direct binding of the Gβγ subunits to either the entire K_G channel subunits (98) or to channel subunit segments (77,78,99,100). Efforts to define minimal channel segments capable of binding Gβ subunits have made use of two approaches: deletion segments (77,100) or chimeras with a Gβγ-insensitive but related channel (78). From such studies, minimal segments exhibiting strong Gβγ binding have been identified in the N-terminus (Gln^{34}–Ile^{86}; Ref. 100) and C-terminus (Glu^{318}–Pro^{462}, Refs. 77,100; Thr^{290}–Tyr^{356}, Ref. 78). Functional studies have been in general agreement with the conclusions drawn from the biochemical studies. Thus, using a chimeric approach, Takao and colleagues (81) implicated a Gβγ-responsive, C-terminal K_{G1} region (Lys^{339}–Thr^{501}), Slesinger and colleagues (80) suggested that Gβγ activation of K_{G1} requires either the N-terminal (Met^{1}–Phe^{85}) or part of the C-terminal (e.g., His^{325}–Thr^{501}), while Kubo and Iizuka (82) suggested that both K_{G4} N-terminus (Met^{1}–Leu^{91}) and C terminus (Met^{186}–Val^{419}) ought to be present for efficient Gβγ signaling. Interestingly, the N-terminal domain of K_{G1} (80) or of K_{G4} (82) facilitated fast rates of activation and deactivation following receptor stimulation. Coexpression of a myristoylated form of the C-terminus of K_{G1} (Lys^{183}–Thr^{501}) with the full-length K_{G1} channel subunit inhibits K_G currents, presumably either by interfering with K_G channel assembly or by providing a sink for free Gβγ subunits (101).

Evidence using the α subunits of neuronal calcium channels, which are Gβγ-modulated, suggests distinct binding sites along the primary sequence (102,103).

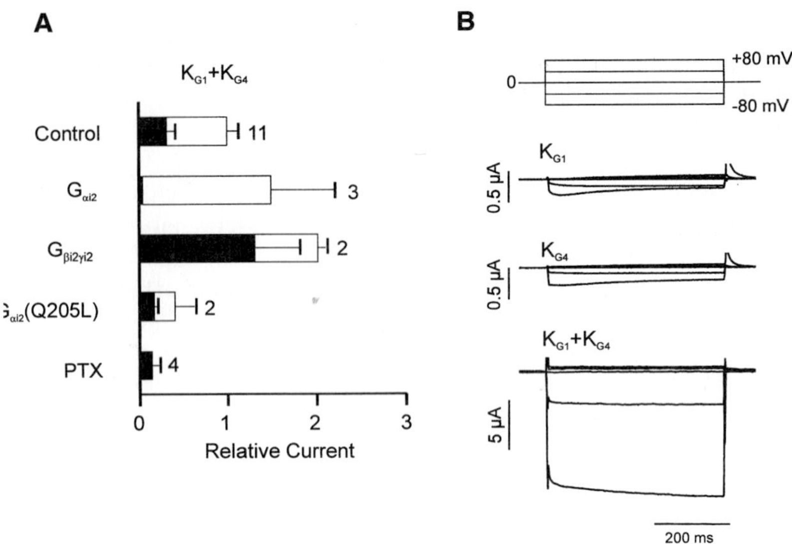

FIG. 7. Average amplitude of basal activity and ACh responses of wild-type heteromeric channels in the presence of different G protein subunits and representative whole-cell currents of control injections. **(A)** For every batch of oocytes, basal and ACh-induced currents were obtained from a number of experiments. To reduce errors arising from batch-to-batch variability in level of expression, these values were normalized to the total current obtained in the presence of ACh under control conditions (the amplitude of total ACh current was on average 12.4 μA, for K_{G1}/K_{G4}). The average of these normalized currents is shown. Numbers indicate the numbers of batches tested in each conditions. **(B)** Representative whole-cell currents obtained with two-electrode voltage clamp of oocytes injected with ~4 ng of K_{G1}, 4 ng of K_{G4}, or 2 ng of each.

Interestingly, the first Gβγ cocrystal with the effector phosducin has revealed five distinct regions of interactions in phosducin, a significant portion of which shows overlap with the Gα sites of interaction (104). Thus, if multiple Gβγ-binding sites are required for proper functional effects, the task of identifying all the sites, their relative importance, and their concerted mechanism of action is certainly not a simple one.

G Protein Subunit Regulation of K_G Channel Activity

Vivaudou and colleagues (105) studied the regulation of recombinant K_G heteromeric channel activity in response to muscarinic stimulation in different conditions expected to perturb the receptor–G protein signaling cascade. They coexpressed different G protein subunits with the m2 receptor and the K_{G1}/K_{G4} heteromer in *Xenopus* oocytes and examined effects on the basal- and receptor-stimulated K^+ channel activity (Fig. 7). Compared to control (K_{G1}/K_{G4}/hm2), addition of $G\alpha_{i2}$ nearly abolished basal activity while enhancing ACh responses twofold. Addition of Gβγ raised basal activity severalfold as first reported by Reuveny et al. (96), but the magnitude of the ACh-induced current was only slightly less than in control conditions. Addition of $G\alpha_{i2}$(Q205L) reduced both basal and ACh-induced activity to 50 and 40% of their control values. Coinjection of the PTX-S1 cRNA encoding the active protomer of PTX thoroughly abolished activation by ACh but only partially inhibited basal activity, in agreement with previous data using incubation of oocytes with purified PTX proteins (87).

In terms of mechanism, the foregoing observations, although consistent with a major role for $G_{\beta\gamma}$ in mediating activation of G protein gated K^+ channels (43), also suggest the existence of additional regulatory pathways involving $G\alpha$ or receptors. In particular, the effects of exogenous $G\alpha_{i2}$(Q205L) suggest, that $G\beta\gamma$ is not the only regulator. The mutation Q205L in $G\alpha_{i2}$ is cognate to the mutation Q227L in G protein subunit α_s, which drastically reduces its GTPase rate without affecting GDP dissociation and, as a consequence, changes the fractional occupancy by GTP from about 5 to 80% (106). Injection of $G\alpha_{i2}$(Q205L) cRNA should therefore produce a large increase in GTP-bound $G\alpha_{i2}$ and a moderate, but still significant, increase in GDP-bound $G\alpha_{i2}$. Since the mutation does not appear to influence the ability of the protein to interact with $G\beta\gamma$ and receptor (106), mutant $G\alpha_{i2}$, like the wild type, will be able to sequester $G\beta\gamma$, but to a lesser extent and should therefore have similar, if not smaller, effects on basal and receptor-dependent activity. $G\alpha_{i2}$(Q205L) did reduce basal activity as expected, but unlike $G\alpha_{i2}$, it did not potentiate receptor-dependent activation and instead reduced it to below control levels. In contrast, equivalent experiments by Lim et al. (107) with $G\alpha_s$(Q227L) and β_2-adrenergic receptors showed an increase in receptor-dependent activity but no effect on basal activity. These results do not readily point to a precise mechanism, although they suggest that specific $G\alpha$ subunits could have different modulatory roles, unrelated to their ability to interact with $G\beta\gamma$. Indeed certain G protein α_{GTP} subunits can directly antagonize channel activation by $G\beta\gamma$ (108), while the α_{GDP} subunits have been shown to bind to K_{G1} (77) and are proposed to participate in the tight link between receptor and channel (80) implied by the "membrane-delimited" pathway of the signaling cascade (33).

Physiological Consequences of K_G Channel Knockouts

The report published in 1995 that the mutant gene responsible for the *weaver (wv)* mouse is a mutation that functionally alters K_{G2} triggered great interest (109). The *wv* mutation results in the degeneration of specific neuronal populations, such as the granule cells of the cerebellum and dopaminergic neurons in the substantial nigra (see reviews in Refs. 110,111). A missense point mutation in the pore-forming (P) region of K_{G2} causes substitution of Gly^{156} to a Ser residue. As a result, K_{G2}(*wv*) exhibits a dual functional alteration, namely, a decrease in K^+ currents and a loss of selectivity to potassium ions, allowing Na ions to readily permeate (112–114). It has been suggested that the decrease in K^+ currents may reflect a loss of sensitivity to $G\beta\gamma$ subunits (115). The loss and gain of selectivity to K^+ and Na ions, respectively, is likely to be due to alteration of the selectivity signature sequence TXXTX**G**YG found in K^+ channels, which acts as the selectivity filter during ion permeation (116,122) (the **G** corresponds to the mutated Gly_{156} residue in K_{G2}).

Is it the loss of K^+ current (114) or the gain in Na^+ current (113) that causes neuronal cell death in *wv* mice? Knockout studies of K_{G2} have addressed this question by showing that in homozygous K_{G2} knockout mice, granule cerebellar as well as dopaminergic neurons are alive, unlike their *wv* counterparts (117). These results led Signorini and colleagues to conclude that the *weaver* phenotypes arise from the gain in Na^+ rather than the loss in K^+ ion permeabilities. Interestingly, the K_{G2} knockout mice developed spontaneous seizures and were more susceptible to pharmacologically induced seizures using a γ-aminobutyric acid antagonist. Thus, these mice provide a useful model in which to study the importance of the absence (or reduction) of K_G channel function in the whole animal. The production of a K_{G4} knockout mouse and its effects on the regulation of heart rate have been reported (118). It will serve as a model in attempts to characterize possible defects in the regulation of cardiac rhythm as well as in pancreatic cell function, where these channels are abundant.

REFERENCES

1. Henderson R, Baldwin, JM, Ceska, TA, Zemlin, F, Beckmann E, Downing KH: Model for the structure of bacteriorhodopsin based on high-resolution electron cryo-microscopy. *J Mol Biol* 1990; 213:899–929.
2. Schertler, GF, Villa C, Henderson R: Projection structure of rhodopsin. *Nature* 1993;362:770–772.
3. Watson N, Linder ME, Druey KM, Kehrl JH, Blumer KJ: RGS family members: GTPase-activating proteins for heterotrimeric G-protein α-subunits. *Nature* 1996;383:172–175.
4. Lambright DG, Sondek J, Bohm A, Skiba N P, Hamm HE, Sigler PB: The 2.0 Å crystal structure of a heterotrimeric G protein. *Nature* 1996;379:311–319.
5. Noel JP, Hamm HE, Sigler PB: The 2.2 Å crystal structure of transducin α complexed with GTPγS. *Nature* 1993;366:654–663.
6. Lambright DG, Noel JP, Hamm HE, Sigler PB: Structural determinants for activation of the α-subunit of a heterotrimeric G protein. *Nature* 1994;369:621–628.
7. Wall MA, Coleman DE, Lee E, Iniguez-Lluhi JA, Posner BA, Gilman AG, Sprang SR: The structure of the G protein heterotrimer Gi$\alpha_1\beta_{1\gamma2}$. *Cell* 1995;83:1047–1058.
8. Sondek J, Bohm A, Lambright DG, Hamm HE, Sigler PB: Crystal structure of a G_A protein β_γ dimer at 2.1 Å resolution. *Nature* 1996;379:369–374.
9. Hamill OP, Marty A, Neher E, Sakmann B, Sigworth FJ. Improved patch–clamp techniques for high-resolution current recording from cells and cell-free membrane patches. *Pfluegers Arch* 1981; 391:85–100.
10. Trautwein W, Dudel J: Zum Mechanismus der Membranwirkung des Acetylcholin an der Herzmuskelfaser. *Pfluegers Arch* 1958;266:324–334.
11. Fosset M, Schmid-Antomarchi H, de Weille JR, Lazdunski M: Somatostatin activates glibenclamide-sensitive and ATP-regulated K^+ channels in insulinoma cells via a G-protein. *FEBS Lett* 1988;242: 94–96.
12. Dunne MJ, Bullett MJ, Li G, Wollheim CB, Petersen, OH: Galanin activates nucleotide-dependent K^+ channels in insulin-secreting cells via a pertussis toxin-sensitive G-protein. *EMBO J* 1989;8: 413–420.
13. Rorsman P, Bokvist K, Ammala C, Arkhammar P, Berggren P-O, Larsson O, Wahlander K: Activation by adrenaline of a low-conductance G protein-dependent K^+ channel in mouse pancreatic B cells. *Nature* 1991;349:77–79.
14. Bauer CK, Meyerhof W, Schwarz JR: An inward-rectifying K^+ current in clonal rat pituitary cells and its modulation by thyrotrophin-releasing hormone. *J Physiol (Lond)* 1990;429:169–189.
15. Einhorn LC, Oxford GS: Guanine nucleotide binding proteins mediate D_2 dopamine receptor activation of a potassium channel in rat lactotrophs. *J Physiol (Lond)* 1993;462:563–578.
16. Takano K, Asano S, Yamashita N: Activation of G protein-coupled K^+ channels by dopamine in human GH-producing cells. *Am J Physiol* 1994;266:E318–E325.

17. Gähwiler BH, Brown DA: $GABA_B$-receptor-activated K^+ current in voltage-clamped CA3 pyramidal cells in hippocampal cultures. *Proc Natl Acad. Sci U S A* 1985;82:1558–1562.
18. Andrade R, Malenka RC, Nicoll RA: A G protein couples serotonin and $GABA_B$ receptors to the same channels in hippocampus. *Science* 1986;234:1261–1265.
19. Colino A, Halliwell JV: Differential modulation of three separate K-conductances in hippocampal CA1 neurons by serotonin. *Nature* 1987;328:73–77.
20. Trussell LO, Jackson MB: Dependence of an adenosine-activated potassium current on a GTP-binding protein in mammalian central neurons. *J Neurosci* 1987;7:3306–3316.
21. VanDongen AM, Codina J, Olate J, Mattera R, Joho R, Birnbaumer L, Brown AM: Newly identified brain potassium channels gated by the guanine nucleotide binding protein G_o. *Science* 1988;242: 1433–1437.
22. Wakamori M, Hidaka H, Akaike N: Hyperpolarizing muscarinic responses of freshly dissociated rat hippocampal CA1 neurones. *J Physiol (Lond)* 1993;463:585–604.
23. Williams JT, Colmers WF, Pan ZZ: Voltage- and ligand-activated inwardly rectifying currents in dorsal raphe neurons *in vitro*. *J Neurosci* 1988;8:3499–3506.
24. Lacey MG, Mercuri NB, North RA: On the potassium conductance increase activated by $GABA_B$ and dopamine D_2 receptor in rat substantia nigra neurones. *J Physiol (Lond)* 1988;401:437–453.
25. North A, Williams JT, Suprenant A, Christie, MJ: μ and δ receptors belong to a family of receptors that are coupled to potassium channels. *Proc Natl Acad Sci U S A* 1987;84:5487–5491.
26. Inoue M, Nakajima S, Nakajima Y: Somatostatin induces an inward rectification in rat locus coeruleus neurones through a pertussis toxin-sensitive mechanism. *J Physiol (Lond)* 1988;407:177–198.
27. Miyake M, Christie MJ, North RA: Single potassium channels opened by opioids in rat locus ceruleus neurons. *Proc Natl Acad Sci U S A* 1989;86:3419–3422.
28. Velimirovic BM, Koyano K, Nakajima S, Nakajima, Y: Opposing mechanisms of regulation of a G-protein-coupled inward rectifier K^+ channel in rat brain. *Proc Natl Acad Sci U S A* 1995;92: 1590–1594.
29. Nakajima Y, Nakajima S, Inoue M: Pertussis toxin-insensitive G protein mediates substance P-induced inhibition of potassium channels in brain neurons. *Proc Natl Acad Sci U S A* 1988;85: 3643–3647.
30. Farkas RH, Nakajima S, Nakajima Y: Neurotensin excites basal forebrain cholinergic neurons: ionic and signal-transduction mechanisms. *Proc Natl Acad Sci U S A* 1994;91:2853–2857.
31. Takano K, Stanfield PR, Nakajima S, Nakajima Y: Protein kinase C-mediated inhibition of an inward rectifier potassium channel by substance P in nucleus basalis neurons. *Neuron* 1995;14:999–1008.
32. Tatsumi H, Costa M, Schimerlik M, North RA: Potassium conductance increased by noradrenaline, opioids, somatostatin, and G-proteins: whole-cell recording from guinea pig submucous neurons. *J Neurosci* 1990;10:1675–1682.
33. Soejima M, and Noma A: Mode of regulation of the ACh-sensitive K^+ channel by the muscarinic receptor in the rabbit atrial cells. *Pfluegers Arch* 1984;400:424–431.
34. Dascal N, Schreibmayer W, Lim NF, Wang W, Chavkin C, DiMagno L, Labarca C, Kieffer BL, Caveriaux-Ruff C, Trollinger D, Lester HA, Davidson N: Atrial G protein-activated K^+ channel: Expression cloning and molecular properties. *Proc Natl Acad Sci U S A* 1993;90:10235–10239.
35. Kubo Y, Reuveny E, Slesinger PA, Jan YN, Jan LY: Primary structure and functional expression of a rat G-protein-coupled muscarinic potassium channel. *Nature* 1993;364:802–806.
36. Kurachi Y, Tung RT, Ito H, Nakajima T: G protein activation of cardiac muscarinic K^+ channels. *Prog Neurobiol* 1992;39:229–246.
37. Clapham DE, Neer EJ: New roles of G-protein $\beta\gamma$ dimers in transmembrane signalling. *Nature* 1993;365:403–406.
38. Yamada M, Terzic A, Kurachi Y: Regulation of potassium channels by G-protein subunits and arachidonic acid metabolites. *Methods Enzymol.* 1994;238:394–422.
39. Wickman K, Clapham DE: Ion channel regulation by G proteins. *Physiol Rev* 1995;75:865–885.
40. Pfaffinger PJ, Martin JM, Hunter DD, Nathanson NM, Hille B: GTP-binding proteins couple cardiac muscarinic receptors to a K^+ channel. *Nature* 1985;317:536–538.
41. Breitwieser G, Szabo G: Uncoupling of cardiac muscarinic and β-adrenergic receptors from ion channels by a guanine nucleotide analogue. *Nature* 1985;317:538–540.
42. Kurachi Y, Nakajima T, Sugimoto T: Acetylcholine activation of K^+ channels in cell-free membrane of atrial cells. *Am J Physiol* 1986;251:H681–H684.
43. Logothetis DE, Kurachi Y, Galper J, Neer EJ, Clapham DE: The $\beta\gamma$ subunits of GTP-binding proteins activate the muscarinic K^+ channel in heart. *Nature* 1987;325:321–326.

44. Clapham DE, Neer EJ: G protein $\beta\gamma$ subunits. *Annu Rev Pharmacol Toxicol.* 1997;37:167–203.
45. Codina J, Yatani A, Grenet D, Brown AM, Birnbaumer L: The α subunit of G_K opens atrial potassium channels. *Science* 1987;236:442–445.
46. Logothetis DE, Kim D, Northup JK, Neer EJ, Clapham DE: Specificity of action of guanine nucleotide-binding regulatory protein subunits on the cardiac muscarinic K^+ channel. *Proc Natl Acad Sci U S A* 1988;85:5814–5818.
47. Wickman KD, Iniquez-Lluhi JA, Davenport PA, Taussig R, Krapivinsky GB, Linder ME, Gilman A.G, Clapham, DE: Recombinant G-protein $\beta\gamma$ subunits activate the muscarinic-gated atrial potassium channel. *Nature* 1994;386:255–257.
48. Chen J, DeVivo M, Dingus J, Harry A, Li J, Sui J, Carty D, Blank JL, Exton J, Stoffel, RH, Inglese J, Lefkowitz RJ, Logothetis DE, Hildebrandt JD, Iyengar R: A region of adenylyl cyclase 2 critical for regulation by G protein $\beta\gamma$ subunits. *Science* 1995;268:1166–1169.
49. Nakajima Y, Nakajima S, Kozasa T: Activation of G protein-coupled inward rectifier K^+ channels in brain neurons requires association of G protein $\beta\gamma$ subunits with cell membrane. *FEBS Lett* 1996; 390:217–220.
50. Kurachi Y, Nakajima T, Sugimoto T: Short-term desensitization of muscarinic K^+ channel current in isolated atrial myocytes and possible role of GTP-binding proteins. *Pfluegers Arch* 1987;410: 227–233.
51. Shui Z, Boyett MR, Zhang W-J, Haga T, Kameyama K: Receptor kinase-dependent desensitization of the muscarinic K^+ current in rat atrial cells. *J Physiol (Lond)* 1995;487:359–366.
52. Heidbüchel H, Callewaert G, Vereecke J, Carmeliet E: ATP-dependent activation of atrial muscarinic K^+ channels in the absence of agonist and G-nucleotides. *Pfluegers Arch* 1990;416:213–215.
53. Kaibara M, Nakajima T, Irisawa H, Giles W: Regulation of spontaneous opening of muscarinic K^+ channels in rabbit atrium. *J Physiol (Lond)* 1991;433:589–613.
54. Kim D: Modulation of acetylcholine-activated K^+ channel function in rat atrial cells by phosphorylation. *J Physiol (Lond)* 1991;437:133–155.
55. Sui JL, Chan, KW, Logothetis DE: Na^+ activation of the muscarinic K^+ channel by a G-protein-independent mechanism. *J Gen Physiol* 1996;108:381–391
56. Sui J-L, Petit-Jacques J, Logothetis DE: Stimulation of the atrial K_{ACh} channel activity by the $\beta\gamma$ subunits of G proteins depends on the presence of phosphatidylinositol phosphates. *Proc Natl Acad Sci U S A* 1978;95:1307–1312.
57. Kurachi Y, Ito H, Sugimoto T, Shimizu I, Miki I, Ui, M: Arachidonic acid metabolites as intracellular modulators of the G-protein-gated cardiac K^+ channel. *Nature* 1986;337:555–557.
58. Kim D, Lewis LL, Graziadei L, Neer EJ, Bar-Sagi D, Clapham, DE: G-protein $\beta\gamma$ subunits activate the cardiac muscarinic K^+ channel via phospholipase A2. *Nature* 1989;337:557–560.
59. Bünemann M, Brandts B, zu-Heringdorf DM, van-Koppen CJ, Jacobs KH, Pott L: Activation of muscarinic K^+ current in guinea-pig atrial myocytes by sphingosine-1-phosphate. *J Physiol (Lond)* 1986;489:701–777.
60. van Koppen C, Meyer-zu-Heringdorf M, Laser KT, Zhang C, Jacobs KH, Bünemann M, Pott L: Activation of a high affinity Gi protein-coupled plasma membrane receptor by sphingosine-1-phosphate. *J Biol Chem* 1996;271:2082–2087.
61. Hilgemann DW, Ball R: Regulation of cardiac Na^+, Ca^{2+} exchange and K_{ATP} potassium channels by PIP_2. *Science* 1996;273:956–959.
62. Fan Z, Makielski, JC: Anionic phospholipids activate ATP-sensitive potassium channels. *J Biol Chem* 1997;272:5388–5395.
63. Doupnik CA, Davidson N, Lester, HA: The inward rectifier potassium channel family. *Curr Opin Neurobiol* 1995;5:268–277.
64. Stoffel M, Espinosa R, Powell KL, Philipson LH, Le-Beau M M, Bell GI: Human G-protein-coupled inwardly rectifying potassium channel (GIRK1) gene (KCNJ3): localization to chromosome 2 and identification of a simple tandem repeat polymorphism. *Genomics* 1994;21: 254–256.
65. Ferrer J, Nichols CG, Makhina, EN, Salkoff L, Bernstein J, Gerhard D, Wasson J, Ramanadham S, Permutt, A: Pancreatic islet cells express a family of inwardly rectifying K^+ channel subunits which interact to form G-protein-activated channels. *J Biol Chem* 1995;270:26086–26091.
66. Karschin C, Schreibmayer W, Dascal N, Lester H, Davidson N, Karschin A: Distribution and localization of a G protein-coupled inwardly rectifying K^+ channel in the rat. *FEBS Lett* 1994;348: 139–144.
67. Kobayashi T, Ikeda K, Ichikawa T, Abe S, Togashi S, Kumanishi T: Molecular cloning of a mouse

G-protein-activated K^+ channel (mGIRK1) and distinct distributions of three GIRK (GIRK1, 2 and 3) mRNAs in mouse brain. *Biochem Biophys Res Commun* 1995;208:1166–1173.
68. Ponce A, Bueno E, Vega-Saenz de Miera E, Chow L, Hillman D, Chen S, Wu MB, Wu X, Zhu L, Rudy B, Thornhill, WB: G protein-gated inward rectifying K^+ channel proteins (GIRK1) are present in the soma and dendrites as well as nerve terminals of specific neurons in the brain. *J Neurosci* 1996;16:1990–2001.
69. Morishige K-I, Inanobe A, Takahashi N, Yoshimoto Y., Kurachi H, Miyake A, Tokunaga Y, Maeda T, Kurachi Y: G protein-gated K^+ channel (GIRK1) protein is expressed presynaptically in the paraventricular nucleus of the hypothalamus. *Biochem Biophys Res Commun* 1996;220: 300–305.
70. Lesage F, Duprat F, Fink M, Guillemare E, Coppola T, Lazdunski M, Hugnot, J-P: Cloning provides evidence for a family of inward rectifier and G-protein coupled K^+ channels in the brain. *FEBS Lett* 1994;353:37–42.
71. Dissmann E, Wischmeyer E, Spauschus A, Pfeil DV, Karschin C, Karschin A: Functional expression and cellular mRNA localization of a G protein-activated K^+ inward rectifier isolated from rat brain. *Biochem Biophys Res Commun* 1996;223:474–479.
72. Ashford MLJ, Bond CT, Blair TA, Adelman, JP: Cloning and functional expression of a rat heart K_{ATP} channel. *Nature* 1994;370:456–459.
73. Krapivinsky G, Gordon EA, Wickman K, Velimirovic B, Krapivinsky L, Clapham DE: The G-protein-gated atrial K^+ channel I_{KACh} is a heteromultimer of two inwardly rectifying K^+ -channel proteins. *Nature* 1995;374:135–141.
74. Chan KW, Langan MN, Sui JL, Kozak JA, Pabon A, Ladias JAA, Logothetis DE: A recombinant inwardly rectifying potassium channel coupled to GTP-binding proteins. *J Gen Physiol* 1996;107: 381–397.
75. Spauschus A, Lentes KU, Wischmeyer E, Dissmann E, Karschin C, Karschin A: A G-protein-activated inwardly rectifying K^+ channel (GIRK4) from human hippocampus associates with other GIRK channels. *J Neurosci* 1996;16:930–938.
76. Hedin KE, Lim NF, Clapham DE: Cloning of a *Xenopus laevis* inwardly rectifying K^+ channel subunit that permits GIRK1 expression of I_{KACh} currents in oocytes. *Neuron* 1996;16:423–429.
77. Huang CL, Slesinger PA, Casey PJ, Jan YN, Jan LY: Evidence that direct binding of $G_{\beta\gamma}$ to the GIRK1 G protein-gated inwardly rectifying K^+ channel is important for channel activation. *Neuron* 1995;15:1133–1143.
78. Kunkel MT, Peralta EG: Identification of domains conferring G protein regulation on inward rectifier potassium channels. *Cell* 1995;83:443–449
79. Tucker SJ, Pessia M, Adelman JP: Muscarine-gated K^+ channel: subunit stoichiometry and structural domains essential for G protein stimulation. *Am J Physiol* 1996;40:H379–H385.
80. Slesinger PA, Reuveny E, Jan YN, Jan LY: Identification of structural elements involved in G protein gating of the GIRK1 potassium channel. *Neuron* 1995;15:1145–1156.
81. Takao K, Yosii M, Kanda A, Kokubun S, Nukada T: A region of the muscarinic-gated atrial K^+ channel critical for activation by G protein $\beta\gamma$ subunits. *Neuron* 1994;13:747–755.
82. Kubo Y, Iizuka M: Identification of domains of the cardiac inward rectifying K^+ channel, CIR, involved in the heteromultimer formation and in the G-protein gating. *Biochem Biophys Res Commun* 1996;227:240–247.
83. Tinker A, Jan YN, Jan LY: Regions responsible for the assembly of inwardly rectifying potassium channels. *Cell* 1996;87:857–868.
84. Lesage F, Guillemare E, Fink M, Duprat F, Heurteaux C, Fosset M, Romey G, Barhanin J, Lazdunski M: Molecular properties of neuronal G-protein-activated inwardly rectifying K^+ channels. *J Biol Chem* 1995;270:28660–28667.
85. Duprat F, Lesage F, Guillemare E, Fink M, Hugnot JP, Bigay J, Lazdunski M, Romey G, Barhanin J: Heterologous multimeric assembly is essential for K^+ channel activity of neuronal and cardiac G-protein-activated inward rectifiers. *Biochem Biophys Res Commun* 1995;212:657–663.
86. Kofuji P, Davidson N, Lester, HA: Evidence that neuronal G-protein-gated inwardly rectifying K^+ channels are activated by $G_{\beta\gamma}$ subunits and function as heteromultimers. *Proc Natl Acad Sci U S A* 1995;92:6542–6546.
87. Chan KW, Langan MN, Sui JL, Kozak JA, Pabon A, Ladias J A A, Logothetis DE: A recombinant inwardly rectifying potassium channel coupled to GTP-binding proteins. *J Gen Physiol* 1996;107: 381–397.

88. Silverman SK, Lester HA, Dougherty DA: Subunit stoichiometry of a heteromultimeric G protein-coupled inward-rectifier K^+ channel. *J Biol Chem* 1996;271:30524–30528.
89. Pessia M, Tucker SJ, Lee K, Bond CT, Adelman JP: Subunit positional effects revealed by novel heteromeric inwardly rectifying K^+ channels. *EMBO J* 1996;15:2980–2987.
90. Kennedy ME, Nemec J, Clapham DE: Localization and interaction of epitope-tagged GIRK1 and CIR inward rectifier K^+ channel subunits. *Neuropharmacology* 1996;35:831–839.
91. Woodward R, Stevens EB, Murrell-Lagnado RD: Molecular determinants for assembly of G-protein-activated inwardly rectifying K^+ channels. *J Biol Chem* 1997; 272:10823–10830.
92. Chan KW, Sui JL, Vivaudou M, Logothetis DE: Control of channel activity through a unique amino acid residue of a G protein-gated inwardly rectifying K^+ channel subunit. *Proc Natl Acad Sci U S A* 1996;93:14193–14198.
93. Chan KW, Sui J-L, Vivaudou M, Logothetis DE: Specific regions of heteromeric subunits involved in the enhancement of G-protein-gated K^+ channel activity. *J Biol Chem* 1997;272:6548–6555.
94. Schreibmayer W, Dascal N, Davidson N, Lester HA: Block of a G-protein activated potassium channel cloned from rat atrium by peptides applied from the cytosolic side. *Biophys J* 1995;68:A35.
95. Luchian T, Dascal N, Dessauer C, Platzer D, Davidson N, Lester HA, Schreibmayer W: A C-terminal peptide of the GIRK1 subunit directly blocks the G protein-activated K^+ channel (GIRK) expressed in *Xenopus* oocytes. *J Physiol (Lond)* 1997;505:13–22.
96. Reuveny E, Slesinger PA, Inglese J, Morales JM, Iniguez-Lluhi JA, Lefkowitz RJ, Bourne HR, Jan YN, Jan, LY: Activation of the cloned muscarinic potassium channel by G protein $\beta\gamma$ subunits. *Nature* 1994;370:143–146.
97. Ito H, Tung RT, Sugimoto T, Kobayashi I, Takahashi K, Katada T, Ui M, Kurachi, Y: On the mechanism of G protein $\beta\gamma$ subunit activation of the muscarinic K^+ channel in guinea pig atrial cell membrane. *J Gen Physiol* 1992;99:961–983.
98. Krapivinsky G, Krapivinsky L, Wickman K, Clapham DE: $G_{\beta\gamma}$ binds directly to the G protein-gated K^+ channel, I_{KACh}. *J Biol Chem* 1995;270:29059–29062.
99. Inanobe A, Morishige K-I, Takahashi N, Ito H, Yamada M, Takumi T, Nishina H, Takahashi K, Kanaho Y, Katada T, Kurachi Y: $G_{\beta\gamma}$ directly binds to the carboxyl terminus of the G protein-gated muscarinic K^+ channel, GIRK1. *Biochem Biophys Res Commu* 1995;212:1022–1028.
100. Huang C-L, Jan Y. N, Jan L Y: Binding of the G protein $\beta\gamma$ subunit to multiple regions of G protein-gated inward-rectifying K^+ channels. *FEBS Lett* 1997;405:291–298.
101. Dascal N, Doupnik CA, Ivanina T, Bausch S, Wang W, Lin C, Garvey J, Chavkin C, Lester HA, Davidson N: Inhibition of function in *Xenopus* oocytes of the inwardly rectifying G-protein-activated atrial K^+ channel (GIRK1) by overexpression of a membrane-attached form of the C-terminal tail. *Proc Natl Acad Sci U S A* 1995;92:6758–6762.
102. De Waard M, Liu H, Walker D, Scott VES, Gurnett CA, Campbell KP: Direct binding of G-protein $\beta\gamma$ complex to voltage-dependent calcium channels. *Nature* 1997;385:446–450.
103. Zamponi GW, Bourinet E, Nelson D, Nargeot J, Snutch TP: Crosstalk between G proteins and protein kinase C mediated by the calcium channel α_1 subunit. *Nature* 1997;385:442–446.
104. Gaudet R, Bohm A, Sigler PB: Crystal structure at 2.4 Å resolution of the complex of transducin $\beta\gamma$ and its regulator, phosducin. *Cell* 1996;87:577–588.
105. Vivaudou M, Chan KW, Sui JL, Logothetis DE: Probing the interactions with G-proteins of GIRK1 and GIRK4, the two subunits of the K_{ACh} channel, using functional homomeric mutants. *Biophys* J 1997;72:A11.
106. Graziano MP, Gilman AG: Synthesis in *Escherichia coli* of GTPase deficient mutants of $G_{s\alpha}$. *J Biol Chem* 1989;264:15475–15482.
107. Lim FN, Dascal N, Labarca C, Davidson N, Lester HA: A G-protein-gated K^+ channel is activated via β_2-adrenergic receptors and $G_{\beta\gamma}$ subunits in *Xenopus* oocytes. *J Gen Physiol* 1995;105: 421–439.
108. Schreibmayer W, Dessauer CW, Vorobiov D, Gilman AG, Lester HA, Davidson N, Dascal N: Inhibition of an inwardly rectifying K^+ channel by G-protein α-subunits. *Nature* 1996;380: 624–627.
109. Patil N, Cox DR, Bhat D, Faham M, Myers RM, Peterson AS: A potassium channel mutation in *weaver* mice implicates membrane excitability in granule cell differentiation. *Nat Genet* 1995;11: 126–129.
110. Hess EJ: Identification of the *weaver* mouse mutation: the end of the beginning. *Neuron* 1996;16: 1073–1076.

111. Herrup K: The *weaver* mouse: a most cantankerous rodent. *Proc Natl Acad Sci U S A* 1996;93: 10541–10542.
112. Slesinger PA, Patil N, Liao YJ, Jan YN, Jan LY, Cox DR: Functional effects of the mouse *weaver* mutation on G protein-gated inwardly rectifying K^+ channels. *Neuron* 1996;16:321–331.
113. Kofuji P, Hofer M, Millen KJ, Millonig JH, Davidson N, Lester HA, Hatten ME: Functional analysis of the *weaver* mutant GIRK2 K^+ channel and rescue of the *weaver* granule cells. *Neuron* 1996;16: 941–952.
114. Surmeier DJ, Mermelstein PG, Goldowitz D: The *weaver* mutation of GIRK2 results in a loss of inwardly rectifying K^+ current in cerebellar granule cells. *Proc Natl Acad Sci U S A* 1996;93: 11191–11195.
115. Navarro B, Kennedy ME, Velimirovic B, Bhat D, Peterson AS, Clapham, DE: Nonselective and $G_{\beta\gamma}$-insensitive *weaver* K^+ channels. *Science* 1996;272:1950–1953.
116. Heginbotham L, Lu Z, Abramson T, MacKinnon, R: Mutations in the K^+ channel signature sequence. *Biophys J* 1994;66:1061–1067.
117. Signorini S, Liao YJ, Duncan SA, Jan LY, Stoffel M: Normal cerebellar development but susceptability to seizures in mice lacking G protein-coupled, inwardly rectifying K^+ channel GIRK2. *Proc Natl Acad Sci U S A* 1997;94:923–927.
118. Wickman K, Nemec J, Gendler SJ, Clapham DE: Abnormal heart rate regulation in GIRK4 knockout mice. *Neuron* 1998;20:103–114.
119. Isomoto S, Kondo C, Kurachi Y: Inwardly rectifying potassium channels: Their molecular heterogeneity and function. *Jpn J Physiol* 1997;47:11–39.
120. Petit-Jacques J, Sui JL, Logothetis DE. The ATP-dependence of K-ACh gating: G-proteins, Na^+ and Mg^{2+} ions. *J Gen Physiology* 1999 (submitted).
121. Huang CL, Feng S, Hilgemann DW: Direct activation of inward rectifier potassium channels by PIP2 and its stabilization by G-betagamma. *Nature* 1998;391:803–806.
122. Doyle DA, Cabral JM, Pfuetzner RA, Kuo A, Gulbis JM, Cohen SL, Chait BT, MacKinnon R: The structure of the potassium channel: Molecular basis of K^+ conduction and selectivity. *Science* 1998; 280:69–77.

Advances in Second Messenger and Phosphoprotein Research, Vol. 33

1040-7952/99 $30.00

9

The Company They Keep: Ion Channels and Their Intracellular Regulatory Partners

Barry D. Johnson

Department of Physiology and Neurobiology, University of Connecticut, Storrs, Connecticut 06269

This chapter highlights progress in determining how ion channels and their regulatory partners are brought together. As in business and real estate, location determines who one interacts with and ultimately one's value to and impact on the overall market. For cells, this means placing ion channels in membrane regions where they will have the most impact in terms of inter- and intracellular signaling. Transmitter and hormone receptors are placed in apposition to sites of secretion or access to the circulatory system, and voltage-gated ion channels are targeted to critical sites of signal transduction. This focus on efficiency also means placing the intracellular channel modulators and targets of modulation in close proximity. When one is working on a time scale of microseconds, diffusion over a distance of micrometers is too slow to be relevant.

Cells generate and maintain their subcellular compartments in relation to the structures known collectively as the cytoskeleton. As more detail of cell structure is uncovered, fewer and fewer key proteins appear to be freely diffusing; instead, they tend to be bound in multienzyme complexes that generally include the substrates and the effector. The large number of proteins and organelles that are bound to the cytoskeleton makes it difficult to define the term "cytoskeletal protein," and the distinction between targeting a protein for the sake of localization and targeting for the purpose of regulating its activity becomes blurred. It is for this reason that one of the greatest challenges in the field of ion channel regulation by "structural" proteins will be to determine which interactions are direct allosteric effects (e.g., binding of actin or actin-linker to channels) and which are indirect effects involving anchoring of signaling pathways to channels by the cytoskeleton.

The chapter is divided along the lines of the primary foci of the studies: (1) the mechanisms involved in clustering and localizing channels to subcellular domains, (2) the mechanisms and proteins that are involved in targeting regulatory enzymes to ion channels, and (3) the role of the primary filaments of the cytoskeleton in regulating ion channel activity. Overlap between these groups is inevitable, since targeting primarily involves protein–protein interactions that, more often than not, influence channel and enzyme activity through allostery. The collection of these examples will hopefully serve to stimulate more interaction between these necessarily related endeavors.

LOCALIZING CHANNELS

Overview

By directing ion channels to particular subcellular domains, cells ensure that intercellular and intracellular signaling is as rapid and efficient as possible. Cell polarization is critical to the function of nearly every cell type, with two prime examples being epithelial cells and neurons. Intercellular communication by chemical synapses in neurons is focused in discrete locations by the targeting of presynaptic machinery, including fusion pores and calcium channels, and postsynaptic machinery, including transmitter receptors, sodium channels, and potassium channels. Integration of synaptic inputs on dendrites and the cell soma is determined by the spatial position of synapses in relation to the site(s) of action potential generation (axon hillock or dendritic "booster"). For example, a neuronal potassium channel (Kv1.2) may be localized to discrete dendritic regions (1) to modulate integration of signals generated by differentially localized, transmitter-gated ion channels (2). The proteins responsible for clustering and localizing ion channels are summarized in Table 1.

Intracellular signals generated and regulated by ion channels, especially those involving intracellular Ca^{2+}, are tightly controlled by channel localization. One sub-

TABLE 1. *Summary of proteins involved in channel localization*

Channel	Localizing protein	Ref.
Nicotinic AChR	Rapsyn	8
AMPA-type	GRIP	17
glutamate receptor	NSF	22
	CRIPT	33
NMDA-type	α-Actinin	29
glutamate receptor	PSD-95	17
	Yotiao	26
	Spectrin	32
	Neurofilament NF-L	34
Kainate-type glutamate receptor	SAP90 and SAP102	18
$GABA_A$ and glycine	Gephyrin	35–38
CFTR Cl^-	EBP50	49
TRP Ca^{2+} channel	INAD	44
shaker K^+ channel	DLG	53
Kv1.1 and Kir 4.1	PSD-95	54,56
L-type Ca^{2+} channel	Sorcin	73
Na^+ channels	Spectrin	75
	Syntrophin and dystrophin	76

type of calcium channels, the L-type Ca^{2+} channel, is well positioned to regulate gene expression, largely because of its localization in the cell soma near the nucleus (3). Even within the soma, calcium channels exhibit different properties depending on their location (4) and relationship to signaling structures, which include Ca^{2+} stores, calcium buffers, and modifying enzymes (kinases, phosphatases, lipases, and G proteins).

In polarized epithelial cells of lung and kidney, vectorial transport depends on the targeting of transporters and ion channels through associations with the cytoskeleton or cholesterol-containing lipid rafts (5). In one recent example, motile Madin–Darby canine kidney cells (MDCK cells) were found to have a polarized distribution of voltage-dependent potassium channels (Kv1.4 type), with greater numbers at the leading edge. Concentration of ion channels and signaling proteins by lipid rafts may be a dominant mechanism for localization in many cell types, including neurons (6), with some overlap with what are now considered to be purely protein–protein interactions. One of the reasons for this overlap is that association with cytoskeletal elements, other possibly direct protein–protein interactions, and the presence of proteins in lipid rafts are partially determined in biochemical experiments by insolubility with the detergent Triton X-100. For the purposes of this chapter, what are now considered to be direct protein interactions should be assumed to include the possibility of a lipid intermediary where direct binding has not been clearly demonstrated.

Ligand-Gated Channels

Nicotinic Acetylcholine Receptor

As the first ion channel purified and cloned (7), the nicotinic acetylcholine receptor (nAChR) has the longest history of study on associated proteins, and more is known about the extracellular signals involved in its clustering than for any other ion channel. Work is now beginning to focus on the intracellular determinants. As reviewed recently by Colledge and Froehner (8), the nAChR of skeletal muscle is concentrated at the neuromuscular junction in part by binding to a 43-kDa protein called rapsyn. Localization of the nAChR is switched by growth factor signaling involving receptor tyrosine kinases (RTKs) and their opposing phosphotyrosine phosphatases (9).

Ionotropic Glutamate Receptors

Whereas study of the localization of the nAChR initially focused on the extracellular signals that cluster the channel, initial work on glutamate receptors focused primarily on the intracellular connections that the channel makes in clustering at postsynaptic densities. Synaptic activity appears to regulate the localization of the NMDA-type glutamate receptor (10) through a pathway that involves Ca^{2+}. AMPA- and NMDA-type receptors are differentially localized to spine and shaft synapses (2) by interacting at some level with actin microfilaments (*see summary of the cytoskel-*

eton, Table 4). Clustering and localization of the AMPA-,NMDA-, and kainate-type glutamate receptors appear to occur by binding of cytoskeleton-associated proteins at the C-terminus of the receptor pore-forming subunits. The consensus primary sequence of these PDZ-binding, C-terminal residues is glutamate-(serine or threonine)-X-valine, where X is any amino acid [*see Table 2* (10a–10e)] and this sequence or a closely related one is found in many ion channel types (11–13). PDZ domains are roughly 90 amino acid domains found in a large family of proteins that get their family name from PSD-95, Discs large, and ZO-1 (14–16).

NMDA-, AMPA-, and kainate-type glutamate receptors bind to the PDZ domain containing proteins PSD-95, GRIP, and SAP90/102, respectively (17,18). These three proteins are concentrated in postsynaptic densities, and immunohistochemical studies have shown that they colocalize with the receptors and studies of heterologously expressed proteins have demonstrated binding (18–20). The case for the involvement of PSD-95 and other family members in receptor and clustering is further supported by experiments in heterologous cells that demonstrate robust clustering with coexpression of NMDA-type glutamate receptors and the PSD-95 family member PSD-93/chapsyn-110 (21). Despite strong evidence for the involvement of PSD-95 in the clustering of the NMDA-type receptor, however, its connection to the cytoskeleton is less certain.

Numerous other proteins also interact with glutamate receptors and may be involved in their localization and binding to the cytoskeleton; these include NSF (22–25) for the AMPA receptor and yotiao (26) and α-actinin for the NMDA-type glutamate receptor (27). Yotiao is a cytoskeleton-associated protein in muscle that colocalizes with the glutamate receptor at the neuromuscular junction and also coimmunoprecipitates with the NMDA receptor subunit NR1. α-Actinin is an actin cross-linking protein (28) whose interaction with the NMDA receptor subunits NR1 and NR2B appears to be competitively antagonized by Ca^{2+} and calmodulin (29). This regulated interaction may explain both the activity-dependent localization of the receptor observed (10) and its physiological inhibition by intracellular calcium (30). Some of the possible mechanisms for Ca^{2+} and calmodulin regulation of channel activity are summarized in Fig. 1.

Given the variety of synaptic locations (e.g., spines, shafts, soma) for even a single type of receptor, different mechanisms for localizing receptors are required (2). A protein called GKAP (***g**uanylate **k**inase -**a**ssociated **p**rotein*) has been found

TABLE 2. *Summary of common interaction motifs*

Binding type	Binding motif	Ref.
PDZ/Discs large	E-(S/T)-X-V (X is any amino acid)	10a
14-3-3	Phosphoserine and threonine	10b
SH2	Phosphotyrosine	10c
SH3	Polyproline	10d

Source: Summarized from Pawson and Scott (10e).

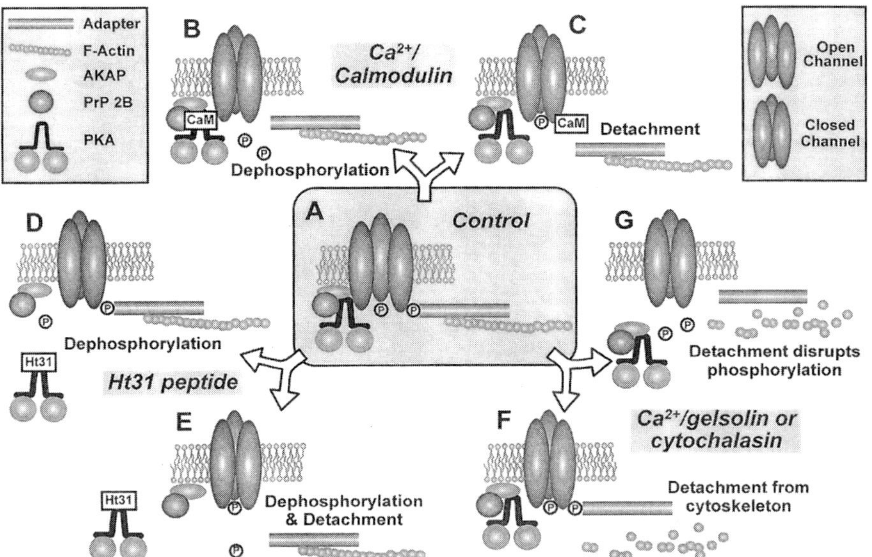

FIG. 1. Alternate pathways of ion channel regulation by phosphorylation and the cytoskeleton. Under control conditions *(A),* an ion channel may be regulated by two mechanisms: (1) phosphorylation and dephosphorylation of the channel or its auxiliary subunits by anchored kinases and phosphatases and (2) allosteric regulation by binding to the cytoskeleton through an adapter protein (e.g., α-actinin). Attachment to the cytoskeleton through an adapter protein may be switched by phosphorylation. *Ca^{2+}/calmodulin:* Ca^{2+} bound to calmodulin activates the phosphatase calcineurin (protein phosphatase 2B) and may also compete with an adapter protein for binding to the channel. Increased intracellular Ca^{2+} acting through a binding protein such as calmodulin may inactivate the channel by two mechanisms: (1) causing channel dephosphorylation alone by turning on a Ca^{2+}-activated phosphatase *(B),* or (2) causing detachment of the channel from the cytoskeleton by binding to the linker protein *(C). Ht31 peptide:* Ht31 peptide binds to the regulatory subunit dimer of PKA and dissociates the complex (consisting of two catalytic subunits bound to the regulatory subunit dimer) from an endogenous cAMP-dependent protein kinase anchoring protein (AKAP). *Ca^{2+}/gelsolin or cytochalasin:* Disruption of kinase anchoring by an anchoring inhibitor peptide such as Ht31 peptide could lead to channel inactivation by causing channel dephosphorylation at a site required for activity *(D)* or by reversing phosphorylation-dependent attachment of the channel to the cytoskeleton *(E).* Detachment might occur by dephosphorylation of the channel or the adapter protein. Disruption of microfilaments (F-actin) by increased intracellular Ca^{2+} acting through an actin-severing protein such as gelsolin, or directly with a destabilizer compound such as cytochalasin B or D, could inactivate the channel by two mechanisms: eliminating an attachment to the cytoskeleton required for channel activity *(F),* or causing channel dephosphorylation through disruption of kinase anchoring *(G).*

to interact with PSD-95 (31) and also colocalizes with the NMDA and AMPA receptor at the postsynaptic density (10). In the same studies, α-actinin was found to colocalize with the NMDA receptor at spiny but not shaft synapses. Spectrin, a member of the same family of actin cross-linking proteins as α-actinin, has also been found to bind the NMDA receptor (32). Another protein, called CRIPT (***c****ysteine-****ri****ch interactor of* ***PDZ t****hree*), may be responsible for connecting glutamate receptors to the microtubule component of the cytoskeleton, since it too binds strongly to PSD-95, immunoprecipitates with tubulin, and is found at the postsynaptic density (33).

Last, since one subtype (NF-L) has been found to bind to the NR1 subunit of the NMDA receptor, the neurofilament component of the cytoskeleton may also be involved in glutamate receptor localization (34).

Glycine and GABA$_A$ Receptors

The protein gephyrin, unrelated in sequence to the proteins that are thought to localize glutamate receptors, interacts with the inhibitory glycine (35,36) and GABA$_A$ receptors (37,38). Gephyrin is a 93-kDa protein that is thought to bind to tubulin and microtubules in a kinase-regulated manner (39) and does not colocalize with NMDA receptor containing synapses (37). Further, localization of GABA$_A$ receptors is distinguished from the localization of glutamate receptors by the lack of disruption of GABA$_A$ receptor clusters by depolymerization of the F-actin component of the cytoskeleton (2). Gephyrin appears to bind to the β subunit of the glycine receptor (40), and the γ subunit is required for clustering of the GABA$_A$ receptor (38).

Second Messenger- and Mechano-Gated Channels

TRP and TRPL Calcium Channels

The *transient receptor potential and transient receptor potential-like* (TRP and TRPL) calcium channels in *Drosophila* photoreceptors are responsible for the light-induced current required for vision, and mutants lacking both channels are blind (41,42). These channels are operated by an unknown second messenger but are closely associated with a host of signaling enzymes as discussed later, through interaction with a protein called INAD for ***i****nactivation* ***n****o* ***a****fterpotential* ***D*** (43,44). INAD, the product of the *InaD* gene, is a 90-kDa protein (45) that binds to the C-terminus of TRP through a PDZ domain and may be associated with the actin component of the cytoskeleton.

Epithelial Sodium and Chloride Channels

The activity of the renal epithelial sodium channel is regulated by PKA and the actin cytoskeleton, making its localization and clustering likely to be linked to actin. The renal amiloride-sensitive sodium channel, Apx (46), which is inhibited by F-actin, may contain its own actin-binding domains that would be responsible for channel localization and regulation (47). In addition, another epithelial sodium channel, rENaC, appears to bind to the actin cross-linking protein spectrin through an SH3 (polyproline)-binding region (48).

The chloride channel believed to be altered in cystic fibrosis, the cystic fibrosis transmembrane conductance regulator (CFTR), may be localized to the apical membrane of epithelial cells by binding to a PDZ protein called ERM-binding phosphoprotein 50 (EBP50) (49,50). ERM proteins get their name from family members ezrin, radixin, and moesin, which link the actin cytoskeleton to the plasma membrane (51).

Voltage-Gated Channels

Potassium Channels

As with the glutamate receptors, much of the work on the clustering and localization of voltage-gated potassium channels has focused on the PDZ proteins. Many potassium channels contain PDZ-binding domains in their C-termini (52), including *shaker* potassium channels and inward rectifiers. Channels appear to be clustered by PSD-95 and other family members by disulfide-linked, head-to-head multimers (11). In *Drosophila,* the ~100-kDa protein DLG is responsible for clustering the *shaker* potassium channel through a PDZ-binding motif in the channel's C-terminus (53). In heterologous expression in *Xenopus* oocytes, the interaction between Kv1.1 and the β subunit alters channel inactivation. PSD-95 family members modulate the α–β interaction and channel inactivation in a process that involves actin (54). In another expression system, PSD-95 and SAP97 cluster potassium channels differentially, suggesting separate mechanisms of interaction with homologous channels (55). In inward rectifier channels, PSD-95/SAP90 is found to cluster and enhance the activity of Kir 4.1 (56) in an interaction that may be switched by PKA phosphorylation of the channel. Phosphorylation would lead to uncoupling and inhibition of channel activity (57). The link to the cytoskeleton may come from a third class of proteins, represented by GKAP, which bind PSD-95 through guanylate kinase–like domains and cocluster with potassium channels (58).

Calcium Channels

Compared to potassium and sodium channels, relatively little is known about the proteins that localize calcium channels. Microfilament and microtubule elements regulate calcium channels (59–62), making either direct connections or indirect connections through linker proteins likely. The structural homology between the voltage-gated channels suggests candidates such as PSD-95, α-actinin, syntrophin, dystrophin, and utrophin in both neurons and muscle. Interactions between presynaptic Ca^{2+} channels and components of the exocytotic machinery, including the SNARE complex proteins syntaxin, SNAP-25, and synaptotagmin, may also help to localize Ca^{2+} channels at neurotransmitter release sites (63–66).

In skeletal and cardiac muscle, calcium channels initiate contraction by triggering Ca^{2+} release from the sarcoplasmic reticulum ryanodine receptor (67), and these channels maintain a close physical relationship with the ryanodine receptor (68). Although calcium channels and ryanodine receptors appear to be targeted separately (69), physical contact between them may play a role in their regulation, especially in skeletal muscle (70–72). One potential intermediary is the 22-kDa protein sorcin, which is thought to bind and regulate skeletal and cardiac Ca^{2+} channels (73).

Sodium Channels

Brain and muscle sodium channels are localized in part through interactions with the extracellular matrix receptor ankyrin (74,75) and the actin-associated proteins spectrin, syntrophin, and dystrophin/utrophin (76). Cardiac and skeletal muscle sodium channels have PDZ recognition domains at their C-termini and coimmunoprecipitate with the peripheral membrane protein syntrophin, which contains multiple PDZ domains (76). Surprisingly, brain sodium channels, which lack a strong PDZ-binding consensus sequence, still coimmunoprecipitate with syntrophin, suggesting that there are multiple pathways for interaction.

TARGETING REGULATORY PROTEINS TO CHANNELS

Overview

The targeting of signaling proteins to ion channels they regulate solves two problems associated with signaling from a distance. First, since the signals generated by the ion channel may be restricted spatially, the rest of the cascade would not be activated if the source and target were separated by any distance. This is the case for Ca^{2+} signaling through voltage-gated Ca^{2+} channels and NMDA-type glutamate receptors, where endogenous intracellular buffers approach the speed and strength of the exogenous buffer BAPTA (77–80). The second solution provided by enzyme targeting is the speed of signaling: if the response to second messengers required diffusion over even short distances, a cascade with multiple steps would take too long to be relevant in most situations.

Caveolins and AKAPs represent a class of proteins that are specialized to target enzymes and multienzyme complexes (81,82). Caveolins are proteins that target multiple enzymes such as G proteins, src, eNOS, and PKC to membrane invaginations in most cell types. AKAPs (**A** **K**inase **a**nchoring **p**roteins) derive their name from their ability to localize cAMP-dependent protein kinase (83), but some family members (e.g., AKAP79) are now known to target multiple enzymes and signaling molecules such as other kinases, phosphatases, and the calcium-binding protein calmodulin (84,85). AKAPs are targeted to specific subcellular locations by binding to cytoskeletal proteins such as actin or tubulin, while others have covalently bound lipid moieties that target them to lipid membranes (86,87). A combination of lipid and protein targeting is used to specify their destination. Proteins such as the ERM proteins (51) and AKAP79 are targeted to both the plasma membrane and microfilaments (83). The mechanisms for targeting enzymes to ion channels are summarized in Table 3.

The role of PKA and AKAPs in regulating ion channel activity is currently tested in physiological experiments by applying a peptide corresponding to the am-

TABLE 3. *Summary of scaffolding proteins associated with ion channels*

Scaffolding protein	Enzymes localized	Channel	Ref.
Caveolins	Gα, Ha-ras, src, eNOS, EGF receptor, PKC	Unknown	82
INAD	PLCβ, $G_q\alpha$, PKC	TRP Ca^{2+} channel	44
AKAP79	PKA, PKC, calcineurin	Glutamate receptors	85
Unknown AKAP	PKA	K_{Ca}	101
AKAP15	PKA	L-type Ca^{2+} channel, brain Na^+ channel	86,108,113

phipathic helical region of a human thyroid AKAP, Ht31 (88). This region of Ht31 is responsible for binding the regulatory subunit of PKA, especially the RII isoform (89) and therefore competes with any endogenous AKAPs for PKA binding (Fig. 1). Although this technique cannot be used to determine which AKAP is responsible for regulating a particular channel, it will indicate whether any interaction with PKA through its regulatory subunit is involved in the process.

Ligand-Gated Channels

Ionotropic Glutamate Receptors

NMDA-, AMPA-, and kainate-type glutamate receptors have been shown to be phosphorylated and modulated by PKA (90), but the physiological role of the association between the receptor and kinase was not fully appreciated until experiments with Ht31, the PKA anchoring inhibitor peptide, were performed. In these experiments, the response of patch-clamped neurons to glutamate application was monitored over time and was found to decline slightly under normal conditions (decline known as rundown or washout). Application of the direct peptide PKA inhibitor PKI(5-24) accelerated the decline of glutamate receptor activity (of the AMPA/kainate type), and application of Ht31 enhanced the loss of receptor activity to an intermediate level (91). The multienzyme anchoring protein AKAP79 is a likely candidate for the regulation of glutamate receptor activity, since it is expressed highly in the same regions containing glutamate receptors. PKA anchoring by AKAP79 is also switched by phosphorylation in the PKA regulatory subunit binding region, allowing an additional mechanism for plasticity of receptor activity (92).

Glycine Receptor

The activity of the inhibitory glycine receptor is stimulated by PKA (93), and an unidentified kinase copurifies with the channel and the putative localizing protein gephyrin (39). This indicates either that gephyrin is also a kinase-anchoring protein or that it is part of the scaffolding in a signaling complex.

Second Messenger- and Mechano-Gated Channels

The cystic fibrosis transmembrane conductance regulator (CFTR) chloride channel is prominently regulated by PKA (94), possibly anchored through binding to an ERM protein/AKAP such as ezrin (95) and the ERM-binding phosphoprotein EBP50, which binds to CFTR (49). cGMP-dependent protein kinase (cGK or PKG) also regulates the activity of this channel, and the membrane targeting of cGK is required for the activation of CFTR (96). cGK may be localized to the membrane by N-terminal myristoylation (97) or by binding to anchoring proteins called GKAPs (98), for **G-k**inase **a**nchoring **p**roteins (not to be confused with the other GKAP, *guanylate kinase associated protein*).

In *Drosophila,* the TRP calcium channel, which is responsible for phototransduced currents in retina, binds to a protein called INAD (43) though a PDZ-binding motif. Although the molecule responsible for channel gating has not been identified, it is believed to be a product of the enzyme complex assembled by INAD (41). This complex includes phospholipase Cβ, the G protein $G_q\alpha$, and PKC (44). In this TRP/INAD complex, both a Ca^{2+} source and Ca^{2+}-dependent enzymes are collected in one place to speed signaling.

The last example of enzyme targeting to a second messenger-gated channel is that of the Ca^{2+}-activated potassium channel. Ca^{2+}-activated potassium channels from brain can be purified and reconstituted in lipid bilayer with an endogenous kinase (PKC-like) and phosphatase (99,100). In tracheal myocytes (101), excised patches containing the K_{Ca} channel also have endogenous PKA activity. Application of Ht31, the PKA anchoring inhibitor peptide, blocked the ability of intracellular ATP to stimulate channel activity. This result suggests that the endogenous PKA in these patches is anchored near the channel by an unidentified AKAP.

Voltage-Gated Channels

Potassium Channels

The inward rectifiers Kir2.1 and Kir2.3 from forebrain and hippocampus are clustered at postsynaptic densities by binding the PDZ domain of PSD-95. Clustering may enhance channel activity by binding the autoinhibitory C-terminus. Stimulation of PKA in intact cells uncouples the channels from PSD-95 and causes their dispersal (57), which would then inhibit channel activity by releasing the autoinhibitory C-terminus from PSD-95 (102).

Calcium Channels

The regulation of L-type Ca^{2+} channels by PKA was one of the first mechanisms of channel modulation to be described in much detail and offers one of the most reproducible demonstrations of the importance of channel modulation in physiology. An elegant demonstration of signaling compartmentalization comes from the work of Jurevicius and Fischmeister (103), who showed in cardiac myocytes that adrenergic signals are spatially restricted by the action of phosphodiesterases. The high concentration of phosphodiesterases requires that there be a functional coupling between the voltage-gated calcium channel, the β-adrenergic receptor, adenylyl cyclase, and PKA.

In skeletal muscle, the L-type Ca^{2+} channel is similarly modulated by the PKA signaling pathway and shows an additional voltage-dependent potentiation that is modulated by PKA (104). Application of Ht31, the PKA anchoring inhibitor peptide (89) (Fig. 1), inhibits skeletal muscle voltage-dependent potentiation to the same extent as direct PKA inhibition with the catalytic inhibitor peptide PKI (105). This PKA-anchoring- and voltage-dependent potentiation has been reconstituted in heterologous cells with channel cDNAs alone (106), suggesting that AKAPs capable of targeting PKA to the channel are not unique to skeletal muscle and may be common elements in cell signaling. A study of Ca^{2+} channel voltage-dependent potentiation in skeletal muscle from an RII knockout mouse revealed that contrary to earlier assumptions, the RI-type regulatory subunit of PKA can anchor to AKAPs and mediate anchoring-dependent modulation (107).

When the AKAP responsible for anchoring PKA near the skeletal muscle Ca^{2+} channel was purified, sequenced, and cloned (81,86,108), it was found to be the smallest yet described (15–18 kDa). AKAP15/18 is targeted to the plasma membrane by virtue of its myristoyl and palimtoyl lipid moieties and may be involved in the β-adrenergic modulation of the cardiac L-type Ca^{2+} channel and insulin secretion in pancreatic beta cells (87,109,110).

Tyrosine kinases, associated with the focal adhesions in smooth muscle, comprise a recently tapped vein in the search for modulators of Ca^{2+} channels. The tyrosine kinases c-SRC and focal adhesion kinase (FAK) stimulate Ca^{2+} channel activity in colonic smooth muscle and may be involved in the stimulation of Ca^{2+} current and contraction induced by platelet-derived growth factor (PDGF) (111). These kinases coimmunoprecipitate with each other, and c-SRC coimmunoprecipitates with the Ca^{2+} channel, indicating that channel and kinase form an association similar to the L-type Ca^{2+} channel and PKA in skeletal muscle.

Sodium Channels

An effort parallel to that on the skeletal muscle L-type Ca^{2+} channel revealed that the rat brain sodium channel forms a close association with AKAP15 as well. The brain sodium channel is inhibited by PKA (112), and purification of the channel carries with it endogenous AKAP and PKA activity (113). Purification and sequencing of this sodium channel-associated AKAP revealed an AKAP of the same sequence as that purified from skeletal muscle. Since AKAP15 is membrane-targeted, specific targeting to ion channels will require that this protein have additional determinants that bring it in even closer proximity to the channel, including direct binding to the pore-forming α subunit, auxiliary β subunits, or localizing proteins such as spectrin.

REGULATION OF CHANNELS BY THE CYTOSKELETON

Overview

The cytoskeleton is composed of three primary types of filament polymerized from protein monomers (summarized in Table 4 from Alberts *et al.*) (114): microfilaments composed of actin monomers, intermediate filaments composed of any number of large intermediate filament proteins (e.g., neurofilaments), and microtubules composed of α and β tubulin dimers. A large number of accessory subunits are responsible for nucleating, capping, severing, and cross-linking each of these filaments.

Many of the same intracellular agents that modulate neurotransmitter receptors and ion channels are physically associated with and modulate the cytoskeleton. These include G proteins (115–117), protein kinases including PKA (118,119), PKC (120,121), CaM kinase II (122), tyrosine kinases (123), and intracellular Ca^{2+} (124,125). Disruption of the cytoskeleton may influence the activity of an ion channel indirectly by altering the activity of other second messenger pathways. Disruption of microtubules and microfilaments has been found to increase cAMP production (126); disruption of microtubules activates the JNK/SAPK tyrosine kinase pathways (127); and the cytoskeleton may be involved in the regulation of intracellular Ca^{2+} (128–130). In addition, ion channels attached to the cytoskeleton may influence the structure of the cytoskeleton and the ability of other interacting pathways to restructure the cytoskeleton (131).

The interactions between these players makes determining whether they are acting directly or indirectly on an ion channel challenging (Fig. 1). What, for example, is responsible for the inhibition of Ca^{2+} channel activity upon removal of intracellular ATP? Is it depolymerization of microfilaments and a disruption of a direct interaction between microfilaments and the Ca^{2+} channel? Inhibition of ATP-dependent second messenger and kinase cascades that normally phosphorylate the Ca^{2+} channel? Or a more indirect mechanism in which depolymerization of microfilaments disrupts the anchoring of a stimulatory kinase, phosphatase, or lipase? Given this potential for uncertainty, definitive demonstration of a direct channel

TABLE 4. *Comparison of the primary filaments of the cytoskeleton*

Filament type	Diameter (nm)	Nucleotide	Monomers	Disrupting agents	Stabilizing agent
Microfilament (F-actin)	8	ATP	Actin	Cytochalasins, DNAse I, latrunculin A	Phalloidin, Jasplakinolide
Intermediate filament	10–12	none	Keratins, vimentin, desmin, GFAP, neurofilaments, nuclear lamins	Acrylamide, tricresyl phosphate	Unknown
Microtubule	25	GTP	α and β tubulin	Colchicine, colcemid, nocadazole, vinblastine, vincristine	Taxol (paclitaxel)

Source: Summarized from Alberts *et al.* (114).

regulation by the cytoskeleton should eliminate the possibility of indirect effects by ruling out the actions of kinases, phosphatases, G proteins, Ca^{2+}, and phospholipids. The studies described in the sections that follow represent the starting point in determining how the cytoskeleton regulates ion channel activity. Some of the channels whose activity is influenced by the cytoskeleton are summarized in Table 5.

Ligand-Gated Channels

Some compounds that act on the cytoskeleton, including cytochalasin B (132) and colchicine, have nonspecific actions that can confound the examination of ion channel regulation by the cytoskeleton. The ligand-gated channels are especially sensitive to colchicine, and direct (competitive) interactions have been demonstrated for nicotinic acetylcholine receptors (133), $GABA_A$ receptors (134,135), and glycine receptors (136).

For the ligand-gated channels, there are currently three examples of receptors regulated by the cytoskeleton. The NMDA-type glutamate receptor was one of the first ligand-gated channels for which cytoskeletal regulation was described (30,137). NMDA-type glutamate receptors generate a significant intracellular Ca^{2+} signal that can lead to depolymerization of microfilaments (138) and receptor inactivation (139). Agents that disrupt microfilaments can limit the Ca^{2+} toxicity caused by receptor activation (140). The mechanism beginning to emerge is that microfilaments enhance the activity of the NMDA receptor through the linker protein α-actinin (29), and this interaction is antagonized by Ca^{2+}/calmodulin and manipulations that disrupt microfilaments (Fig. 1). These disruptions include cytochalasins (30) and knockout of the actin-severing protein gelsolin (141). Microtubules may also regulate the NMDA

TABLE 5. *Summary of channels regulated by the cytoskeleton*

Channel	Cytoskeletal filament	Ref.
NMDA-type glutamate receptor	Microfilaments	30
$GABA_A$	Microtubules	143
$P2X_1$	Microfilaments	144
ENaC	Microfilaments	147
K^+ (CCD, muscaranic, ATP sensitive)	Microfilaments	155,157,161,165
Cl^-	Microfilaments	168,169,174–176,178
Voltage-gated H^+	Microfilaments and microtubules	131a
Voltage-gated Na^+ (squid axon)	Microtubules	181
Voltage-gated Na^+ (DRG)	Microfilaments	59
Voltage-gated Ca^{2+}	Microtubules	59,62,193,194
Voltage-gated Ca^{2+}	Microfilaments	141,196,197
Voltage-gated Ca^{2+}	Microfilaments and microtubules	60,61
Voltage-gated K^+	Microfilaments	12,164,165,200,201

receptor in some systems, inasmuch as they and their associated proteins, MAPs (**m**icrotubule-**a**ssociated **p**roteins), have been found to interact with the receptor (142).

Two other ligand-gated channels regulated by the cytoskeleton are the $GABA_A$ receptor (143) and the ATP-gated cation channel (144). The $GABA_A$ receptor, studied in cerebral cortical microsacs, proteoliposomes, and in *Xenopus* oocytes, is inhibited by disruption (colchicine, nocodazole, vinblastine) or stabilization (taxol) of microtubules (143). The ATP-gated cation channel ($P2X_1$ receptor) cloned from rat vas deferens initially exhibits more rapid kinetics than the native channel when expressed in HEK 293 cells. As the heterologous cells take on a flatter morphology, channel kinetics slow to normal time courses, and this normalization is prevented by treating the cells with the microfilament disrupters cytochalasin B and D (144). Normal $P2X_1$ receptor function apparently requires an attachment to intact microfilaments.

Second Messenger- and Mechano-Gated Channels

Cation Channels

Mechanosensitive channels are common transducers of physical force on the cell membrane and are found throughout the plant and animal kingdom (145). Their physiology and molecular biology are advancing at a rapid pace, but the examination of their molecular interactions with the cytoskeleton are at an early stage. From a physiological point of view, it is clear that the renal epithelial sodium channel is regulated by actin microfilaments and that this regulation may play a role in mechanosensitivity and regulatory volume decrease (RVD) in response to osmotic challenges (146). The activity of the epithelial sodium channel (ENaC), is up-regulated by PKA, but this up regulation appears to be indirect, through short actin filaments (47,147–149). Stabilization of microfilaments inhibits channel activity (147), and heterologous expression of one ENaC type (Apx) was possible only in a cell line lacking an actin-stabilizing protein (47). PKA appears to activate the channel by phosphorylating microfilaments and thereby shortening them (148). In pneumocytes, G protein modulation of the channel also required short actin filaments (150). The actin regulation of the epithelial sodium channel has been studied extensively both in native cells and in the amphibian renal cell line, A6, where identical regulation by PKA (151), actin (152), and actin-stabilizing proteins (153) have been observed.

Other primarily nonmechanosensitive cation channels that appear to be regulated by the cytoskeleton include a nuclear pore ion channel (154), as well as a variety of potassium channels including a potassium channel in cortical collecting duct (155), a potassium channel in human melanoma cells (156), a muscarinic potassium channel, Kv2.3 (157), Ca^{2+} -activated potassium channels (158,159), and ATP-sensitive potassium channels (160–164). In most of these examples, microfilaments impart a mechanosensitivity to the potassium channel (156) where stable microfilaments decrease channel activity. For example, in cardiac myocytes the sulfonylurea receptor (SUR) inhibits the activity of the ATP-sensitive potassium channel, and this interaction is mediated by microfilaments. Disruption of microfilaments causes disinhibition and an increase in the potassium current (161,165).

Anion Channels

Like the cation channels, chloride channels are primarily regulated by the microfilament component of the cytoskeleton. Microtubules are involved in the stimulated secretion of chloride by epithelial cells, but the mechanism appears to be at the level of trafficking and inserting additional chloride channels into the plasma membrane (166). Secretion of chloride is critical for cell volume regulation and transport of water in epithelial cells (146), and kinase pathways including PKA and MAPK appear to be involved (167).

The relationship between the kinase pathways and microfilaments follow the same lines as that of the epithelial sodium channels. Cell shape changes due to cell swelling or associated with the cell cycle modulate chloride channel activity, with high levels of F-actin generally inhibiting channel activity (168). Cell swelling, PKA activation, and direct treatment with cytochalasins disrupt actin filaments and stimulate channel activity (168–170).

The multiple chloride channel types are undoubtedly regulated by different mechanisms (171), and because of the regulation of actin dynamics by phosphorylation, and actin's role in enzyme anchoring, it has been difficult to determine whether phosphorylation is the dominant activator of Cl^- channels (172,173) or whether microfilaments play a primary role (174–176).

The chloride channel whose mutation is responsible for cystic fibrosis, the cystic fibrosis transmembrane conductance regulator (CFTR), appears to be regulated both by direct phosphorylation by PKA (177) and through an interaction with actin filaments (178,179). Aside from the similarities in their activation by short actin filaments, CFTR and the epithelial sodium channel share another interaction involving actin, whereby CFTR tonically inhibits the activity of ENaC (180).

Voltage-Gated Channels

Sodium Channels

The earliest descriptions of voltage-gated ion channel modulation by the cytoskeleton were those of Matsumoto *et al.* (181–184). In these experiments on squid giant axon, Na^+ channels were found to require intact microtubules to function; thus treatment of axons with colchicine suppressed the sodium current and action potential, and restoration of excitability was achieved by reintroduction of tubulin and other molecules that supported the polymerization of microtubules (184). At about the same time, Fukuda (59) demonstrated that action potentials were suppressed in guinea pig dorsal root ganglion neurons treated with the microfilament disrupter cytochalasin B, but not by disruption of microtubules with colchicine. Application of colchicine to the dorsal root ganglion neurons instead inhibited calcium spikes.

Modulation of voltage-gated sodium channels has been reexamined recently, this time in cardiac myocytes, where Na^+ channel properties appear to be influenced by microfilaments and microtubules (185). Disruption of F-actin was found to reduce the peak current by reducing open probability, thereby slowing channel inactivation and shifting activation toward more positive potentials. Stabilization of microtubules shifted activation to more negative potentials (186).

Calcium Channels

Building on Matsumoto's and Fukuda's early experiments (59,181), the idea that voltage-gated Ca^{2+} channels are regulated by the cytoskeleton was proposed as a solution to a perplexing problem in channel modulation (60). Voltage-gated Ca^{2+} channels show a dependence on intracellular ATP called metabolic dependence, rundown, or washout (187), and are inactivated when intracellular Ca^{2+} rises (188). The rapid inactivation of channels observed after photorelease of intracellular Ca^{2+} from caged compounds (77,189,190) appeared to be too rapid to be mediated by a purely diffusion-limited, enzymatic mechanism. Work by Chad, Eckert, and Armstrong demonstrated convincingly that metabolic dependence and Ca^{2+}-dependent inactivation in *Helix* neurons and GH3 cells were mediated by phosphorylation and dephosphorylation of the Ca^{2+} channel (191,192). Attempts at confirming an enzymatic mechanism in *Lymnaea* neurons had not met with success (60).

Since the cytoskeleton was known to require nucleotides for polymerization and to be sensitive to intracellular Ca^{2+}, a channel dependence on polymerized microfilaments or microtubules could serve as an alternate mechanism for both channel metabolic dependence and Ca^{2+}-dependent inactivation. Application of cytochalasin B or colchicine reduced the ability of nucleotides to maintain channel activity, and stabilizers of microfilaments and microtubules reduced Ca^{2+}-dependent inactivation of Ca^{2+} channels in snail neurons and rat hippocampal pyramidal neurons (60,61). Subsequently, inactivation of a cardiac myocyte L-type Ca^{2+} channel was found to be modulated by microtubules (62), and the involvement of microfilaments in Ca^{2+} channel metabolic dependence was confirmed in a study of neurons from a mouse knockout of a microfilament disrupting protein, gelsolin (141) (Fig. 1). Inhibition of Ca^{2+} channel activity by microtubules has been observed in a plant cell (193,194).

The recent discovery that the critical enzymes involved in Ca^{2+} channel Ca^{2+}-dependent inactivation and metabolic dependence (PKA and calcineurin) may be collected in a complex near the channel by an AKAP such as AKAP79 makes an enzymatic mechanism more tenable. If Ca^{2+} channels had to bind to the cytoskeleton to function, Ca^{2+} channel activity and intracellular Ca^{2+} signals would be prevented except where they were needed and directed by the cell, such as near intracellular stores and sites of transmitter release. An attachment of voltage-gated Ca^{2+} channels to the cytoskeleton also allows for a dual use as mechanosensors. For example, voltage-gated Ca^{2+} channels in smooth muscle cells are regulated by membrane stretch and volume changes (195–197), and hyperosmotic stress reduces Ca^{2+} current in pituitary somatotrophs (198). Since phosphorylation is known to regulate some channel interactions with the cytoskeleton, both enzymatic and cytoskeletal mechanisms could regulate channel activity in concert and may have more or less influence over Ca^{2+} channel Ca^{2+}-dependent inactivation and metabolic dependence in different systems.

Potassium Channels

The inwardly rectifying potassium channels (IRKs) of cardiac ventricular myocytes derive their voltage-dependent rectification from block by Ca^{2+} and Mg^{2+} ions, or from polyamines (199). In cardiac myocytes, application of cytochalasin D to inside-out patches containing IRKs sped the loss of rectification associated with Ca^{2+} ions, but not rectification due to Mg^{2+}. These results suggest that microfilaments may constitute part of the channel's mechanism of rectification (200). In retinal bipolar cells, a voltage-gated potassium channel also appears to be tonically inhibited by an interaction with microfilaments (201), as evidenced by the activation of the normally quiescent channel that follows application of cytochalas in D. Direct application of actin to excised patches had the same activating effect as cytochalasin D.

The last two examples of voltage-gated potassium channels regulated by the cytoskeleton involve the classical delayed rectifier potassium channel. The first is in skeletal muscle myotubes from a mouse model for Duchenne muscular dystrophy, the *mdx* mouse (202). Individuals with Duchenne muscular dystrophy, like the *mdx* mouse, lack the protein dystrophin, which links microfilaments to the plasma membrane (203). Although an earlier study from the same laboratory did not report any changes in the gating of the voltage-gated sodium channel (204), examination of the voltage-dependence of the delayed rectifier potassium channel in cell-attached patches revealed a significant shift toward more positive potentials in *mdx* muscle (164). The conspicuous absence of the ATP-sensitive potassium channel in *mdx,* but not control muscle, also points to its likely regulation by microfilaments.

The potassium channel Kv1.1 from rat brain exhibits delayed-rectifier-type slow gating when expressed alone in *Xenopus* oocytes. Coexpression with the subunit $\beta_{1.1}$ imparts a fast *shaker-A*-type inactivation on the channel that is enhanced by phosphorylation of the channel and by disruption of microfilaments in a nonadditive manner (12). Phosphorylation of the channel by PKA may reduce binding of microfilaments, leading to enhanced β modulation of gating kinetics. This last example highlights the complex relationship between ion channels, the cytoskeleton, and associated signaling pathways, as well as the difficulties in separating the effects of one from the other.

CONCLUSION

With limited exceptions, the majority of work on ion channel localization has failed to address the functional consequences of binding to cytoskeletal components on the activity of the ion channel. Likewise, work on the functional consequences of disrupting the cytoskeleton has not necessarily been related to discerning which components of the cytoskeleton might be localizing channels. Future work will be required to relate these closely linked phenomena. Protein–protein interactions often have effects on activity, especially for such complex molecules as ion channels.

As noted at the outset, one of the greatest challenges in the field of ion channel regulation by "structural" proteins will be to determine which interactions are direct allosteric effects (e.g., binding of actin or actin-linker to channels) and which interactions are indirect effects involving anchoring of signaling pathways to channels. As more channel-interacting partners are identified by knockout, biochemical, and library screen (e.g., two-hybrid) approaches, we may begin to distinguish direct from indirect actions of the proteins that link the cytoskeleton, regulatory enzymes, and ion channels.

ACKNOWLEDGMENTS

I would like to thank Shalini Gera, Lou Byerly, and Angel de Blas for comments on the manuscript.

REFERENCES

1. Southan AP, Robertson B: Patch–clamp recordings from cerebellar basket cell bodies and their presynaptic terminals reveal an asymmetric distribution of voltage-gated potassium channels. *J Neurosci* 1998;18:948–955.
2. Allison DW, Gelfand VI, Spector I, Craig AM: Role of actin in anchoring postsynaptic receptors in cultured hippocampal neurons: differential attachment of NMDA versus AMPA receptors. *J Neurosci* 1998;14:2423–2436.
3. Hell JW, Westenbroek RE, Warner C *et al:* Identification and differential subcellular localization of the neuronal class C and class D L-type calcium channel alpha 1 subunits. *J Cell Biol* 1993;123: 949–962.
4. Thompson S, Coombs J: Spatial distribution of Ca currents in molluscan neuron cell bodies and regional differences in the strength of inactivation. *J Neurosci* 1988;8:1929–1939.
5. Harder T, Simons K: Caveolae, DIGs, and the dynamics of sphingolipid–cholesterol microdomains. *Curr Opin Cell Biol* 1997;9:534–542.
6. Ledesma MD, Simons K, Dotti CG: Neuronal polarity: essential role of protein–lipid complexes in axonal sorting. *Proc Natl Acad Sci U S A* 1998;95:3966–3967.
7. Noda M, Takahashi H, Tanabe T *et al:* Primary structure of alpha-subunit precursor of *Torpedo californica* acetylcholine receptor deduced from cDNA sequence. *Nature* 1998;299:793–797.
8. Colledge M, Froehner SC: Signals mediating ion channel clustering at the neuromuscular junction. *Curr Opin Neurobiol* 1998;8:357–363.
9. Dai Z, Peng HB: A role of tyrosine phosphatase in acetylcholine receptor cluster dispersal and formation. *J Cell Biol* 1998;141:1613–1624.
10. Rao A, Craig AM: Activity regulates the synaptic localization of the NMDA receptor in hippocampal neurons. *Neuron* 1997;19:801–812.

10a. Sheng M: PDZs and receptor/channel clustering: rounding up the latest suspects. *Neuron* 1996;17: 575–578.

10b. Muslin AJ, Tanner JW, Allen PM, Shaw AS: Interaction of 14-3-3 with signaling proteins is mediated by the recognition of phosphoserine. *Cell* 1996;84:889–897.

10c. Moran MF, Koch CA, Anderson D *et al:* Src homology region 2 domains direct protein–protein interactions in signal transduction. *Proc Natl Acad Sci U S A* 1990;87:8622–8626.

10d. Ren R, Mayer BJ, Cicchetti P, Baltimore D: Identification of a ten-amino acid proline-rich SH3 binding site. *Science* 1993;259:1157–1161.
10e. Pawson T, Scott JD: Signaling through scaffold, anchoring, and adaptor proteins. *Science* 1997;278: 2075–2080.
11. Hsueh YP, Kim E, Sheng M: Disulfide-linked head-to-head multimerization in the mechanism of ion channel clustering by PSD-95. *Neuron* 1997;18:803–814.
12. Levin G, Chikvashvili D, Singer Lahat D *et al:* Phosphorylation of a K^+ channel α subunit modulates the inactivation conferred by a β subunit. Involvement of cytoskeleton. *J Biol Chem* 1996;271: 29321–29328.
13. Songyang Z, Fanning AS, Fu C *et al:* Recognition of unique carboxyl-terminal motifs by distinct PDZ domains. *Science* 1997;275:73–77.
14. Niethammer M, Kim E, Sheng M: Interaction between the C terminus of NMDA receptor subunits and multiple members of the PSD-95 family of membrane-associated guanylate kinases. *J Neurosci* 1996;16:2157–2163.
15. Sheng M, Wyszynski M: Ion channel targeting in neurons. *BioEssays* 1997;19:847–853.
16. Sheng M, Kim E: Ion channel associated proteins. *Curr Opin Neurobiol* 1996;6:602–608.
17. O'Brien RJ, Lau LF, Huganir RL: Molecular mechanisms of glutamate receptor clustering at excitatory synapses. *Curr Opin Neurobiol* 1998;8:364–369.
18. Garcia EP, Mehta S, Blair LAC *et al:* SAP90 binds and clusters kainate receptors causing incomplete desensitization. *Neuron* 1998;21:727–739.
19. Kornau HC, Schenker LT, Kennedy MB, Seeburg PH: Domain interaction between NMDA receptor subunits and the postsynaptic density protein PSD-95. *Science* 1995;269:1737–1740.
20. Dong H, O'Brien RJ, Fung ET *et al:* GRIP: a synaptic PDZ domain–containing protein that interacts with AMPA receptors. *Nature* 1997;386:279–284.
21. Kim E, Cho KO, Rothschild A, Sheng M: Heteromultimerization and NMDA receptor-clustering activity of Chapsyn-110, a member of the PSD-95 family of proteins. *Neuron* 1996;17:103–113.
22. Lin JW, Sheng M: NSF and AMPA receptors get physical. Neuron 1998;21:267–270.
23. Nishimune A, Isaac JT, Molnar E *et al:* NSF binding to GluR2 regulates synaptic transmission. *Neuron* 1998;21:87–97.
24. Osten P, Srivastava S, Inman GJ *et al:* The AMPA receptor GluR2 C terminus can mediate a reversible, ATP-dependent interaction with NSF and alpha- and beta-SNAPs. *Neuron* 1998;21: 99–110.
25. Song I, Kamboj S, Xia J *et al:* Interaction of the *N*-ethylmaleimide-sensitive factor with AMPA receptors. *Neuron* 1998;21:393–400.
26. Lin JW, Wyszynski M, Madhavan R *et al:* Yotiao, a novel protein of neuromuscular junction and brain that interacts with specific splice variants of NMDA receptor subunit NR1. *J Neurosci* 1998; 18:2017–2027.
27. Wyszynski M, Kharazia V, Shanghvi R *et al:* Differential regional expression and ultrastructural localization of alpha-actinin-2, a putative NMDA receptor–anchoring protein, in rat brain. *J Neurosci* 1998;18:1383–1392.
28. Matsudaira P: Actin crosslinking proteins at the leading edge. *Semin Cell Biol* 1994;5:165–174.
29. Wyszynski M, Lin J, Rao A *et al:* Competitive binding of α-actinin and calmodulin to the NMDA receptor. *Nature* 1997;385:439–442.
30. Rosenmund C, Westbrook GL: Calcium-induced actin depolymerization reduces NMDA channel activity. *Neuron* 1993;10:805–814.
31. Naisbitt S, Kim E, Weinberg RJ *et al:* Characterization of guanylate kinase–associated protein, a postsynaptic density protein at excitatory synapses that interacts directly with postsynaptic density-95/synapse-associated protein 90. *J Neurosci* 1997;17:5687–5696.
32. Wechsler A, Teichberg VI: Brain spectrin binding to the NMDA receptor is regulated by phosphorylation, calcium and calmodulin. *EMBO J* 1998;17:3931–3939.
33. Niethammer M, Valtschanoff JG, Kapoor TM *et al:* CRIPT, a novel postsynaptic protein that binds to the third PDZ domain of PSD-95/SAP90. *Neuron* 1998;20:693–707.
34. Ehlers MD, Fung ET, OBrien RJ, Huganir RL: Splice variant–specific interaction of the NMDA receptor subunit NR1 with neuronal intermediate filaments. *J Neurosci* 1998;18:720–730.
35. Kirsch J, Betz H: The postsynaptic localization of the glycine receptor–associated protein gephyrin is regulated by the cytoskeleton. *J Neurosci* 1995;15:4148–4156.
36. Feng G, Tintrup H, Kirsch J *et al:* Dual requirement for gephyrin in glycine receptor clustering and molybdoenzyme activity. *Science* 1998;282:1321–1324.

37. Craig AM, Banker G, Chang WR, McGrath ME, Serpinskaya AS: Clustering of gephyrin at GABAergic but not glutamatergic synapses in cultured rat hippocampal neurons. *J Neurosci* 1996;16: 3166–3177.
38. Essrich C, Lorez M, Benson JA, Fritschy JM, Luscher B: Postsynaptic clustering of major $GABA_A$ receptor subtypes requires the gamma 2 subunit and gephyrin. *Nat Neurosci* 1998;1:563–571.
39. Langosch D, Hoch W, Betz H: The 93 kDa protein gephyrin and tubulin associated with the inhibitory glycine receptor are phosphorylated by an endogenous protein kinase. *FEBS Lett* 1992;298: 113–117.
40. Meyer G, Kirsch J, Betz H, Langosch D: Identification of a gephyrin binding motif on the glycine receptor beta subunit. *Neuron* 1995;15:563–572.
41. Scott K, Zuker C: TRP, TRPL and trouble in photoreceptor cells. *Curr Opin Neurobiol* 1998;8:383–388.
42. Niemeyer BA, Suzuki E, Scott K, Jalink K, Zuker CS: The *Drosophila* light-activated conductance is composed of the two channels TRP and TRPL. *Cell* 1996;85:651–659.
43. Montell C: TRP trapped in fly signaling web. *Curr Opin Neurobiol* 1998;8:389–397.
44. Shieh BH, Zhu MY: Regulation of the TRP Ca^{2+} channel by INAD in *Drosophila* photoreceptors. *Neuron* 1996;16:991–998.
45. Shieh BH, Niemeyer B: A novel protein encoded by the *InaD* gene regulates recovery of visual transduction in *Drosophila. Neuron* 1995;14:201–210.
46. Staub O, Verrey F, Kleyman TR *et al:* Primary structure of an apical protein from *Xenopus laevis* that participates in amiloride-sensitive sodium channel activity. *J Cell Biol* 1992;119:1497–1506.
47. Prat AG, Holtzman EJ, Brown D *et al:* Renal epithelial protein (Apx) is an actin cytoskeleton–regulated Na^+ channel. *J Biol Chem* 1996;271:18045–18053.
48. Rotin D, Bar Sagi D, O'Brodovich H *et al:* An SH3 binding region in the epithelial Na^+ channel (alpha rENaC) mediates its localization at the apical membrane. *EMBO J* 1994;13:4440–4450.
49. Short DB, Trotter KW, Reczek D *et al:* An apical PDZ protein anchors the cystic fibrosis transmembrane conductance regulator to the cytoskeleton. *J Biol Chem* 1998;273:19797–19801.
50. Reczek D, Berryman M, Bretscher A: Identification of EBP50: A PDZ-containing phosphoprotein that associates with members of the ezrin–radixin–moesin family. *J Cell Biol* 1997;139:169–179.
51. Vaheri A, Carpen O, Heiska L *et al:* The ezrin protein family: membrane–cytoskeleton interactions and disease associations. *Curr Opin Cell Biol* 1997;9:659–666.
52. Kim E, Niethammer M, Rothschild A, Jan YN, Sheng M: Clustering of *shaker*-type K^+ channels by interaction with a family of membrane-associated guanylate kinases. *Nature* 1995;378:85–88.
53. Tejedor FJ, Bokhari A, Rogero O *et al:* Essential role for *dlg* in synaptic clustering of Shaker K^+ channels in vivo. *J Neurosci* 1997;17:152–159.
54. Jing J, Peretz T, Singer-Lahat D *et al:* Inactivation of a voltage-dependent K^+ channel by bβ subunit. Modulation by a phosphorylation-dependent interaction between the distal C terminus of α subunit and cytoskeleton. *J Biol Chem* 1997;272:14021–14024.
55. Kim E, Sheng M: Differential K^+ channel clustering activity of PSD-95 and SAP97, two related membrane-associated putative guanylate kinases. *Neuropharmacology* 1996;35:993–1000.
56. Horio Y, Hibino H, Inanobe A *et al:* Clustering and enhanced activity of an inwardly rectifying potassium channel, Kir4.1, by an anchoring protein, PSD-95/SAP90. *J Biol Chem* 1997;272: 12885–12888.
57. Cohen NA, Brenman JE, Snyder SH, Bredt DS: Binding of the inward rectifier K^+ channel Kir 2.3 to PSD-95 is regulated by protein kinase A phosphorylation. *Neuron* 1996;17:759–767.
58. Kim E, Naisbitt S, Hsueh YP *et al:* GKAP, a novel synaptic protein that interacts with the guanylate kinase–like domain of the PSD-95/SAP90 family of channel clustering molecules. *J Cell Biol* 1997; 136:669–678.
59. Fukuda J, Kameyama M, Yamaguchi K: Breakdown of cytoskeletal filaments selectively reduces Na and Ca spikes in cultured neurones. *Nature* 1981;294:82–85.
60. Johnson BD, Byerly L: A cytoskeletal mechanism for Ca^{2+} channel metabolic dependence and inactivation by intracellular Ca^{2+}. *Neuron* 1993;10:797–804.
61. Johnson BD, Byerly L: Ca^{2+} Channel Ca^{2+}-dependent inactivation in a mammalian central neuron involves the cytoskeleton. *Pfluegers Arch* 1994;429:14–21.
62. Galli A, DeFelice L: Inactivation of L-type Ca channels in embryonic chick ventricle cells: dependence on the cytoskeletal agents colchicine and taxol. *Biophys J* 1994;67:2296–2304.
63. Sheng ZH, Westenbroek RE, Catterall WA: Physical link and functional coupling of presynaptic calcium channels and the synaptic vesicle docking/fusion machinery. *J Bioenerg Biomembr* 1998; 30:335–345.

64. Stanley EF, Mirotznik RR: Cleavage of syntaxin prevents G-protein regulation of presynaptic calcium channels. *Nature* 1997;385:340–343.
65. Bezprozvanny I, Scheller RH, Tsien RW: Functional impact of syntaxin on gating of N-type and Q-type calcium channels. *Nature* 1995;378:623–626.
66. Martin-Moutot N, Charvin N, Leveque C *et al:* Interaction of SNARE complexes with P/Q-type calcium channels in rat cerebellar synaptosomes. *J Biol Chem* 1996;271:6567–6570.
67. Ríos E, Stern MD: Calcium in close quarters: microdomain feedback in excitation–contraction coupling and other cell biological phenomena. *Annu Rev Biophys Biomol Struct* 1997;26:47–82.
68. Flucher BE, Franzini-Armstrong C: Formation of junctions involved in excitation-contraction coupling in skeletal and cardiac muscle. *Proc Natl Acad Sci U S A* 1996;93:8101–8106.
69. Powell JA, Petherbridge L, Flucher BE: Formation of triads without the dihydropyridine receptor alpha subunits in cell lines from dysgenic skeletal muscle. *J Cell Biol* 1996;134:375–387.
70. El-Hayek R, Ikemoto N: Identification of the minimum essential region in the II–III loop of the dihydropyridine receptor $\alpha 1$ subunit required for activation of skeletal muscle-type excitation-contraction coupling. *Biochemistry* 1998;37:7015–7020.
71. Leong P, MacLennan DH: The cytoplasmic loops between domains II and III and domains III and IV in the skeletal muscle dihydropyridine receptor bind to a contiguous site in the skeletal muscle ryanodine receptor. *J Biol Chem* 1998;273:29958–29964.
72. Nakai J, Tanabe T, Konno T, Adams B, Beam KG: Localization in the II–III loop of the dihydropyridine receptor of a sequence critical for excitation–contraction coupling. *J Biol Chem* 1998;273: 24983–24986.
73. Meyers MB, Puri TS, Chien AJ *et al:* Sorcin associates with the pore-forming subunit of voltage-dependent L-type Ca^{2+} channels. *J Biol Chem* 1998;273:18930–18935.
74. Srinivasan Y, Lewallen M, Angelides KJ: Mapping the binding site on ankyrin for the voltage-dependent sodium channel from brain. *J Biol Chem* 1992;267:7483–7489.
75. Srinivasan Y, Elmer L, Davis J, Bennett V, Angelides K: Ankyrin and spectrin associate with voltage-dependent sodium channels in brain. *Nature* 1988;333:177–180.
76. Gee SH, Madhavan R, Levinson SR *et al:* Interaction of muscle and brain sodium channels with multiple members of the syntrophin family of dystrophin-associated proteins. *J Neurosci* 1998;18: 128–137.
77. Johnson BD, Byerly L: Photo-released intracellular Ca^{2+} rapidly blocks Ba^{2+} current in *Lymnaea* neurons. *J Physiol (London)* 1993;462:321–347.
78. Berlin JR, Bassani JW, Bers DM: Intrinsic cytosolic calcium buffering properties of single rat cardiac myocytes. *Biophys J* 1994;67:1775–1787.
79. Chard PS, Bleakman D, Christakos S, Fullmer CS, Miller RJ: Calcium buffering properties of calbindin D28k and parvalbumin in rat sensory neurones. *J Physiol (London)* 1993;472:341–357.
80. Tse A, TSE FW, Hille B: Calcium homeostasis in identified rat gonadotrophs. *J Physiol (London)* 1994;477:511–525.
81. Gray PC, Scott JD, Catterall WA: Regulation of ion channels by cAMP-dependent protein kinase and A-kinase anchoring proteins. *Curr Opin Neurobiol* 1998;8:330–334.
82. Okamoto T, Schlegel A, Scherer PE, Lisanti MP: Caveolins, a family of scaffolding proteins for organizing "preassembled signaling complexes" at the plasma membrane. *J Biol Chem* 1998;273: 5419–5422.
83. Dell'Acqua ML, Scott JD: Protein kinase A anchoring. *J Biol Chem* 1997;272:12881–12884.
84. Faux MC, Scott JD: Regulation of the AKAP79-protein kinase C interaction by Ca^{2+}/calmodulin. *J Biol Chem* 1997;272:17038–17044.
85. Klauck TM, Faux MC, Labudda K *et al:* Coordination of three signaling enzymes by AKAP79, a mammalian scaffold protein. *Science* 1996;271:1589–1592.
86. Gray PC, Johnson BD, Westenbroek RE *et al:* Primary structure and function of an A kinase anchoring protein associated with calcium channels. *Neuron* 1998;20:1017–1026.
87. Fraser ID, Tavalin SJ, Lester LB *et al:* A novel lipid-anchored A-kinase anchoring protein facilitates cAMP-responsive membrane events. *EMBO J* 1998;17:2261–2272.
88. Carr DW, Hausken ZE, Fraser ID, Stofko-Hahn RE, Scott JD: Association of the type-II cAMP-dependent protein-kinase with a human thyroid RII–anchoring protein—cloning and characterization of the RII-binding domain. *J Biol Chem* 1992;267:13376–13382.
89. Carr DW, Stofko-Hahn RE, Fraser ID *et al:* Interaction of the regulatory subunit (RII) of cAMP-dependent protein kinase with RII-anchoring proteins occurs through an amphipathic helix binding motif. *J Biol Chem* 1991;266:14188–14192.

90. Raymond LA, Tingley WG, Blackstone CD, Roche KW, Huganir RL: Glutamate receptor modulation by protein phosphorylation. *J Physiol (Paris)* 1994;88:181–192.
91. Rosenmund C, Carr DW, Bergeson SE *et al:* Anchoring of protein kinase A is required for modulation of AMPA/kainate receptors on hippocampal neurons. *Nature* 1994;368:853–856.
92. Dell'Acqua ML, Faux MC, Thorburn J, Thorburn A, Scott JD: Membrane-targeting sequences on AKAP79 bind phosphatidylinositol-4, 5-bisphosphate. *EMBO J* 1998;17:2246–2260.
93. Song YM, Huang LY: Modulation of glycine receptor chloride channels by cAMP-dependent protein kinase in spinal trigeminal neurons. *Nature* 1990;348:242–245.
94. Berger HA, Travis SM, Welsh MJ: Regulation of the cystic fibrosis transmembrane conductance regulator Cl^- channel by specific protein kinases and protein phosphatases. *J Biol Chem* 1993;268: 2037–2047.
95. Dransfield DT, Bradford AJ, Smith J *et al:* Ezrin is a cyclic AMP-dependent protein kinase anchoring protein. *EMBO J* 1997;16:35–43.
96. Vaandrager AB, Smolenski A, Tilly BC *et al:* Membrane targeting of cGMP-dependent protein kinase is required for cystic fibrosis transmembrane conductance regulator Cl^- channel activation. *Proc Natl Acad Sci U S A* 1998;95:1466–1471.
97. Vaandrager AB, Ehlert EM, Jarchau T, Lohmann SM, de Jonge HR: N-terminal myristoylation is required for membrane localization of cGMP-dependent protein kinase type II. *J Biol Chem* 1996; 271:7025–7029.
98. Vo NK, Gettemy JM, Coghlan VM: Identification of cGMP-dependent protein kinase anchoring proteins (GKAPs). *Biochem Biophys Res Commun* 1998;246:831–835.
99. Reinhart PH, Chung S, Martin BL, Brautigan DL, Levitan IB: Modulation of calcium-activated potassium channels from rat brain by protein kinase A and phosphatase 2A. *J Neurosci* 1991;11: 1627–1635.
100. Reinhart PH, Levitan IB: Kinase and phosphatase activities intimately associated with a reconstituted calcium-dependent potassium channel. *J Neurosci* 1995;15:4572–4579.
101. Wang ZW, Kotlikoff MI: Activation of KCa channels in airway smooth muscle cells by endogenous protein kinase A. *Am J Physiol* 1996;271:L100–L105.
102. Wischmeyer E, Karschin A: Receptor stimulation causes slow inhibition of IRK1 inwardly rectifying K^+ channels by direct protein kinase A–mediated phosphorylation. *Proc Natl Acad Sci U S A* 1996;93:5819–5823.
103. Jurevicius J, Fischmeister R: cAMP compartmentalization is responsible for a local activation of cardiac Ca^{2+} channels by β-adrenergic agonists. *Proc Natl Acad Sci U S A* 1997;93:295–299.
104. Sculptoreanu A, Scheuer T, Catterall WA: Voltage-dependent potentiation of L-type Ca^{2+} channels due to phosphorylation by cAMP-dependent protein kinase. *Nature* 1993;364:240–243.
105. Johnson BD, Scheuer T, Catterall WA: Voltage-dependent potentiation of L-type Ca^{2+} channels in skeletal muscle cells requires anchored cAMP-dependent protein kinase. *Proc Natl Acad Sci U S A* 1994;91:11492–11496.
106. Johnson BD, Brousal JP, Peterson BZ *et al:* Modulation of the cloned skeletal muscle L-type Ca^{2+} channel by anchored cAMP-dependent protein kinase. *J Neurosci* 1997;17:1243–1255.
107. Burton KA, Johnson BD, Hausken ZE *et al:* Type II regulatory subunits are not required for the anchoring-dependent modulation of Ca^{2+} channel activity by cAMP-dependent protein kinase. *Proc Natl Acad Sci U S A* 1997;94:11067–11072.
108. Gray PC, Tibbs VC, Catterall WA, Murphy BJ: Identification of a 15-kDa cAMP-dependent protein kinase–anchoring protein associated with skeletal muscle L-type calcium channels. *J Biol Chem* 1997;272:6297–6302.
109. Lester LB, Langeberg LK, Scott JD: Anchoring of protein kinase A facilitates hormone-mediated insulin secretion. *Proc Natl Acad Sci U S A* 1997;94:14942–14947.
110. Gao T, Yatani A, Dell'Acqua ML *et al:* cAMP-dependent regulation of cardiac L-type Ca^{2+} channels requires membrane targeting of PKA and phosphorylation of channel subunits. *Neuron* 1997; 19:185–196.
111. Hu XQ, Singh N, Mukhopadhyay D, Akbarali HI: Modulation of voltage-dependent Ca^{2+} channels in rabbit colonic smooth muscle cells by c-Src and focal adhesion kinase. *J Biol Chem* 1998;273: 5337–5342.
112. Li M, West JW, Lai Y, Scheuer T, Catterall WA: Functional modulation of brain sodium channels by cAMP-dependent phosphorylation. *Neuron* 1992;8:1151–1159.
113. Tibbs VC, Gray PC, Catterall WA, Murphy BJ: AKAP15 anchors cAMP-dependent protein kinase to brain sodium channels. *J Biol Chem* 1998;273:25783.

114. Alberts B, Bray D, Lewis J *et al: Molecular biology of the cell.* New York: Garland, 1994.
115. Rodbell M: The role of GTP-binding proteins in signal transduction: from the sublimely simple to the conceptually complex. *Curr Top Cell Regul* 1992;32:1–47.
116. Narumiya S, Ishizaki T, Watanabe N: Rho effectors and reorganization of actin cytoskeleton. *FEBS Lett* 1997;410:68–72.
117. Wu HC, Huang PH, Lin CT: G protein beta subunit is closely associated with microtubules. *J Cell Biochem* 1998;70:553–562.
118. Li Y, Ndubuka C, Rubin CS: A kinase anchor protein 75 targets regulatory (RII) subunits of cAMP-dependent protein kinase II to the cortical actin cytoskeleton in non-neuronal cells. *J Biol Chem* 1996;271:16862–16869.
119. Cheek TR, Burgoyne RD: Cyclic AMP inhibits both nicotine-induced actin disassembly and catecholamine secretion from bovine adrenal chromaffin cells. *J Biol Chem* 1987;262:11663–11666.
120. Blobe GC, Stribling DS, Fabbro D, Stabel S, Hannun YA: Protein kinase C βII specifically binds to and is activated by F-actin. *J Biol Chem* 1996;271:15823–15830.
121. Masson-Gadais B, Salers P, Bongrand P, Lissitzky JC: PKC regulation of microfilament network organization in keratinocytes defined by a pharmacological study with PKC activators and inhibitors. *Exp Cell Res* 1997;236:238–247.
122. Vallano ML, Goldenring JR, Lasher RS, Delorenzo RJ: Association of calcium/calmodulin-dependent kinase with cytoskeletal preparations: phosphorylation of tubulin, neurofilament, and microtubule-associated proteins. *Ann N Y Acad Sci* 1986;466:357–374.
123. Schlaepfer DD, Hunter T: Integrin signalling and tyrosine phosphorylation: just the FAKs? *Trends Cell Biol* 1998;8:151–157.
124. Forscher P: Calcium and polyphosphoinositide control of cytoskeletal dynamics. *Trends Neurosci* 1989;12:468–474.
125. Neely MD, Nicholls JG: Electrical activity, growth cone motility and the cytoskeleton. *J Exp Biol* 1995;198:1433–1446.
126. Shin HY, Lee EH, Shin TY, Kim HM: Effect of cytochalasin D on systemic and local anaphylaxis in a murine model. *Pharmacol Res* 1997;36:141–146.
127. Wang TH, Wang HS, Ichijo H *et al:* Microtubule-interfering agents activate c-Jun N-terminal kinase/stress-activated protein kinase through both *ras* and apoptosis signal-regulating kinase pathways. *J Biol Chem* 1998;273:4928–4936.
128. Tseng S, Kim R, Kim T, Morgan KG, Hai CM: F-actin disruption attenuates agonist-induced $[Ca^{2+}]$, myosin phosphorylation, and force in smooth muscle. *Am J Physiol* 1997;41:C1960-C1967.
129. Ribeiro CM, Reece J, Putney JW Jr. Role of the cytoskeleton in calcium signaling in NIH 3T3 cells. An intact cytoskeleton is required for agonist-induced $[Ca^{2+}]i$ signaling, but not for capacitative calcium entry. *J Biol Chem* 1997;272:26555–26561.
130. Undrovinas AI, Maltsev VA: Cytochalasin D alters kinetics of Ca^{2+} transient in rat ventricular cardiomyocytes: an effect of altered actin cytoskeleton? *J Mol Cell Cardiol* 1998;30:1665–1670.
131. Vaaraniemi J, Huotari V, Lehto VP, Eskelinen S: Effect of PMA on the integrity of the membrane skeleton and morphology of epithelial MDCK cells is dependent on the activity of amiloride-sensitive ion transporters and membrane potential. *Eur J Cell Biol* 1997;74:262–272.
131a. Klee R, Heinemann U, Eder C: Changes in proton currents in murine microglia induced by cytoskeletal disruptive agents. *Neurosci Lett* 1998;247:191–194.
132. Ebstensen RD, Plagemann PG: Cytochalasin B: inhibition of glucose and glucosamine transport. *Proc Natl Acad Sci U S A* 1972;69:1430–1434.
133. McKay DB, Aronstam RS, Schneider AS: Interactions of microtubule-active agents with nicotinic acetylcholine receptors. Relationship to their inhibition of catecholamine secretion by adrenal chromaffin cells. *Mol Pharmacol* 1985;28:10–16.
134. Bueno OF, Leidenheimer NJ: Colchicine inhibits $GABA_A$ receptors independently of microtubule depolymerization. *Neuropharmacology* 1998;37:383–390.
135. Weiner JL, Buhler AV, Whatley VJ, Harris RA, Dunwiddie TV: Colchicine is a competitive antagonist at human recombinant $GABA_A$ receptors. *J Pharmacol Exp Ther* 1998;284:95–102.
136. Machu TK: Colchicine competitively antagonizes glycine receptors expressed in *Xenopus* oocytes. *Neuropharmacology* 1998;37:391–396.
137. Paoletti P, Ascher P: Mechanosensitivity of NMDA receptors in cultured mouse central neurons. *Neuron* 1992;13:645–655.
138. Shorte SL: N-Methyl-D-aspartate evokes rapid net depolymerization of filamentous actin in cultured rat cerebellar granule cells. *J Neurophysiol* 1997;78:1135–1143.

139. Zhang S, Ehlers MD, Bernhardt JP, Su CT, Huganir RL: Calmodulin mediates calcium-dependent inactivation of N-Methyl-D-aspartate receptors. *Neuron* 1998;21:443–453.
140. Furukawa K, Mattson MP: Cytochalasins protect hippocampal neurons against amyloid beta-peptide toxicity: evidence that actin depolymerization suppresses Ca^{2+} influx. *J Neurochem* 1995;65: 1061–1068.
141. Furukawa K, Fu WM, Li Y *et al:* The actin-severing protein gelsolin modulates calcium channel and NMDA receptor activities and vulnerability to excitotoxicity in hippocampal neurons. *J Neurosci* 1997;17:8178–8186.
142. Sanchez C, Ulloa L, Montoro RJ, Lopez-Barneo J, Avila J: NMDA-glutamate receptors regulate phosphorylation of dendritic cytoskeletal proteins in the hippocampus. *Brain Res* 1997;765: 141–148.
143. Whatley VJ, Mihic SJ, Allan AM, McQuilkin SJ, Harris RA: $GABA_A$ receptor function is inhibited by microtubule depolymerization. *J Biol Chem* 1994;269:19546–19552.
144. Parker KE: Modulation of ATP-gated non-selective cation channel ($P2X_1$ receptor) activation and desensitization by the actin cytoskeleton. *J Physiol (London)* 1998;510:19–25.
145. Sackin H: Mechanosensitive channels. *Annu Rev Physiol* 1995;57:333–353.
146. Cantiello HF: Role of actin filament organization in cell volume and ion channel regulation. *J Exp Zool* 1997;279:425–435.
147. Cantiello HF, Stow JL, Prat AG, Ausiello DA: Actin filaments regulate epithelial Na^+ channel activity. *Am J Physiol* 1991;261:C882–C886.
148. Prat AG, Bertorello AM, Ausiello DA, Cantiello HF: Activation of epithelial Na^+ channels by protein kinase A requires actin filaments. *Am J Physiol* 1993;265:C224-C233.
149. Berdiev BK, Prat AG, Cantiello HF *et al:* Regulation of epithelial sodium channels by short actin filaments. *J Biol Chem* 1996;271:17704–17710.
150. Berdiev BK, Shlyonsky VG, Senyk O *et al:* Protein kinase A phosphorylation and G protein regulation of type II pneumocyte Na^+ channels in lipid bilayers. *Am J Physiol* 1997;272: C1262–C1270.
151. Marunaka Y, Shintani Y, Downey GP, Niisato N: Activation of Na^+-permeant cation channel by stretch and cyclic AMP-dependent phosphorylation in renal epithelial A6 cells. *J Gen Physiol* 1997; 110:327–336.
152. Rehn M, Weber WM, Clauss W: Role of the cytoskeleton in stimulation of Na^+ channels in A6 cells by changes in osmolality. *Pfluegers Arch* 1998;436:270–279.
153. Smith PR, Stoner LC, Viggiano SC, Angelides KJ, Benos DJ: Effects of vasopressin and aldosterone on the lateral mobility of epithelial Na^+ channels in A6 renal epithelial cells. *J Membr Biol* 1995; 147:195–205.
154. Prat AG, Cantiello HF: Nuclear ion channel activity is regulated by actin filaments. *Am J Physiol* 1996;270:C1532–C1543.
155. Wang WH, Cassola A, Giebisch G: Involvement of actin cytoskeleton in modulation of apical K channel activity in rat collecting duct. *Am J Physiol* 1994;267:F592–F598.
156. Cantiello HF, Prat AG, Bonventre JV *et al:* Actin-binding protein contributes to cell volume regulatory ion channel activation in melanoma cells. *J Biol Chem* 1993;268:4596–4599.
157. Ji S, John SA, Lu YJ, Weiss JN: Mechanosensitivity of the cardiac muscarinic potassium channel—A novel property conferred by Kir3.4 subunit. *J Biol Chem* 1998;273:1324–1328.
158. Benz I, Meyer DK, Kohlhardt M: Properties and the cytoskeletal control of Ca^{2+}-independent large conductance K^+ channels in neonatal rat hippocampal neurons. *J Membr Biol* 1998;161:275–286.
159. Ehrhardt AG, Frankish N, Isenberg G: A large-conductance K^+ channel that is inhibited by the cytoskeleton in the smooth muscle cell line DDT1 MF-2. *J Physiol (London)* 1996;496:663–676.
160. Terzic A, Kurachi Y: Actin microfilament disrupters enhance KATP channel opening in patches from guinea-pig cardiomyocytes. *J Physiol (London)* 1996;492:395–404.
161. Yokoshiki H, Katsube Y, Sunugawa M, Seki T, Sperelakis N: Disruption of actin cytoskeleton attenuates sulfonylurea inhibition of cardiac ATP-sensitive K^+ channels. *Pfluegers Arch* 1997;434: 203–205.
162. Furukawa T, Yamane T, Terai T, Katayama Y, Hiraoka M: Functional linkage of the cardiac ATP-sensitive K^+ channel to the actin cytoskeleton. *Pfluegers Arch* 1996;431:504–512.
163. Mauerer UR, Boulpaep EL, Segal AS: Regulation of an inwardly rectifying ATP-sensitive K^+ channel in the basolateral membrane of renal proximal tubule. *J Gen Physiol* 1998;111:161–180.
164. Hocherman SD, Bezanilla F: A patch–clamp study of delayed rectifier currents in skeletal muscle of control and *mdx* mice. *J Physiol (London)* 1996;493:113–128.

165. Brady PA, Alekseev AE, Aleksandrova LA, Gomez LA, Terzic A: A disrupter of actin microfilaments impairs sulfonylurea-inhibitory gating of cardiac K_{ATP} channels. *Am J Physiol* 1996;271: H2710–H2716.
166. Fuller CM, Bridges RJ, Benos DJ: Forskolin- but not ionomycin-evoked Cl^- secretion in colonic epithelia depends on intact microtubules. *Am J Physiol* 1994;266:C661–C668.
167. Crepel V, Panenka W, Kelly MEM, MacVicar BA: Mitogen-activated protein and tyrosine kinases in the activation of astrocyte volume-activated chloride current. *J Neurosci* 1998;18:1196–1206.
168. Ullrich N, Sontheimer H: Cell cycle-dependent expression of a glioma-specific chloride current: proposed link to cytoskeletal changes. *Am J Physiol* 1997;42:C1290–C1297.
169. Zhang JP, Larsen TH, Lieberman M: F-actin modulates swelling-activated chloride current in cultured chick cardiac myocytes. *Am J Physiol* 1997;42:C1215–C1224.
170. Schwiebert EM, Mills JW, Stanton BA: Actin-based cytoskeleton regulates a chloride channel and cell volume in a renal cortical collecting duct cell line. *J Biol Chem* 1994;269:7081–7089.
171. Levitan I, Almonte C, Mollard P, Garber SS: Modulation of a volume-regulated chloride current by F-actin. *J Membr Biol* 1995;147:283–294.
172. Fischer H, Illek B, Machen TE: The actin filament disrupter cytochalasin D activates the recombinant cystic fibrosis transmembrane conductance regulator Cl^- channel in mouse STS fibroblasts. *J Physiol (London)* 1995;489:745–754.
173. Villaz M, Cinniger JC, Moody WJ: A voltage-gated chloride channel in ascidian embryos modulated by both the cell cycle clock and cell volume. *J Physiol (London)* 1995;488:689–699.
174. Lascola CD, Nelson DJ, Kraig RP: Cytoskeletal actin gates a Cl^- channel in neocortical astrocytes. *J Neurosci* 1998;18:1679–1692.
175. Haussler U, Rivet-Bastide M, Fahlke C *et al:* Role of the cytoskeleton in the regulation of Cl^- channels in human embryonic skeletal muscle cells. *Pfluegers Arch* 1994;428:323–330.
176. Suzuki M, Miyazaki K, Ikeda M, Kawaguchi Y, Sakai O: F-actin network may regulate a Cl^- channel in renal proximal tubule cells. *J Membr Biol* 1993;134:31–39.
177. Seibert FS, Tabcharani JA, Chang XB *et al:* cAMP-dependent protein kinase-mediated phosphorylation of cystic fibrosis transmembrane conductance regulator residue Ser-753 and its role in channel activation. *J Biol Chem* 1995;270:2158–2162.
178. Cantiello HF: Role of the actin cytoskeleton in the regulation of the cystic fibrosis transmembrane conductance regulator. *Exp Physiol* 1996;81:505–514.
179. Prat AG, Xiao Y, F, Cantiello HF: cAMP-independent regulation of CFTR by the actin cytoskeleton. *Am J Physiol* 1995;268:C1552–C1561.
180. Ismailov II, Berdiev BK, Shlyonsky VG *et al:* Role of actin in regulation of epithelial sodium channels by CFTR. *Am J Physiol* 1997;272:C1077–C1086.
181. Matsumoto G, Sakai H: Microtubules inside the plasma membrane of squid giant axons and their possible physiological function. *J Membr Biol* 1979;50:1–14.
182. Matsumoto G, Ichikawa M, Tasaki A, Murofushi H, Sakai H: Axonal microtubules necessary for generation of sodium current in squid giant axons. I. Pharmacological study on sodium current and restoration of sodium current by microtubule proteins and 260K protein. *J Membr Biol* 1984;77: 77–91.
183. Matsumoto G, Ichikawa M, Tasaki A: Axonal microtubules necessary for generation of sodium current in squid giant axons. II. Effect of colchicine upon asymmetrical displacement current. *J Membr Biol* 1984;77:93–99.
184. Matsumoto G, Sakai H: Restoration of membrane excitability of squid giant axons by reagents activating tyrosine–tubulin ligase. *J Membr Biol* 1979;50:15–22.
185. Undrovinas AI, Shander GS, Makielski JC: Cytoskeleton modulates gating of voltage dependent sodium channel in heart. *Am J Physiol* 1995;269:H203–H214.
186. Maltsev VA, Undrovinas AI: Cytoskeleton modulates coupling between availability and activation of cardiac sodium channel. *Am J Physiol* 1997;42:H1832–H1840.
187. Eckert R, Chad JE: Inactivation of Ca channels. *Prog Mol Biol* 1984;44:215–267.
188. Hagiwara SL, Byerly L: Calcium channel. *Annu Rev Neurosci* 1981;4:69–125.
189. Morad M, Davies NW, Kaplan JH, Lux HD: Inactivation and block of calcium channels by photoreleased Ca in dorsal root ganglion neurons. *Science* 1988;241:842–844.
190. Fryer MW, Zucker RS: Ca^{2+}-dependent inactivation of Ca^{2+} current in *Aplysia* neurons: kinetic studies using photolabile Ca^{2+} chelators. *J Physiol (London)* 1993;464:501–528.
191. Chad JE, Eckert R: An enzymatic mechanism for calcium current inactivation in dialysed *Helix* neurones. *J Physiol (London)* 1986;378:31–51.

192. Armstrong D, Eckert R: Voltage-activated calcium channels that must be phosphorylated to respond to membrane depolarization. *Proc Natl Acad Sci U S A* 1987;84:2518–2522.
193. Thion L, Mazars C, Nacry P *et al:* Plasma membrane depolarization-activated calcium channels, stimulated by microtubule-depolymerizing drugs in wild-type *Arabidopsis thaliana* protoplasts, display constitutively large activities and a longer half- life in *ton 2* mutant cells affected in the organization of cortical microtubules. *Plant J* 1998;13:603–610.
194. Thion L, Mazars C, Thuleau P *et al:* Activation of plasma membrane voltage-dependent calcium-permeable channels by disruption of microtubules in carrot cells. *FEBS Lett* 1996;393:13–18.
195. McCarron JG, Crichton CA, Langton PD, MacKenzie A, Smith GL: Myogenic contraction by modulation of voltage-dependent calcium currents in isolated rat cerebral arteries. *J Physiol (London)* 1997;498:371–379.
196. Xu WX, Kim SJ, So I *et al:* Effect of stretch on calcium channel currents recorded from the antral circular myocytes of guinea-pig stomach. *Pfluegers Arch* 1996;432:159–164.
197. Xu WX, Kim SJ, So I, Kim KW: Role of actin microfilament in osmotic stretch-induced increase of voltage-operated calcium channel current in guinea-pig gastric myocytes. *Pfluegers Arch* 1997;434: 502–504.
198. Matzner O, Ben Tabou S, Nussinovitch I: Hyperosmotic regulation of voltage-gated calcium currents in rat anterior pituitary cells. *J Neurophysiol* 1996;75:1894–1900.
199. Taglialatela M, Ficker E, Wible BA, Brown AM: C-terminus determinants for Mg^{2+} and polyamine block of the inward rectifier K^+ channel IRK1. *EMBO J* 1995;14:5532–5541.
200. Mazzanti M, Assandri R, Ferroni A, DiFrancesco D: Cytoskeletal control of rectification and expression of four substates in cardiac inward rectifier K^+ channels. *FASEB J* 1996;10:357–361.
201. Maguire G, Connaughton V, Prat AG, Jackson GR Jr., Cantiello HF: Actin cytoskeleton regulates ion channel activity in retinal neurons. *NeuroReport* 1998;9:665–670.
202. Sunada Y, Campbell KP: Dystrophin–glycoprotein complex: molecular organization and critical roles in skeletal muscle. *Curr Opin Neurol* 1995;8:379–384.
203. Ahn AH, Kunkel LM: The structural and functional diversity of dystrophin. *Nat Genet* 1993;3: 283–291.
204. Mathes C, Bezanilla F, Weiss RE: Sodium current and membrane potential in EDL muscle fibers from normal and dystrophic (*mdx*) mice. *Am J Physiol* 1991;261:C718–C725.

Part III

Second Messengers

Advances in Second Messenger and Phosphoprotein Research, Vol. 33

1040-7952/99 $30.00

10

Cyclic Nucleotide Gated Channels

Martin Biel, Xiangang Zong, and Franz Hofmann

Institut für Pharmakologie und Toxikologie der Technischen Universität München, 80802 Munich, Germany

INTRODUCTION

Cyclic nucleotides are important second messengers that trigger a variety of cellular processes including visual and olfactory transduction (1,2), regulation of smooth muscle tone (3), intestinal chloride and and water secretion (4), and synaptic plasticity (5). There are three main classes of receptors for cGMP or cAMP: the cyclic nucleotide dependent protein kinases (6), the cyclic nucleotide regulated phosphodiesterases (7) and the cyclic nucleotide gated (CNG) cation channels. The first identified members of the CNG channel family were the cGMP activated cation channels of rod (8,9) and cone (10,11) photoreceptors. Subsequently, a CNG channel that is activated roughly equally by cGMP and cAMP was detected in the membrane of olfactory neurons (12). For some years it was thought that CNG channel expression was limited to sensory cells and that the basic function of CNG channels was to generate receptor currents by allowing influx of Na^+ upon activation by cAMP or cGMP. However, the work of several groups has provided evidence that this physiological function might be restricted to vertebrate photoreceptors. It has become clear that CNG channels are present in many cell types and that they represent receptor-operated calcium channels that allow neurotransmitter-controlled calcium entry.

THE CNG CHANNEL MULTI-GENE FAMILY

CNG channels are key elements in the visual (13,14) and olfactory (2,15) signal transduction. These signaling cascades differ from each other in the intrinsic second messenger and the type of CNG channel, respectively. Visual transduction of rod and cone photoreceptors is controlled by the concentration of cGMP in the outer segment of both photoreceptor types (13,16). In the dark, a high concentration of cGMP maintains the photoreceptor CNG channel in the open state, leading to a membrane depolarization and a steady release of glutamate from the synaptic terminal. Photoactivated rhodopsin activates a phosphodiesterase, which catalyzes the hydrolysis of cGMP. The resulting decrease in the cGMP concentration closes the photoreceptor CNG channel and causes a hyperpolarization of the photoreceptor membrane. This hyperpolarization reduces the synaptic transmitter release from the photoreceptor to second-order visual neurons. Interestingly, photoreceptors of the lizard parietal eye respond to light by a depolarization instead of a hyperpolarization (17). The underlying transduction cascade involves the light-induced synthesis of cGMP followed by the activation of a cGMP-activated CNG channel that may be related to rod and cone photoreceptor CNG channels.

Olfactory signal transduction is coupled to the odor-dependent synthesis of cAMP, which is catalyzed by the adenylyl cyclase of olfactory cilia (2,15). The increase in cAMP concentration opens a cAMP-gated channel. This then leads to a depolarization of the olfactory neuron. It has been suggested that inositol 1,4,5-triphosphate (IP_3) may also be involved in transduction of certain odors (15,18). However, at least in mammals the IP_3-mediated pathway of olfaction may be less important or even absent, since the deletion of the cAMP-gated channel gene in mice resulted in a complete loss of excitatory olfactory signal transduction (19).

The molecular characterization of CNG channels became possible with the purification of the photoreceptor channel from bovine rod outer segment (20). Subsequently, the cDNA of the channel (CNG1) has been cloned and expressed functionally (21). The cDNAs of other CNG channels have been isolated by homology screening with CNG1-derived probes or by PCR-based cloning (Fig. 1A; Table 1 and references cited there). Native CNG channels are oligomeric complexes consisting of distinct, yet homologous subunits. Presently, the CNG channel family comprises five different members in mammals (CNG1–5). The primary sequences of CNG1–5 are homologous to each other, indicating that these subunits have evolved from a common ancestral CNG channel. The CNG channel subunits can be subdivided into two different groups on the basis of their ability (α subunits) or inability (β subunits) to form functional homomeric channels in expression systems like *Xenopus* oocytes or human embryonic kidney (HEK) cells (Fig. 1A).

The α subunits of the rod photoreceptor (CNG1), the cone photoreceptor (CNG3), and the olfactory (CNG2) channel contain 640–740 amino acid residues and have overall sequence identities of 60–70%. The calculated molecular masses range from 75 to 85 kDa. The molecular mass of the rod photoreceptor channel α subunit as deduced from the cDNA (79 kDa) differs from the mass of the purified protein from rod outer segment (63 kDa) as estimated by SDS-PAGE. This difference can be attributed to the post-translational cleavage of the cytoplasmic portion of the N-terminus (22). There is evidence that the cone photoreceptor channel (CNG3) is also truncated in the N-terminal region (23,24). A CNG channel that may represent a splice variant of the cone photoreceptor channel has been identified in taste buds of rat tongue (25), suggesting that CNG channels are also implicated in taste chemoreception.

Although CNG channel β subunits (CNG4 and CNG5) are structurally related to CNG1–3 they fail to induce the formation of cGMP- or cAMP-gated channels in heterologous expression systems. However, recent studies indicate that CNG5 may

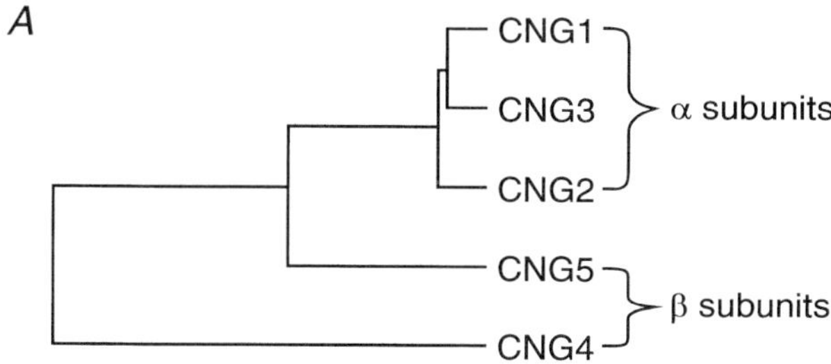

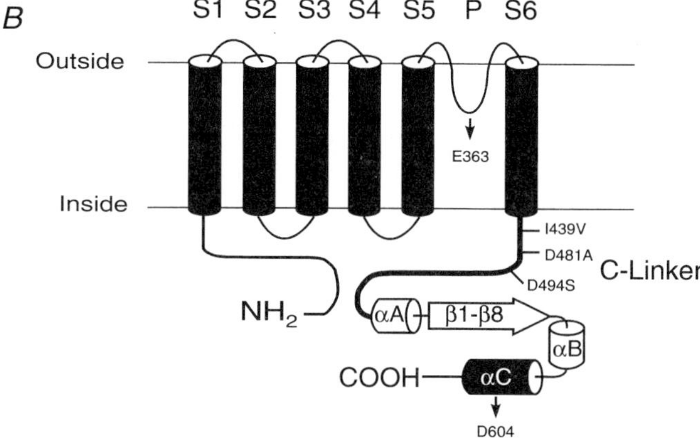

FIG. 1. **(A)** Phylogenetic tree of cyclic nucleotide gated channel α and β subunits. The tree was calculated on the basis of sequence alignments with the transmembrane domain and the cyclic nucleotide binding pocket of the respective subunits. **(B)** Two-dimensional model of the CNG channel: S1–S6, α helical transmembrane segments; P, pore region. The eight β strands (β_1–β_8) and the three α helices (α_A–α_C) that form the cyclic nucleotide binding domain (CNBD) are schematically shown in the C-terminus of the channel. The C-linker that connects S6 with the CNBD is illustrated by a thick line. The three amino acids in the C-linker of CNG3 and CNG2 that determine cAMP efficacy, as well as the positions of D604 and E363 that are important for ligand selectivity and Ca^{2+} block, respectively, are indicated.

form a homomeric, NO-activated channel in neurons of the vomeronasal organ, which is important for the sensory transduction of pheromones (26,27). CNG channel β subunits impart to the respective α subunits specific properties that are also found in native channels. For this reason, β subunits have been also designated as modulatory subunits of CNG channels. Two types of β subunits were originally cloned from photoreceptors (CNG4) and olfactory neurons (CNG5). There is evidence that a specific CNG4 isoform (CNG4.3) is also expressed in olfactory neurons and is likely to represent a subunit of the native olfactory channel (28).

Hydropathicity analysis and sequence alignments indicate that CNG4 and CNG5, like the CNG channel α subunits, contain six transmembrane segments, an ion-

TABLE 1. *Family of CNG channel subunits*

Gene[a]	Source	Species	Reference
CNG1	Retina	Bovine	21
		Mouse	134
		Rat	135
		Human	134, 136
		Dog	137
		Chicken	23
	Kidney	Mouse	123
	Heart	Porcine	138
CNG2	Olfactory epithelium	Rat	139
		Bovine	140
		Catfish	72
	Heart	Mouse	141
	Aorta	Rabbit	142
CNG3	Retina	Chicken	23
		Human	143, 144
	Kidney	Bovine	54
	Testis	Bovine	24
	Pineal gland	Chicken	111
CNGgust	Taste buds	Rat	25
hCNG4.1	Retina	Human	61
hCNG4.2		Human	61
hCNG4.3		Human	90, 91
bCNG4.1		Bovine	62
bCNG4.2	Testis	Bovine	44
bCNG4.3		Bovine	44
bCNG4.4		Bovine	44
rCNG4.1	Pineal gland	Rat	112
rCNG4.2		Rat	112
rCNG4.3	Olfactory epithelium	Rat	28
bGARP	Retina	Bovine	62, 91, 92
hGARP		Human	91, 93
rGARP	Pineal gland	Rat	112
CNG5	Olfactory epithelium	Rat	43, 85
DmCNG	Eye, antenna	*D. melanogaster*	29
tax-2	Genomic DNA	*C. elegans*	30
tax-4	Genomic DNA	*C. elegans*	31

[a]The nomenclature of the CNG4 isoforms has been changed with respect to the previous naming (44). The new name consists of a lowercase letter that indicates the species (e.g, "h" for human) and the number of the isoform of the gene in order of its appearance in the literature (e.g., hCNG4.1 is the first human CNG4 isoform to be described). In this nomenclature the names of the isoforms of CNG4 cloned from bovine testis are bCNG4.2, bCNG4.3, bCNG4.4 instead of the old names CNG4c, CNG4d, and CNG4e.

conducting pore, and a putative binding pocket for cGMP in the C-terminus. Despite this general structural relationship, the overall sequence identity between the β subunits and the α subunits (30–50%) is significantly lower than the homology between the different α subunits (60–70%). This more distant relationship indicates an early evolutionary separation of α and β subunits from an ancestral CNG channel. In contrast to the α subunits and the CNG5 subunit, the primary transcript of the CNG4 gene is extensively spliced giving rise to several isoforms (Table 1; Fig. 2), which are expressed in a tissue- and species-specific manner.

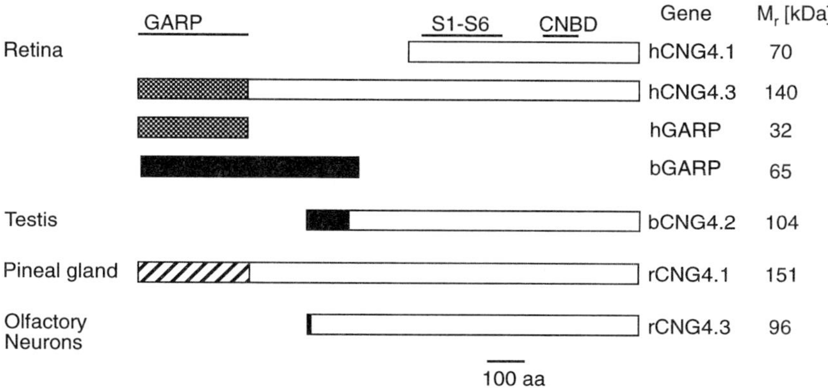

FIG. 2. Schematic representation of CNG4 and GARP isoforms cloned from various tissues. The proteins are illustrated by boxes with the N-terminus on the left and the C-terminus at the right. The location of the transmembrane segments (S1–S6), the cyclic nucleotide binding domain (CNBD) and the glutamic acid rich domain are shown. The protein region of CNG4 displaying high sequence identity between the isoforms is represented by open boxes, whereas the largely diverging N-termini of the isoforms are shown as dotted, solid, or hatched boxes. The calculated molecular masses (M_r) as deduced from the cDNAs are indicated. For nomenclature of isoforms see Table 1.

CNG channels have also been cloned from invertebrates like *Drosophila* (DmCNG) and *Caenorhabditis elegans* (tax-2 and tax-4). The DmCNG channel (29) is expressed in both photoreceptor and olfactory sensillae. Its sequence has roughly equal homology to vertebrate photoreceptor and olfactory channels. DmCNG represents also in its affinities to cyclic nucleotides an intermediate between photoreceptor- and olfactory-type channels. The functional role of the *Drosophila* CNG channel is not clear yet, since there is no report of a light-induced change of the cGMP concentration in invertebrate photoreceptor cells (1). The tax-2 (30) and tax-4 (31) proteins of *C. elegans* are coexpressed in olfactory, gustatory, and thermosensory neurons, implicating that these subunits form a single heteromeric channel, which is involved in multiple sensory modalities.

CHANNEL STRUCTURE AND STOICHIOMETRY

Functionally, CNG channels belong to the class of ligand-gated cation channels (32), since they are activated by the binding of a ligand (cGMP or cAMP) to the cyclic nucleotide binding pocket. However, inspection of the primary sequences of the cloned channels revealed that CNG channels belong structurally to the superfamily of voltage-gated cation channels (21,33). The proposed two-dimensional model of the CNG channel is shown in Fig. 1B. Like voltage-gated potassium channels, CNG channels contain six transmembrane segments (S1–S6). The ion-conducting pore is formed by a "hairpin" structure between the S5 and S6 regions. Both N- and C-termini are localized on the cytoplasmic site of the cell. The presence of six

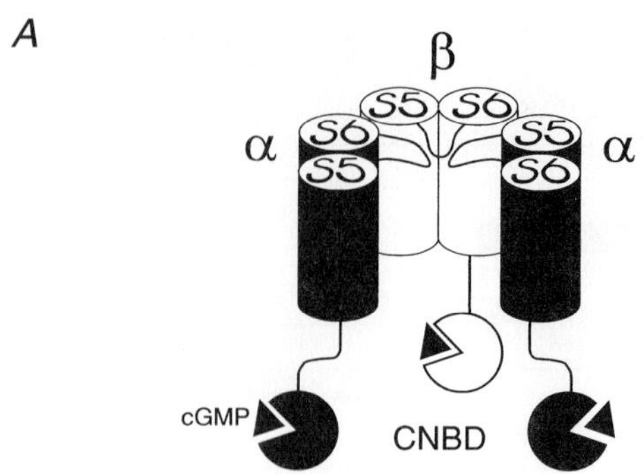

B

Ca^{2+}

CNG1	346	K Y V Y S L Y W S T L T L T T I G . . E T P P P V R	369
CNG2	323	E Y I Y C L Y W S T L T L T T I G . . E T P P P V K	346
CNG3	370	K Y I Y S L Y W S T L T L T T I G . . E T P P P V K	393
CNG4	489	S Y I R C Y Y W A V K T L I T I G . . G L P D P R T	512
CNG5	217	Q Y L Y S F Y F S T L I L T T V G . . D T P L P D R	240
Shaker	428	S I P D A F W W A V V T M T T V G Y G D M T P V G V	453
α_{1Cb}	1424	F Q T F P Q A V L L L F R C A T G . . E A W Q D I M	1447

FIG. 3. (A) Scheme of the pore region and the cyclic nucleotide binding domain (CNBD) of the heteromeric CNG channel. Each of four subunits contributes a single P loop to form the selectivity filter of the channel. For clarity, only three out of four subunits are shown; the number and order of α and β subunits within the tetrameric complex has been arbitrarily fixed in this model because the subunit stoichiometry of native channels is not known yet. **(B)** Sequence alignment of the pore regions of CNG1–5 (for references see Table 1), *Shaker* A channel (145), and the α subunit (α_{1Cb}) of the smooth muscle L-type calcium channel (146). Amino acid residues identical in at least five sequences are boxed. The YG motif present in K^+ channels and absent in CNG and calcium channels is displayed in white against black. The glutamate residue interacting with Ca^{2+} in CNG and calcium channels is marked.

transmembrane segments has been experimentally verified in the bovine CNG1 subunit by a gene fusion approach using the bacterial reporter enzymes alkaline phosphatase and β-galactosidase (34).

All CNG channel subunits contain the positively charged "voltage sensor" motif in S4 (21,35). However, the number of regularly spaced positive charges (3–4 basic residues at every third position) is reduced in comparison with the S4 segment of potassium channels (5–7 basic residues). This reduced number of positive residues is possibly responsible for the very weak voltage dependence of CNG channel opening (13,36). A chimeric *Drosophila ether-à-gogo* K^+ channel containing the voltage sensor of the rat olfactory channel is still activated by voltage (37), indicating that the S4 segment of the CNG channel is functionally intact but may be inactivated by other CNG channel domains. It has been speculated that the S4 segment already was present in the core structure of an ancestor gene that is common to channels activated by both voltage and cyclic nucleotides (38,39). According to this hypothesis, the ancestral S4 segment evolved into the voltage sensor of voltage-activated cation channels, whereas it was conserved as an important structural element necessary for the maintainance of the general channel architecture in the CNG channels.

Recent results from the heterologous expression of tandem dimers of CNG channels (40,41) demonstrate that CNG channels, like the related voltage-gated potassium channels, are tetrameric complexes. Native CNG channels are likely to contain a combination of α and β subunits (Fig. 3A). There is also evidence that homomeric CNG channels consisting only of α subunits may exist in some tissues (42–44).

PERMEATION OF MONO- AND DIVALENT CATIONS

The region of highest homology between CNG channels and voltage-gated cation channels corresponds to the pore (P) region, which is localized between the transmembrane segments S5 and S6 (Fig. 3; see also ref. 45). This sequence is also the most conserved structure within the different members of the CNG channel family. The work of several groups has verified the identity of the postulated region with the CNG channel pore. The expression of chimeric channels in which the pore was exchanged between K^+ channels and CNG channels (46) or between different CNG channels (47) revealed that the specific permeation properties of the channels are mainly determined by the respective pore regions. In addition, rod and olfactory CNG channels interact with the "ball peptide" that produces rapid inactivation in *Shaker*-type K^+ channels (48) by binding to a region within the pore (49). A study using cysteine-scanning mutagenesis (50) indicated that the P region of the CNG channel, like that of K^+ channels, does not dip into and out of the membrane as initially proposed but extends toward the central axis of the channel forming the blades of an irislike structure. If one assumes a tetrameric channel, the S5 and S6 segments of the four subunits form the large-diameter internal vestibule of the channel, whereas the P regions may form the narrowest part of the pore that serves as the selectivity filter (Fig. 3; see also ref. 45).

CNG channels conduct a variety of monovalent cations like Na^+ and K^+ (51–53). The influx of Na^+, which leads to membrane depolarization, has been considered as the archetypal function of CNG channels. However, there is now strong evidence that at least in some cells CNG channels actually work as cyclic nucleotide gated calcium channels (Fig. 4). Measurement of the permeability ratios between Ca^{2+} and Na^+ revealed that CNG channels are more permeable for Ca^{2+} than for Na^+. Especially the CNG3 channel (24,54,55) and the CNG2 channel (55) strongly select Ca^{2+} over Na^+. At physiological concentrations of Ca^{2+}, the current through the heterologously expressed CNG2 channel is almost entirely carried by Ca^{2+} (55).

Site-directed mutagenesis of amino acid residues within the P region has identified the amino acid residues crucially important for the specific permeation properties. A YG-motif, which is present in the pore of the K^+ channel but not in those of CNG channels (Fig. 3B), determines the selectivity for K^+ over Na^+ (45,56). When these residues are deleted in the K^+ channel, the resulting channel resembles CNG chan-

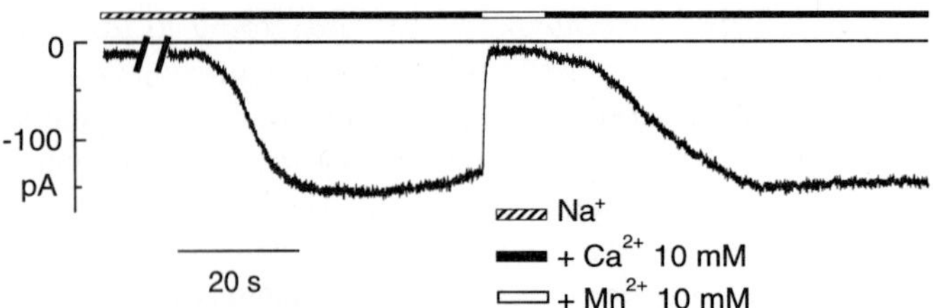

FIG. 4. Whole-cell measurement of the calcium current through the homomeric CNG3 channel expressed in HEK 293 cells. The cell was clamped to 0 mV and dialyzed with a pseudo-intracellular solution (140 mM K-aspartate, 8 mM NaCl, 10 mM EGTA, 10 mM Na-HEPES pH 7.2) supplemented with 1 mM cGMP. After initial superfusion with a Ca^{2+}-free Ringer's solution (140 mM NaCl, 5 mM KCl, 10 mM HEPES, 1 mM NaEGTA pH 7.2) the cell was superfused with Ringer's solutions supplemented with 10 mM Ca^{2+} as indicated. The NaCl concentration was reduced according to the respective concentrations of Ca^{2+}; NaEGTA was omitted. The resulting inward current was carried by Ca^{2+} and was reversibly blocked when extracellular Ca^{2+} was exchanged by 10 mM Mn^{2+}.

nels in that it shows little selectivity among monovalent cations (46). Ca^{2+} efficiently permeates CNG channels but also partially blocks them in a voltage-dependent manner (55,57,58). A glutamate residue within the P region of CNG1–3 (E363 in CNG1; Fig. 1B) is a major determinant of this block. Neutralization of the glutamate residue reduces the block by divalent cations (59,60). The CNG4 subunit has a glycine residue replacing the glutamate residue, and this may explain why heteromeric channels containing the CNG4 subunit show less block by Ca^{2+} than homomeric CNG1–3 channels (28,61,62).

There is evidence that amino acid residues outside the P region also influence ion permeation. The P region of the *Drosophila* CNG channel (DmCNG) is nearly identical to the pore of CNG1–3 in vertebrates and also contains the crucial glutamate residue. However, whereas maximal potency is similar for the blockade of DmCNG and vertebrate CNG channels by Ca^{2+}, Ca^{2+} permeability and voltage dependence of the block by Ca^{2+} are entirely different for the two channels (29). The ion permeation of CNG channels resembles that of voltage-dependent calcium channels. Voltage-dependent Ca^{2+} channels conduct monovalent cations in the absence of divalent cations, whereas in the presence of Ca^{2+} the conductivity of monovalent cations is blocked (63). Interestingly, the pore of calcium channels also lacks the YG motif and contains a set of conserved glutamate residues that determine the ion selectivity (64). Similar to calcium channels, the permeation properties of CNG channels are consistent with a multi-ion occupancy of the pore region (65).

CHANNEL ACTIVATION AND GATING

CNG channels contain a cyclic nucleotide–binding domain (CNBD) in the C-terminal part that bears structural homology to those of the cAMP- and cGMP-dependent protein kinases and to that of the catabolite gene activator protein (CAP) of *E. coli* (36,66). In analogy to the X-ray-resolved structure of the binding domain of CAP (67), the CNBD of CNG channels may consist of an eight-stranded β roll and three α helices (68). Although the CNBD is highly conserved between photoreceptor and olfactory channels (sequence identity of 80–90%), both channel types reveal remarkable differences in their response to cGMP and cAMP. Olfactory channels are 30- to 50-fold more sensitive to cGMP, and up to 1000-fold more sensitive to cAMP than photoreceptor channels (36,52). In addition, cAMP acts as a partial agonist on photoreceptor channels, activating only a fraction of the current induced by cGMP, whereas it fully activates olfactory channels. Studies with mutated and chimeric CNG channels have elucidated within the primary structure of the channel key structural elements that determine these differences. Initially, it was proposed that an invariant threonine located on the β_7 roll of CNG channels (T560 in CNG1) controls the relative ligand specificity of CNG channels (69). An equivalent threonine in the binding pockets of cGMP-dependent protein kinases confers selectivity of cGMP over cAMP, whereas an alanine residue that is present in cAMP-dependent protein kinases confers selectivity of cAMP over cGMP (70). However, the major effect seen when T560 is mutated into alanine in CNG1 is a decrease in the apparent affinities of the channel for cGMP and cAMP (69). Although the decrease is more prominent for cGMP than for cAMP, which would argue for a role of T560 in ligand discrimination, the exchange does not substantially alter the relative agonist efficacy of the channel, which is determined by the maximal current at saturating ligand concentrations (71). In addition, a threonine equivalent to T560 is present in the catfish olfactory channel, which has roughly equal apparent affinities for cGMP and cAMP (72). Thus, T560 may be important for the initial binding of the ligand but is not likely to be a major determinant of agonist selectivity. Mutational studies point to an important role of the the putative α helix (α_C helix) at the C-terminal end of the CNBD for the control of relative ligand specificity (71,73). Substituting of an aspartate residue within the α_C helix of CNG1 (D604 in bovine CNG1; Fig. 1B) by a nonpolar residue inverts CNG1 from a channel that is highly selective for cGMP to a channel with greater cAMP sensitivity (71).

Interestingly, the molecular structures that affect the apparent affinity for cGMP or cAMP are not restricted to the CNBD but are distributed throughout the channel sequence (74). The N-terminal segment (73,74) and the linker peptide (C-linker) connecting the S6 transmembrane segment and the CNBD (Fig. 1B) are especially influential in determining apparent agonist affinity and channel gating. The importance of the C-linker is underlined by several findings.

1. The exchange of three amino acid residues in the C-linker of the cone photoreceptor channel by the respective residues of the olfactory channel (I439V, D481A, and D494S; Fig. 1B) dramatically increases the cAMP-efficacy of the cone channel (75).
2. Histidine residues identified in the C-linker of the CNG1 channel (H420) and the CNG2 channel (H396) confer potentiation or inhibition of the corresponding currents by micromolar concentrations of Ni^{2+} (74,76–78).

3. An additional histidine that is present in all α subunits (H468 in bovine CNG1) has been found to be involved in channel gating (79).

4. Donors of nitric oxide, as well as other agents that modify free SH groups, activate native and expressed olfactory CNG channels in the absence of cGMP or cAMP, presumably by covalently binding to a conserved cysteine residue, also located in the C-linker (26,80).

5. Mild oxidants like copper phenanthroline potentiate activation of the rod photoreceptor channel (81) by facilitating the formation of a disulfide bond between two cysteine residues present in the C-linker (C481) and the N-terminus (C35), respectively.

In summary, CNG channel activation may require a concerted conformational change of various protein domains. Overlay assays suggest that the cytoplasmic N-terminal domain of CNG channels directly binds to domains of the C-terminus including the C-linker and part of CNBD (81,82). This interdomain interaction, which may occur within the same subunit or between different channel subunits of the tetrameric complex, is likely to constitute a crucial part of the channel gating machinery.

Based on the observation that CNG channels reveal spontaneous openings in the absence of any cyclic nucleotide an allosteric model of channel activation has been proposed (83). According to this model, cyclic nucleotides stabilize the channel in the open state, thereby promoting a dramatic increase of the open probability. However, it was demonstrated in 1997 that partially liganded CNG channels can move freely between up to three distinct subconductance states (84). This finding clearly contrasts with the simple closed–open transition postulated in cyclic allosteric models, indicating that a more complex model will be necessary to precisely describe CNG channel activation.

FUNCTIONAL PROPERTIES OF MODULATORY CHANNEL SUBUNITS

Two types of CNG channel β subunits have been cloned, CNG4 and CNG5 (Fig. 1A and 2; Table 1). Although the sequence identity is rather low between CNG4 and 5 (25%), these subunits reveal functional similarities. When coexpressed with the respective α subunit, they form heteromeric complexes and induce channel properties that are present in the native channels but absent from the channels formed by the α subunits alone (28,43,44,61,62,85). The "new" but physiological properties include single channel flickering, current blockade by L-*cis*-diltiazem, weakening of the current outward rectification induced by external calcium, and increase of the apparent affinity for cAMP. The latter effect seems to be especially important in the case of the olfactory channel, since cAMP is the natural second messenger of olfaction (2,86). The participation of CNG4 in agonist binding was confirmed by photoaffinity labeling with an analog of cGMP (87) that identified a peptide corresponding to the predicted CNBD of CNG4.

So far, only a single transcript has been identified for CNG5. In contrast, the cloning of CNG4 yielded multiple transcripts that encode a whole family of isoforms for this subunit (Table 1 and Fig. 2). Initially, two isoforms of CNG4 with calculated molecular masses of 70 kDa (hCNG4.1) and 102 kDa (hCNG4.2) were cloned from human retina (61). In contrast to these molecular weights, only a 240-kDa protein was purified together with the 63-kDa α subunit from rod outer segment of different vertebrates (20,88,89). These discrepancies have been recently unraveled by the cloning of the 240-kDa protein from bovine (bCNG4.1; Ref. 62) and human (hCNG4.3; Refs. 90,91) retina. Northern analysis indicated that the bCNG4.1 and hCNG4.3 transcripts are the predominant forms in retina, whereas the hCNG4.1 transcript is much less abundant (90). This might explain why the 70-kDa hCNG4.1 protein is not detected in the purified channel. Analysis of the intron–exon structure of the human CNG4 gene indicated that the reported hCNG4.2 sequence does not represent a naturally occurring mRNA but rather is derived from a truncated mRNA that is not translated into a mature protein (90).

The proteins bCNG4.1 and hCNG4.3 have an unusual bipartite structure consisting of a core region including the transmembrane segments and the CNBD that is homologous to other CNG channel subunits, and an extended N-terminal domain, that is characterized by a large number of glutamate residues. The N-terminal domain is also expressed as a distinct soluble protein (the glutamic acid rich protein or GARP), which has been purified and cloned from bovine (62,91,92) and human (91,93) retina. At present, the function of the GARP domain is not known. Alternative splicing of the GARP domain gives rise to several CNG4 isoforms, which are expressed in a tissue- and species-dependent manner (Fig. 2, Table 1). The CNG4 subunits cloned from bovine testis (bCNG4.2, bCNG4.3, and bCNG4.4) contain partial deletions within the GARP domain (44). Interestingly, a specific CNG4 isoform (rCNG4.3) is highly expressed in rat olfactory neurons (28). Coexpression of CNG4.3 with the olfactory α subunit, CNG2, resulted in a channel with a significantly increased affinity for cAMP, suggesting that heteromeric CNG4.3/CNG2 channels may also exist in olfactory neurons. Comparison of the primary structures of different CNG4 isoforms expressed in rat, bovine, and human tissues revealed a surprisingly unequal distribution of sequence homology. The core region of CNG4 including the transmembrane domain and the CNBD is highly conserved between the different species (sequence identity >90%). However, the corresponding extended N-terminal domain and the soluble GARP protein strongly diverge with homologies of only 50–60%. At present, there is no plausible explanation for such a large divergence among mammals for the "same" gene.

CHANNEL MODULATION

Recent studies have focused on the modulation of CNG channels by Ca^{2+}, which influences the activity of CNG channels via different pathways. The calcium ion efficiently permeates CNG channels (Fig. 4) but also inhibits these channels from both the extracellular and intracellular surface in a strongly voltage-dependent manner (57,58,94). As stated earlier, the extracellular block is conferred by high affinity binding of Ca^{2+} to the conserved glutamate residue (E363 in CNG1; Fig. 1B) within the P region of the CNG channel (59,60). The block of the channel from the cytoplasmic site requires 50- to 100-fold higher concentrations of Ca^{2+} or Mg^{2+} than the extracellular block (58,94,95). The amino residues that form this low affinity site have not yet been determined. The voltage-dependent block by divalent cations leads to a prominent reduction of the apparent single channel conductance from 20 to 60 pS in the absence of divalents to less than 0.1 pS in their presence (for reviews see Refs. 2,13). At least for CNG channels expressed in sensory neurons, the reduction of the single channel conductance by Ca^{2+} or Mg^{2+} is of crucial physiological significance because it improves the signal-to-noise ratio of sensory transduction (2,13). This is because the current generated through the gating of many low conductance channels will be less noisy than the current induced by a few high conductance channels.

In addition to its direct interaction with CNG channels, Ca^{2+} causes a decrease in the apparent cyclic nucleotide affinity of both photoreceptor and olfactory channels by activation of calmodulin (96–98). This inhibitory action of Ca^{2+}/calmodulin (CaM) is a key component of olfactory adaptation (97,99) and may also contribute to light adaption (96), although to a lesser degree. The modulation by CaM does not involve the activation of a kinase but is mediated by the direct binding of the CaM to the channel. The CaM modulation of the native rod photoreceptor channel is conferred by the β subunit (62,100–102). Two distinct CaM-binding sides have been identified in the primary sequence of the CNG4 subunit (101,102). These sides are localized in the cytoplasmic N-terminus preceding the first transmembrane segment and in the C-terminus downstream of the CNBD, respectively. Interestingly, coexpression of CNG1/CNG4 heteromers revealed that deletion of the N-terminal CaM-binding side abolishes current inhibition by CaM, whereas deletion of the C-terminal side has no major effect on current modulation by CaM. In contrast to the rod channel, CaM controls the activity of the olfactory channel (103) by binding to a single site situated on the CNG2 α subunit. The inhibition of the olfactory channel by CaM is much more pronounced (up to 50-fold shift of *K*a for cGMP) than that found with rod and cone photoreceptor channels (two- to three-fold shift of *K*a). The CaM-binding site of the CNG2 subunit has been mapped to its N-terminus (103). It is formed by a 26 amino acid stretch that, unlike both CaM-binding sites identified in the CNG4 subunit, corresponds to the consensus sequence of CaM-binding sites found in other CaM-binding proteins (104). Since the binding site for CaM constitutes an autoexcitatory channel domain that binds to the C-terminus of the CNG channel (82), CaM may inhibit channel activity by targeting this interdomain interaction.

Interestingly, inhibition of channel activity by direct binding of CaM has also been described for the NR1 subunit of the NMDA receptor (105). In this case CaM reduces the open probability of the channel by binding to two separate sites at the C-terminus. Direct CaM binding might thus be a general mechanism for Ca^{2+}-dependent regulation of ion channel properties: Ca^{2+} might modulate CNG channels by interaction with other factors that are different from calmodulin. There is evidence for an endogenous Ca^{2+}-binding protein that induces calmodulin-like inhibition of the frog rod photoreceptor channel. This protein has apparently different Ca^{2+} dependence and/or higher affinity for the channel than calmodulin (98).

The activity of CNG channels is also regulated by protein kinases and phosphatases. The apparent cGMP affinities of both native and expressed rod photoreceptor channels are modulated by tyrosin phosphorylation, with cGMP sensitivity being low in the phosphorylated state and high in the dephosphorylated state (106). Additionally, the native rod channel is regulated by Ser/Thr phosphorylation (107). It has not yet been determined, whether modulation of rod channels involves direct phosphorylation of the channel α or β subunits or is indirect, with phosphorylation occurring on unidentified proteins that may be closely associated with the channel. Protein kinases may also control activity of the olfactory channel. Activation of protein kinase C increases the ligand affinity of the heterologously expressed CNG2 subunit (108). The effect is mediated by a serine located in the N-terminal region near the CaM-binding site of the channel. The close proximity of both regulatory sites in the N-terminus gives further evidence that this channel domain plays a critical role in the gating of CNG channels.

CNG CHANNELS IN NONSENSORY TISSUES

Originally, CNG channels were considered to be restricted to sensory transduction. However, it has become clear that these channels have a fairly widespread tissue distribution. A cGMP-activated channel that may be involved in the regulation of melatonin secretion has been functionally characterized in chick (109) and rat (110) pinealocytes. PCR analysis and cloning revealed that chick pineal gland expresses both rod photoreceptor (CNG1) and cone photoreceptor (CNG3) α subunits (111). Similarily, a rod photoreceptor CNG channel consisting of the CNG1 α subunit and the CNG4 β subunit has been cloned from rat pineal (112). Photoreceptor-type channels have also been detected by a combination of immunocytochemistry, PCR amplification, and excised-patch recordings in retinal On-bipolar cells (113,114), retinal ganglion cells (115), and sympathetic neurons (116).

CNG channels may be functionally important for development and normal function of the mammalian brain (117). Both olfactory and rod photoreceptor channels have been detected in CA1–3 hippocampal neurons (118,119). It has been speculated that CNG channels regulate synaptic efficacy in these neurons by modulating Ca^{2+} levels in response to changes in cyclic nucleotide concentrations (117).

CNG channels are not restricted to the central nervous system but are also found in various peripheral tissues including heart, colon, and aorta (120,121). The amount of CNG channel specific mRNA is quite low in most of these tissues. In contrast, a CNG channel is abundantly expressed in testis. Two channel subunits have been identified, the CNG3 α subunit (24,54) and the CNG4 β subunit (44). Western blotting, immunohistochemistry, and patch–clamp recordings have shown that the channel protein is specifically localized in the sperm cells (24).

Photoreceptor-type CNG channels have also been cloned from bovine (CNG3, ref. 54) and mouse kidney (CNG1; refs. 122,123). Interestingly, the modulatory CNG4 subunit seems to be expressed in a species-dependent fashion. CNG4 was detected in rat kidney, whereas no evidence for its presence was found in bovine kidney, suggesting that the bovine renal CNG channel might either be a homomer or contain other not yet identified channel subunits (44).

The functional role of CNG channels in the kidney is unknown. It has been speculated that the CNG channel protein constitutes a subunit of the cGMP-inhibited, amiloride-sensitive cation channel of inner medullary collecting duct (124,125). The latter channel belongs to the heterogeneous group of cyclic nucleotide modulated cation channels that do not obligatory require cyclic nucleotides to open but are modulated by them (for review see Ref. 1). First members of this class of channels, the hyperpolarization-activated cation (HAC) channels, have been cloned recently from mammals (126,127) and sea urchin (128). HAC channels comprise a multigene family containing at least four different members (HAC1–4). HAC channels reveal a very complex gating behavior. They become activated by membrane hyperpolarization; however, their opening kinetics and current amplitude are enhanced by direct binding of cyclic nucleotides. Thus, HAC channels constitute a functional link between voltage- and ligand-gated cation channels. HAC channels are likely to fullfil a variety of functions in different cell types, the most important one being the control of pacemaker activity in spontaneously active neurons and heart cells (for detailed reviews see Refs. 129 and 130).

The functional role of CNG channels is not clear yet for most nonsensory cells. The high permeability of these channels for Ca^{2+} suggests that they work as cyclic nucleotide operated calcium channels that transduce a hormonally induced rise in the intracellular cGMP/cAMP level into an influx of Ca^{2+}. The signaling cascade that triggers sperm chemotaxis may provide a system that exemplifies the role of CNG channels in a nonsensory cell. In sea urchin, the regulation of sperm motility involves the activation of the membrane-bound guanylyl cyclase by peptides secreted from sea urchin eggs. This then leads to an increase in the cGMP concentration and an internal rise of Ca^{2+} (131). Although the ability of mammalian sperm to detect and respond to chemical cues has not been demonstrated unequivocally, the presence of odorant receptors (132), rhodopsin kinases (133), arrestin (133), and CNG channel subunits (24,44) suggests that a signal transduction pathway similar to those of photoreceptors and olfactory cells may also exist in mammalian sperm. Since CNG channels are highly permeable for Ca^{2+}, they could serve as the terminal component that links the rises in cGMP and Ca^{2+}.

CONCLUSION

Since the first description of a CNG channel in rod outer segment in the mid-1980s, our knowledge about this class of ion channels has increased tremendeously. Whereas it was initially assumed that CNG channels are specifically expressed in visual and olfactory neurons, it is now generally accepted that these channels are distributed throughout the body and may have important roles in many cell types. Molecular cloning revealed that CNG channels comprise a multigene family of at least five members. In addition, there exists a whole variety of cyclic nucleotide modulated channels whose primary structure is not known in most cases. Most importantly, it has become clear that the major function of CNG channels is to provide a second messenger–triggered influx pathway for Ca^{2+}. Depending on the cell type, a rise of intracellular Ca^{2+} could trigger such diverse functions as cell motility, secretion, and neural plasticity.

ACKNOWLEDGMENTS

The research conducted in the author's laboratory was supported by grants from Fonds der Chemischen Industrie, BMBF, and Deutsche Forschungsgemeinschaft.

REFERENCES

1. Finn JT, Grunwald ME, Yau KW: Cyclic nucleotide-gated ion channels: an extanded family with diverse functions. *Annu Rev Physiol* 1996;58:395–426.
2. Zufall F, Firestein S, Shepherd GM. Cyclic nucleotide-gated ion channels and sensory transduction in olfactory receptor neurons. *Annu Rev Biophys Biomol Struct* 1994;23:577–607.
3. Hofmann F, Ludwig A, Pfeifer A: Cyclic GMP and the control of airways smooth muscle tone. In: Raeburn D, Giembycz MA, eds. *Airways smooth muscle: Biochemical control of contraction and relaxation.* Basel: Birkhäuser Verlag, 1994;253–269.
4. Vaandrager AB, De Jonge HR. Effect of cyclic GMP on intestinal transport. *Adv Pharmacol* 1994; 26:252–282.
5. Arancio O, Kandel ER, Hawkins RD: Activity-dependent long-term enhancement of transmitter release by presynaptic 3',5'-cyclic GMP in cultured hippocampal neurons, *Nature* 1995;376:74–80.
6. Hofmann F, Dostmann W, Keilbach A, Landgraf W, Ruth P: Structure and physiological role of cGMP-dependent protein kinase. *Biochim Biophys Acta* 1992;1135:51–60.
7. Beavo JA: Cyclic nucleotide phosphodiesterases: functional implications of multiple isoforms. *Physiol Rev* 1995;75:725–748.
8. Fesenko EE, Kolesnikov SS, Luyubarsky AL: Induction by cyclic GMP of cationic conductance in plasma membrane of retinal rod outer segments. *Nature* 1985;313:310–313.
9. Yau KW, Nakatani K: Light-suppressible, cyclic GMP-sensitive conductance in the plasma membrane of a truncated rod outer segment. *Nature* 1985;317:252–255.
10. Cobbs WH, Barkdoll AE III, Pugh EN Jr: Cyclic GMP increases photocurrent and light sensitivity of retinal cones. *Nature* 1985;317:64–66.
11. Haynes LW, Yau KW: Cyclic GMP-sensitive conductance in outer segment membrane of the catfish cones. *Nature* 1985;317:61–64.
12. Nakamura T, Gold GH: A cyclic nucleotide-gated conductance in olfactory receptor cilia. *Nature* 1987;325:442–444.
13. Yau KW, Baylor DA: Cyclic GMP activated conductance of retinal photoreceptor cells. *Annu Rev Neurosci* 1989;12:289–327.
14. Baylor D: How photons start vision. *Proc Natl Acad Sci U S A* 1996;93:560–565.
15. Breer H, Raming K, Krieger J: Signal recognition and transduction in olfactory neurons. *Biochim Biophys Acta* 1994;1224:277–287.
16. Haynes LW, Yau KW: Single-channel measurement from the cyclic GMP-activated conductance of catfish retinal cones. *J Physiol (Lond)* 1990;429:454–481.
17. Finn JT, Solessio EC, Yau KW: A cGMP-gated cation channel in depolarizing photoreceptors of the lizard parietal eye. *Nature* 1997;385:815–819.
18. Hatt H, Ache BW: Cyclic nucleotide- and inositol phosphate-gated ion channels in lobster olfactory neurons. *Proc Natl Acad Sci U S A* 1994;91:6264–6268.
19. Brunet LJ, Gold GH, Ngai J: General anosmia caused by a targeted disruption of the mouse olfactory cyclic nucleotide-gated cation channel. *Neuron* 1996;17:681–693.

20. Cook NJ, Hanke W, Kaupp UB: Identification, purification, and functional reconstitution of the cyclic GMP-dependent channels from rod photoreceptors. *Proc Natl Acad Sci U S A* 1987;84: 585–589.
21. Kaupp UB, Niidome T, Tanabe T et al: Primary structure and functional expression from complementary DNA of the rod photoreceptor cyclic GMP-gated channel. *Nature* 1989;342:762–766.
22. Molday RS, Molday LL, Dose A et al: The cGMP-gated channel of the rod photoreceptor cell: characterization and orientation of the amino terminus. *J Biol Chem* 1991;266:21917–21922.
23. Bönigk W, Altenhofen W, Müller F et al: Rod and cone photoreceptor cells express distinct genes for cGMP-gated channels. *Neuron* 1993;10:865–877.
24. Weyand I, Godde M, Frings S et al: Cloning and functional expression of a cyclic nucleotide-gated channel from mammalian sperm. *Nature* 1994;368:859–863.
25. Misaka T, Kusakabe Y, Emori Y, Gonoi T, Arai S, Abe K: Taste buds have a cyclic nucleotide-activated channel, CNGgust. *J Biol Chem* 1997;272:22623–22629.
26. Broillet MC, Firestein S: β Subunits of the olfactory cyclic nucleotide-gated channel form a nitric oxide activated Ca^{2+} channel. *Neuron* 1997;18:951–958.
27. Berghard A, Buck LB, Liman ER: Evidence for distinct signaling mechanism in two mammalian olfactory sense organs. *Proc Natl Acad Sci U S A* 1996;93:2365–2369.
28. Sautter A, Zong XG, Hofmann F, Biel M: An isoform of the rod photoreceptor cyclic nucleotide-gated channel β subunit expressed in olfactory neurons. *Proc Natl Acad Sci U S A* 1998;95:4696–4701.
29. Baumann A, Frings S, Godde M, Seifert R, Kaupp UB: Primary structure and functional expression of a *Drosophila* cyclic nucleotide-gated channel present in eyes and antennae. *EMBO J* 1994;13: 5040–5050.
30. Coburn CM, Bargmann CI: A putative cyclic nucleotide-gated channel is required for sensory development and function in *C. elegans*. *Neuron* 1996;17:695–706.
31. Komatsu H, Mori I, Rhee JS, Akaike N, Ohshima Y: Mutations in a cyclic nucleotide-gated channel lead to abnormal thermosensation and chemosensation in *C. elegans*. *Neuron* 1996;17:707–718.
32. Barnard EA: Receptor classes and the transmitter-gated ion channels. *Trends Biochem Sci* 1992;17: 2623–2635.
33. Guy HR, Durell SR, Warmke J, Drysdale R, Ganetzky B: Similarities in amino acid sequences of *Drosophila eag* and cyclic nucleotide-gated channels. *Science* 1991;254:730.
34. Henn DK, Baumann A, Kaupp UB: Probing the transmembrane topology of cyclic nucleotide-gated ion channels with a gene fusion approach. *Proc Natl Acad Sci U S A* 1995;92:7425–7429.
35. Goldstein SAN: A structural vignette common to voltage sensors and conduction pores: canaliculi. *Neuron* 1996;16:717–722.
36. Zagotta WN, Siegelbaum SA: Structure and function of cyclic nucleotide-gated channels. *Annu Rev Neurosci* 1996;19:235–263.
37. Tang CY, Papazian DM: Transfer of voltage independence from a rat olfactory channel to the *Drosophila* ether-à-go-go K^+-channel. *J Gen Physiol* 1997;109:301–311.
38. Jan LY, Jan YN: A superfamily of ion channels. *Nature* 1990;345:672.
39. Jan LY, Jan YN: Structural elements involved in specific K^+ channel functions. *Annu Rev Physiol* 1992;54:537–555.
40. Gordon SE, Zagotta WN: Subunit interactions in coordination of Ni^{2+} in cyclic nucleotide-gated channels. *Proc Natl Acad Sci U S A* 1995;92:10222–10226.
41. Liu DT, Tibbs GR, Siegelbaum SA: Subunit stoichiometry of cyclic nucleotide-gated channels and effects of subunit order on channel function. *Neuron* 1996;16:983–990.
42. Torre V, Straforini M, Sesti F, Lamb TD: Different channel-gating properties of two classes of cyclic GMP-activated channel in vertebrate photoreceptors. *Proc R Soc Lond Ser B* 1992;250: 209–215.
43. Bradley J, Li J, Davidson N, Lester HA, Zinn K: Heteromeric olfactory cyclic nucleotide-gated channels: a new subunit that confers increased sensitivity to cAMP. *Proc Natl Acad Sci U S A* 1994; 91:8890–8894.
44. Biel M, Zong X, Ludwig A, Sautter A, Hofmann F: Molecular cloning and expression of a modulatory subunit of the cyclic nucleotide-gated cation channel. *J Biol Chem* 1996;271:6349–6355.
45. Doyle DA, Cabral JM, Pfuetzner RA et al: The structure of the potassium channel: molecular basis of K^+ conduction and selectivity. *Science* 1998;280:69–77.
46. Heginbotham L, Abramson T, MacKinnon R: A functional connection between the pores of distantly related ion channels as revealed by mutant K^+ channels. *Science* 1992;258:1152–1155.

47. Goulding EH, Tibbs GR, Liu D, Siegelbaum SA: Role of H5 domain in determining pore diameter and ion permeation through cyclic nucleotide-gated channels. *Nature* 1993;364:61–64.
48. Hoshi T, Zagotta WN, Aldrich RW: Biophysical and molecular mechanism of *Shaker* potassium channel inactivation. *Science* 1990;250:533–538.
49. Kramer RH, Goulding E, Siegelbaum SA: Potassium channel inactivation peptide blocks cyclic nucleotide–gated channels by binding to the conserved pore domain. *Neuron* 1994;12:655–662.
50. Sun ZP, Akabas MH, Goulding EH, Karlin A, Siegelbaum SA: Exposure of residues in the cyclic nucleotide-gated channel pore: P region structure and function in gating. *Neuron* 1996;16:141–149.
51. Menini A: Currents carried by monovalent cations through cyclic GMP-activated channels in excised patches from salamander rods. *J Physiol (Lond)* 1990;424:167–185.
52. Frings S, Lynch JW, Lindemann B: Properties of cyclic nucleotide-gated channels mediating olfactory transduction: activation, selectivity, and blockage. *J Gen Physiol* 1992;100:45–67.
53. Haynes LW: Permeation of internal and external monovalent cations through the catfish cone photoreceptor cGMP-gated channel. *J Gen Physiol* 1995;106:485–505.
54. Biel M, Zong X, Distler M et al: Another member of the cyclic nucleotide-gated channels family expressed in testis, kidney and heart. *Proc Natl Acad Sci U S A* 1994;91:3505–3509.
55. Frings S, Seifert R, Godde M, Kaupp UB: Profoundly different calcium permeation and blockage determine the specific function of distinct cyclic nucleotide-gated channels. *Neuron* 1995;15: 169–179.
56. Choe S, Robinson R: An ingenious filter: the structural basis for ion channel selectivity. *Neuron* 1998;20:821–823.
57. Zufall F, Firestein S: Divalent cations block the cyclic nucleotide-gated channel of olfactory receptor neurons. *J Neurophysiol* 1993;69:1758–1768.
58. Haynes LW: Permeation and block by internal and external cations of the catfish cone photoreceptor cGMP-gated channel. *J Gen Physiol* 1995;106:507–523.
59. Root MJ, MacKinnon R: Identification of an external divalent cation-binding site in the the pore of a cGMP-activated channel. *Neuron* 1993;11:459–466.
60. Eismann E, Müller F, Heinemann SH, Kaupp UB: A single negative charge within the pore region of a cGMP-gated channel controls rectification, Ca^{2+} blockage, and ion selectivity. *Proc Natl Acad Sci U S A* 1994;91:1109–1113.
61. Chen TY, Peng YW, Dhallan RS, Ahamed B, Reed RR, Yau KW: A new subunit of the cyclic nucleotide-gated cation channel in retinal rods. *Nature* 1993;362:764–767.
62. Körschen HG, Illing M, Seifert R et al: A 240 kDa protein represents the complete β subunit of the cyclic nucleotide-gated channel from rod photoreceptor. *Neuron* 1995;15:627–636.
63. Hess P, Tsien RW: Mechanism of ion permeation through calcium channels. *Nature* 1984;309: 453–456.
64. Yang J, Ellinor PT, Sather WA, Zhang JF, Tsien RW: Molecular determinants of Ca^{2+} selectivity and ion permeation of L-type Ca^{2+} channels. *Nature* 1993;366:158–161.
65. Sesti F, Eismann E, Kaupp UB, Nizzari M, Torre V: The multi-ion nature of the cGMP-gated channel from vertebrate rods. *J Physiol (Lond)* 1995;487:17–36.
66. Shabb JB, Corbin JD: Cyclic nucleotide-binding domains in proteins having diverse functions. *J Biol Chem* 1992;267:5723–5726.
67. McKay DB, Steitz TA: Structure of catabolite gene activator protein at 2.9 Å resolution suggests binding to left-handed B-DNA. *Nature* 1981;290:744–749.
68. Kumar VD, Weber IT: Molecular model of the cyclic GMP-binding domain of the cyclic GMP-gated ion channel. *Biochemistry* 1992;31:4643–4649.
69. Altenhofen W, Ludwig J, Eismann E, Kraus W, Bönigk W, Kaupp, UB: Control of ligand specificity in cyclic nucleotide-gated channels from rod photoreceptors and olfactory epithelium. *Proc Natl Acad Sci U S A* 1991;88:9868–9872.
70. Shabb JB, Ng L, Corbin JD: One amino acid change produces a high affinity cGMP-binding site in cAMP-dependent protein kinase. *J Biol Chem* 1990;265:16031–16034.
71. Varnum MD, Black KD, Zagotta WN: Molecular mechanism for ligand discrimination of cyclic nucleotide-gated channels. *Neuron* 1995;15:619–625.
72. Goulding EH, Ngai J, Kramer RH, Colicos S, Axel R, Siegelbaum SA, Chess A: Molecular cloning and single-channel properties of the cyclic nucleotide-gated channel from catfish olfactory neurons. *Neuron* 1992;8:45–58.
73. Goulding EH, Tibbs GR, Siegelbaum SA: Molecular mechanism of cyclic nucleotide-gated channel activation. *Nature* 1994;372:369–374.

74. Gordon SE, Zagotta WN: Localization of regions affecting an allosteric transition in cyclic nucleotide-activated channels. *Neuron* 1995;14:857–864.
75. Zong X, Zucker H, Hofmann F, Biel M: Three amino acids in the C-linker are major determinants of gating in cyclic nucleotide-gated channels. *EMBO J* 1998;17:353–362.
76. Idlefonse M, Bennett N: Single-channel study of the cGMP-dependent conductance of retinal rods from incorporation of native vesicles into planar lipid bilayers. *J Membr Biol* 1991;123: 133–147.
77. Karpen JW, Brown RL, Stryer L, Baylor DA: Interaction between divalent cations and the gating machinery of cyclic GMP-activated channels in salamander rods. *J Gen Physiol* 1993;101:1–25.
78. Gordon SE, Zagotta WN: A histidine residue associated with the gate of the cyclic nucleotide-activated channels in rod photoreceptors. *Neuron* 1995;14:177–183.
79. Gordon SE, Oakley JC, Varnum MD, Zagotta WN: Altered ligand specificity by protonation in the ligand binding domain of cyclic nucleotide-gated channels. *Biochemistry* 1996;35:3994–4001.
80. Broillet MC, Firestein S: Direct activation of the olfactory cyclic nucleotide-gated channel through modification of sulfhydryl groups by NO compounds. *Neuron* 1996;16:377–385.
81. Gordon SE, Varnum MD, Zagotta WN: Direct interaction between amino- and carboxyl-terminal domains of cyclic nucleotide-gated channels. *Neuron* 1997;19:431–441.
82. Varnum MD, Zagotta WN: Interdomain interactions underlying activation of cyclic nucleotide-gated channels. *Science* 1997;278:110–113.
83. Tibbs GR, Goulding EH, Siegelbaum SA: Allosteric activation and tuning of ligand efficacy in cyclic nucleotide-gated channels. *Nature* 1997;386:612–615.
84. Ruiz ML, Karpen JW: Single cyclic nucleotide-gated channels locked in different ligand-bound states. *Nature* 1997;389:389–392.
85. Liman ER, Buck LB: A second subunit of the olfactory cyclic nucleotide-gated channel confers high sensitivity to cAMP. *Neuron* 1994;13:611–621.
86. Firestein S: Electric signals in olfactory transduction. *Curr Opin Neurobiol* 1992;2:444–448.
87. Brown RL, Gramling R, Bert RJ, Karpen JW: Cyclic GMP contact points within the 63-kDa subunit and a 240-kDa associated protein of retinal rod cGMP-activated channels. *Biochemistry* 1995;34: 8365–8370.
88. Molday LL, Cook NJ, Kaupp UB, Molday RS: The cGMP-gated cation channel of bovine rod photoreceptor cells is associated with a 240 kDa protein exhibiting immunochemical cross-reactivity with spectrin. *J Biol Chem* 1990;265:18690–18695.
89. Hsu YT, Molday RS: Interaction of calmodulin with the cyclic GMP-gated channel of rod photoreceptor cells. *J Biol Chem* 1994;269:29765–29770.
90. Ardell MD, Aragon I, Oliveira L, Porche GE, Burke E, Pittler SJ: The β subunit of human rod photoreceptor cGMP-gated cation channel is generated from a complex transcription unit. *FEBS Lett* 1996; 389:213–218.
91. Colville CA, Molday RS: Primary structure and expression of the human β-subunit and related proteins of the rod photoreceptor cGMP-gated channel. *J Biol Chem* 1996;271:32968–32974.
92. Sugimoto Y, Yatsunami K, Tsujimoto M, Khorana HG, Ichikawa A: The amino acid sequence of a glutamic acid-rich protein from bovine retina as deduced from the cDNA sequence. *Proc Natl Acad Sci U S A* 1991;88:3116–3119.
93. Ardell MD, Makhija AK, Olivera L, Miniou P, Viegas-Péquignot E, Pittler SJ: cDNA, gene structure, and chromosomal localisation of human GAR1 (CNCG3L), a homolog of the third subunit of bovine photoreceptor cGMP-gated channel. *Genomics* 1995;28:32–38.
94. Zimmerman AL, Baylor DA: Cation interactions within the cyclic GMP-activated channel of retinal rods from the tiger salamander. *J Physiol (Lond)* 1992;449:759–783.
95. Colamartino G, Menini A, Torre V: Blockage and permeation of divalent cations through the cyclic GMP-activated channel from tiger salamander retinal rods. *J Physiol (Lond)* 1991;440:189–206.
96. Hsu YT, Molday RS: Modulation of the cGMP-gated channel of rod photoreceptor cells by calmodulin. *Nature* 1993;361:76–79.
97. Chen TY, Yau KW: Direct modulation by Ca^{2+}-calmodulin of cyclic nucleotide-activated channel of rat olfactory receptor neurons. *Nature* 1994;368:545–548.
98. Gordon SE, Downing-Park J, Zimmermann AL: Modulation of the cGMP-gated ion channel in frog rods by calmodulin and an endogenous inhibitory factor. *J Physiol (Lond)* 1995;486:533–546.
99. Kurahashi T, Menini A: Mechanism of odorant adaptation in the olfactory receptor cell. *Nature* 1997;385:725–729.

100. Chen TY, Illing M, Hsu YT, Yau KW, Molday RS: Subunit 2 (or β) of retinal rod cGMP-gated cation channel is a component of the 240-kDa channel-associated protein and mediates Ca^{2+}-calmodulin modulation. *Proc Natl Acad Sci U S A* 1994;91:11757–11761.
101. Grunwald ME, Yu WP, Yu HH, Yau KW: Identification of a domain on the β-subunit of the rod cGMP-gated cation channel that mediates inhibition by calcium–calmodulin. *J Biol Chem* 1998; 273:9148–9157.
102. Weitz D, Zoche M, Müller F et al: Calmodulin controls the rod photoreceptor CNG channel through an unconventional binding site in the N-terminus of the β-subunit. *EMBO J* 1998;17:2273–2284.
103. Liu M, Chen TY, Ahamed B, Li J, Yau KW: Calcium–calmodulin modulation of the olfactory cyclic nucleotide-gated cation channel. *Science* 1994;266:1348–1354.
104. Ikura M, Clore GM, Gronenborn AM, Zhu G, Klee CB, Bax A: Solution structure of a calmodulin-target peptide complex by multidimensional NMR. *Science* 1992;256:632–638.
105. Ehlers MD, Zhang S, Bernhardt JP, Huganir RL: Inactivation of NMDA receptors by direct interaction of calmodulin with the NR1 subunit. *Cell* 1996;84:745–755.
106. Molokanova E, Trivedi B, Savchenko A, Kramer RH: Modulation of rod photoreceptor cyclic nucleotide-gated channels by tyrosine phosphorylation. *J Neurosci* 1997;17:9068–9076.
107. Gordon SE, Brautigan DL, Zimmerman AL: Protein phosphatases modulate the apparent agonist affinity of the light-regulated ion channel in retinal rods. *Neuron* 1992;9:739–748.
108. Müller F, Bönigk W, Sesti F, Frings S: Phosphorylation of mammalian olfactory cyclic nucleotide-gated channels increases ligand sensitivity. *J Neurosci* 1998;18:164–173.
109. Dryer S, Henderson D: A cyclic GMP-activated channel in dissociated cells of chick pineal gland. *Nature* 1991;353:756–758.
110. Schaad NC, Vanecek J, Rodriguez IR, Klein DC, Holtzclaw L, Russel JT: Vasoactive intestinal peptide elevates pinealocyte intracellular calcium concentrations by enhancing influx: evidence for involvement of a cyclic GMP-dependent mechanism. *Mol Pharmacol* 1995;47:923–933.
111. Bönigk W, Müller F, Middendorff R, Weyand I, Kaupp UB: Two alternatively spliced forms of the cGMP-gated channel α-subunit from cone photoreceptor are expressed in the chick pineal organ. *J Neurosc* 1996;16:7458–7468.
112. Sautter A, Biel M, Hofmann F: Molecular cloning of cyclic nucleotide-gated cation channel subunits from rat pineal gland. *Mol Brain Res* 1997;48:171–175.
113. Shiells RA, Falk G: Glutamate receptors of rod bipolar cells are linked to a cyclic GMP cascade via a G-protein. *Proc R Soc Lond Ser B* 1990;242:91–94.
114. Nawy S, Jahr CE: cGMP-gated conductance in retinal bipolar cells is suppressed by the photoreceptor transmitter. *Neuron* 1991;7:677–683.
115. Ahmad I, Leinders-Zufall T, Kocsis JD, Shepherd GM, Zufall F, Barnstable CJ: Retinal ganglion cells express a cGMP-gated cation conductance activatable by nitric oxide donors. *Neuron* 1994;12: 155–166.
116. Thompson SH: Cyclic GMP-gated channels in a sympathetic neuron cell line. *J Gen Physiol* 1997; 110:155–164.
117. Zufall F, Shepherd GM, Barnstable CJ: Cyclic nucleotide-gated channels as regulator of CNS development and plasticity. *Curr Opin Neurobiol* 1997;7:404–412.
118. Bradley J, Zhang Y, Bakin R, Lester HA, Ronnett GV, Zinn K: Functional expression of the heteromeric "olfactory" cyclic nucleotide-gated channel in hippocampus: a potential effector of synaptic plasticity in brain neurons. *J Neurosc* 1997;17:1993–2005.
119. Kingston PA, Zufall F, Barnstable CJ: Rat hippocampal neurons express genes for both rod retinal and olfactory cyclic nucleotide-gated channels: novel targets for cAMP/cGMP function. *Proc Natl Acad Sci U S A* 1996;93:10440–10445.
120. Distler M, Biel M, Flockerzi V, Hofmann F: Expression of cyclic nucleotide-gated cation channels in non-sensory tissues and cells. *Neuropharmacology* 1994;33:1275–1282.
121. Ding C, Potter ED, Qiu W, Coon SL, Levine MA: Guggino SE. Cloning and widespread distribution of the rat rod-type cyclic nucleotide-gated cation channel. *Am J Physiol* 1997;272:C1335–C1344.
122. Ahmad I, Korbmacher C, Segal AS, Cheung P, Boulpaep EL, Barnstable CJ: Mouse cortical collecting duct cells show nonselective cation channel activity and express a gene related to the cGMP-gated rod photoreceptor channel. *Proc Natl Acad Sci U S A* 1992;89:10262–10266.
123. Karlson KH, Ciampolillo-Bates F, McCoy DE, Kizer NL, Stanton BA: Cloning of a cGMP-gated cation channel from mouse kidney inner medullary collecting duct. *Biochim Biophys Acta* 1995; 1236:197–200.

124. Light DB, Corbin JD, Stanton BA: Dual ion-channel regulation by cyclic GMP and cyclic GMP-dependent protein kinase. *Nature* 1990;344:336–339.
125. McCoy DE, Guggino SE, Stanton BA: The renal cGMP-gated cation channel: its molecular structure and physiological role. *Kidney Int* 1995;48:1125–1133.
126. Ludwig A, Zong X, Jeglitsch M, Hofmann F, Biel M: A family of hyperpolarization-activated mammalian cation channels. *Nature* 1998;393:587–591.
127. Santoro B, Liu DT, Yao H et al: Identification of a gene encoding a hyperpolarization-activated pacemaker channel of brain. *Cell* 1998;93:717–729.
128. Gauss R, Seifert R, Kaupp UB: Molecular identification of a hyperpolarization-activated channel in sea urchin sperm. *Nature* 1998;393:583–687.
129. DiFrancesco D: Pacemaker mechanisms in cardiac tissue. *Annu Rev Physiol* 1993;55:455–472.
130. Pape HC: Queer current and pacemaker: the hyperpolarization-activated cation current in neurons. *Annu Rev Physiol* 1996;58:299–327.
131. Garbers DL: Molecular basis of fertilization. *Annu Rev Biochem* 1989;58:719–742.
132. Parmentier M, Libert F, Schurmans S et al: Expression of members of the putative olfactory receptor gene family in mammalian germ cells. *Nature* 1992;355:453–455.
133. Dawson TM, Arriza JL, Jaworsky DE et al: beta-Adrenergic receptor kinase-2 and beta-arrestin-2 as mediators of odorant-induced desensitization. *Science* 1993;259:825–829.
134. Pittler SJ, Lee AK, Altherr MR et al: Primary structure and chromosomal localization of human and mouse rod photoreceptor cGMP-gated cation channel. *J Biol Chem* 1992;267:6257–6262.
135. Barnstable CJ, Wei JY: Isolation and characterization of the α-subunit of the rat rod photoreceptor cGMP-gated cation channel. *J Mol Neurosci.* 1996;6:289–302.
136. Dhallan RS, Macke JP, Eddy RL et al: Human rod photoreceptor cGMP-gated channel: amino acid sequence, gene structure, and functional expression. *J Neurosci* 1992;12:3248–3256.
137. Zhang Q, Pearce-Kelling S, Acland GM, Aguirre GD, Ray K: Canine rod photoreceptor cGMP-gated channel protein α-subunit: studies on the expression of the gene and characterization of the cDNA. *Exp Eye Res* 1997;65:301–309.
138. Ratcliffe CF, Conley EC, Brammar WJ: Distribution of cGMP-gated cation channel expression in bovine and porcine cardiovascular tissue. *Biochem Soc Trans* 1995;23:441S
139. Dhallan RS, Yau KW, Schrader KA, Reed RR: Primary structure and functional expression of a cyclic nucleotide-activated channel from olfactory neurons. *Nature* 1990;347:184–187.
140. Ludwig J, Margalit T, Eismann E, Lancet D, Kaupp UB: Primary structure of cAMP-gated channel from bovine olfactory epithelium. *FEBS Lett* 1990;270:24–29.
141. Ruiz ML, London B, Nadal-Ginard B: Cloning and characterization of an olfactory cyclic nucleotide-gated channel expressed in mouse heart. *J Mol Cell Cardiol* 1996;28:1453–1461.
142. Biel M, Altenhofen W, Hullin R et al: Primary structure and functional expression of a cyclic nucleotide-gated channel from rabbit aorta. *FEBS Lett* 1993;329:134–138.
143. Yu WP, Grunwald ME, Yau KW: Molecular cloning, functional expression and chromosomal localization of a human homolog of the cyclic nucleotide-gated ion channel of retinal cone photoreceptors. *FEBS Lett* 1996;393:211–215.
144. Wissinger B, Müller F, Weyand I et al: Cloning, chromosomal localization and functional expression of the gene encoding the α-subunit of the cGMP-gated channel in human cone photoreceptors. *Eur J Neurosci* 1997;9:2512–2521.
145. Schwarz TL, Temple BI, Papazian DM, Jan YN, Jan LY: Multiple potassium channel components are produced by alternative splicing at the *Shaker* locus in *Drosophila. Nature* 1988;331:137–142.
146. Biel M, Ruth P, Bosse E et al: Primary structure and functional expression of a high voltage activated calcium channel from rabbit lung. *FEBS Lett* 1990;269:409–412.

Advances in Second Messenger and Phosphoprotein Research, Vol. 33

1040-7952/99 $30.00

11

Cyclic GMP and Ion Channel Regulation

Richard E. White

Department of Physiology and Biophysics, Wright State University School of Medicine, Dayton, Ohio 45435

INTRODUCTION AND HISTORICAL BACKGROUND

Thirty years of research into the cellular effects of cyclic GMP have finally overcome "the notion that this cyclic nucleotide is merely a poor substitute for cyclic AMP" (1). Although approximately 10,000 articles have explained much concerning the nature, regulation, and biological effects of cyclic GMP, we are still only beginning to understand the role played by this nucleotide in cellular physiology and pathophysiology. One reason for the unfortunate latency in appreciating the overall importance of cyclic GMP was that, in contrast to cyclic AMP, no extracellular "first messenger" signal could be demonstrated to stimulate cyclic GMP synthesis. The importance of cyclic AMP in regulating cellular activity was recognized in the seminal studies of Sutherland, Rall, and their associates, who identified this nucleotide as the critical second messenger mediating the metabolic effects of glycogenolytic hormones (2), but it was not until 6 years later that cyclic GMP was first isolated from a biological source (3). Although structural and biochemical similarities to cyclic AMP suggested that cyclic GMP might also be an important regulator of cellular function, the enzyme responsible for cyclic GMP synthesis, guanylyl cyclase, was discovered to exist primarily as a cytosolic enzyme (4). At the time these studies cast doubt on whether any extracellular hormone could stimulate cyclic GMP synthesis via a cyclic AMP–like plasma membrane transduction system. Therefore, cyclic GMP became one of a number of interesting, yet largely unappreciated cellular biochemicals. Shortly thereafter, however, findings from Goldberg's group (5) demonstrated that perfusing isolated hearts with acetylcholine elevated the levels of cyclic GMP in cardiac tissue, suggesting that an extracellular signal might indeed stimulate cyclic GMP production to antagonize the excitatory effects of β-adrenergic agonists and cyclic AMP in the heart. Subsequent studies strove to establish cyclic GMP as the physiological antagonist of cyclic AMP in other organ systems as well; however, a "yin–yang" model of reciprocal cyclic nucleotide actions could not be clearly established in all tissues. Once again, interest in cyclic GMP as an important signaling molecule began to wane.

Several important discoveries during the early to mid–1970s sparked a renewed interest in cyclic GMP research. Kuo and Greengard (6) discovered a protein kinase that was activated by cyclic GMP selectively; however, this cyclic GMP dependent protein kinase (PKG) was identified in extracts of lobster, cockroach, moth, and only later from a mammalian source (7,8). On the other hand, mammalian tissues, most of which contained the cyclic AMP dependent protein kinase (PKA), appeared to have a dearth of PKG activity, if any at all. Thus, cyclic GMP was again consigned to the background of research into signal transduction mechanisms in mammalian biology. Fortunately, not all research into the effects of cyclic GMP was dormant. Another clue into a possible cellular function for cyclic GMP was revealed when this nucleotide was found to regulate cyclic AMP hydrolysis, thus providing further evidence for a physiological antagonism between these cyclic nucleotides (9). In addition, retinal rod outer segments were found to contain high levels of cyclic GMP which declined in response to light exposure, suggesting that cyclic GMP might play a role in visual transduction (10).

Within the next several years important discoveries were made that proposed a link between cyclic GMP and relaxation of smooth muscle. Nitrovasodilator drugs, such as nitroglycerin, had been used clinically for nearly a century to treat angina and hypertensive crises, but it was not until the mid to late 1970s that these agents were found to enhance production of cyclic GMP in vascular and other smooth muscle (11–13). Subsequent studies from the laboratories of Murad and Ignarro revealed that nitric oxide (NO) released from nitrovasodilators was actually the active agent that stimulated cyclic GMP production (14,15); however, mammalian cells were thought to be incapable of synthesizing NO. Thus, the effects of NO on cyclic GMP production were considered more of a pharmacological curiosity than a relevant physiological mechanism, and potential effects of NO on cyclic GMP production were largely ignored—that is, until the early 1980s, when the prejudices of the past were forgotten upon the new discovery of an old endogenous cyclic GMP stimulator.

In 1980 Furchgott and Zawadzki (16) demonstrated that acetylcholine relaxed arterial strips when care was taken to avoid damaging the endothelial cell layer, but this vasodilatory response could be abolished by rubbing the intimal surface of the preparation to destroy the endothelium. These and other studies suggested that vascular endothelial cells release a diffusible substance that induces relaxation of the underlying vascular smooth muscle, and very shortly a number of other endogenous or exogenous agents were shown to produce an "endothelium-dependent" relaxation of blood vessels. This "endothelium-derived relaxing factor" (EDRF) was the first endogenous substance found to directly activate guanylyl cyclase, and thus to function as a physiological stimulator of cyclic GMP production. Because of the lability of EDRF, 7 years of intensive investigation passed before nitric oxide was finally identified as the elusive EDRF (17,18). Since that time, our knowledge of the biological roles of the NO–cyclic GMP signaling system have expanded beyond cardiovascular regulation to include a wide array of physiological and pathophysiological effects such as respiratory and gastrointestinal smooth muscle relaxation, penile erection, neuronal long-term potentiation and possibly cerebellar long-term depression, and septic shock (for recent comprehensive reviews, see Refs. 19–21). Finally, the role of cyclic GMP as an important regulator of mammalian cellular biology has been firmly established.

Another important discovery in the early 1980s also stimulated a resurgence of research into cyclic GMP. In 1956 Kisch (22) observed that mammalian atrial cells contained membrane-bound storage granules and a more highly developed Golgi complex and rough endoplasmic reticulum compared to their ventricular counterparts. During that same year Henry et al. (23) used a balloon to increase left atrial pressure in dog hearts and discovered that urine output was increased markedly. Twenty years later Marie et al. (24) reported that the density of these atrial granules could be altered depending on the intake of salt and water. These findings were consistent with the idea that atrial granules contained a substance that played a physiological role in salt and water balance, and this hypothesis was later proven by deBold et al. (25), who injected crude atrial extracts into rats and observed a 30-fold increase in sodium and chloride excretion, and a 10-fold increase in urine output. Subsequent purification of this substance revealed a 28- amino acid peptide with potent natriuretic and vasorelaxant properties, and this peptide was named atrial natriuretic factor (ANF or ANP; Refs. 26–28). Interestingly, ANP was found to elevate cyclic GMP levels in vascular smooth muscle by stimulating a membrane-bound guanylyl cyclase (29). This particulate enzyme was not activated by NO, as was its soluble counterpart. ANP could stimulate guanylyl cyclase activity in broken cell preparations, indicating that guanylyl cyclase could indeed be regulated by extracellular signals; so now cyclic GMP could be considered a genuine second messenger. More recent studies have established that cyclic GMP production by natriuretic peptides is a powerful negative feedback system to counteract hypervolemia. Besides their direct relaxant effect on the vasculature, these peptides provide a functional antagonism of the renin–angiotensin–aldosterone system to promote diuresis and natriuresis. In addition to protecting the body against fluid overload and hypertension, ANP functions as an important modulator of secretory activity (30).

In summary, for nearly 20 years cyclic GMP was an intracellular signal in search of an endogenous activator. With the discoveries of NO and ANP, the short-sightedness of the past was replaced by an avid interest in understanding this complex yet vital cellular communication process. The complexity of the system resides in the multifaceted means by which cells synthesize, process, and respond to cyclic GMP. Guanylyl cyclase exists as both a soluble and a membrane-associated enzyme, and each form has its own endogenous ligand(s). Once synthesized, cyclic GMP may interact with various effector molecules such as ion channels, phosphodiesterases, and PKG; and PKG adds another layer of complexity insofar as this kinase phosphorylates a number of different substrate proteins that induce a wide variety of cell-specific responses. Some excellent reviews have described the biochemistry and molecular biology of cyclic GMP metabolism and its overall importance as a signal transduction molecule (19,31,32), and I will not attempt an exhaustive overview of this important area. In contrast, there are few articles that focus primarily on how cyclic GMP regulates ion channel activity. This chapter endeavors to address this specific area of interest by discussing how cyclic GMP dependent mechanisms regulate the ionic basis of cellular excitability via direct and indirect modulation of ion channel gating mechanisms.

CYCLIC GMP: SYNTHESIS AND SUBSTRATES

Guanylyl Cyclases

Given the diverse effects of cyclic GMP on a number of different tissues, it is not surprising that this nucleotide has a rather ubiquitous distribution among various cell types, and that the enzymes that catalyze its formation (guanylyl cyclase) or metabolism (phosphodiesterase) have been identified in virtually all tissue sources examined (31). Guanylyl cyclases (GC) catalyze the formation of cyclic GMP from GTP, and there are two forms of this enzyme: soluble and particulate. Adenylyl cyclase, in contrast, exists only in a membrane-bound form. Although both isoforms of GC are expressed in practically all cells, the relative proportions of these enzymes can vary depending on cell type. For example, soluble GC predominates in vascular smooth muscle, whereas endothelial and intestinal epithelial cells express a greater proportion of the particulate enzyme. Purified soluble GC is a heterodimer consisting of a 70-kDa α subunit and a 82-kDa β subunit, and coexpression of both subunits is necessary to reconstitute significant GC activity (33). In addition, divalent ions are required for GC activity, with Mn^{2+} apparently being the preferred ion; however, Mg^{2+} is probably the important physiological species. In the presence of Mg^{2+}, NO stimulates GC activity by at least 40- to 50-fold (34), but this stimulation requires the presence of an enzyme-bound heme. Ignarro et al. (35) have demonstrated that NO and nitrovasodilators activate heme-reconstituted GC, but not the heme-deficient variety.

Particulate GC is a polypeptide transmembrane receptor containing an extracellular ligand binding domain, a single membrane-spanning region, and an intracellular cyclase domain that contains a significant amount of homology among all isoforms. At least five different isoforms of particulate GC have been cloned. GC-A (ANP-R1 or type I receptor) binds ANP and brain natriuretic peptide (BNP, 32 amino acids) and is located primarily in the cardiovascular system. GC-B is located in neural tissue, where it specifically binds the related C-natriuretic peptide (CNP, 22 amino acids), which may function as a neurotransmitter. GC-C appears to be limited to intestinal epithelial cells; its endogenous ligand is guanylin, which increases chloride secretion and decreases fluid reabsorption, and it is also activated by heat-stable enterotoxin ST_a (36). In addition, a retinal GC may also be a particulate isoform, with the Ca^{2+}-binding peptide recoverin as its ligand (37). Another ANP receptor (ANP-R2 or type II receptor) does not catalyze cyclic GMP synthesis but probably binds and removes ANP from the circulation via a receptor-mediated endocytosis mechanism; this protein has been termed the "clearance" or C receptor. Interestingly,

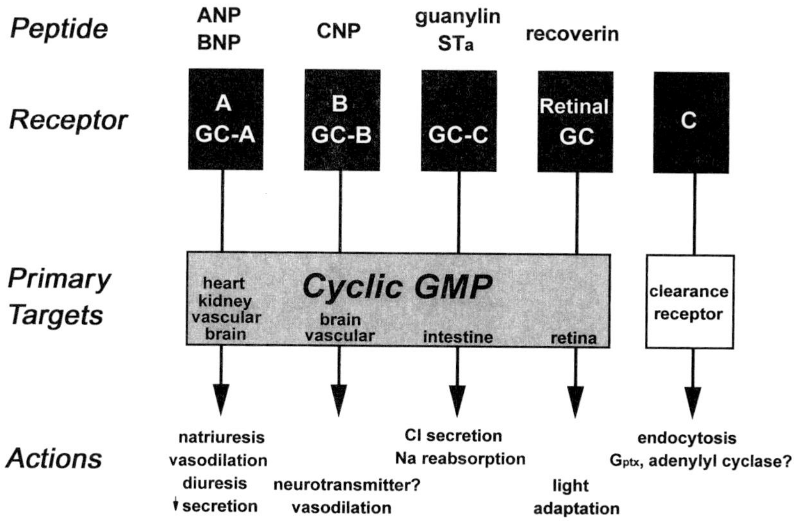

FIG. 1. Natriuretic peptides: a summary of their receptors, targets, and physiological effects.

activation of the clearance receptor depresses neurotransmitter release from neurosecretory (PC12) cells by a cyclic GMP independent, pertussis toxin sensitive mechanism, suggesting that ANP may attenuate sympathetic outflow by inhibiting adenylyl cyclase activity (38). This unfortunate confusion regarding terminology and classification of natriuretic peptides and their receptors may be clarified somewhat by the organization presented in Fig. 1. In light of this evidence, it is becoming increasingly apparent that the family of natriuretic peptides does far more than simply adjust salt and water balance.

Targets of Cyclic GMP

Underlying much of the diversity of cyclic GMP action is the interaction of cyclic GMP with at least three classes of cellular proteins: cyclic GMP regulated phosphodiesterase, cyclic GMP gated ion channels, and cyclic GMP dependent protein kinase. Furthermore, multiple effector molecules for cyclic GMP may also exist within the same cell. In contrast, the vast majority of physiological effects produced by cyclic AMP are mediated through activation of PKA, but PKA-induced phosphorylation is at most only half the story. For cyclic AMP dependent phosphorylation to be a true regulatory mechanism, a reciprocal mechanism must also exist to "undo" or dephosphorylate the substrates of PKA; otherwise, the effects of cyclic AMP would be irreversible. Nature appears to have assigned much of this important task to cyclic GMP, which we now know can antagonize the effects of cyclic AMP dependent phosphorylation by both direct and indirect mechanisms. The direct dephosphorylation mechanism (i.e., stimulation of phosphoprotein phosphatase activ-

TABLE 1. *Cyclic GMP as a regulator and substrate of phosphodiesterase activity*

PDE[a]	Classification	Substrate
Cyclic GMP stimulated PDE	Type II	Cyclic GMP (micromolar concentrations) stimulates hydrolysis of cyclic AMP (and cyclic GMP)
Cyclic GMP inhibited PDE	Type III	Cyclic GMP (submicromolar concentrations) inhibits cyclic AMP hydrolysis
Cyclic GMP specific PDE	Type V	Hydrolyze cyclic GMP
	Type VI	Hydrolyze cyclic GMP for phototransduction
Ca^{2+}/calmodulin PDE	Type I	Hydrolyze cyclic AMP and cyclic GMP
Cyclic AMP specific PDE	Type IV	Hydrolyze cyclic AMP
Unknown	Type VII	Unknown

[a]Phosphodiesterase.

ity) is discussed later in regard to modulation of ion channel activity. An important indirect mechanism was discovered first, and it involves cyclic GMP dependent regulation of cyclic AMP levels via modulation of phosphodiesterase activity, which catalyzes the hydrolysis of cyclic nucleotides to their respective inactive 5'-nucleotide monophosphates.

Phosphodiesterases

More than 30 different phosphodiesterase (PDE) isozymes, representing at least seven gene families, are currently recognized, and have been designated as types I–VII according to substrate specificity and effector molecules (31,39). Table 1 illustrates the important regulatory control exerted by cyclic GMP on PDE activity. Types I and V PDE are important for reducing cyclic GMP levels in most cell types, with type VI PDE hydrolyzing cyclic GMP in retinal photoreceptor cells. These enzymes mediate a very rapid (sometimes millisecond) hydrolysis of cyclic GMP, providing an efficient means of adjusting intracellular cyclic GMP levels for precise control. On the other hand, activity of PDE types II and III is regulated by cyclic GMP, and activity of these PDEs adjusts the magnitude of cyclic AMP dependent phosphorylation by increasing or lowering levels of this nucleotide. Type II PDE is stimulated by cyclic GMP to hydrolyze both cyclic AMP and cyclic GMP, but because the former is usually present in higher concentration, degredation of cyclic AMP is often the predominant effect. Still, cyclic GMP induced hydrolysis of cyclic GMP can be a negative feedback mechanism to prevent excessive accumulation of cyclic GMP. Thus, type II PDE is an important regulatory enzyme underlying the physiological antagonism of cyclic AMP by cyclic GMP, as discussed shortly when we consider how cyclic GMP modulates cardiac calcium currents. In contrast, type III PDE is inhibited by cyclic GMP, and thus serves to augment the effects of PKA-induced phosphorylation.

Cyclic GMP Dependent Protein Kinase

PKG is a serine/threonine protein kinase that is stimulated selectively, but not exclusively, by cyclic GMP, having a 20-fold higher affinity for this nucleotide than for cyclic AMP *in vitro;* but because cyclic AMP levels are normally much higher than those of cyclic GMP *in situ,* it is quite possible that PKG may be activated by cyclic AMP under normal conditions. This cross-activation has been reported in arterial smooth muscle (40,41) and could account for the somewhat puzzling finding that cyclic AMP and cyclic GMP may be "physiological agonists" in vascular smooth muscle, since both nucleotides mediate vasodilatory responses. There are at least two PKG gene products: type I PKG is a homodimer of 78-kDa subunits, displays alternate mRNA splicing to produce two isoforms, Iα and Iβ, and is expressed in a wide range of mammalian cells, especially vascular smooth muscle, platelets, and Purkinje cells; type II PKG possesses 86-kDa subunits and is expressed principally in intestinal epithelial cells and brain. Since PKG phosphorylates a number of substrates, elucidation of its precise mechanism of action is complex (42,43). The most well-known model for studying PKG action is probably smooth muscle, which is a rich source of this kinase. Felbel et al. (44) were the first to demonstrate that PKG could regulate intracellular concentrations of Ca^{2+}, $[Ca^{2+}]_i$, in smooth muscle when they introduced purified PKG into isolated tracheal smooth muscle cells and observed a decrease in cytosolic calcium. Subsequent studies in vascular smooth muscle proposed that PKG induces relaxation by inhibition of phospholipase C activity, stimulation of Ca^{2+}–ATPase activity, inhibition of receptors for inositol 1,4,5-trisphosphate, decreased sensitivity of contractile proteins, inhibition of Ca^{2+} channels, and/or stimulation of K^+ channel activity. Indeed, there are pleotropic effects of PKG (43). In light of the evidence for these mechanisms, two things are very clear concerning the inhibitory effects of PKG on cell excitability: (1) redundancy (i.e., multiple effector mechanisms for PKG probably exist in a single cell) and (2) heterogeneity. The relative contribution of each mechanism(s) to PKG-induced modulation will depend on the specific cell type involved. Of the various mechanisms proposed to mediate the effects of PKG, one that is receiving significant attention is the effect of cyclic GMP dependent phosphorylation on ion channel activity, and the remainder of this chapter focuses primarily on findings that are beginning to shed some light on this important mechanism of cyclic GMP transduction.

CYCLIC GMP AND ION CHANNEL GATING

In contrast to the first 20 years of cyclic GMP research, the next 10 years witnessed an increasing shift of emphasis from the biochemist's bench to the electrophysiologist's platform. Cyclic GMP is now known to produce many of its physiological effects by regulating ion channel activity either directly or indirectly. In fact, ion channels are important effector molecules for cyclic GMP action in each of the three known intracellular "receptors" for cyclic GMP (ion channels, phosphodiesterases,

TABLE 2. *Summary of ion channels modulated by cyclic GMP*

Cyclic GMP action	Type of channel[a]	Location	Reference
Cyclic GMP gated channels	(+) Cation	Rod outer segments	45
		Cone outer segments	52
		Olfactory epithelia	114
		Kidney and testis	115
	(−) Cation	Kidney epithelia	53, 54
	(+) K^+	Sperm cells	55
Metabolic regulation by cyclic GMP	(−) Ca^{2+}	Frog ventricle	58
		Chick heart	61
		Rat ventricle	60
		Guinea pig ventricle	59, 63
		Rabbit portal vein	116
		Rat pituitary tumor cells	64
		Human fetal ventricle	117
		Guinea pig hippocampus	62
	(+) Ca^{2+}	Human atria	67
		Snail neurons	68, 69
		Rat sympathetic neurons	119
	(+) Ca^{2+}-activated K^+	Rat pituitary tumor cells	64
		Dog coronary artery	94
		Rabbit basilar artery	95
		Rat pulmonary artery	96
		Pig coronary artery	97, 98
		Bovine trachea	65, 99
		Human mesangial cells	86
		Rat aorta	87
	(+) K	Rat collecting duct	118
		Human fetal ventricle	117
		Rabbit corneal epithelium	120
	(+) K_{ATP}	Rat aorta	87

[a](+) indicates stimulation of channel activity or increase in current, whereas (−) indicates decreased activity or current.

and PKG). This is not to imply that regulation of ion channel activity mediates all of cyclic GMP's diverse actions. Rather, what follows is meant to present some of the ever increasing evidence that biological transducers like ion channels are among the favorite targets of NO, cyclic GMP, PKG, and cyclic GMP stimulating drugs. A primary focus is on the mechanisms by which cyclic GMP controls cellular excitability via modulation of ion channel gating. A brief summary of cyclic GMP effects on ion channels is presented in Table 2.

Cyclic GMP Gated Ion Channels

Nearly all the responses to cyclic AMP, and probably the majority of those produced by cyclic GMP, are mediated via stimulation of protein kinase activity. It was quite a surprise then in 1985 when Fesenko and his colleagues demonstrated that cyclic GMP stimulated an ion channel directly (45). Cyclic GMP was applied to the cytoplasmic surface of inside-out patches of retinal rod outer segments, and a cationic conductance that did not require the presence of ATP was generated. These studies demonstrated that cyclic GMP, without stimulating PKG-induced phosphorylation, could interact directly with ion channels to modulate their gating, and gave birth to one of the more rapidly developing areas of cyclic GMP research.

The role of cyclic GMP in visual transduction remains possibly the most well-understood mechanism of cyclic GMP function in cellular communication. Over 30 years ago it was discovered that light causes the membranes of retinal rod and cone cells to hyperpolarize (46), and subsequent studies demonstrated that this hyperpolarization results from hydrolysis of cyclic GMP. In the absence of light, cyclic GMP binds to and opens a cation-selective ion channel that is permeable to calcium and sodium, but it is sodium that carries approximately 70% of this "dark current" because it is the predominant extracellular cation. Influx of this positive charge under resting conditions maintains rod cells in a partially depolarized state near -40 mV, and at this voltage neurotransmitter is continuously released from the synaptic terminal. Upon exposure to light, photons are captured by rhodopsin, a transmembrane protein containing the prosthetic group 11-*cis*-retinal, which is converted into all-*trans*-retinal by a single photon. This isomerization produces a conformational change that activates transducin, a specific heterotrimeric guanine nucleotide binding (G) protein. Activated transducin can now stimulate phosphodiesterase (type VI) activity to hydrolyze approximately 106 cyclic GMP molecules, resulting in closure of the cation channel and hyperpolarization of the rod cell (47–49). The greater the hyperpolarization, the fewer action potentials are produced by the rod bipolar cell, and the less neurotransmitter is released.

Light-induced closure of the cation channel also reduces the entry of Ca^{2+} in the rod cell, and continued activity of the Na^{+}/Ca^{2+} exchanger lowers cellular calcium levels in the outer segment. As Ca^{2+} levels fall into the low nanomolar range, the GC activity of rod outer segments is increased 5- to 20-fold (50). Interestingly, Ca^{2+} does not interact directly with GC, but is bound by a Ca^{2+}-sensitive regulatory protein named recoverin (51). Unlike calmodulin, bound Ca^{2+} inhibits the activity of recoverin; but when Ca^{2+} levels fall in the absence of light, bound Ca^{2+} is released, and recoverin directly stimulates GC activity to "recover" cyclic GMP and keep the visual transduction system operable by preventing depletion of cyclic GMP. Thus, it is light that affects the "life cycle" of cyclic GMP in the visual system: PDE-induced hydrolysis predominates during light exposure, whereas cyclic GMP levels are higher when GC is active and PDE activity is low during dark periods. Similar cyclic GMP gated cation channels have also been characterized in cone photoreceptors (52). The cyclic GMP-gated channel contains no consensus sequences for phosphorylation by PKG or PKA, and the protein is not phosphorylated in vitro by these kinases (49); therefore, phosphorylation appears to play little, if any, role in regulating the activity of these channels.

Cyclic nucleotide gated channels are also present in other systems. In olfactory cells cyclic nucleotide gated channels can be activated by either cyclic GMP or cyclic AMP, but the latter nucleotide may be the more important physiological regulator of channel activity. Many known odorant receptors have been cloned. These proteins normally consist of seven transmembrane-spanning regions and are coupled to heterotrimeric G proteins (e.g., G_{olf}) that interact with and stimulate adenylyl cyclase activity. In addition, there is a nonselective cation channel expressed in a renal collecting duct cell line that appears to be very similar to the rod photoreceptor cation channel, but is inhibited, rather than activated, by cyclic GMP (53,54), and spermatozoa express a K^+ channel that is activated by the egg peptide speract, which stimulates GC activity (55). Because sperm cells probably do not express PKG in significant amounts, the effect of cyclic GMP on this channel is likely due to direct interaction with the channel complex.

Why do these and other cells express cyclic GMP gated channels? One possibility could be to enhance efficiency of signaling. Direct interaction of cyclic GMP with a channel is a more rapid means of altering cellular conditions (e.g., membrane potential) than other more complex mechanisms involving additional regulatory molecules such as kinases and phosphatases, and possibly this is one reason for the frequent presence of cyclic nucleotide gated channels in the cells of sensory organs (e.g., olfactory, visual), which should be able to respond to acute changes in environmental conditions.

Phosphodiesterases, Cyclic GMP, and Ion Channel Regulation

It has been suggested that probably all kinds of channels can be phosphorylated, and cyclic AMP dependent phosphorylation is known to modulate the gating behavior of a variety of channels (56). For many cells then, at least those expressing type II or type III PDE, cyclic GMP could antagonize or enhance the effects of cyclic AMP on ion channel activity by either stimulating or inhibiting the enzymes controlling cyclic AMP metabolism. Probably the classic example of cyclic nucleotide antagonism is found in the heart. β-Adrenergic agonists stimulate cardiac activity by enhancing production of cyclic AMP, whereas depression of this activity is induced by the parasympathetic neurotransmitter acetylcholine (ACh). With the studies of George et al. (5), ACh was found to increase cyclic GMP accumulation in the heart, and a number of studies since that time have endeavored to determine the physiological basis of cyclic GMP induced inhibition.

Within the last 10 years it has become apparent that the focal point of this reciprocal interaction is probably the cardiac dihydropyridine-sensitive (L-type) calcium channel. The slow inward Ca^{2+} current through these channels, which is responsible for the plateau phase of the cardiac action potential, influences the rate and strength of cardiac contraction. It is known that cyclic AMP stimulating agents increase this inward Ca^{2+} current, whereas cyclic GMP decreases ^{45}Ca influx and contractile force, and shortens the duration of the action potential (57). Cyclic GMP was found to depress Ca^{2+} currents in cardiac cells from frog (58), guinea pig (59), rat (60), or chicken (61); but in all species except chicken, there was no significant effect on unstimulated currents. Thus, the primary effect of cyclic GMP appeared to be reversal of cyclic AMP dependent phosphorylation, rather than a direct effect on the Ca^{2+} channel itself.

The first direct evidence for a mechanism of cyclic GMP action on cardiac Ca^{2+} currents was obtained from frog ventricular myocytes. In these experiments Hartzell and Fischmeister (58) dialyzed cells with cyclic GMP and found a reversible inhibition of the stimulatory effect of cyclic AMP; however, cyclic GMP was ineffective on currents stimulated by a nonhydrolyzable analog of cyclic AMP, ruling out a direct effect of cyclic GMP on channel proteins. Furthermore, the action of cyclic GMP was blocked by an inhibitor of PDE activity. It was concluded, therefore, that cyclic GMP antagonized the excitatory effects of cyclic AMP by stimulating PDE activity to promote cyclic AMP hydrolysis, probably via the cyclic GMP stimulated (type II) PDE. A similar mechanism of cyclic GMP action was also described in guinea pig hippocampal cells, where 8-bromo-cyclic GMP depressed whole-cell Ca^{2+} currents, and this effect was mimicked by cyclic IMP, a potent phosphodiesterase activator, but was not affected by inhibitors of protein kinase activity (62). Now there was a complete signaling process that could explain the physiological antagonism of cyclic GMP and cyclic AMP in excitable cells; however, additional mechanisms of cyclic GMP action were soon discovered.

In contrast to amphibian cells, cyclic GMP reversed the stimulatory action of 8-bromo-cyclic AMP or the PDE inhibitor isobutylmethylxanthine on Ca^{2+} currents in mammalian heart (59); thus, a distinct, PDE-independent mechanism of cyclic GMP action exists in these cells. These experiments in guinea pig heart, and subsequent studies of rat heart cells demonstrating that the proteolytic active fragment of purified PKG mimicked the effect of cyclic GMP, suggested that the effect of cyclic GMP in mammalian heart involved stimulation of cyclic GMP dependent phosphorylation (60). Similar effects have recently been observed in guinea pig ventricular cells, where a nitric oxide donor, 3-morpholino-sydnonymine (SIN-1), inhibited isoproterenol-stimulated Ca^{2+} current, and this effect was prevented by inhibiting PKG activity with KT5823 (63). In light of these studies, it appears that in mammalian cardiac tissue cyclic GMP dependent phosphorylation is able to reverse the effects of cyclic AMP dependent phosphorylation by a mechanism distal to synthesis or hydrolysis of cyclic AMP. Although the precise nature of this cardiac inhibitory mechanism is not known, one possibility could be PKG-induced stimulation of phosphoprotein phosphatase activity. In neurosecretory cells, ANP or cyclic GMP was found to inhibit L-type Ca^{2+} currents, and the effect of PKG in these cells appeared to involve stimulation of serine/threonine phosphatase 2A to induce dephosphorylation of ion channels (64). A similar mechanism could account for the ability of PKG to reverse cyclic AMP dependent phosphorylation in mammalian heart, although this pathway has not been characterized in cardiac cells. Interestingly though, PKG-stimulated phosphatase activity has recently been demonstrated to modulate K^+ channel activity in tracheal smooth muscle (65), suggesting that this dephosphorylation mechanism might be an important regulator of ion channel activity in a variety of cell types.

In summary, the actions of ACh on the heart are complex, and are known to involve cyclic GMP independent pathways such as inhibition of pacemaker channels in nodal cells and stimulation of an inwardly rectifying K^+ channel by activation of GTP-binding proteins (57,66); however, it is likely that the inhibitory effects of ACh and other cyclic GMP producing agents such as ANP and NO (e.g., cytokines, NO donors) on slow inward Ca^{2+} currents involve stimulation of PKG-dependent phosphorylation in mammalian cardiac tissue. On the other hand, NO can also stimulate Ca^{2+} currents in both frog and human cardiac cells by a mechanism probably involving inhibition of type III (cyclic GMP inhibited) PDE activity: low nanomolar concentrations of SIN-1 stimulated Ca^{2+} current in frog ventricular cells (60), and a more recent study has described a similar phenomenon in human atrial myocytes (67). In the latter study SIN-1 had no effect on currents in the presence of milrinone, an inhibitor of the cyclic GMP inhibited PDE. Therefore, it was proposed that low levels of NO released from SIN-1 stimulated cyclic GMP production, which then depressed cyclic AMP hydrolysis by inhibiting PDE activity, thereby indirectly enhancing PKA stimulation of Ca^{2+} current; however, these experiments did not exclude a role for PKG in this mechanism. Interestingly, in the majority of atrial cells higher (micromolar) concentrations of SIN-1 reduced Ca^{2+} currents, suggesting a dual effect of NO–cyclic GMP on Ca^{2+} currents in human atria mediated by both type III PDE and PKG.

Thus, it seems clear that cyclic GMP exerts an important regulatory influence on the activity of Ca^{2+} channels through either stimulation of PKG activity or interaction with specific PDE enzymes; however, it appears equally clear that the response to cyclic GMP and the signal transduction mechanism(s) involved in this response exhibit a heterogeneity that is dependent on cell type. For example, intracellular injection of PKG has been found to increase inward currents in snail neurons (68,69), and it has also been suggested that NO may produce cyclic GMP independent effects on ion channel activity in some smooth muscle cells (70).

Cyclic GMP Dependent Phosphorylation and K^+ Channels

The primary transduction mechanism mediating the effects of cyclic GMP in most cells is phosphorylation via PKG, and this mechanism also appears to account for the influence of cyclic GMP on ion channel activity in excitable cells. Stimulation of cyclic GMP dependent phosphorylation is believed to evoke a wide array of physiological responses including depression of cardiac activity, decreased peripheral vascular resistance, decreased blood pressure, inhibition of secretion from pituitary, renal, adrenal, or neuronal cells, and relaxation of respiratory or gastrointestinal smooth muscle; it may also play a role in learning and memory. Therefore, cyclic GMP dependent phosphorylation exerts a major influence on the activity of a variety of excitable cells by ultimately influencing the concentration of activator Ca^{2+} that regulates cellular function. Several known substrates for PKG are involved in modulating $[Ca^{2+}]_i$. For example, nitrovasodilators have been reported to inhibit inositol phosphate accumulation possibly by PKG-mediated inhibition of phospholipase C activity directly (71,72) or by interacting with pertussis toxin sensitive G protein α subunits in renal cells (53); however, there is apparently no substantial evidence that PKG can phosphorylate phospholipase C directly, and G proteins are poor substrates for PKG-dependent phosphorylation (73). In contrast, the Ca^{2+}–ATPase and possibly the Na^+, K^+, Cl^- cotransporter of smooth muscle, and the cerebellar inositol 1,4,5-trisphosphate receptor are likely substrates for PKG (74–76).

In excitable cells the precarious balance between calcium entry and exit is regulated to a large degree by membrane voltage, and voltage-dependent ion channels are the primary players in this homeostatic mechanism. Besides depression of Ca^{2+} currents, studies have implicated stimulation of K^+ channels as a powerful mediator of PKG-induced reduction of $[Ca^{2+}]_i$. Opening of these channels hyperpolarizes membrane potential and closes voltage-sensitive Ca^{2+} channels, and thereby allows the various Ca^{2+} resequestration/elimination mechanisms to exert a dominant influence on $[Ca^{2+}]_i$. Cyclic GMP elevating agents are reported to stimulate distinct K^+ channels (Table 2), but a favorite target appears to be the calcium-activated K^+ channel. Although cytosolic concentrations of Ca^{2+} reach high nanomolar levels during stimulation, the actual Ca^{2+} concentration near the inner aspect of the plasma membrane is probably greater (77), making calcium-sensitive ion channels important effector molecules involved in regulatory feedback mechanisms.

Calcium-activated potassium channels serve such a role because opening these channels is one way Ca^{2+} can "turn off" Ca^{2+} entry to prevent excess or deleterious accumulation. All excitable cells, with the apparent exception of cardiac myocytes, express calcium- and voltage-activated potassium channels, which represent a diverse group of proteins that can be broadly categorized vis-à-vis their ionic conductance. The first group of these channels to be identified exhibited a high conductance for K^+, was expressed at high density (sometimes >10,000 per cell), and is referred to as large conductance, calcium- and voltage-activated potassium (BK or maxi K) channels. In addition, both small (SK) and intermediate (IK) conductance channels exist, but their expression is less ubiquitous than that of the BK channel. Because of their high conductance and close proximity to membrane calcium channels, BK channels constitute a sensitive repolarizing mechanism that can respond to cytoplasmic or localized changes in calcium concentrations.

Within the last several years it has become increasingly clear that BK channels are far more than just passive brakes that help maintain cells in a resting state. In addition to membrane voltage and $[Ca^{2+}]_i$ BK channels are subject to metabolic control via phosphorylation, and the presence or absence of a covalently linked organic phosphate group(s) can influence the gating behavior of BK channels dramatically. It is also apparent that a number of hormones, neurotransmitters, and pharmacological agents regulate cellular activity by either stimulating or inhibiting BK channel activity, and the majority of these compounds are believed to work by modulating channel phosphorylation. Interestingly, regulation of BK channel activity is functionally heterogeneous in that phosphorylation may either enhance or depress channel activity depending on the specific cells involved. For example, injection of the catalytic subunit of PKA into snail neurons increased Ca-dependent K^+ currents (78,79), whereas PKA-induced phosphorylation depresses these currents in photoreceptors (80) or neuroendocrine cells (81). Moreover, the BK channel has been identified as an important effector molecule mediating PKG-induced inhibition of cellular excitability, and much of this exciting evidence involves research into the signaling mechanisms of the same two molecules that reopened the field of cyclic GMP research: ANP and NO.

Atrial Natriuretic Peptide

Since its discovery in the early 1980s, much research has gone into elucidating the cellular basis of ANP action. An important clue was, of course, the intrinsic guanylyl cyclase activity exhibited by the ANP receptor, which means that the major effects of ANP should be mediated via cyclic GMP. Thus, ANP could conceivably modulate cyclic GMP gated ion channels or phosphodiesterase action, or stimulate PKG activity, depending on the specific cell type. At present there is little evidence that cyclic nucleotide gated ion channels are expressed to a significant degree in electrically excitable cells such as cardiac and vascular smooth muscle or endocrine tissue; ANP, however, exerts a powerful inhibitory influence on these cells. Similarly, ANP is a physiological antagonist for the renin–angiotensin system and inhibits the cyclic AMP independent effects of angiotensin II. Thus, modulation of cyclic nucleotide gated ion channels or PDE activity appears not to be a primary mechanism of ANP action in excitable cells, so another mechanism is necessary to explain how natriuretic peptides inhibit cellular excitability. The molecular target of cyclic GMP with the greatest potential to explain the effects of ANP is stimulation of PKG, and we have investigated the role of cyclic GMP-dependent phosphorylation in ANP-induced depression of cell excitability (64).

Secretion is triggered by calcium-stimulated exocytosis. Although the precise mechanisms by which Ca^{2+} stimulates this process is unclear, it is clear that many neurotransmitters or hormones that raise intracellular Ca^{2+} levels augment secretion. In contrast, inhibitory neurohormones reduce cytosolic Ca^{2+} levels in neuroendocrine cells, and thereby attenuate secretory activity. The ionic basis of secretory regulation in pituitary cells was initially thought to involve phosphorylation/dephosphorylation of voltage-sensitive Ca^{2+} channels or possibly direct effects of hormone-stimulated G protein subunits on channel proteins (82); however, studies (81) employing the perforated patch technique have identified BK channels as important targets of regulatory hormones and neurotransmitters in a cell line derived from the rat anterior pituitary (GH cells; Ref. 83). In addition to increasing Ca^{2+} current, cyclic AMP inhibited BK channel activity, and thereby diminished the hyperpolarizing "breaking effect" of K^+ efflux through these channels; conversely, any agent that opened BK channels would tend to inhibit secretory activity. ANP was known to be a important inhibitor of secretion from the anterior pituitary, and our studies revealed that this peptide is indeed a potent stimulator of BK channel activity.

In contrast, ANP suppresses Ca^{2+} currents to a much lesser degree. The action on BK channels involves stimulation of phosphoprotein phosphatase activity to antagonize the effects of cyclic AMP dependent phosphorylation (64). Studies employing perfused cell-free patches demonstrated that BK channel activity was high under "dephosphorylating" conditions (i.e., bathing the cytoplasmic face of the membrane with an ATP-free solution) (Fig. 2). When ATP and cyclic AMP were then added to

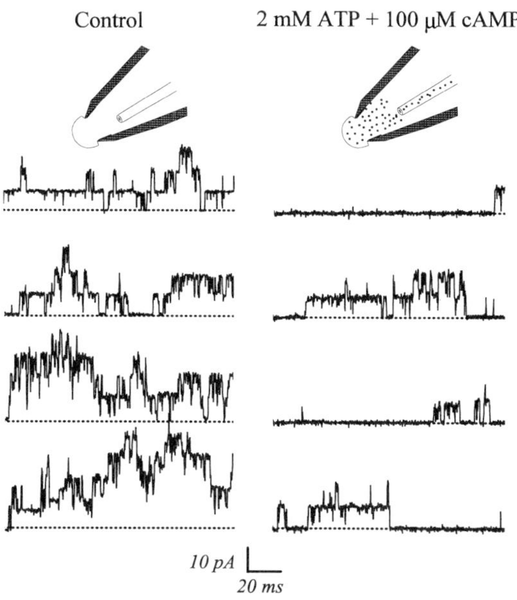

FIG. 2. BK channel activity in GH_4C_1 cells is inhibited by cyclic AMP dependent phosphorylation. Representative records of unitary potassium currents from a cell-free patch in the outside-out configuration at +40mV. Perfusion of the pipette interior with ATP and cyclic AMP dramatically reduced the activity of the large-conductance channels. The maximal number of simultaneously active BK channels decreased from 5 to 2. Channel openings are upward deflections, and the broken line represents the baseline closed state. (Redrawn with permission from *Nature,* Ref. 81. Copyright 1991 Macmillan Magazines Limited.)

this solution, BK channel activity was inhibited significantly. These experiments indicated that cyclic AMP dependent phosphorylation, presumably through stimulation of PKA activity, exerts an inhibitory effect on channel gating by phosphorylating either the BK channel or a regulatory molecule that remains intimately associated with the channel complex under cell-free conditions. In these studies ANP was shown to enhance cyclic GMP accumulation in GH_4C_1 cells and to produce its effects on BK channels via stimulation of PKG activity, as cyclic GMP or 8-bromo-cyclic GMP mimicked the effect of ANP, whereas blockers of PKG activity were inhibitory. Furthermore, perfusing purified PKG onto the cytoplasmic face of cell-free patches enhanced BK channel-open probability dramatically. Blocking PDE activity had no effect on the response, and cyclic GMP did nothing in the absence of ATP, thus making involvement of PDE activity or cyclic GMP gated channels unlikely. Instead, inhibition of phosphoprotein phosphatase activity with a low concentration (10 nM) of okadaic acid reversed cyclic GMP induced stimulation (Fig. 3), suggesting that the effects of ANP or cyclic GMP on BK channel activity involved stimulation of phosphatase activity to dephosphorylate the channels. An earlier study had demonstrated that another inhibitory neuropeptide, somatostatin, also stimulates BK channel activity in these cells, but through a cyclic GMP independent mechanism (81).

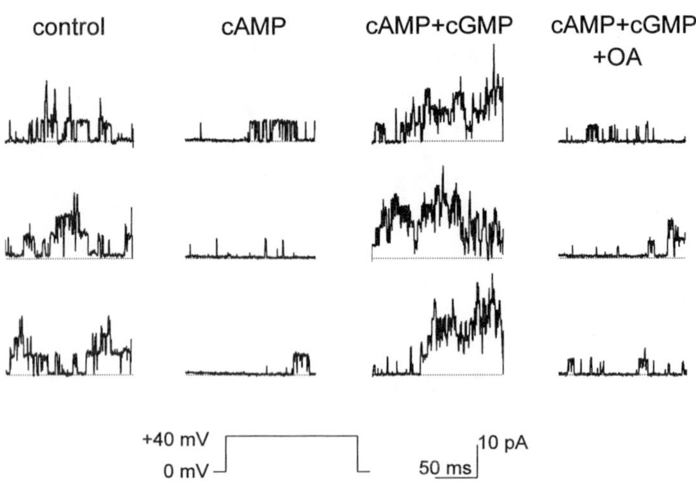

FIG. 3. Cyclic GMP reverses the inhibitory effect of cyclic AMP on BK channel activity in GH_4C_1 cells through protein dephosphorylation. Representative records of unitary BK currents from a cell-free patch in the outside-out configuration at +40mV. Perfusion of the cytoplasmic surface of the patch with 100 μM cyclic AMP inhibited channel activity, but addition of 100 μM cyclic GMP to the perfusate reversed this effect. Okadaic acid (OA; 10 nM) completely blocked the stimulation of channel activity produced by cyclic GMP. Channel openings are upward deflections, and the broken line represents the baseline closed state. (Redrawn with permission from *Nature,* Ref. 64. Copyright 1993 Macmillan Magazines Limited.)

Other laboratories have presented further evidence for a stimulatory effect of ANP on K^+ channel activity. For example, ANP was found to inhibit secretion of aldosterone from adrenal glomerulosa cells, and this inhibition was associated with a depression of $[K^+]_i$ involving stimulation of BK channel activity (84). In addition, ANP, BNP, or 8-bromo-cyclic GMP (but not CNP) was found to hyperpolarize mesangial cells, which are an important renal target for natriuretic peptides; because the effect of the cyclic GMP analog was blocked by Ba^{2+}, it was concluded that K^+ channels were involved in the response to ANP (85). Furthermore, a patch–clamp study of mesangial cells identified stimulation of the BK channel as an important repolarizing mechanism for ANP action (86). In myocytes from rat aorta ANP or cyclic GMP stimulated both the BK and the ATP-sensitive (K_{ATP}) K^+ channel (87), and studies of *Xenopus* oocytes have also indicated a stimulatory effect of ANP on glibenclamide-sensitive (K_{ATP}?) channels (88,89). These studies employed 8-bromo-cyclic GMP as a membrane-permeable cyclic GMP analog, and because this compound is a more selective stimulator of PKG than cyclic GMP, it is likely that these effects of ANP on K^+ channels were also mediated via PKG activity.

In summary, there is increasing evidence that many, if not most, of the physiological effects of natriuretic peptides are mediate by cyclic GMP dependent phosphorylation mechanisms, particularly those involving PKG-stimulated activity of K^+ channels. The fall in $[Ca^{2+}]_i$ brought about by this hyperpolarizing effect is a potent mechanism for inducing depression of secretory activity or muscle contraction, and can account for the effects of ANP on a variety of cell types.

Nitric Oxide

In contrast to many systems in which cyclic GMP was believed to be only a mechanism for reversing the effects of cyclic AMP, there is accumulating evidence that cyclic GMP dependent phosphorylation is actually the more important endogenous mechanism regulating vascular smooth muscle tone. For example, both cyclic GMP and cyclic AMP mediate vasodilation, and recent evidence strongly suggests that, in at least some arteries, the transduction mechanisms of both nucleotides converge on a common biochemical mediator, PKG (40,41). A similar cross-activation of PKG can also occur in gastric smooth muscle (90). The clearest illustration of the importance of cyclic GMP dependent phosphorylation in mediating a physiological response is probably the vasodilatory effect of endogenous or exogenous NO. Although NO has been reported to produce cyclic GMP independent effects in large conduit arteries (70), the majority of studies have implicated cyclic GMP dependent phosphorylation as the primary mechanism of NO-induced vasodilation. PKG has multiple substrates in vascular smooth muscle, as indicated earlier, and phosphorylation of one or more of these proteins could contribute to vascular relaxation. In fact, it seems likely that there are multiple mechanisms mediating the effects of cyclic GMP existing in the same smooth muscle cell. Nonetheless, there is no doubt that stimulation of PKG activity opens BK channels in arterial smooth muscle, and that a number of important vasodilators open this channel.

Studies of intact smooth muscle preparations indicate that blockade of BK channels can inhibit the relaxation response to various agents. For example, charybdotoxin, a BK channel antagonist, prevented the induction of relaxation of tracheal smooth muscle, but not rabbit aorta, by ANP or NO donors (91); however, a subsequent study from the same laboratory revealed that either charybdotoxin or iberiotoxin, a highly selective BK channel antagonist, significantly attenuated nitroglycerin-induced relaxation of rabbit aorta without affecting cyclic GMP production or contractile proteins directly (92). In contrast, blockers of other K^+ channels (i.e., glibenclamide, apamin) had no effect. Additional studies determined that the effect of NO donors on this artery was mediated via activation of PKG (93). Taken together, these studies from intact tissues suggest that a primary mechanism of nitrovasodilator-induced relaxation of smooth muscle is stimulation of BK channel activity via PKG-dependent phosphorylation.

Studies on single smooth muscle myocytes have established the importance of this ionic mechanism directly. Taniguchi et al. (94) demonstrated that application of purified PKG to the cytoplasmic surface of inside-out membrane patches of dog coronary artery smooth muscle cells enhanced the open probability of BK channels dramatically; furthermore, the magnitude of the relaxant effect of ANP was related inversely to extracellular concentrations of K^+ and was blocked by charybdotoxin, suggesting that ANP-induced relaxation of coronary arteries occurred via opening of

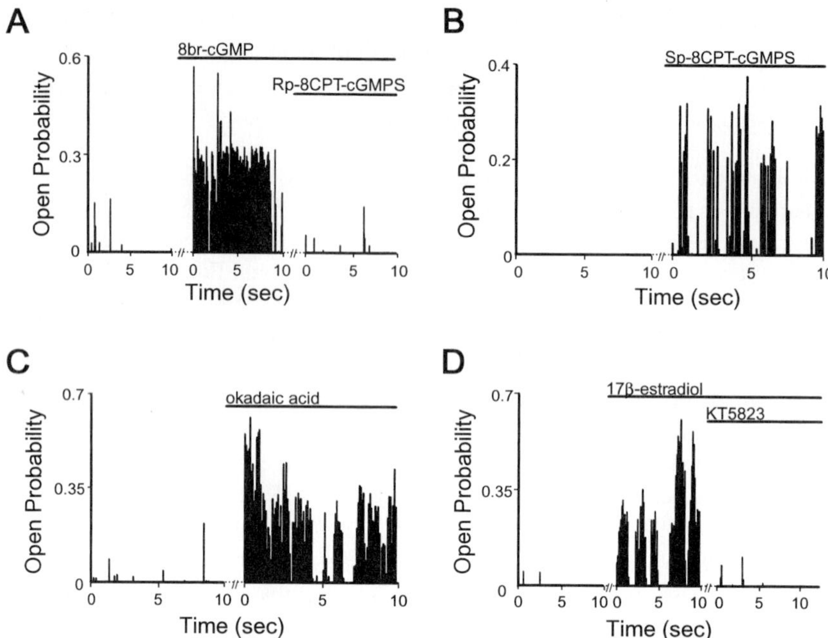

FIG. 4. Cyclic GMP dependent phosphorylation stimulates BK channel activity in myocytes isolated from porcine coronary arteries. Each panel represents a plot of channel activity from a different cell-attached patch during 100-ms intervals at +40mV. To conserve space on the computer's hard disk, recording was interrupted at the times indicated by the broken lines on the time axis. Period of drug exposure is indicated by the bars. **(A)** BK channel activity was stimulated by 1 mM 8-bromo-cyclic GMP (30 min), whereas inhibition of PKG activity with 100 μM Rp-8CPT-cGMPS (20 min) reversed the effect. **(B)** Stimulation of PKG activity with Sp-8CPT-cGMPS (100 μM, 30 min) increased BK channel open probability. **(C)** Inhibition of phosphoprotein phosphatase activity with okadaic acid (500 nM, 20 min) stimulated BK channel activity. **(D)** BK channel activity was stimulated by a 20-minute exposure to 10 μM 17β-estradiol, but inhibition of PKG activity with 300 nM KT5823 (7 min) reversed the effect of estrogen. (Reprinted with permission from Ref. 98.)

BK channels. Stimulatory effects of PKG-mediated phosphorylation on BK channel activity also were observed in subsequent studies of myocytes from rabbit basilar arteries (95), rat pulmonary arteries (96), porcine coronary arteries (97,98), and tracheal smooth muscle (99,100). Thus, it is clear that a significant portion of the relaxation response to NO derived from the endothelium or exogenous donors, ANP, and in some cases β-adrenergic agonists is mediated by stimulation of BK channel activity involving cyclic GMP dependent phosphorylation.

In addition, we have reported that low nanomolar–micromolar levels of exogenous estrogen stimulate BK channel activity in myocytes from porcine coronary arteries via cyclic GMP dependent phosphorylation (Fig. 4) and that inhibiting BK channels with iberiotoxin nearly abolishes the endothelium-independent relaxant effect of estrogen in these vessels (98). In light of these and other studies, it is becoming increasingly apparent that PKG-induced modulation of BK channel activity is an important regulatory mechanism that helps mediate many physiological and pharmacological responses in the cardiovascular system, including endothelial control of vascular tone and blood flow, the vasodilatory effects of ANP and other endogenous agents, mechanisms of nitrovasodilator action and antihypertensive therapy, and antianginal treatment. Furthermore, this mechanism may also underlie much of the protection against cardiovascular disease exhibited by women during their childbearing years and the efficacy of postmenopausal estrogen replacement therapy to restore this protection. Still, it should be remembered that this ionic mechanism is not the only cyclic GMP effector process in smooth muscle.

Mechanisms of Channel Modulation

While there is little doubt that PKG-mediated phosphorylation stimulates BK channel activity in various cell types, the specific substrate for this phosphorylation has not been identified. Several studies have claimed that demonstrating a stimulatory effect of purified PKG on BK channels in cell-free patches proves direct channel activation by the kinase. This, however, is not the case because studies on purified channel complexes inserted into artificial bilayers reveal that regulatory kinases and/or phosphatases can copurify with BK channel proteins, as discussed later. Furthermore, findings from White et al. (64) suggested that phosphatases are also present in excised patches and that these enzymes mediate the stimulatory effects of cyclic GMP dependent phosphorylation on BK channel activity in neurosecretory cells; even though purified PKG stimulated channel activity, the BK channel was not the substrate for PKG. Given the uncertainty regarding the actual substrate of PKG in cell-free systems, most investigators have retreated to making the disclaimer that PKG acts on the channel or a "closely associated regulatory molecule."

Studies on cell-free patches or artificial lipid bilayers have suggested that there is an intimate association between BK channels and their regulatory molecules (i.e., kinases and phosphatases). For example, BK channels prepared from rat brain plasma membrane vesicles and inserted into bilayers are stimulated simply by adding 50–500 μM ATP to the cytoplasmic surface of the membrane (101,102). Because this stimulation required Mg^{2+} and was not mimicked by nonhydrolyzable ATP derivatives, it was concluded that channel opening was due to phosphorylation by an endogenous protein kinase that was intimately associated with the channel complex. In addition, ATP or ATPγS stimulated the activity of channels in cell-free patches from oocytes injected with cRNA for calcium-activated K^+ channels cloned from *Drosophila melanogaster (dSlo),* and this stimulation was blocked by inhibitors of PKA activity (103). These findings suggested that a PKA-like kinase remains functionally associated with the channel in detached membrane patches. BK channels in cell-free patches from gastric smooth muscle (104) or nerve terminals (105) were stimulated by ATP, but not by nonhydrolyzable analogs, and this effect could be reversed by washout of ATP.

The foregoing experiments provide further evidence for a channel-associated or at least membrane-delimited kinase, and also suggest that a phosphatase may be a part of the channel complex because of the reversibility of ATP-induced activity in the absence of exogenous phosphatase. Further evidence that phosphatases are intimately associated with BK channels was obtained from studies demonstrating that inhibition of phosphatase activity mimicked the effect of cyclic AMP on BK channels in excised patches from neuroendocrine cells (64), or enhanced the effect of ATP on neuronal BK channels in lipid bilayers (102) or cell-free patches from smooth muscle (104). Given this evidence obtained from cell-free systems, the simplest explanation appears to be that BK channels are modulated by kinases and phosphatases that form a regulatory complex with the channel protein, or are at least functionally associated with the channel. Recent studies indicate that enzymes such as these can be localized to their protein substrate by anchoring proteins. One such protein, the A kinase anchoring protein (AKAP79), is known to localize protein kinases and phosphoprotein phosphatases to the same subcellular site to provide an efficient, compartmentalized phosphorylation–dephosphorylation system to regulate the activity of a single substrate protein (106), and a similar scaffolding system could account for the functional association of ion channel regulatory molecules with their substrates.

The Integral Role of Phosphoprotein Phosphatases

Although it is clear that BK channel activity depends on a precisely regulated balance between kinase and phosphatase activity, understanding the interaction between these two enzymes and how they influence channel activity is becoming increasingly complex. What is clear is that regulation of BK channel activity is functionally heterogeneous in that phosphorylation may either enhance or depress channel activity depending on the specific cells involved. Incorporation of rat brain BK channels into artificial lipid bilayers has actually revealed two distinct types: one that was stimulated by PKA and another that was inhibited by PKA (107). Further complicating our understanding of these regulatable systems is the once overlooked fact that phosphorylated channels must also be dephosphorylated by phosphoprotein phosphatases.

Protein kinases have been the subject of intense investigation during the past decade. In comparison, much less was known about protein phosphatases until recently. While ion channel regulation by kinases has been recognized for a number of years, only comparatively recent studies that have examined the equally important role of phosphoprotein phosphatases in regulating K^+ channel activity (64,81,101, 102,105,107–110). Studies from Armstrong and his collaborators have clearly indicated that it is dephosphorylation that stimulates BK channel gating in GH pituitary cells, whether stimulated by PKG or a mechanism involving a pertussis toxin sensitive G protein (64,81,110). In these studies low concentrations of the phosphatase inhibitor okadaic acid inhibited or reversed the effects of PKG-mediated phosphorylation, suggesting a novel target of PKG action: stimulation of protein dephosphorylation, possibly via phosphoprotein phosphatase 2A. In the pituitary, dephosphorylation would provide a powerful mechanism exerting control over secretion. For example, secretagogues stimulate PKA activity to phosphorylate (activate)

Ca^{2+} channels and (inhibit) BK channels. In this manner phosphorylation not only stimulates cellular activity, but also "removes the brake" by inhibiting K^+ efflux. Conversely, inhibitors of secretion need only to stimulate phosphatase activity to dephosphorylate the same channels and depress secretion by lowering the levels of stimulatory Ca^{2+}. Thus, cyclic AMP and cyclic GMP exert a powerful reciprocal mechanism to control the activity of these neurosecretory cells by influencing the phosphorylation state of the channel.

Although it may seem superficially paradoxical that a kinase stimulates protein dephosphorylation, such behavior helps explain many of the physiological effects of cyclic GMP. For example, by activating a protein phosphatase, ANP could reverse the effects of both angiotensin II and vasopressin that are mediated by distinct protein kinases. Furthermore, dephosphorylation of Ca^{2+} and K^+ channels could also explain the inhibitory effects of ANP and cyclic GMP on mammalian cardiac cells, smooth muscle cells, and secretory cells. Although direct effects of PKG on phosphatases have not been firmly established, other studies have confirmed the importance of this mechanism in a variety of cell types. PKG activity opens BK channels in cell-free patches from tracheal smooth muscle, and this response was inhibited by the protein phosphatase inhibitors microcystin or okadaic acid (65); moreover, the catalytic subunit of phosphatase 2A mimicked the effect of PKG. In addition, the β-amyloid precursor protein opens K^+ channels in hippocampal neurons, and this effect can be inhibited by blocking PKG or phosphatase activity (111). Thus, these studies implicate PKG-induced dephosphorylation as an important regulatory mechanism for inhibition of secretion, neuronal excitability, and smooth muscle contraction. Furthermore, PKG stimulation of phosphatase activity could also explain why cyclic GMP reverses the effects of cyclic AMP dependent phosphorylation in a number of tissues.

The potential role for phosphoprotein phosphatases in regulating BK channels of smooth muscle is not completely understood. In contrast to the conclusions of Zhou et al. (65), Alioua et al. (99) have demonstrated that the α subunit of the BK channel from tracheal smooth muscle can be phosphorylated by PKG-Iα and that purified PKG stimulates the activity of these BK channels inserted into artificial lipid bilayers; however, these results remain to be confirmed in intact cells. That is, if direct phosphorylation of channel proteins is involved, inhibition of phosphatase activity should stimulate rather than inhibit PKG-induced opening of BK channels. Interestingly, both whole-cell and single channel patch–clamp studies have demonstrated that high concentrations of okadaic acid (>100 nM) increase the activity of BK channels in myocytes from gastric (104) or arterial smooth muscle (Refs. 96,98; Fig. 4C). These findings are consistent with a model predicting a stimulatory effect of phosphorylation on smooth muscle BK channels; however, it is also clear that phosphorylation mediated by C kinase *inhibits* the activity of vascular smooth muscle BK channels in cultured coronary artery myocytes (112) or BK channels in neuroendocrine cells (113). These studies raise the possibility of a more complex model of BK channel regulation involving multiple sites of phosphorylation/dephosphorylation on the channel protein or on a number of regulatory kinases and/or phosphatases. Clearly, further investigation will be needed to unravel the mysteries of how channel proteins and modulatory molecules work together to control phosphorylation/dephosphorylation of BK and other ion channels, hence to regulate cellular excitability.

FINAL SYNOPSIS

Although 40 years have passed since cyclic nucleotides were discovered to be cellular second messengers, our knowledge of the specific molecular mechanisms stimulated by these compounds is far from complete; and this is particularly true for cyclic GMP. What is perfectly clear now is that cyclic GMP is not a simple "yang" molecule that serves only to reverse the effects of cyclic AMP. Indeed, in many cases it is the cyclic GMP signal transduction cascade that exerts the primary influence upon cellular biology to control a diversity of important physiological and pathophysiological processes. The recent discoveries that ion channels serve as primary effector molecules governing much of the pharmacodynamics of cyclic GMP have again opened an exciting new area of investigation into how second messengers regulate cellular excitability. It is anticipated that much future effort will be directed into understanding how cyclic GMP turns these proteins on or off, either by direct interaction or by more indirect, but no less effective, phosphorylation/dephosphorylation mechanisms. Once these mechanisms are understood, it is not overly optimistic to anticipate the development of novel or improved therapeutic measures to treat a variety of complications. Not bad for a simple molecule that was once envisioned to be just a "poor substitute" for an important intracellular biochemical.

ACKNOWLEDGMENTS

The author would like to acknowledge support received from the American Heart Association (Ohio affiliate), the American Federation for Aging Research, and the National Institutes of Health (NHLBI). The editorial assistance of Ms. M. White is also greatly appreciated.

REFERENCES

1. Goldberg ND, Haddox MK, Nicol SE et al: Biologic regulation through opposing influences of cyclic GMP and cyclic AMP: the yin–yang hypothesis. *Adv Cyclic Nucl Res* 1975;5:307–330.
2. Sutherland EW, Rall TW: The properties of an adenine ribonucleotide produced with cellular particles, ATP, Mg^{2+}, and epinephrine or glucagon. *J Am Chem Soc* 1957;79:3608–3611.
3. Ashman DR, Lipton R, Melicow MM, Price TD: Isolation of adenosine 3',5'-monophosphate and guanosine 3',5'-monophosphate from rat urine. *Biochem Biophys Res Commun* 1963;11:330–334.
4. Hardman JG, Sutherland EW: Guanylate cyclase, an enzyme catalyzing the formation of guanosine 3':5'-monophosphate from guanosine triphosphate. *J Biol Chem* 1969;244:6363–6370.
5. George WJ, Polson JB, O'Toole AG, Goldberg ND: Elevation of guanosine 3',5'-monophosphate in rat heart after perfusion with acetylcholine. *Proc Natl Acad Sci U S A* 1970;66:398–403.
6. Kuo JF, Greengard P: Cyclic nucleotide-dependent protein kinases. IV. Isolation and partial purification of a protein kinase activated by guanosine 3',5'-monophosphate. *J Biol Chem* 1970;245: 2493–2498.
7. Hoffmann F, Sold G: A protein kinase activity from rat cerebelllum stimulated by guanosine-3',5'-monophosphate. *Biochem Biophys Res Commun* 1972;49:1100–1107.

8. Lincoln TM, Corbin JD: Characterization and biological role of the cGMP-dependent protein kinase. *Adv Cyclic Nucl Res* 1983;15:139–192.
9. Beavo JA, Hardman, JG, Sutherland EW: Stimulation of adenosine 3',5'-monophosphate hydrolysis by guanosine 3',5'-monophosphate. *J Biol Chem* 1971;246:3841–3846.
10. Miki N, Keirns, JJ, Marcus FR, Freeman J, Bitensky MW: Regulation of cyclic nucleotide concentrations in photoreceptors: an ATP-dependent stimulation of cyclic nucleotide phosphodiesterase by light. *Proc Natl Acad Sci U S A* 1973;70:3820–3824.
11. Diamond J, Holmes TG: Effects of potassium chloride and smooth muscle relaxants on tension and cyclic nucleotide levels in rat myometrium. *Can J Physiol Pharmacol* 1975;53:1099–1107.
12. Katsuki S, Arnold WP, and Murad F: Effects of sodium nitroprusside, nitroglycerin, and sodium azide on levels of cyclic nucleotides and mechanical activity of various tissues. *J Cyclic Nucl Res* 1977;3:239–247.
13. Schultz KD, Schultz K, Shultz G: Sodium nitroprusside and other smooth muscle relaxants increase cyclic GMP levels in rat ductus deferens. *Nature* 1977;265:750–751.
14. Arnold WP, Mittal C, Katsuki S, Murad F: Nitric oxide activates guanylate cyclase and increases guanosine 3',5'-cyclic monophosphate levels in various tissue preparations. *Proc Natl Acad Sci U S A* 1977;74:3202–3207.
15. Gruetter CA, Barry BK, McNamara DB, Kadowitz PJ, Ignarro LJ: Coronary arterial relaxation and guanylate cyclase activation by cigarette smoke, *N'*-nitrosonornicotine and nitric oxide. *J Pharmacol Exp Ther* 1980;214:9–15.
16. Furchgott RF, Zawadzki JV: The obligatory role of endothelial cells in the relaxation of arterial smooth muscle by acetylcholine. *Nature* 1980;288:373–376.
17. Ignarro LJ, Buga GM, Wood KS, Byrns RE, Chaudhuri G: Endothelium-derived relaxing factor produced and released from artery and vein is nitric oxide. *Proc Natl Acad Sci U S A* 1987;84: 9265–9269.
18. Palmer RMJ, Ferrige AG, Moncada S: Nitric oxide release accounts for the biological activity of endothelium-derived relaxing factor. *Nature* 1987;327:524–526.
19. Lincoln TM: *Cyclic GMP: Biochemistry, physiology and pathophysiology.* Austin: R. G. Landes Co., 1994.
20. Ignarro L, Murad F: Nitric oxide: biochemistry, molecular biology, and therapeutic implications. *Ad Pharmacol* 1995;34.
21. Packer L: Nitric oxide: physiological and pathological processes. *Methods Enzymol* 1996;269.
22. Kisch B: Electron microscopy of the atrium of the heart. *Exp Med Surg* 1956;14:99–112.
23. Henry JP, Gauer OH, Reeves JL: Evidence of the atrial location of receptors influencing urine flow. *Circ Res* 1956;4:85–90.
24. Marie JP, Guillemot H, Hatt PY: Le degré de granulation des cardiocytes auriculaires. *Pathol Biol* 1976;24:549–554.
25. deBold AF, Borenstein HB, Veress AT, Sonnenberg H: A rapid and potent natriuretic response to intravenous injection of atrial myocardial extract in rats. *Life Sci* 1981;28:89–94.
26. Flynn TG, deBold ML, deBold AJ: The amino acid sequence of an atrial peptide with potent diuretic and natriuretic properties. *Biochem Biophys Res Commun* 1983;117:859–865.
27. Thibault G, Garcia R, Cantin M et al: Atrial natriuretic factor. Characterization and partial purification. *Hypertension* 1983;5(Suppl I):75–80.
28. Kangawa K, Matsuo H: Purification and complete amino acid sequence of α-human atrial natriuretic polypeptide (α-hANP). *Biochem Biophys Res Commun* 1984;118:131–139.
29. Winquist RJ, Faison EP, Waldman SA et al: Atrial natriuretic factor elicits an endothelium-independent relaxation and activates particulate guanylate cyclase in vascular smooth muscle. *Proc Natl Acad Sci U S A* 1984;81:7661–7664.
30. Armstrong DL, White RE: Natriuretic peptides and receptors. *Sci Am Sci Med* 1994;1:34–43.
31. MacFarland RT: Molecular aspects of cyclic GMP signaling. *Zool Sci* 1995;12:151–163.
32. Imai S: Cyclic GMP as a second messenger in the cardiovascular system. *Jpn Heart J* 1995;36: 127–177.
33. Nakane M, Arai K, Saheki S, Kuno T, Buechler W, Murad F: Molecular cloning and expression of cDNAs coding for soluble guanylate cyclase from rat lung. *J Biol Chem* 1990;265:16841–16849.
34. Hobbs AJ, Ignarro LJ: Nitric oxide–cyclic GMP signal transduction system. *Methods Enzymol* 1996; 209:134–148.
35. Ignarro LJ, Degnan JN, Baricos WH, Kadowitz PJ, Wolin MS: Activation of purified guanylate cyclase by nitric oxide requres heme. *Biochim Biophys Acta* 1982;718:49–59.
36. Leitman DC, Waldman SA, Murad F: Regulation of particulate guanylate cyclase by natriuretic peptides and *Escherichia coli* heat-stable enterotoxin. *Adv. Pharmacol.* 1994;26:67–86.

37. Schmidt HHHW, Pollock JS, Nakane M, Forstermann U, Murad F: The nitric oxide and cGMP signal transduction system: Regulation and mechanism of action. *Biochim Biophys Acta* 1993;1178: 153–175.
38. Drewett JG, Ziegler RJ, Trachte GJ: Neuromodulatory effects of atrial natriuretic peptides correlate with an inhibition of adenylate cyclase but not an activation of guanylate cyclase. *J Pharmacol Exp Ther* 1992;260:689–696.
39. Kelly RA, Smith TW: Pharmacological treatment of heart failure. In: Hardman JG, Limbird LE., eds. *The pharmacological basis of therapeutics.* New York: McGraw-Hill, 1996;809–838.
40. Francis SH, Noblett BD, Todd BW, Wells JN, Corbin JD: Relaxation of vascular and tracheal smooth muscle by cyclic nucleotide analogs that preferentially activate purified cGMP-dependent protein kinase. *Mol Pharmacol* 1988;34:506–517.
41. Lincoln TM, Cornwell TL, Taylor AE: cGMP-dependent protein kinase mediates the reduction of Ca^{2+} by cAMP in vascular smooth muscle cells. *Am J Physiol* 1990;158:C399-C407.
42. Lincoln TM, Cornwell TL: Intracellular cyclic GMP receptor proteins. *FASEB J* 1993;7:328–342.
43. Lincoln TM, Komalavilas P, Cornwell TL: Pleiotropic regulation of vascular smooth muscle tone by cyclic GMP-dependent protein kinase. *Hypertension* 1994;23:1141–1147.
44. Felbel J, Trockur B, Ecker T et al: Regulation of cytosolic calcium by cAMP and cGMP in freshly isolated smooth muscle cells from bovine brachea. *J Biol Chem* 1988;263:16764–16771.
45. Fesenko EE, Kolenikov SS, Lyubarsky AL: Induction by cyclic GMP of cationic conductance in plasma membranes of retinal rod outer segment. *Nature* 1985;313:310–313.
46. Bortoff A: Localization of slow potential responses in the *Necturus* retina. *Vision Res* 1964;4: 627–635.
47. Miki N, Baraban JM, Keirns JJ et al: Purification and properties of the light activated cyclic nucleotide phosphodiesterase from rod outer segments. *J Biol Chem* 1975;250:6320–6327.
48. Yau K-W, Baylor DA: Cyclic GMP-activated conductance of retinal photoreceptor cells. *Annu Rev Neurosci* 1989;12:289–327.
49. Kaupp UB, Koch K-W: Role of cGMP and Ca^{2+} in vertebrate photoreceptor excitation and adaptation. *Annu Rev Physiol* 1992;54:153–175.
50. Koch K-W, Stryer L: Highly cooperative feedback control of retinal rod guanylate cyclase by calcium ions. *Nature* 1988;334:64–66.
51. Dizhoor AM, Ray S, Kumar S et al: Recoverin: a calcium sensitive activator of retinal rod guanylate cyclase. *Science* 1991;251:915–918.
52. Haynes LW, Yau K-W: Cyclic GMP-sensitive conductance in outer segment membrane of catfish cones. *Nature* 1985;317:661–664.
53. Light DB, Corbin JD, Stanton BA: Dual ion-channel regulation by cyclic GMP and cyclic GMP-dependent protein kinase. *Nature* 1990;344:336–339.
54. Ahmad I, Korbmacher C, Segal AS et al: Mouse cortical collecting duct cells show nonselective cation channel activity and express a gene related to the cGMP-gated rod photoreceptor channel. *Proc Natl Acad Sci U S A* 1992;89:10262–10266.
55. Cook SP, Babcock DF: Selective modulation by cGMP of the K^+ channel activated by speract. *J Biol Chem* 1993;268:22401–22407.
56. Hille B: *Ionic channels of excitable membranes.* Sunderland, MA: Sinauer Associates, 1992; 234–235.
57. Hartzell HC: Regulation of cardiac ion channels by catecholamines, acetylcholine, and 2nd messenger systems. *Prog Biophys Mol Biol* 1988;52:165–247.
58. Hartzell HC, Fischmeister R: Opposite effects of cyclic GMP and cyclic AMP on Ca^{2+} current in single heart cells. *Nature* 1986;323:273–275.
59. Levi RC, Alloatti G, Fischmeister R: Cyclic GMP regulates the Ca-channel current in guinea pig ventricular myocytes. *Pfluegers Arch* 1989;413:685–687.
60. Mery P-F, Pavoine C, Belhassen L et al: Nitric oxide regulates cardiac Ca^{2+} current—involvement of cGMP-inhibited and cyclic stimulated phosphodiesterases through guanylyl cyclase activation. *J Biol Chem* 1993;268:26286–26295.
61. Wahler GM, Rusch JF, Sperelakis N: 8-Bromo-cyclic GMP inhibits the calcium channel current in embryonic chick ventricular myocytes. *Can J Physiol* 1990;68:531–534.
62. Doerner D, Alger BE: Cyclic GMP depresses hippocampal Ca^{2+} current through a mechanism independent of cGMP-dependent protein kinase. *Neuron* 1988;1:693–699.
63. Wahler GM, Dollinger SJ: Nitric oxide donor SIN-1 inhibits mammalian cardiac calcium current through cGMP-dependent protein kinase. *Am J Physiol* 1995;268:C45–C54.

64. White RE, Lee AB, Shcherbatko AD, Lincoln TM, Schonbrunn A, Armstrong DL: Potassium channel stimulation by natriuretic peptides through cGMP-dependent dephosphorylation. *Nature* 1993;361:263–266.
65. Zhou X, Ruth P, Schlossmann J, Hofmann F, Korth M: Protein phosphatase 2A is essential for the activation of Ca^{2+}-activated K^+ currents by cGMP-dependent protein kinase in tracheal smooth muscle and Chinese hamster ovary cells. *J Biol Chem* 1996;271:19760–19767.
66. Lohmann SM, Fischmeister R, Walter U: Signal transductin by cGMP in heart. *Basic Res Cardiol* 1991;86:503–514.
67. Kirstein M, Rivet-Bastide M, Hatem S et al: Nitric oxide regulates the calcium current in isolated human atrial myocytes. *J Clin Invest* 1995;95:794–802.
68. Paupardin-Tritsch D, Hammond C, Gerschenfeld HM et al: cGMP-dependent protein kinase enhances Ca^{2+} current and potentiates the serotonin-induced Ca^{2+} current increase in snail neurones. *Nature* 1986;323:812–814.
69. Ichinose M, McAdoo DJ: the cyclic GMP-induced inward current in neuron R14 of *Aplysia californica:* similarity to a FMRFamide-induced inward current. *J Neurobiol* 1989;20:10–24.
70. Bolotina VM, Najibi S, Palacino JJ et al: Nitric oxide directly activates calcium-dependent potassium channels in vascular smooth muscle. *Nature* 1994;368:850–853.
71. Rapoport RM: Cyclic guanosine monophosphate inhibition of contraction may be mediated through inhibiton of phosphatidylinositol hydrolysis in rat aorta. *Circ Res* 1986;58:407–410.
72. Takai Y, Kaibuchi K, Matsubara T et al: Inhibitory action of guanosine 3',5'-monophosphate on thrombin-induced phosphatidylinositol turnover and protein phosphorylation in human platelets. *Biochem Biophys Res Commun* 1981;101:61–67.
73. Lincoln TM: Pertussis toxin-sensitive and insensitive guanine nucleotide binding proteins (G-proteins) are not phosphorylated by cyclic GMP-dependent protein kinase. *Second Messenger Phosphoproteins* 1991;13:99–109.
74. Sarcevic G, Bookes V, Martin TJ et al: Atrial natriuretic peptide-dependent phosphorylation of smooth muscle cell particulate fraction proteins is mediated by cGMP-dependent protein kinase. *J Biol Chem* 1989;264:20648–20654.
75. Komalavilas P, Lincoln TM: Phosphorylation of the inositol 1,3,4-inositol trisphosphate receptor by cyclic GMP-dependent protein kinase. *J Biol Chem* 1994;269:8701–8707.
76. O'Donnell ME, Owen NE: Role of cyclic GMP in atrial natriuretic factor stimulation of Na^+,K^+,Cl^- cotransport in vascular smooth muscle cells. *J Biol Chem* 1986;261:15461–15466.
77. Carl A, Lee HK, Sanders KM: Regulation of ion channels in smooth muscles by calcium. *Am J Physiol* 1996;271:C9–C34.
78. DePeyer J, Cachelin A, Levitan IB, Reuter H: Ca^{2+}-activated K^+ conductance in internally perfused snail neurons is enhanced by protein phosphorylaton. *Proc Natl Acad Sci U S A* 1982;79: 4207–4211.
79. Ewald D, Williams A, Levitan IB: Modulation of single Ca^{2+}-dependent K^+ channel activity by protein phosphorylation. *Nature* 1985;315:503–506.
80. Alkon D, Kubota M, Neary J, Naito S, Coulter D, Rasmusssen J: C-kinase activation prolongs Ca^{2+}-dependent inactivation of K^+ currents. *Biochem Biophys Res Commun* 1986;134:1245–1253.
81. White RE, Schonbrunn A, Armstrong D: Somatostatin stimulates Ca^{2+}-activated K^+ channels through protein dephosphorylation. *Nature* 1991;351:570–573.
82. Lewis DL, Weight FF, Luini A: A guanine nucleotide-binding protein mediates the inhibition of voltage-dependent calcium current by somatostatin in a pituitary cell line. *Proc Natl Acad Sci U S A* 1986;83:9035–9039.
83. Tashjian AH: Clonal strains of hormone-producing pituitary cells. *Methods Enzymol* 1979;58: 527–535.
84. Ganz MG, Nee JJ, Isales CM et al: Atrial natriuretic peptide enhances activity of potassium conductance in adrenal glomerulosa cells. *Am J Physiol* 1994;266:C1357–C1365.
85. Cermak R, Kleta R, Forssmann WG et al: Natriuretic peptides increase a K^+ conductance in rat mesangial cells. *Pfluegers Arch* 1996;431:571–577.
86. Stockand JD, Sansom SC: Role of large Ca^{2+}-activated K^+ channels in regulation of mesangial contraction by nitroprusside and ANP. *Am J Physiol* 1996;270:C1773–C1779.
87. Kubo M, Nakaya Y, Matsuoka S et al: Atrial natriuretic factor and isosorbide dinitrate modulate the gating of ATP-sensitive K^+ channels in cultured vascular smooth muscle cells. *Circ Res* 1994;74: 471–476.

88. Sakuta H, Okamoto K, Watanabe Y: Modification by cGMP of glibenclamide-sensitive K^+ currents in *Xenopus* oocytes. *Jpn J Pharmacol* 1993;61:259–262.
89. Sakuta H, Okamoto K, Tandai M: Atrial natriuretic factor potentiates glibenclamide-sensitive K^+ currents via the activation of receptor guanylate cycliase in follicle-enclosed *Xenopus* oocytes. *Eur J Pharmacol* 1994;267:281–287.
90. Murthy KS, Makhlouf GM: Interaction of cA-kinase and cG-kinase in mediating relaxation of dispersed smooth muscle cells. *Am J Physiol* 1995;268:C171–C180.
91. Hamaguchi M, Ishibashi T, Imai S: Involvement of charybdotoxin-sensitive K^+ channel in the relaxation of bovine tracheal smooth muscle by glyceryl trinitrate and sodium nitroprusside. *J Pharmacol Exp Ther* 1991:262:263–270.
92. Ishibashi T, Kawada T, Kato K et al: Contribution of activation of K^+ channels to glyceryl trinitrate-induced relaxation of rabbit aorta. *Gen Pharmacol* 1995;26:543–552.
93. Nakazawa M, Imai S. Rp-8-Br-Guanosine-3',5'-cyclic monophosphorothioate inhibits relaxation elicited by nitroglycerin in rabbit aorta. *Eur J Pharmacol* 1994;253:179–181.
94. Taniguchi J, Furukawa K, Shigekawa M: Maxi K^+ channels are stimulated by cyclic guanosine monophosphate-dependent protein kinase in canine coronary artery smooth muscle cells. *Pfluegers Arch* 1993;423:167–172.
95. Robertson, BE, Schubert, R, Hescheler J, Nelson MT: cGMP-Dependent protein kinase activates Ca-activated K channels in cerebral artery smooth muscle cells. *Am J Physiol* 1993;265: C299–C303.
96. Archer SL, Huang JMC, Hampl V, Nelson DP, Shultz PJ, Weir EK: Nitric oxide and cGMP cause vasorelaxation by activation of a charybdotoxin-sensitive K channel by cGMP-dependent protein kinase. *Proc Natl Acad Sci U S A* 1994;91:7583–7587.
97. Fujino K, Nakaya S, Wakatsuki T et al: Effects of nitroglycerine of ATP-induced Ca^{++}-mobilization, Ca^{++}-activated K channels and contraction of cultured smooth muscle cells of porcine coronary artery. *J Pharmacol Exp Ther* 1991;256:371–377.
98. White RE, Darkow DJ, Falvo Lang JL: Estrogen relaxes coronary arteries by opening BKCa channels through a cGMP-dependent mechanism. *Circ Res* 1995;77:936–942.
99. Alioua A, Huggins JP, Rousseau E: PKG-Iα phosphorylates the α-subunit and upregulates reconstituted GKCa channels from tracheal smooth muscle. *Am J Physiol* 1995;268:L1057–L1063.
100. Yamakage M, Hirshman, CA, Croxton TL. Sodium nitroprusside stimulates Ca^{2+}-activated K^+ channels in porcine tracheal smooth muscle cells. *Am J Physiol* 1996;270:L338–L345.
101. Chung S, Reinhart PH, Martin BL, Brautigan D, Levitan IB: Protein kinase activity closely associated with a reconstituted calcium-activated potassium channel. *Science* 1991;253:560–562.
102. Reinhart PH, Levitan IB: Kinase and phosphatase activities intimately associated with a reconstituted calcium-dependent potassium channel. *J Neurosci* 1995;15:4572–4579.
103. Esguerra M, Wang J, Foster CD, Adelman JP, North RA, Levitan IB: Cloned Ca^{2+}-dependent K^+ channel modulated by a functionally associated protein kinase. *Nature* 1994;369:563–565.
104. Lee M, Bang H, Lim I, Uhm D, Rhee S: Modulation of large conductance calcium-activated K^+ channel by membrane-delimited protein kinase and phosphatase activities. *Pfluegers Arch* 1994; 429:150–152.
105. Bielefeldt K, Jackson MB: Phosphorylation and dephosphorylation modulate a Ca^{2+}-activated K^+ channel in rat peptidergic nerve terminals. *J Physiol (Lond)* 1994;475.2:241–254.
106. Faux MC, Scott JD: Molecular glue: kinase anchoring and scaffold proteins. *Cell* 1996;85:9–12.
107. Reinhart PH, Chung S, Martin BL, Brautigan DL, Levitan IB: Modulation of calcium-activated potassium channels from rat brain by protein kinase A and phosphatase 2A. *J Neurosci* 1991;11: 1627–1635.
108. Ichinose M, Byrne JH: Role of protein phosphatases in the modulation of neuronal membrane currents. *Brain Res* 1991;549:146–150.
109. Endo S, Critz SD, Byrne JH, Shenolikar S: Protein phosphatase-1 regulates outward K^+ currents in sensory neurons of *Aplysia californica. J Neurochem* 1995;64:1833–1840.
110. Duerson K, White RE, Jiang F, Schonbrunn A, Armstrong DL: Somatostatin stimulates BK_{Ca} channels in rat pituitary tumor cells through lipoxygenase metabolites of arachidonic acid. *Neuropharmacology* 1996;35:949–961.
111. Furukawa K, Barger SW, Blalock EM et al: Activation of K^+ channels and suppression of neuronal activity by secreted beta-amyloid-precursor protein. *Nature* 1996;379:74–78.
112. Minami K, Fukuzawa K, Nakaya Y: Protein kinase C inhibits the Ca^{2+}-activated K^+ channel of cultured porcine coronary artery smooth muscle cells. *Biochem Biophys Res Commun* 1993;190: 263–269.

113. Shipston MJ, Armstrong DL: Activation of protein kinase C inhibits large conductance calcium- and voltage-activated potassium (BK) channels in clonal (GH_4C_1) rat anterior pituitary cells. In press.
114. Nakamura T, Gold GH: A cyclic nucleotide-gated conductance in olfactory receptor cilia. *Nature* 1987;325:442–444.
115. Biel M, Altenhofer W, Hullin R et al: Primary structure and functional expression of a cyclic nucleotide-gated channel from rabbit aorta. *FEBS Lett* 1993;329:134–138.
116. Ishikawa T, Hume JR, Keef KD: Regulation of Ca^{2+} channels by cAMP and cGMP in vascular smooth muscle cells. *Circ Res* 1993;73:1128–1137.
117. Bkaily G, Perron N, Wang S et al: Atrial natriuretic factor blocks the high-threshold Ca^{2+} current and increases K^+ current in fetal single ventricular cells. *J Mol Cell Cardiol* 1993;25:1305–1316.
118. Hirsch J, Schlatter E: K^+ channels in the basolateral membrane of rat cortical collecting duct are regulated by a cGMP-dependent protein kinase. *Pfluegers Arch* 1995;429:338–344.
119. Chen C, Schofield GG: Nitric oxide donors enhanced Ca^{2+} currents and blocked noradrenaline-induced Ca^{2+} current inhibition in rat sympathetic neurons. *J Physiol (Lond)* 1995;482:521–531.
120. Farrugia G, Rae JL: Regulation of a potassium-selective current in rabbit corneal epithelium by cyclic GMP, carbachol and diltiazem. *J Membr Biol* 1992;129:99–107.

Advances in Second Messenger and Phosphoprotein Research, Vol. 33

1040-7952/99 $30.00

12

Store-Operated Calcium Channels

Richard S. Lewis

Department of Molecular and Cellular Physiology, Stanford University School of Medicine, Stanford, California 94305

INTRODUCTION

The regulated entry of Ca^{2+} across the plasma membrane is an essential and ubiquitous signaling mechanism in both electrically excitable and nonexcitable cells. Ca^{2+} entry has long been implicated in an impressive range of cellular behaviors, ranging from exocytosis and contraction to gene expression and cell differentiation. Ca^{2+} signaling in electrically excitable and nonexcitable cells differs in a fundamental way: whereas excitable cells in many cases use voltage-gated Ca^{2+} channels to regulate Ca^{2+} entry, nonexcitable cells must rely on voltage-independent Ca^{2+} channels. Several classes of voltage-independent Ca^{2+} channels have been discovered, including receptor-operated, ligand-operated, G protein coupled, and store-operated channels (reviewed in refs. 1–3). Store-operated channels (SOCs), defined as channels that open in response to depletion of the endoplasmic reticulum (ER) Ca^{2+} stores, represent one of the most ubiquitous mechanisms for triggering Ca^{2+} influx in nonexcitable cells.

Several techniques have dominated the study of these channels and their regulation, each with its own set of strengths and weaknesses. Fluorescent dye based measurements of Ca^{2+} or Mn^{2+} influx are well suited to detect the activity of SOCs, which in many cases conduct small currents that are difficult to study electrophysiologically, and the use of fluorescence methods has provided much information about the tissue distribution and dynamic behavior of store-operated pathways. However, Ca^{2+} or Mn^{2+} influx only indirectly reflects SOC activity, since the influx rate can also be modulated by membrane potential, buffering, pumping, sequestration, and mitochondrial activity. Thus, changes in any of these parameters can confound the interpretation of imaging experiments. Voltage–clamp measurements of the Ca^{2+}-activated Cl^- current, an approach that has seen wide use as an indirect assay of I_{SOC} in *Xenopus* oocytes, circumvent problems associated with membrane potential changes. However, it appears that at least two distinct Cl^- channels with different voltage and Ca^{2+} sensitivities contribute to $I_{Cl(Ca)}$, and this may also complicate the interpretation of such experiments (4). Direct measurement of SOC currents with patch–clamp techniques (5) offers distinct advantages to the detailed characterization of these channels and their regulation. While important cytosolic factors can be lost during conventional whole-cell recording by diffusion into the recording pipette, perforated-patch recording has largely removed this drawback (6). Much of the progress made over the past 5–10 years in our understanding of SOCs has come about through a combination of fluorescence- and patch–clamp-based studies that have exploited the strengths of each approach, and these have been complemented most recently by molecular biology studies. It is now clear that SOCs are a diverse family of channels that serve important functions, and their identification at a molecular level looks promising. This chapter focuses on the properties, modulation, and possible molecular basis of store-operated channels as determined primarily through electrophysiological studies. Several other reviews have recently covered aspects of store-operated Ca^{2+} entry from a more general perspective (7–10).

HISTORICAL OVERVIEW

A large number of cell surface receptors are coupled through G proteins or tyrosine kinases to the activation of phospholipase C (PLC) and consequently the generation of inositol 1,4,5-trisphosphate (IP_3) (11). It has long been recognized that an almost universal consequence of the production of IP_3 is an increase in the concentration of intracellular Ca^{2+} ($[Ca^{2+}]_i$) resulting from the release of Ca^{2+} from intracellular stores by IP_3 followed by influx across the plasma membrane (7,10,11). While the molecular mechanism of Ca^{2+} release is understood to result from IP_3 acting on a receptor/Ca^{2+} channel in the ER (11,12), the mechanism underlying Ca^{2+} entry is considerably more mysterious. Perhaps the simplest mechanism would involve a direct action of IP_3 on Ca^{2+} channels in the plasma membrane, in much the same way that it activates Ca^{2+} channels in the ER. However, several studies revealed that the Ca^{2+} permeability of the plasma membrane was more closely related to the degree of filling of internal stores than to either agonist receptor occupancy or IP_3 levels. Following a transient activation of PLC-coupled receptors in the absence of extracellular Ca^{2+}, the increased Ca^{2+} permeability of the plasma membrane was maintained long after IP_3 or its metabolites had returned to baseline levels (7). These and other results led Putney to propose that store depletion rather than IP_3 was responsible for triggering influx (13). This process was called "capacitative Ca^{2+} entry" (CCE), to reflect the initial idea that Ca^{2+} entered the cytoplasm from outside by traversing the ER (the "capacitor"), which was assumed at that time to have a privileged connection to the external milieu. Subsequent experiments showed that Ca^{2+} enters the cytoplasm before it is taken up into the stores, leading to a revision of the original hypothesis (7). In its currently accepted form, the CCE hypothesis states that depletion of stores sends an activating signal of some kind to channels in the plasma membrane, causing them to open and promote Ca^{2+} entry, creating an autoregulatory loop that ensures efficient refilling of the stores (Fig. 1).

Converging evidence obtained through imaging and patch–clamp studies has confirmed many of the key aspects of the CCE hypothesis and has led to an intensive

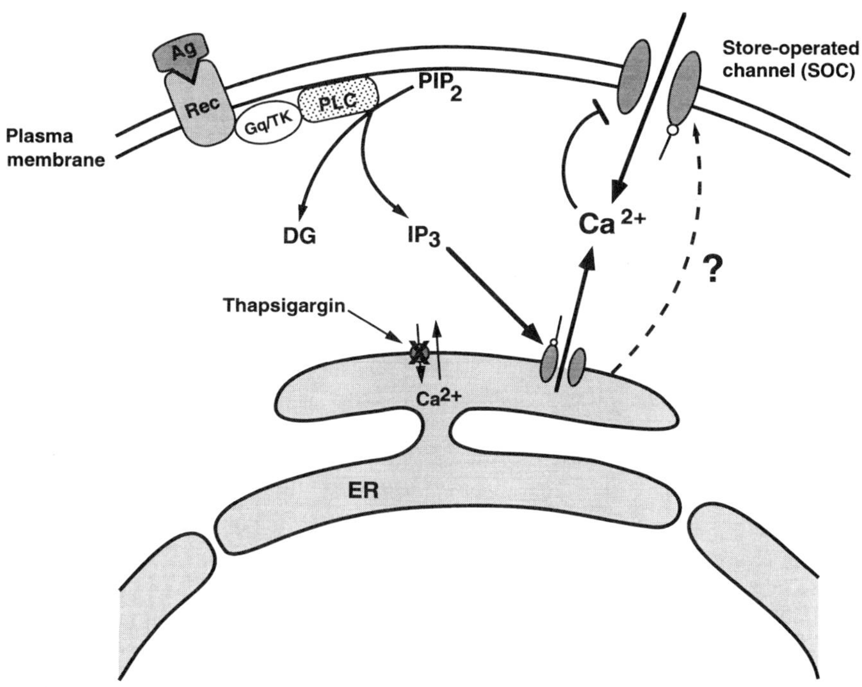

FIG. 1. The store-operated ("capacitative") Ca^{2+} entry pathway. In this generic model, agonist (Ag) binding to a receptor (Rec) triggers activation of PLC via a G protein or protein tyrosine kinase intermediate (Gq/TK). Cleavage of PIP_2 yields IP_3, which opens Ca^{2+} channels in the endoplasmic reticulum (ER). Depletion of the ER activates SOCs in the plasma membrane via an unknown mechanism (dashed line) that may involve a diffusible messenger or conformational coupling between proteins in the ER and plasma membrane. Elevated Ca^{2+} in many cases feeds back to inhibit SOC activity. Thapsigargin can be used to deplete stores without generating IP_3 by inhibiting Ca^{2+} uptake via SERCA Ca^{2+}-ATPases.

study of its underlying mechanism. An important step was the demonstration of a Ca^{2+} conductance activated by store depletion. The original discovery of this current was made in mast cells and T lymphocytes by means of a combination of Ca^{2+} imaging and patch–clamp techniques. In a study of second messenger–activated Ca^{2+} entry mechanisms in rat peritoneal mast cells, Penner et al. (14) noted that intracellular dialysis with IP_3 in whole-cell recordings activated an extremely small (<1–2 pA) and noise-free inward current in parallel with a sustained rise in $[Ca^{2+}]_i$, and suggested that this current underlies the Ca^{2+} influx induced in these cells by IP_3 and agonist. Similarly, during a search for the mitogen-activated Ca^{2+} current in Jurkat human leukemic T cells, Lewis and Cahalan (15) found a small Ca^{2+}-selective inward current that developed in a delayed fashion following the onset of whole-cell recordings with low buffered $[Ca^{2+}]_i$. Several aspects of the current, later to be identified as I_{CRAC}, were characterized, including a lack of voltage-dependent gating, an inwardly rectifying current–voltage relation, blockade by Ni^{2+} and Cd^{2+}, and feedback inhibition by Ca^{2+} such that intracellular Ca^{2+} buffers were needed to maximize the size of the current (15). At the time, the cause of its spontaneous activation was unknown; however, in perforated-patch recordings (in which spontaneous activation did not occur), the current could be activated by phytohemagglutinin, a lectin that elevates intracellular IP_3 through activation of the T-cell antigen receptor (TCR) (15). McDonald et al. (16) later showed that a similar current was activated in Jurkat cells by photolysis of intracellular caged IP_3, providing direct evidence for the role of IP_3 in the activation process. Together, these studies established that Ca^{2+} currents were activated in some way by IP_3 in mast cells and T cells, but the mechanism of their activation was not understood at the time.

Subsequent Ca^{2+} imaging studies in lymphocytes led to the notion that IP_3-triggered Ca^{2+} entry in these cells occurred through a store depletion mechanism. Rapid progress in understanding the role of depletion in triggering Ca^{2+} entry ensued following the discovery that thapsigargin (TG), a plant-derived sesquiterpene lactone, depletes IP_3-sensitive Ca^{2+} stores through its ability to inhibit Ca^{2+}-ATPases in the ER, without generating IP_3 (17,18). Several groups showed that thapsigargin depleted stores in T cells and also activated Ca^{2+} influx that was similar to but not additive with that evoked by TCR stimulation (19–21). These imaging results were subsequently extended by patch–clamp recordings in mast cells and T cells. Hoth and Penner described a Ca^{2+} current in mast cells, presumably the same that had been reported previously, that could be activated by several treatments that were expected to deplete intracellular stores: extracellular ionomycin or intracellular infusion of Ca^{2+} buffers or IP_3 (22,23). They named the current I_{CRAC}, for Ca^{2+} release-activated Ca^{2+} current. In Jurkat T cells, Zweifach and Lewis (24) showed that thapsigargin was able to activate a Ca^{2+} current whose properties matched those of I_{CRAC} and the previously described current that was activated by TCR stimulation. These studies demonstrated the existence of currents that were activated by depletion rather than directly by IP_3. Subsequent work (25–36) greatly extended the characterization of this current and established I_{CRAC} as a paradigm for Ca^{2+}-selective SOCs. A variety of SOCs have now been reported in many different cells, and their distinguishing characteristics are discussed next.

THE DIVERSITY OF STORE-OPERATED CHANNELS

Over the past several years, electrophysiological and molecular biological studies have demonstrated the existence of a diverse family of store-operated channels. The salient features of the best-characterized SOCs are summarized in Table 1. The essential defining feature of these channels is activation by a variety of stimuli that deplete stores, independently of the Ca^{2+} that is released during the depletion process. These stimuli include agonists to phosphoinositide-linked receptors, intracellular IP_3, Ca^{2+} ionophores, inhibitors of SERCA-type Ca^{2+}-ATPases, and intracellular dialysis with buffered solutions containing low concentrations of Ca^{2+} (<50 nM). Because the content of intracellular Ca^{2+} stores is at present difficult to measure and correlate with current activation, the identification of a current as being store-operated relies entirely on indirect inference. For this reason, it is essential to test as many independent store depletion methods as possible. Activation by IP_3 suggests that the agonist receptor is not directly coupled to channel activation. Activation by ionophores and SERCA inhibitors, which deplete stores without generating IP_3, implies that the channels respond to depletion rather than directly to IP_3 (although sensitivity to more than one activator should not be excluded; see ref. 37). However, because most of these agents (agonists, IP_3, ionophores, and SERCA inhibitors) increase the net Ca^{2+} flux from stores, it is also important to demonstrate current activation under conditions of constant $[Ca^{2+}]_i$. For example, strong evidence for a store-dependent mechanism is obtained from current activation by intracellular Ca^{2+} chelators, which are thought to deplete stores by reducing $[Ca^{2+}]_i$ and interrupting a futile cycle of ongoing Ca^{2+} leak and reuptake by stores. These activation criteria have been met for a number of store-operated channels listed in Table 1.

A second feature common to all SOCs described to date is a lack of voltage-dependent gating, although this need not be a defining feature. Despite the lack of voltage-dependent gating, the membrane potential nevertheless plays several important roles in regulating Ca^{2+} influx through SOCs. First, depolarization inhibits Ca^{2+} entry by reducing the driving force for Ca^{2+} entry. Second, hyperpolarization promotes rapid inactivation of CRAC channels by Ca^{2+}, presumably by raising the local $[Ca^{2+}]_i$ at the inactivation sites (22,23,32). Finally, hyperpolarization can enhance CRAC channel activity through effects on Ca^{2+}-dependent potentiation (36). Inactivation and potentiation are discussed in greater detail later. As shown in Table 1, the different SOCs are most easily distinguished from one another by their ionic selectivity and unitary conductance, and to a limited extent by their pharmacological sensitivities.

Hallmarks of the CRAC Channel

The most intensively studied of the SOCs is the Ca^{2+} release-activated Ca^{2+} (CRAC) channel. CRAC channels are easily distinguished from other SOCs by their extremely high selectivity for Ca^{2+} over monovalent cations and by their extremely small unitary conductance. Channels with these characteristics have been positively identified in T cells, mast cells, megakaryocytes, and macrophages (see Table 1). While the true extent of CRAC channel distribution is not known, it is intriguing that all the positively identified CRAC channels to date have appeared in cells of the hematopoietic lineage. An interesting two-part question is whether CRAC channels are expressed in all hematopoietic cells and whether their expression is lineage restricted. Detailed studies of the ion selectivity and unitary conductance of store-operated channels in additional cell types will be needed to address this point.

Ion Selectivity

CRAC channels show an extremely high selectivity for Ca^{2+} over monovalent cations. Relative permeabilities have been difficult to determine based on reversal

TABLE 1. *The family of store-operated calcium channels*

Cell type	Activator[a]	γ (pS)	Selectivity[b]	Selected blockers	References
CRAC-type channels					
Mast cells	IP_3 48/80 Ionomycin Low $[Ca^{2+}]_i$	<1	$Ca^{2+}>Ba^{2+}\sim Sr^{2+}>Mn^{2+}>>Na^+$ $P_{Ca}/P_M > 1000$ G_{Na} in 0 Ca^{2+}, Mg^{2+}	La^{3+}, $Zn^{2+}>Cd^{2+}>Be^{2+}$, Co^{2+}, $Mn^{2+}>Ni^{2+}>Ba^{2+}$, Sr^{2+} SKF 96365 Econazole	22, 23, 25, 26, 29
RBL cells	Low $[Ca^{2+}]_i$ Ionomycin		$Ca^{2+}>Ba^{2+}\sim Sr^{2+}>Mn^{2+}>>Na^+$ $P_{Ca}/P_M > 1000$ G_{Na} in 0 Ca^{2+}, Mg^{2+}	$La^{3+}>Ni^{2+}$	29, 35
Jurkat T cells	TCR IP_3 TG, CPA, tBHQ Ionomycin Low $[Ca^{2+}]_i$	0.01 (2 mM Ca^{2+}) 0.024 (110 mM Ca^{2+})	$Ca^{2+}>Ba^{2+}\sim Sr^{2+}>>Mn^{2+}>>Na^+$ $P_{Ca}/P_M > 1000$ G_{Na} in 0 Ca^{2+}, Mg^{2+}	$La^{3+}>Cd^{2+}>Co^{2+}$, Ni^{2+} Econazole [Not diltiazem, nifedipine, D600, δ-conotoxin, aga-IVa]	16, 24, 28, 29, 34, 122
Human T cells (blood)	TCR IP_3 TG Ionomycin	~0.008	$Ca^{2+}>>Na^+$	La^{3+}, Cd^{2+}, Zn^{2+}	27, 123
Macrophages	IP_3 TG Low $[Ca^{2+}]_i$		$Ca^{2+}>Sr^{2+}>>Mg^{2+}$, Mn^{2+}, Ni^{2+}, Na^+		99
Megakaryocytes	ATP_o TG low $[Ca^{2+}]_i$	<1[c]	$Ca^{2+}>Ba^{2+}>>Mg^{2+}$, Na^+		124

Non-CRAC-type channels					
A431 cells	TG Low $[Ca^{2+}]_i$	2 (200 mM Ca^{2+}) 16 (160 mM Ba^{2+})	$Ba^{2+} > Ca^{2+} >> Mn^{2+}$, Na^+		46
Endothelial cells (BAEC)	Bradykinin ATP tBHQ IP_3	5–11	$P_{Ca}/P_{Na} > 10$	La^{3+} Heparin	37, 45
dTRP	TG	3?[d]	$Ca^{2+} > Na^+$ $G_{Na} > G_{Ca}$ $P_{Ca}/P_{Cs} = 40$[d]	La^{3+}	53, 54
hTRPC3, bTRP	IP_3 TG	16	$P_{Ca}/P_{Na} = 7$ $G_{Na} > G_{Ca}$	Gd^{3+}	56, 57
Pancreatic acinar cells (mouse)	ACh IP_3 tBHQ Ionomycin Low $[Ca^{2+}]_i$	40–45?	$P_{Na}/P_{Ca} = 13$	[not La^{3+},Gd^{3+},Cd^{2+},Mn^{2+}]	47
MDCK cells	ATP IP_3 TG Ionomycin		$Ca^{2+} >> Ba^{2+}$, Mn^{2+} $P_{Ca}/P_{Na} = 26$	La^{3+} [not SKF 96365, econazole]	48

[a]Activation by low $[Ca^{2+}]_i$ was achieved using BAPTA or EGTA to chelate intracellular Ca^{2+}.

[b]The steady state conductance of CRAC channels to Ba^{2+} (and probably Sr^{2+}) may be artificially low because these ions are unable to fully potentiate CRAC channel activity (see text).

[c]M. Mahaut-Smith, personal communication

[d]Values obtained from photoreceptor currents; all other dTRP data is from expression in Sf9 cells.

Abbreviations: IP_3, inositol 1,4,5-trisphosphate; tBHQ, *tert*-butylhydroquinone; CPA, cyclopiazaonic acid; TG, thapsigargin; TCR, T-cell receptor cross-linking; ACh, acetylcholine; dTRP; *Drosophila* TRP; hTRPC3, human TRPC3; bTRP, bovine TRP.

potential measurements because a clearly defined reversal potential is often not observed (I_{CRAC} often reverses above +50 mV, where other conductances can also contribute). The best estimate of Ca^{2+} selectivity has been obtained using a fura-2/patch–clamp method (38) to estimate the fraction of current carried by Ca^{2+} (23,29). With this technique, a high concentration of fura-2 is applied inside the cell to outcompete endogenous buffers and capture all incoming Ca^{2+}. A comparison is then made between the total amount of Ca^{2+} entering the cell during a step increase in I_{CRAC} (given by the change in fura-2 fluorescence) and the total charge entering the cell (given by the integral of I_{CRAC}). In this way Hoth and Penner determined that CRAC channels are about as selective as voltage-gated Ca^{2+} channels of excitable cells: more than >98% of the currents through both channels is carried by Ca^{2+} in the presence of 10 mM Ca^{2+} and 150 mM Cs^{+} (23,29). Since the permeability ratios ($P_{Ca}/P_{monovalent}$) for voltage-gated Ca^{2+} channels are 1000:1 or greater, a similarly high selectivity is implied for CRAC channels. Thus, CRAC channels are among the most Ca^{2+}-selective channels known.

Recent evidence suggests that CRAC and voltage-gated Ca^{2+} channels may achieve high selectivity for Ca^{2+} in similar ways, in that binding of Ca^{2+} within the pore appears to prevent the permeation of monovalent cations. Several groups have reported that after extracellular divalent cations are removed from cells in which I_{CRAC} is activated, large Na^{+} currents appear (16,23,28). A critical question, made difficult by the lack of specific CRAC channel blockers, is whether this reflects a loss of CRAC channel selectivity or is simply due to an increase in leak current. By periodically switching between Ca^{2+}-containing and Ca^{2+}-free solutions to measure I_{CRAC} and I_{Na}, respectively, Lepple-Wienhues and Cahalan (34) showed that I_{CRAC} slowly activates and inactivates with the same time course as I_{Na}, demonstrating that the Na^{+} current most likely flows through CRAC channels. Extracellular Ca^{2+} blocked the monovalent current with an IC_{50} of 3 μM. Thus, CRAC channels, like voltage-gated Ca^{2+} channels, can conduct monovalent cations in the absence of divalent cations. In voltage-gated channels, this behavior has been traced to the interactions of Ca^{2+} with a quartet of glutamate residues within the pore (39,40); electrostatic repulsion between Ca^{2+} ions within the pore is thought to be responsible for the ability of these channels to attain high selectivity for Ca^{2+} while maintaining a relatively high conduction rate (in isotonic Ca^{2+} or Ba^{2+}, γ is 5–10 pS for L-type channels). Although CRAC channels are comparably selective, their unitary conductance as estimated from noise analysis is about two orders of magnitude lower, as discussed later.

Another distinctive feature of CRAC channels is that, unlike many other types of Ca^{2+} channels, they conduct Ca^{2+} better than Sr^{2+} or Ba^{2+}. In Jurkat cells, substitution of 2 mM Ba^{2+} or Sr^{2+} for Ca^{2+} reduces the size of the current at steady state by approximately half (24,28). However, these measurements are complicated by several factors that have only recently come to light. First, CRAC channel gating can be affected by the permeant ion, complicating the interpretation of steady state permeation measurements. Following substitution of Ba^{2+} for Ca^{2+}, CRAC channel activity declines over seconds (29,32) as a result of the reversal of Ca^{2+}-dependent potentiation (33,36). Thus, steady-state measurements of Ba^{2+} or Sr^{2+} currents may underestimate the true ability of these ions to permeate; in fact, several studies of I_{CRAC} utilizing rapid Ba^{2+} perfusion shows that permeation by Ba^{2+} may rival that of Ca^{2+} (29,33,35). A second complication is that permeation by Ba^{2+} appears to be voltage and cell dependent (29). In rat basophilic leukemia (RBL) and mast cells, the shape of the I–V relation is altered by Ba^{2+}, such that Ba^{2+} conducts less well than Ca^{2+} at potentials positive to ~ -90 mV but better than Ca^{2+} below this potential. Interestingly, this behavior is not as pronounced in Jurkat cells, suggesting that subtly different CRAC channel subtypes may exist (29). Clearly, divalent permeation through CRAC channels is complex, and further structure–function studies will be needed to understand the molecular basis for selectivity.

Stimulated entry of Mn^{2+} and quenching of fura-2 fluorescence is commonly used to monitor CCE in many cells including T cells and mast cells (41). Thus, confusion arose following early reports that Mn^{2+} (10 mM) blocked CRAC channels (23) and was unable to carry current in these two cell types (22,28). However, later studies demonstrated small currents carried by 100 mM Mn^{2+} in mast cells (25) and by 8 mM Mn^{2+} in RBL cells (35) following activation of I_{CRAC}. Thus, depletion-induced Mn^{2+} fluxes in mast cells and T cells are consistent with CRAC channel activity. It remains an open question whether the permeability of CRAC channels to Mn^{2+} is quantitatively sufficient to account for the measured flux, or whether additional pathways also contribute.

Unitary Conductance

A second hallmark of the CRAC channel is its extremely small apparent conductance. From the earliest published recordings, it was noted that an increase in current noise was undetectable by eye as the macroscopic current was slowly induced (14,15). Jurkat T cells are a particularly favorable system for studying I_{CRAC} noise because the current can be isolated from all other currents in the cell at negative potentials. Zweifach and Lewis analyzed whole-cell current fluctuations in these cells as I_{CRAC} was activated by thapsigargin or by cross-linking the T-cell receptor (TCR) with phytohemagglutinin (24). Current variance (σ^2) increased linearly with current magnitude (I), yielding estimates (based on σ^2/I) of 1–4 fA for the single channel current in the presence of 2–110 mM Ca^{2+} at -80 mV (Fig. 2). Assuming a reversal potential of $+80$ mV, these values imply a unitary chord conductance of 10–24 fS, more than 100-fold smaller than other known Ca^{2+} channels.

The channel current amplitude of 1–4 fA corresponds to a flux rate of merely 3000–12,000 ions per second, well within the capabilities of the fastest ion carriers (42). Thus, these results raise the question of whether I_{CRAC} is carried by a channel or a single ion transporter. Several characteristics of I_{CRAC} suggest a channel mechanism. I_{CRAC} exhibits anomalous mole fraction behavior in the presence of mixtures of Ca^{2+} and Ba^{2+} (29), and its selectivity for Ca^{2+} over monovalents is dependent on micromolar amounts of extracellular Ca^{2+} (34); both behaviors are characteristic of multi-ion Ca^{2+} pores (42). Moreover, the I_{CRAC} noise spectrum is fit by a

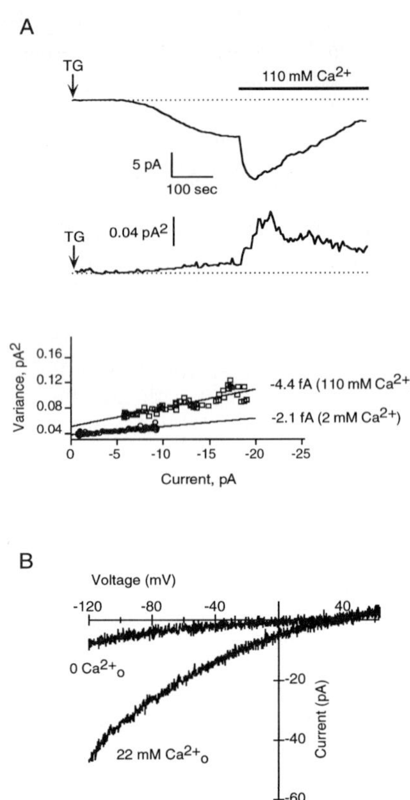

FIG. 2. Hallmarks of the CRAC channel: unitary conductance and selectivity. **(A)** Activation of I_{CRAC} in a Jurkat T cell by TG (current, top trace) is accompanied by a small increase in current noise (variance, bottom trace). Increasing the extracellular concentration $[Ca^{2+}]_o$ from 2 mM to 110 mM causes a transient enhancement of both current and associated noise. In the graph below, current and variance values from this experiment yield an estimate for the unitary conductance (σ^2/I) of -2 to -4 fA. (Reproduced from Ref. 24, with permission. Copyright 1993 National Academy of Science, U.S.A.) **(B)** Following depletion of Ca^{2+} stores in another Jurkat cell, ramp currents were recorded in the absence of extracellular Ca^{2+} and shortly following readdition of 22 mM Ca^{2+}. The current–voltage relation shows a characteristic inward rectification with no clear reversal of current up to +60 mV, a reflection of the high Ca^{2+} selectivity of I_{CRAC}. (Reproduced from Ref. 31 with permission.)

Lorentzian function, declining with the square of frequency (24), another typical feature of noise from ion channels but not of single ion carriers (43; but see Ref. 44). Finally, noise analysis of the Na^+-mediated CRAC current in the absence of extracellular divalents yields a unitary conductance of 2.6 pS, well beyond the transport rates of known carriers (34). Thus, the consensus view at present is that the CRAC transporter is a channel. However, noise analysis may well underestimate the unitary conductance. Such errors can arise if the recording bandwidth is insufficient, or if the channel open probability is very high (42). A more exact estimate of the channel conductance must therefore await single-channel recordings of I_{CRAC}.

Modulation by Calcium

Although CRAC channels are not directly voltage dependent, hyperpolarization indirectly inhibits channel activity through its ability to promote rapid Ca^{2+}-dependent inactivation (22,23,32). In addition, hyperpolarization can enhance I_{CRAC} over a period of seconds by promoting Ca^{2+}-dependent potentiation (CDP) (33,36). These and other forms of Ca^{2+}-dependent modulation are distinctive features of I_{CRAC} in mast cells, RBL cells, and Jurkat T cells, and are discussed in detail in the sections that follow. It is not known whether these characteristics also apply to many of the other SOCs that are distinct from the CRAC channel.

Other Store-Operated Channels

Imaging studies have shown that the capacitative Ca^{2+} entry mechanism operates in a wide variety of cells (8), but limited information is available from such studies about the full range of channel subtypes involved. Patch-clamp studies have identified a number of additional store-operated channels that differ from CRAC channels primarily in two respects: they have a much reduced selectivity for Ca^{2+} over monovalent cations and a higher unitary conductance. The properties of many of the known SOCs are compared in Table 1. Because of their propensity to carry significant amounts of Na^{+} into the cell, these channels would be expected to depolarize the membrane more than the selective CRAC-type SOCs. In most cases the contribution of such channels to changes in intracellular Ca^{2+} concentration has not been measured directly (e.g., with combined measurements of current and fura-2 fluorescence).

Store-operated channels have been identified at the single-channel level in several preparations. In cell-attached recordings from bovine aortic endothelial cells, bradykinin, ATP and *tert*-butylhydroquinone (tBHQ; a SERCA inhibitor) activate an 11-pS channel (in 10 mM Ca^{2+}) with weak selectivity for Ca^{2+} ($P_{Ca}/P_{Na} > 10$) (45). Interestingly, the channel tends to run down within a few minutes of patch excision, but IP_3 applied to the intracellular face partially restores activity for a few more minutes (37). An important question is whether the effect of IP_3 is direct on the channel (implying dual regulation by store depletion and IP_3) or is mediated through depletion of IP_3-sensitive stores that are attached to the patch.

A depletion-activated channel found in A431 epidermal tumor cells is distinguished by a high apparent sensitivity to inactivation by divalent ions (46). Single-channel divalent ion currents were activated in cell-attached patches by intracellular dialysis with BAPTA through a second pipette. These channels conducted Ba^{2+} better than Ca^{2+} ($\gamma = 16$ or 2 pS with 160 mM Ba^{2+} or 200 mM Ca^{2+}, respectively) and were not detectable in the absence of divalents. A high sensitivity to inactivation by divalent ions was suggested based on several observations. First, channel activity usually disappeared within 1–2 minutes of cell-attached recording, and this rundown was slowed by depolarizing the patch to reduce the driving force for Ba^{2+} entry. Second, intracellular BAPTA was necessary to allow channel activation by TG or tBHQ. Third, the whole-cell Ba^{2+} current activated by dialysis with EGTA was immediately inhibited by addition of 5 mM Ca^{2+} to the bath. The rapid inactivation of single-channel and whole-cell currents, particularly by Ca^{2+}, suggests that it may be involved in generating brief $[Ca^{2+}]_i$ transients in intact cells.

Possibly the least Ca^{2+}-selective SOC to date has been reported in mouse pancreatic acinar cells. A study (47) describes a cation current (relative monovalent: Ca^{2+} permeability of 13:1) that is activated in whole-cell recordings by a variety of store-depleting stimuli (ACh, IP_3, ionomycin, tBHQ) or by passive depletion by EGTA or BAPTA in the pipette. Despite its low Ca^{2+} permeability, this current is proposed to underlie CCE because both are inhibited by a subset of tyrosine kinase inhibitors and flufenamic acid but not by La^{3+}, Gd^{3+}, Cd^{2+}, or Mn^{2+}. The insensitivity to La^{3+} is surprising, since CCE in pancreatic acinar cells from rats is blocked by La^{3+}, suggesting a species-specific difference in the channels mediating CCE in these cells. Further electrophysiological studies may help to address this issue more directly. Interestingly, 40- to 45-pS channels were detected in patches from cells stimulated with ACh or tBHQ, suggesting that they may produce the whole-cell current; a pharmacological profile for the channel could be used to test this hypothesis.

Several additional weakly selective depletion-activated currents have been described at the whole-cell level in MDCK kidney epithelial cells (48), HT_{29} colonic carcinoma cells (49), and pancreatic β cells (50). While Ca^{2+} can be demonstrated to carry current in the absence of Na^+, these currents carry mostly Na^+ under physiological conditions. A highly Ca^{2+}-selective inward current is activated in rat hepatocytes by ATP, vasopressin, or TG; while it is proposed that the current is I_{CRAC} (51), more extensive information about selectivity and unitary conductance is needed to confirm this.

Finally, the TRP protein from *Drosophila* photoreceptors may function as a weakly selective store-operated channel. While it has been difficult to demonstrate this for the native protein in the photoreceptor (52), Schilling and colleagues have shown that heterologous overexpression of TRP in Sf9 cells is associated with an increase in TG-activated current (53). The TRP current is distinguished by its lower Ca^{2+} selectivity from the endogenous SOC current; unlike the endogenous SOCs, TRP channels conduct large Na^+ currents in the nominal absence of Ca^{2+} (53). In photoreceptors, observations that the Ca^{2+} selectivity of the light-induced current is largely lost in *trp* null mutants suggests that TRP-containing channels are moderately Ca^{2+} selective (P_{Ca}/P_{Cs} ~40) (54,55). Recent reports of mammalian TRP homologs also suggest a low selectivity for Ca^{2+} in heterologous expression systems (56,57); these are discussed later.

Several questions arise from these studies indicating diversity of the SOC family.

1. Do these various SOCs differ in their regulation, and do they serve distinct functions? A comparison between the inactivation of CRAC channels and A431 channels suggests an extreme difference in their sensitivity to inactivation by Ca^{2+}, but the basis for this difference and the consequences for cell function are not understood.

2. Are the SOCs related by way of an extended gene family? This might be expected based on analogy to other channels, and studies discussed shortly have in fact revealed a family of mammalian homologs of the *Drosophila trp* gene that may encode SOCs.

3. What is the tissue distribution of the various SOCs, and can multiple types of SOCs coexist in the same cell? In T cells and mast cells, it appears that the CRAC channel is the only SOC, and in other cells, only a single type of depletion-activated current has been reported. However, considering the difficulty of the measurements, the failure to detect a current, particularly one as small as I_{CRAC}, must be taken with a grain of salt. Moreover, the ability to discriminate multiple currents is limited by the set of properties that comprise the channel fingerprint, and so far this is not very extensive.

PHYSIOLOGIC FUNCTIONS OF SOCs

Store-operated channels are present in a large number of nonexcitable cell types and are expected to serve a corresponding diversity of functions. Most of the inferences regarding function are indirect; for example, a behavior is known to be triggered by an IP_3-linked receptor and is dependent on extracellular Ca^{2+}, and since SOCs are activated by receptor agonists and elevate $[Ca^{2+}]_i$, they are implicated in the cell's response. In several cases, the link between SOCs and cell function has been further strengthened by showing that receptor stimulation activates the same current as store depletion per se. For example, cross-linking of the IgE receptor in RBL cells by antigen activates a current that is indistinguishable from I_{CRAC} activated in the same cells by TG (30,58). Based on these results, it is likely that I_{CRAC} is responsible for the Ca^{2+} influx that underlies the secretion of histamine and serotonin from RBL cells and mast cells during allergic reactions (59,60). Another example is provided by vascular endothelial cells, in which another type of SOC is activated by bradykinin and by tBHQ (45). In this case, agonist-stimulated Ca^{2+} influx through these channels may contribute to the control of vascular permeability and tone.

Overall, thorough investigations of SOC functions have been hampered by the lack of specific pharmacological blockers. The imidazole antimycotic compounds (including SKF 96365 and econazole) have seen wide use as blockers of CCE and I_{SOC}, but their inhibitory actions extend to other channels (26), as well as to intracellular enzymes such as cytochrome P450 (61), which may have multiple confounding effects on the cell. Similar limitations apply to polyvalent cations such as La^{3+} and Gd^{3+}, which block a variety of Ca^{2+}-selective and -nonselective channels as well as

Ca^{2+} pumps. This problem has been partially circumvented by genetic approaches. For example, null mutations of the *trp* gene in *Drosophila* have been used to implicate the TRP protein in invertebrate phototransduction (52,54,55). In T cells, spontaneous and induced mutagenesis of the CRAC channel pathway have provided evidence for the role of CRAC channels in store refilling and T-cell activation.

T-Cell CRAC Mutants: Store Refilling and T-Cell Activation

The most widely ascribed function for SOCs is to serve as a conduit for the refilling of intracellular Ca^{2+} stores. This idea arose from the observation that store depletion acts to open the channels, and in many cells store refilling requires extracellular Ca^{2+}. Fanger et al. (62) used mutant Jurkat T cells bearing defects in the capacitative Ca^{2+} entry pathway to test the role of CRAC channels in store refilling. In a mutant line (CJ-1) displaying only ~15% of the control level of I_{CRAC}, store content was normal in the steady state condition, but refilling following a transient store depletion with cyclopiazonic acid occurred at about half the rate seen in parental control cells. This result shows directly that I_{CRAC} contributes to refilling, although it does not rule out additional contributions from other Ca^{2+} sources. Generation of null mutants with a complete absence of I_{CRAC} may be useful in addressing this question.

CRAC channels have also been implicated in T-cell activation (63). Antigenic stimulation of T cells is known to generate IP_3 and trigger a rise in $[Ca^{2+}]_i$; to commit cells to an activation program that involves entry into the cell cycle, proliferation, and differentiation into immunologically reactive cells, the Ca^{2+} signal must be sustained for a period of tens of minutes to hours (64). Because of the finite capacity of intracellular stores, Ca^{2+} influx is necessary to sustain this signal. Several lines of evidence indicate that CRAC channels provide the sole pathway for this Ca^{2+} entry (but see ref. 65). First, a close comparison of TCR-activated and depletion-activated Ca^{2+} currents (elicited by TG, intracellular EGTA or BAPTA, or ionomycin) has shown that the two are indistinguishable based on ion permeation, gating, and unitary conductance (24,28). Furthermore, following maximal stimulation with TG, no additional current or Ca^{2+} influx is evoked by TCR cross-linking (20,21,28). SKF 96365 inhibits I_{CRAC} and interleukin 2 production with similar efficacy, further strengthening the link between I_{CRAC} and T cell activation (66).

Finally, perhaps the most compelling evidence comes from two studies in which I_{CRAC} was genetically lesioned. Choquet and colleagues discovered that I_{CRAC} was absent in T cells from human patients with a severe immunodeficiency (27). These cells also failed to respond to TCR cross-linking in terms of either a sustained $[Ca^{2+}]_i$ rise or other markers of cell activation (67). Similarly, in mutant Jurkat T cells selected specifically for the absence of store-operated Ca^{2+} entry, TCR cross-linking also failed to elicit Ca^{2+} entry required for interleukin-2 reporter gene expression (62). Together, these studies provide strong evidence that CRAC channels are solely responsible for Ca^{2+} influx during T-cell activation, and the devastating immunodeficiency associated with their absence underscores their importance in immune system function. In this and other examples, the signaling ability of SOCs clearly extends beyond the rather mundane duty of refilling stores.

REGULATION OF STORE-OPERATED CHANNELS

Probably the most engaging aspect of store-operated channels is the question of how store depletion leads to channel opening. However, despite intense investigation of this question, the only universal conclusion thus far is that no universal mechanism has been found. Some progress has been made in understanding inactivation of SOCs, particularly the effects of Ca^{2+} on CRAC channel inactivation. The following sections discuss primarily patch–clamp studies relevant to activation and inactivation of SOCs.

Activation of Store-Operated Channels

According to the CCE hypothesis, several basic elements are required to couple store depletion to the activation of Ca^{2+} influx. First, the ER must have a sensor of some kind that tracks the luminal concentration of Ca^{2+}. Two groups have suggested that calreticulin may play such a role. Overexpression of calreticulin inhibits TG-induced Ca^{2+} and Mn^{2+} influx in murine L cells despite nearly complete depletion of stores (68). In *Xenopus* oocytes, calreticulin overexpression inhibits IP_3-induced Ca^{2+} waves, possibly by modulating Ca^{2+}-ATPases or IP_3 receptors in the ER (69). The high affinity, low capacity binding domain of calreticulin has been proposed to account for this effect, based on results of deletion mutagenesis experiments. However, no study to date has shown that Ca^{2+}-free and Ca^{2+}-bound forms of calreticulin send different signals, as they must if calreticulin is the ER Ca^{2+} sensor. In principle, much could be learned about the sensor if the concentration of Ca^{2+} in ER could be related quantitatively to SOC activity. Future developments of intraorganellar Ca^{2+} indicators and caged compounds may enable such studies.

A second postulate of the CCE hypothesis is that the luminal Ca^{2+} sensor must control or generate an activator that leads to Ca^{2+} channel opening. Much debate over the past several years has centered on the nature of the activator, yielding an enormous variety of proposed mechanisms. These mechanisms can be grouped loosely into those involving a diffusible messenger and those involving physical contact between proteins in the ER and plasma membrane. The proposals include Ca^{2+} influx factor (CIF), cGMP, heterotrimeric and small G proteins, tyrosine and serine/threonine kinases, cytochrome P-450 and coupling between IP_3 and IP_4 receptors (reviewed in refs. 2,8,9,70). The subsection that follows discusses a subset of these proposals, involving patch-clamp studies of store-operated currents; for a more comprehensive discussion of proposed mechanisms, the reader is referred to the reviews already cited.

Diffusible Messenger or Physical Coupling?

Randriamampita and Tsien (71) originally proposed that a soluble factor (CIF, for Ca^{2+} influx factor) is released from stores upon Ca^{2+} depletion to activate SOCs. Their proposal was based in large part on the ability of cytosolic extracts from stimulated Jurkat cells to trigger Ca^{2+} influx when applied extracellularly to several other cell types. Subsequently it was shown that CIF contains several activities (72,73), including an ability to open nonselective (possibly Ca^{2+}-permeable) channels when applied extracellularly but not intracellularly (74). Hanley and coworkers have focused on fractionation of crude CIF to identify a molecule capable of triggering Ca^{2+} influx without store release when applied inside *Xenopus* oocytes (75). The most recent results suggest that CIF also augments IP_3 receptor function and may be a modulator rather than an activator of SOCs (76); such a role would be consistent with the observation that intracellular CIF by itself does not activate I_{CRAC} in Jurkat cells but appears to enhance I_{CRAC} induced by TG (74).

Cyclic GMP has also been proposed as a soluble mediator of SOC activation, based in part on observations from pancreatic acinar cells that cGMP levels rise after store depletion and that pharmacological inhibitors of guanylyl cyclase inhibit Ca^{2+} entry (77,78). In pancreatic acinar cells (79) and N1E-115 neuroblastoma cells (80,81), cGMP and intracellular IP_3 (applied directly or through muscarinic stimulation) activate similar Ca^{2+} currents. In both systems, reduction of $[Ca^{2+}]_i$ with Ca^{2+} chelators also activates the currents, supporting a mechanism based on store depletion (79,82). However, more recent work suggests that production of cGMP in these cells may be driven by elevated $[Ca^{2+}]_i$ rather than by store depletion itself (83). Mathes and Thompson (81) have described a cascade in neuroblastoma cells by which Ca^{2+} release from the ER (triggered by IP_3 or TG) raises $[Ca^{2+}]_i$ to activate NO synthase, NO activates guanylate cyclase, and the resulting rise in cGMP directly activates channels in the plasma membrane. This mechanism does not easily explain the ability of intracellular Ca^{2+} chelation to activate I_{Ca} under conditions of reduced $[Ca^{2+}]_i$, but two additional possibilities might help to explain this discrepancy: chelators may remove a tonic inhibition of cGMP-gated channels by the resting level of $[Ca^{2+}]_i$, or an I_{SOC} activated through store depletion may provide the initial $[Ca^{2+}]_i$ increase needed to activate the NO/cGMP cascade. Increases in cGMP have been uncoupled from CCE in pancreatic acinar cells and Jurkat cells, further supporting the notion that cGMP may be a product rather than a cause of increased $[Ca^{2+}]_i$ (84,85). These studies highlight the potential complexity of interacting signal transduction pathways and the ambiguities in interpretation that can result.

A GTP-dependent step has been implicated in SOC activation based on the ability of intracellular GTPγS to inhibit I_{CRAC} in RBL cells (86) and TG-induced Ca^{2+} entry in lacrimal acinar cells (87) and HL-60 granulocytes (88). GTPγS tonically activates heterotrimeric G proteins but inhibits small G proteins, while GDPβS should inhibit both; thus far, there is no consensus on which variety of G protein is responsible for the GTPγS effect, since in some cases GDPβS is also inhibitory (87) and in others it is not (86,88). Interestingly, GTPγS inhibits the induction of I_{CRAC} in RBL cells but not its maintenance, since it has no effect when applied simultaneously with ionomycin or intracellular IP_3. Citing the role of small G proteins in vesicle trafficking, Mahaut-Smith and colleagues have proposed that I_{CRAC} may be activated by store-regulated insertion of vesicles containing CRAC channels. This argument is based on the ability of primaquine and low temperature, both inhibitors of vesicular transport, to inhibit I_{CRAC} activation in megakaryocytes (89,90). More direct tests of these hypotheses are needed to rule out possible effects of GTPγS, primaquine, and temperature on other cellular pathways that may be only indirectly related to the control of capacitative Ca^{2+} entry.

The concept that SOC activation may involve direct physical contact between proteins in the ER and plasma membrane dates back to a proposal by Irvine (91) in which IP_3 receptors in the ER membrane were hypothesized to interact with IP_4 receptors in the plasma membrane, analogous to the coupling of ryanodine and dihydropyridine receptors in skeletal muscle. A generic form of this "conformational coupling" model is appealing because it would explain how I_{CRAC} can be sustained for many minutes in the whole-cell configuration, despite escape of diffusible cytosolic constituents into the pipette. Irvine's original hypothesis has undergone a series of transformations as evidence has accumulated against involvement of specific proteins; the most recent elaboration of the model as described by Berridge (9) involves coupling between type III IP_3 receptors in the ER and CRAC/SOC channels in the plasma membrane. Indirect evidence in support of such a model includes the colocalization of CCE and Ca^{2+} release in *Xenopus* oocytes (92), the enhancement of CCE in oocytes by overexpression of type III IP_3 receptors (93), and the identification of ankyrin-like domains in TRP homologs and the ankyrin-binding ability of IP_3 receptors (94,95). Further progress in validating a physical coupling model for CCE will clearly depend on the molecular identification of store-operated channels (discussed later) and the proteins with which they interact in the cell.

Additional Requirements for CRAC Channel Activation

In addition to the store depletion signal, several permissive factors affect the optimal activation of CRAC channels: ADP/ATP, extracellular pH, and extracellular Ca^{2+}. ATP depletion inhibits capacitative Ca^{2+} entry in a variety of cells (96,97). A patch–clamp study of I_{CRAC} in RBL cells by Innocenti et al. (98) has shown that this effect is due to a rise in ADP concentration, inasmuch as 0.5 mM ADP in the recording pipette inhibits I_{CRAC} independently of the concentration of ATP (0–2 mM), while the absence of ATP itself has no effect. The mechanism of inhibition by ADP is intriguing; inhibition requires low intracellular Ca^{2+} buffering, activation of I_{CRAC} within 40 seconds of the start of internal dialysis, and physiological temperature (37 °C). These characteristics hint at a mechanism involving Ca^{2+} and a diffusible molecule that is lost during whole-cell recording. Further studies of ADP-dependent inhibition may shed light on the role of metabolic stress in regulating Ca^{2+} signaling as well as mechanisms of I_{CRAC} activation and inactivation.

I_{CRAC} in macrophages is enhanced by alkaline extracellular pH, increasing by about threefold between pH 6.5 and 9 (99). The effect can be described by a single inhibitory site with an apparent pK_a of 8.2. The mechanism of this effect is not known, but protonation of a site on the channel may be responsible. pH changes may modulate I_{CRAC} on the order of $<40\%$ during macrophage invasion of abcesses or tumors *in vivo,* where pH_o tends to be acidic (99).

Recently, two groups have found that extracellular Ca^{2+} is required to promote full activation of CRAC channels by store depletion, a process we refer to as Ca^{2+}-dependent potentiation, or CDP (33,36). This effect of Ca^{2+} is revealed upon readdition of Ca^{2+} to cells whose stores have been depleted in the absence of extracellular Ca^{2+}. Under these conditions, Ca^{2+} readdition causes I_{CRAC} to appear in two stages: a rapid current increase due to channel activity that was already present in the absence of Ca^{2+}, followed by a severalfold exponential increase in current over the next 10–20 seconds due to CDP. The site of action for CDP does not appear to be intracellular, since intracellular Ca^{2+} chelators do not affect CDP, and Ni^{2+} can replace Ca^{2+} in this role without permeating the channels to a significant extent. Interestingly, Ba^{2+} appears to be unable to potentiate CRAC channel activity, which may explain why Ba^{2+} currents through CRAC channels are smaller than Ca^{2+} currents in the steady state (29) (as discussed earlier in connection with ion selectivity). Interestingly, the extent of CDP increases with hyperpolarization, raising the possibility that the binding site may reside within the pore. An important unanswered question is whether CDP affects the kinetics or amplitude of I_{CRAC} under physiological conditions (i.e., constant $[Ca^{2+}]_o$). One interesting possibility is that CRAC channels normally open in two stages, and initial opening of the channel by the depletion signal may be needed to allow extracellular Ca^{2+} to access the CDP site and further augment the open probability. Further experiments will be needed to address the mechanism and physiological role of CDP.

CRAC Channel Inactivation

Negative feedback by Ca^{2+} is well established as a potent mechanism for modulating the activity of voltage-gated Ca^{2+} channels (see Chapter 2 this volume). Similarly, negative regulation of SOCs is likely to play important roles in controlling the amplitude and timing of Ca^{2+} signals in nonexcitable cells. Evidence for Ca^{2+}-dependent regulation of store-operated channels dates back to the earliest reports of CRAC channels in T cells and mast cells (15,16,22), as well as studies of SOCs in endothelial (37,45) and A431 cells (46). As described next, kinetic studies of CRAC channels have distinguished several types of inactivation by Ca^{2+}, with widely differing time courses and underlying mechanisms.

Fast Inactivation

After entering the cell, Ca^{2+} feeds back to inhibit CRAC channels over tens of milliseconds (22,23,32,99). Fast inactivation is most easily measured as a time-dependent decrease from the peak current elicited by hyperpolarizing voltage pulses (Fig. 3A). Several lines of evidence suggest that this inactivation results from the binding of intracellular Ca^{2+} to sites located extremely close to the mouth of the channel. First, the extent of fast inactivation produced by brief hyperpolarizations is dependent on the single channel current amplitude but is independent of global $[Ca^{2+}]_i$ (up to several micromolar) and the number of open channels in the cell (32). In addition, inactivation is slowed by intracellular BAPTA (10–12 mM), a Ca^{2+} buffer that binds Ca^{2+} rapidly, but not by EGTA (1.2–12 mM), a buffer with a comparable affinity for Ca^{2+} but a slower binding rate (23,32; but see Ref. 99) (Fig. 3A). Zweifach and Lewis (32) estimated the location of the inactivation binding site using a simple model that describes Ca^{2+} diffusion from a point source in the presence of an excess of highly mobile buffer (100). As shown in Fig. 3B, the model predicts the steady state distribution of $[Ca^{2+}]_i$ at different distances from the channel given the single channel current, the buffer concentration, and the binding rate; because of its faster rate of Ca^{2+} binding, BAPTA will create a steeper gradient of $[Ca^{2+}]_i$ near the open channel than will EGTA. The location of the Ca^{2+}-binding site for inactivation was determined by finding the distance at which the apparent $[Ca^{2+}]_i$ dependence of inactivation was the same in the presence of BAPTA or EGTA (32) (Fig. 3C). The steepness of the $[Ca^{2+}]_i$ dependence suggests the involvement of at least two sites positioned several nanometers from the mouth of the channel. Thus, the binding sites could reside on the channel itself or on a closely associated regulatory subunit. Interestingly, recent evidence indicates that TRP channels in the photoreceptor exist in a complex with PLC, PKC, INAD, and calmodulin, and the feedback regulation of TRP channels by Ca^{2+} has been proposed to occur via this signaling complex (101,102).

Slow Ca^{2+}-Dependent Inactivation

Intracellular Ca^{2+} also inhibits I_{CRAC} on a thousandfold slower time scale of tens of seconds. Slow inhibition by Ca^{2+} is an expected consequence of the capacitative Ca^{2+} entry model, in that increased $[Ca^{2+}]_i$ promotes store refilling, which should in turn deactivate Ca^{2+} influx. Slow inactivation of I_{CRAC} has been observed in Jurkat cells through simultaneous measurements of whole-cell current and $[Ca^{2+}]_i$ (16,31). We studied the Ca^{2+} dependence of this behavior by passively depleting stores with intracellular EGTA in the absence of extracellular Ca^{2+}, then raising $[Ca^{2+}]_o$ to trigger an increase in I_{CRAC} (31). When the intracellular concentration of EGTA was relatively low (1.2 mM), its buffering capacity was eventually overwhelmed by Ca^{2+} influx, leading to a rise in $[Ca^{2+}]_i$ and a concomitant decline in I_{CRAC} to a level near zero over ~100 seconds. However, higher levels of $[EGTA]_i$ (12 mM) prevented both the rise in $[Ca^{2+}]_i$ and the fall in I_{CRAC}, suggesting that unlike fast inactivation, slow inactivation is driven by a global increase in $[Ca^{2+}]_i$ (31).

Surprisingly, slow inactivation is not due solely to store refilling. This has been shown using a high dose of TG ($\pm$ IP_3) to prevent reuptake of Ca^{2+} by stores; under these conditions, I_{CRAC} still inactivates (by about 50%) following a rise in $[Ca^{2+}]_i$

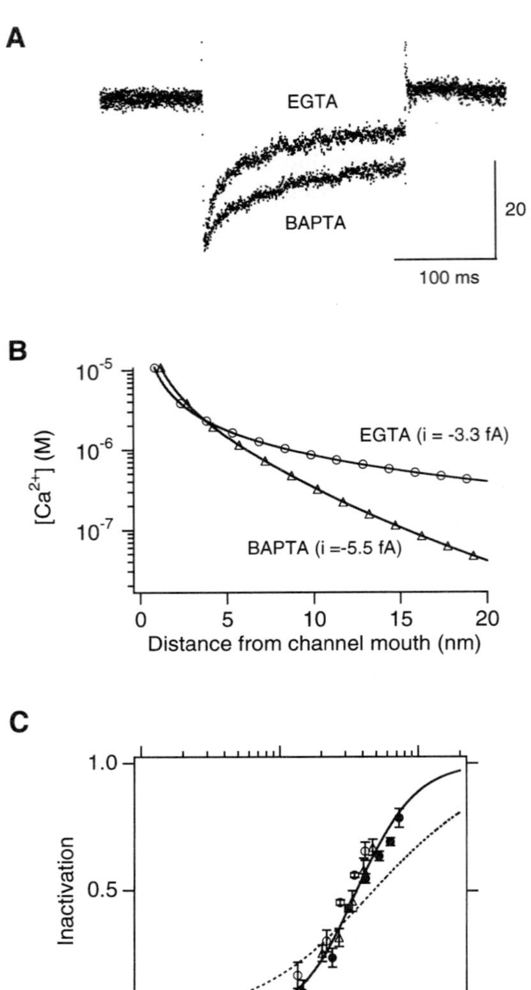

FIG. 3. Rapid inactivation of I_{CRAC} by local binding of intracellular Ca^{2+}. **(A)** Fast inactivation evoked during hyperpolarizing pulses from −12 mV to −142 mV is slowed by intracellular BAPTA (12 mM), a rapid Ca^{2+} buffer, but not by EGTA (12 mM), a slower buffer. **(B)** Predicted $[Ca^{2+}]_i$ gradients near CRAC channels in the presence of 12 mM intracellular BAPTA or EGTA. Single channel I_{CRAC} values of −5.5 fA (BAPTA) or −3.3 fA (EGTA) were inferred from experimental conditions that produced the same degree of inactivation. The data obtained with BAPTA (triangles) and EGTA (circles) predict the same Ca^{2+} dependence of inactivation if and only if the site is assumed to be 3–4 nm from the pore. (See text and Ref. 32 for details). **(C)** The extent of inactivation during pulses like those in A plotted against the predicted $[Ca^{2+}]_i$ at a distance of 3 nm from the mouth of the CRAC channel. Data were obtained at a range of voltages and $[Ca^{2+}]_o$, and the estimated single channel current value for each experiment was used to predict $[Ca^{2+}]_i$. The solid curves show best fits of the Hill equation with $n_H = 1$ (dashed line) or 2 (solid line). [Reproduced from The Journal of General Physiology by copyright permission of The Rockefeller University Press.] (Ref. 32).

(31). Thus, a store-independent mechanism of slow inactivation operates in parallel with the store-dependent one. Both are triggered by a rise of global $[Ca^{2+}]_i$, but their relative sensitivities to Ca^{2+} have not yet been quantitated. The underlying mechanism of store-independent (i.e., TG-insensitive) inactivation is also unclear. Although it is inhibited by okadaic acid as well as by pharmacological doses of cyclosporin A, inhibition by 1-norokadaone and resistance to calyculin A and FK506 is not consistent with a role for protein phosphatases 1, 2A, 2B, 2C, or 3 (31). The broad spectrum kinase inhibitor H-7 appears to prevent recovery from store-independent slow inactivation, suggesting that a kinase may be involved in recovery from the inactivated state (103). A role for calmodulin in SOC regulation was also proposed based on the ability of intracellular dialysis with Ca^{2+}/CaM to evoke delayed inactivation of I_{SOC} in TG-treated vascular endothelial cells (104). The possible roles of calmodulin and specific kinases and phosphatases in slow inactivation of I_{CRAC} remain to be tested.

The operation of two parallel pathways for I_{CRAC} inactivation may have important physiological consequences. This apparent redundancy may ensure feedback control of Ca^{2+} entry under conditions of both weak stimulation (in which IP_3 levels are relatively low and stores can refill efficiently) and intense stimulation (when IP_3 levels are high and store depletion is maintained). Slow store-independent inactivation may contribute to $[Ca^{2+}]_i$ oscillations. For example, the ability of TG to trigger $[Ca^{2+}]_i$ oscillations in parotid acinar cells and HeLa cells with completely empty stores has been attributed to periodic store-independent inactivation of SOC channels by Ca^{2+} (105–107). In contrast, $[Ca^{2+}]_i$ oscillations in T lymphocytes arise from *partial* depletion of stores (108); in this case, the slow kinetics of both store-dependent and -independent inactivation may contribute to the oscillations by creating a lag between the rise in $[Ca^{2+}]_i$ and the consequent decline in Ca^{2+} influx. Modeling studies may provide additional insights into the special functions of these two modes of inactivation.

Inactivation by PKC

Protein kinase C has been shown in imaging studies to inhibit capacitative Ca^{2+} or Mn^{2+} entry in a number of cell types. However, it is difficult in many cases to distinguish effects on the Ca^{2+} entry mechanism itself from those mediated through inhibition of PI-coupled receptors, acceleration of pump activity, or depolarization. More direct evidence for an inhibitory action of PKC was obtained in a patch–clamp study of I_{CRAC} in RBL cells (58). Following depletion with intracellular IP_3, slow I_{CRAC} inactivation was enhanced by ATP, ATPγS, and to a larger extent by PMA, but was inhibited by two PKC antagonists (staurosporine and bisindolylmaleimide). This pharmacological profile led Parekh and Penner to propose that a serine/threonine kinase, most likely PKC, may inhibit I_{CRAC}. This conclusion is at odds with that of an imaging study of RBL cells, in which PMA had no affect on TG-induced influx (109). The differences might be related to the different types of stimulation that were used (IP_3 vs. TG) or to the dialysis of cells with EGTA in the patch–clamp study. PMA also does not appear to inhibit CCE in rat thymocytes, although the phosphatase 1/2A antagonists okadaic acid and calyculin A did, suggesting that a non–PKC serine/threonine kinase may in fact inhibit CRAC channels (110).

Interestingly, another study (31) showed that the same phosphatase 1/2A antagonists did not alter peak I_{CRAC} magnitude in Jurkat cells; instead, okadaic acid reduced the extent of store-independent slow inactivation of I_{CRAC}, which would be expected to *increase* the rate of Ca^{2+} influx. This discrepancies need to be addressed; one obvious possibility is that modulation of I_{CRAC} by kinases is cell specific or is altered by loss of cytoplasmic molecules during whole-cell recording. A challenging goal for future studies will be to determine whether a candidate kinase is directly involved in activation of CCE or is merely providing a permissive environment within which the CCE mechanism can operate. Further progress in this area will largely depend on the identification of specific molecules that form the CCE pathway, including the store-operated channels themselves.

MOLECULAR IDENTIFICATION OF SOCs

Clearly one of the major obstacles in the study of SOC functions and regulatory mechanisms is that none of the molecules underlying capacitative Ca^{2+} entry have been identified. The lack of potent and selective pharmacological antagonists as well as the ubiquitous nature of the pathway have undermined attempts to biochemically isolate proteins in the pathway or to use expression cloning to isolate their genes. Thus, it is widely hoped that the recent identification of the *Drosophila* protein TRP and its mammalian homologs as potential SOCs (reviewed in Refs. 111,112) will lead to significant advances in understanding SOC function.

TRP: A Prototypic SOC?

The *trp* gene was first identified in *Drosophila* mutants whose photoreceptors failed to generate a sustained receptor potential in response to intense sustained light. The gene sequence (113) suggested a channel-like structure based on its hydrophobicity profile, and the *trp* mutation was found to reduce the Ca^{2+} selectivity of the light response (54). These results led to the notion that *trp* may encode a light-responsive Ca^{2+} channel, and the absolute dependence of phototransduction in *Drosophila* on phospholipase C prompted further speculation that the TRP channel may be depletion activated (52,54).

Evidence that TRP functions as a SOC in the photoreceptor has been hard to come by. Light does not release detectable amount of Ca^{2+} from stores, and TG fails to elicit a current in the photoreceptors, nor does it interfere acutely with phototransduction (114). Arguments to explain these discrepancies, have included proposals of stores with unusual properties or a requirement for cytosolic Ca^{2+} as a cofactor in channel activation (115), but several inconsistencies remain unresolved (52,115). A recent genetic study (55) has shown that both TRP and a homologous protein, TRPL (116), contribute to the light response, and although each can function independently in the absence of the other, heteromultimers are difficult to rule out at this point (117). TRP also appears to exist in a complex with several other signaling proteins, including rhodopsin (Rh1), phospholipase Cβ (NorpA), protein kinase C (InaC), InaD (necessary for Ca^{2+}-dependent inactivation), and calmodulin (101,102). Thus, regulation of TRP-containing channels in the photoreceptor may be quite complex, perhaps reflecting specializations that adapt them to the special requirements of phototransduction.

The best evidence that TRP can function as a store-operated channel has been provided by Schilling and colleagues through overexpression of TRP in Sf9 cells (53). In these cells, TRP overexpression caused the appearance of a TG-activated current that was distinguishable from the endogenous SOC current (and I_{CRAC}) by its low Ca^{2+} selectivity. Schilling's group exploited the differences between TRP and TRPL to define functional domains in these proteins. TRPL is homologous to TRP over much of its sequence but differs significantly in the carboxy-terminal putative intracellular domain (116), and overexpression of TRPL in Sf9 cells elicits a constitutively active, nonselective current that does not respond to TG (53). Studies of chimeras in which the C-terminal intracellular regions of TRP and TRPL were swapped indicated that sensitivity to TG, and therefore presumably sensitivity to the depletion signal, is dependent on the C-terminal region of TRP, while ion selectivity is determined within other regions of the two proteins (118). Interestingly, none of the mammalian TRP homologs described next show significant homology to TRP in the C-terminal region.

Mammalian TRP Homologs

The search for mammalian homologs of TRP/TRPL has yielded a family of closely related genes from human (56,95,119,120), murine (121,120), bovine (57), and *Xenopus* (121) cDNA libraries, and the list of homologs is still expanding. Northern analysis suggests that one of the human homologs, TRPC1, is expressed in a wide variety of tissues, including high levels in heart, brain, ovary, and testis, as well as lower levels in several other tissues. In contrast, human TRPC3 appears to be more restricted to brain (120). Significantly, at least two of the homologs are expressed in tissues in which CCE has been demonstrated; bovine TRP appears most highly expressed in adrenal gland (57), and TRPC1 mRNA is present at low levels in human lymphoid tissues (spleen, thymus) (95,119), where CRAC channels are known to function.

What is the relationship of these homologs to capacitative Ca^{2+} entry and SOCs? As a first step in this direction, Zhu et al. (120) overexpressed human TRPC1 (Htrp1) or TRPC3 (Htrp3) together with the M5 muscarinic receptor in COS-M6 cells. CCE was assayed as the $[Ca^{2+}]_i$ rise produced by Ca^{2+} readdition following store depletion with carbachol in Ca^{2+}-free medium. Compared to control cells, CCE was enhanced by 50% or ~200% in TRPC1 or TRPC3 transfectants, respectively. TRPC3 also enhanced Ca^{2+} entry following store depletion with TG; interestingly, the influx was less sensitive to block by La^{3+} and Ni^{2+} than the endogenous CCE, suggesting that the endogenous SOCs are not composed of TRPC3. Two electrophysiological studies support the view that TRP homologs may function as SOCs in heterologous expression systems. In CHO cells transfected with *TRPC1* (56) and HEK-293 cells transfected with *bTRP* (57), intracellular IP_3 or TG activates a cation-nonselective but Ca^{2+}-permeable whole-cell current. Further work will be needed to more precisely define the properties (selectivity, conductance, inactivation, etc.) of these currents and to establish their role in the formation of endogenous SOCs.

A critical goal of these studies is is to identify the constituents of endogenous SOCs such as the CRAC channel, for which a reasonable amount of functional information is already known. One approach to this question has been to suppress endogenous SOC activity with antisense TRP mRNA. Taking this approach, Zhu et al. were able to suppress CCE in a majority (6/9) of murine L-cell clones stably transfected with a combination of six antisense murine *TRP* sequences (120); however, it has not been shown that these effects are due to the down regulation of endogenous TRPs rather than to nonspecific effects of antisense expression. Nevertheless, the results thus far are promising, and further experiments of this kind, combined with immunochemical studies of TRP homologs, may ultimately reveal the subunit composition of store-operated channels.

CONCLUSIONS

In the past 8 years we have come a long way toward understanding the molecular basis of store-operated calcium entry. Several store-operated channels have been characterized in enough detail to suggest that a family of related proteins rather than a single SOC may underlie CCE in different tissues. At least one class of SOC, the CRAC channel, is quite distinctive in its high Ca^{2+} selectivity, and further biophysical studies may lead to new insights into the molecular mechanisms of Ca^{2+} permeation through these channels. While progress is encouraging, clearly several major gaps in our understanding of store-operated channels remain. First and perhaps most importantly, how is store depletion communicated to SOCs to trigger their activation? What mechanisms underlie the feedback regulation of SOC activity, and what role do these play in generating dynamic Ca^{2+} signaling patterns and cell responses? Can a specific SOC antagonist be found?

These and many other questions will be difficult to tackle until more molecules in the store-operated Ca^{2+} entry pathway are brought to light. In this respect, the work on mammalian TRP homologs looks especially promising, although this area also contributes its share of outstanding questions. Do the mammalian TRPs function endogenously as SOCs? Do they encode any of the nonselective SOCs that are known to exist, or do they combine with additional subunits to form Ca^{2+}-selective ones? Further application of molecular approaches, particularly immunochemical and yeast two-hybrid techniques, in combination with electrophysiological studies in well-defined systems, may soon enable the identification of proteins essential for the formation and activation of store-operated channels.

ACKNOWLEDGMENTS

Work in the author's laboratory is supported by grants from the National Institutes of Health, the American Heart Association, and Zeneca Pharmaceuticals Group.

REFERENCES

1. Penner R, Fasolato C, Hoth M: Calcium influx and its control by calcium release. *Curr Opin Neurobiol* 1993;3:368–374.
2. Fasolato C, Innocenti B, Pozzan T: Receptor-activated Ca^{2+} influx: how many mechanisms for how many channels? *Trends Pharmacol Sci* 1994;15:77–83.
3. Felder CC, Singer-Lahat D, Mathes C: Voltage-independent calcium channels. Regulation by receptors and intracellular calcium stores. *Biochem Pharmacol* 1994;48:1997–2004.
4. Hartzell HC: Activation of different Cl currents in *Xenopus* oocytes by Ca liberated from stores and by capacitative Ca influx. *J Gen Physiol* 1996;108:157–175.
5. Hamill OP, Marty A, Neher E, Sakmann B, Sigworth FJ: Improved patch-clamp techniques for high-resolution current recording from cells and cell-free membrane patches. *Pfluegers Arch* 1981;391: 85–100.
6. Marty A, Neher E: Tight-seal whole-cell recording. In: Sakmann B, Neher E, eds. *Single-channel recording*. New York: Plenum, 1995;31–52.
7. Putney JW Jr: Capacitative calcium entry revisited. *Cell Calcium* 1990;11:611–624.
8. Putney JW Jr, Bird GSJ: The inositol phosphate–calcium signaling system in nonexcitable cells. *Endocr Rev* 1993;14:610–631.
9. Berridge MJ: Capacitative calcium entry. *Biochem J* 1995;312:1–11.
10. Clapham DE: Calcium signaling. *Cell* 1995;80:259–268.
11. Berridge MJ: Inositol trisphosphate and calcium signalling. *Nature* 1993;361:315–325.
12. Streb H, Irvine RF, Berridge MJ, Schulz I: Release of Ca^{2+} from a nonmitochondrial intracellular store in pancreatic acinar cells by inositol-1,4,5-trisphosphate. *Nature* 1983;306:67–69.
13. Putney JW Jr: A model for receptor-regulated calcium entry. *Cell Calcium* 1986;7:1–12.
14. Penner R, Matthews G, Neher E: Regulation of calcium influx by second messengers in rat mast cells. *Nature* 1988;334:499–504.
15. Lewis RS, Cahalan MD: Mitogen-induced oscillations of cytosolic Ca^{2+} and transmembrane Ca^{2+} current in human leukemic T cells. *Cell Regul* 1989;1:99–112.
16. McDonald TV, Premack BA, Gardner P: Flash photolysis of caged inositol 1,4,5-trisphosphate activates plasma membrane calcium current in human T cells. *J Biol Chem* 1993;268:3889–3896.
17. Thastrup O, Cullen PJ, Drobak BK, Hanley MR, Dawson AP: Thapsigargin, a tumor promoter, discharges intracellular Ca^{2+} stores by specific inhibition of the endoplasmic reticulum Ca^{2+}-ATPase. *Proc Natl Acad Sci U S A* 1990;87:2466–2470.
18. Lytton J, Westlin M, Hanley MR: Thapsigargin inhibits the sarcoplasmic or endoplasmic reticulum Ca-ATPase family of calcium pumps. *J Biol Chem* 1991;266:17067–17071.
19. Gouy H, Cefai D, Christensen SB, Debré P, Bismuth G: Ca^{2+} influx in human T lymphocytes is induced independently of inositol phosphate production by mobilization of intracellular Ca^{2+} stores. A study with the Ca^{2+} endoplasmic reticulum-ATPase inhibitor thapsigargin. *Eur J Immunol* 1990; 20:2269–2275.
20. Mason MJ, Mahaut-Smith MP, Grinstein S: The role of intracellular Ca^{2+} in the regulation of the plasma membrane Ca^{2+} permeability of unstimulated rat lymphocytes. *J Biol Chem* 1991;266: 10872–10879.
21. Sarkadi B, Tordai A, Homolya L, Scharff O, Gárdos G: Calcium influx and intracellular calcium release in anti-CD3 antibody-stimulated and thapsigargin-treated human T lymphoblasts. *J Membr Biol* 1991;123:9–21.
22. Hoth M, Penner R: Depletion of intracellular calcium stores activates a calcium current in mast cells. *Nature* 1992;355:353–356.
23. Hoth M, Penner R: Calcium release-activated calcium current in rat mast cells. *J Physiol (Lond)* 1993;465:359–386.

24. Zweifach A, Lewis RS: Mitogen-regulated Ca^{2+} current of T lymphocytes is activated by depletion of intracellular Ca^{2+} stores. *Proc Natl Acad Sci U S A* 1993;90:6295–6299.
25. Fasolato C, Hoth M, Penner R: Multiple mechanisms of manganese-induced quenching of fura-2 fluorescence in rat mast cells. *Pfluegers Arch* 1993;423:225–231.
26. Franzius D, Hoth M, Penner R: Non-specific effects of calcium entry antagonists in mast cells. *Pfluegers Arch* 1994;428:433–438.
27. Partiseti M, Le Diest F, Hivroz C, Fischer A, Korn H, Choquet D: The calcium current activated by T cell receptor and store depletion in human lymphocytes is absent in a primary immunodeficiency. *J Biol Chem* 1994;51:32327–32335.
28. Premack BA, McDonald TV, Gardner P: Activation of Ca^{2+} current in Jurkat T cells following the depletion of Ca^{2+} stores by microsomal Ca^{2+}-ATPase inhibitors. *J Immunol* 1994;152:5226–5240.
29. Hoth M: Calcium and barium permeation through calcium release–activated calcium (CRAC) channels. *Pfluegers Arch* 1995;430:315–322.
30. Zhang L, McCloskey MA: Immunoglobulin E receptor-activated calcium conductance in rat mast cells. *J Physiol (Lond)* 1995;483:59–66.
31. Zweifach A, Lewis RS: Slow calcium-dependent inactivation of depletion-activated calcium current. Store-dependent and -independent mechanisms. *J Biol Chem* 1995;270:14445–14451.
32. Zweifach A, Lewis RS: Rapid inactivation of depletion-activated calcium current (I_{CRAC}) due to local calcium feedback. *J Gen Physiol* 1995;105:209–226.
33. Christian EP, Spence KT, Togo JA, Dargis PG, Patel J: Calcium-dependent enhancement of depletion-activated calcium current in Jurkat T lymphocytes. *J Membr Biol* 1996;150:63–71.
34. Lepple-Wienhues A, Cahalan MD: Conductance and permeation of monovalent cations through depletion-activated Ca^{2+} channels (I_{CRAC}) in Jurkat T cells. *Biophys J* 1996;71:787–794.
35. Schofield GG, Mason MJ: A Ca^{2+} current activated by release of intracellular Ca^{2+} stores in rat basophilic leukemia cells (RBL-1). *J Membr Biol* 1996;153:217–231.
36. Zweifach A, Lewis RS: Calcium-dependent potentiation of store-operated calcium channels in T lymphocytes. *J Gen Physiol* 1996;107:597–610.
37. Vaca L, Kunze DL: IP_3-activated Ca^{2+} channels in the plasma membrane of cultured vascular endothelial cells. *Am J Physiol* 1995;269:C733–C738.
38. Neher E: The use of fura-2 for estimating Ca buffers and Ca fluxes. *Neuropharmacology* 1995;34: 1423–1442.
39. Kim M-S, Morii T, Sun L-X, Imoto K, Mori Y: Structural determinants of ion selectivity in brain calcium channel. *FEBS Lett* 1993;318:145–148.
40. Yang J, Ellinor PT, Sather WA, Zhang J-F, Tsien RW: Molecular determinants of Ca^{2+} selectivity and ion permeation in L-type Ca^{2+} channels. *Nature* 1993;366:158–161.
41. Meldolesi J, Clementi E, Fasolato C, Zacchetti D, Pozzan T: Ca^{2+} influx following receptor activation. *Trends Pharmacol Sci* 1991;12:289–292.
42. Hille B: *Ionic channels of excitable membranes.* Sunderland, MA: Sinauer Associates, 1992.
43. Neher E, Stevens CF: Conductance fluctuations and ionic pores in membranes. *Annu Rev Biophys Bioeng* 1977;6:345–381.
44. Hilgemann DW: Unitary cardiac Na^+,Ca^{2+} exchange current magnitudes determined from channel-like noise and charge movements of ion transport. *Biophys J* 1996;71:759–768.
45. Vaca L, Kunze DL: Depletion of intracellular Ca^{2+} stores activates a Ca^{2+}-selective channel in vascular endothelium. *Am J Physiol* 1994;267:C920–C925.
46. Lückhoff A, Clapham DE: Calcium channels activated by depletion of internal calcium stores in A431 cells. *Biophys J* 1994;67:177–182.
47. Krause E, Pfeiffer F, Schmid A, Schulz I: Depletion of intracellular calcium stores activates a calcium conducting nonselective cation current in mouse pancreatic acinar cells. *J Biol Chem* 1996; 271:32523–32528.
48. Delles C, Haller T, Dietl P: A highly calcium-selective cation current activated by intracellular calcium release in MDCK cells. *J Physiol (Lond)* 1995;486:557–569.
49. Kerst G, Fischer K-G, Normann C, Kramer A, Leipziger J, Greger R: Ca^{2+} influx induced by store release and cytosolic Ca^{2+} chelation in HT_{29} colonic carcinoma cells. *Pfluegers Arch* 1995;430: 653–665.
50. Worley JFI, McIntyre MS, Spencer B, Dukes ID: Depletion of intracellular Ca^{2+} stores activates a maitotoxin-sensitive nonselective cationic current in β-cells. *J Biol Chem* 1994;269:32055–32058.
51. Duszynski J, Elensky M, Cheung JY, Tillotson DL, LaNoue KF: Hormone-regulated Ca^{2+} channel in rat hepatocytes revealed by whole cell patch clamp. *Cell Calcium* 1995;18:19–29.

52. Hardie RC, Minke B: Phosphoinositide-mediated phototransduction in *Drosophila* photoreceptors: the role of Ca^{2+} and *trp*. *Cell Calcium* 1995;18:256–274.
53. Vaca L, Sinkins WG, Hu Y, Kunze DL, Schilling WP: Activation of recombinant *trp* by thapsigargin in Sf9 insect cells. *Am J Physiol* 1994;267:C1501–C1505.
54. Hardie RC, Minke B: The *trp* gene is essential for a light-activated Ca^{2+} channel in *Drosophila* photoreceptors. *Neuron* 1992;8:643–651.
55. Niemeyer BA, Suzuki E, Scott K, Jalink K, Zuker CS: The *Drosophila* light-activated conductance is composed of the two channels TRP and TRPL. *Cell* 1996;85:651–659.
56. Zitt C, Zobel A, Obukhov AG et al: Cloning and functional expression of a human Ca^{2+}-permeable cation channel activated by calcium store depletion. *Neuron* 1996;16:1189–1196.
57. Philipp S, Cavalié A, Freichel M et al: A mammalian capacitative calcium entry channel homologous to *Drosophila* TRP and TRPL. *EMBO J* 1996;15:6166–6171.
58. Parekh AB, Penner R: Depletion-activated calcium current is inhibited by protein kinase in RBL-2H3 cells. *Proc Natl Acad Sci U S A* 1995;92:7907–7911.
59. Beaven MA, Rogers J, Moore JP, Hesketh TR, Smith GA, Metcalfe JC: The mechanism of the calcium signal and correlation with histamine release in 2H3 cells. *J Biol Chem* 1984;259:7129–7136.
60. Mohr FC, Fewtrell C: Depolarization of rat basophilic leukemia cells inhibits calcium uptake and exocytosis. *J Cell Biol* 1987;104:783–792.
61. Alvarez J, Montero M, García-Sancho J: Cytochrome P-450 may link intracellular Ca^{2+} stores with plasma membrane Ca^{2+} influx. *Biochem J* 1991;274:193–197.
62. Fanger CM, Hoth M, Crabtree GR, Lewis RS: Characterization of T cell mutants with defects in capacitative calcium entry: genetic evidence for the physiological roles of CRAC channels. *J Cell Biol* 1995;131:766–667.
63. Lewis RS, Cahalan MD: Potassium and calcium channels in lymphocytes. *Annu Rev Immunol* 1995; 13:623–653.
64. Crabtree GR: Contingent genetic regulatory events in T lymphocyte activation. *Science* 1989;243: 355–361.
65. Densmore JJ, Haverstick DM, Szabo G, Gray LS: A voltage-operable current is involved in Ca^{2+} entry in human lymphocytes whereas I_{CRAC} has no apparent role. *Am J Physiol* 1996;271:C1494–C1503.
66. Chung SC, McDonald TV, Gardner P: Inhibition by SK&F 96365 of Ca^{2+} current, IL-2 production and activation in T lymphocytes. *Br J Pharmacol* 1994;113:861–868.
67. Le Deist F, Hivroz C, Partiseti M et al: A primary T-cell immunodeficiency associated with defective transmembrane calcium influx. *Blood* 1995;85:1053–1062.
68. Mery L, Mesaeli N, Michalak M, Opas M, Lew DP, Krause K-H: Overexpression of calreticulin increases intracellular Ca^{2+} storage and decreases store-operated Ca^{2+} influx. *J Biol Chem* 1996; 271:9332–9339.
69. Camacho P, Lechleiter JD: Calreticulin inhibits repetitive intracellular Ca^{2+} waves. *Cell* 1995;82: 765–771.
70. Favre CJ, Nüsse O, Lew DP, Krause K-H: Store-operated Ca^{2+} influx: What is the message from the stores to the membrane? *J Lab Clin Med* 1996;128:19–26.
71. Randriamampita C, Tsien RY: Emptying of intracellular Ca^{2+} stores releases a novel small messenger that stimulates Ca^{2+} influx. *Nature* 1993;364:809–814.
72. Gilon P, Bird GSJ, Bian X, Yakel JL, Putney JW Jr: The Ca^{2+}-mobilizing actions of a Jurkat cell extract on mammalian cells and *Xenopus laevis* oocytes. *J Biol Chem* 1995;270:8050–8055.
73. Thomas D, Hanley MR: Evaluation of calcium influx factors from stimulated Jurkat T-lymphocytes by microinjection into *Xenopus* oocytes. *J Biol Chem* 1995;270:6429–6432.
74. Premack BA, Thomas DW, Hanley MR, Gardner P: Ca^{2+} channel modulation by "calcium influx factors" (CIF's) produced following the depletion of intracellular Ca^{2+} stores. *Biophys J* 1995;68: A54.
75. Kim HY, Thomas D, Hanley MR: Chromatographic resolution of an intracellular calcium influx factor from thapsigargin-activated Jurkat cells. *J Biol Chem* 1995;270:9706–9708.
76. Thomas D, Kim HY, Hanley MR: Regulation of inositol trisphosphate-induced membrane currents in *Xenopus* oocytes by a Jurkat cell calcium influx factor. *Biochem J* 1996;318:649–656.
77. Pandol SJ, Schoeffield-Payne MS: Cyclic GMP mediates the agonist-mediated increase in plasma membrane calcium entry in the pancreatic acinar cell. *J Biol Chem* 1990;265:12846–12853.

78. Xu X, Star RA, Tortorici G, Muallem S: Depletion of intracellular Ca^{2+} stores activates nitric-oxide synthase to generate cGMP and regulate Ca^{2+} influx. *J Biol Chem* 1994;269:12645–12653.
79. Bahnson TD, Pandol SJ, Dionne VE: Cyclic GMP modulates depletion-activated Ca^{2+} entry in pancreatic acinar cells. *J Biol Chem* 1993;268:10808–10812.
80. Mathes C, Thompson SH: Calcium current activated by muscarinic receptors and thapsigargin in neuronal cells. *J Gen Physiol* 1994;104:107–121.
81. Mathes C, Thompson SH: The nitric oxide/cGMP pathway couples muscarinic receptors to the activation of Ca^{2+} influx. *J Neurosci* 1996;16:1702–1709.
82. Mathes C, Thompson SH: The relationship between depletion of intracellular Ca^{2+} stores and activation of Ca^{2+} current by muscarinic receptors in neuroblastoma cells. *J Gen Physiol* 1995;106: 975–993.
83. Thompson SH, Mathes C, Alousi AA: Calcium requirement for cGMP production during muscarinic activation of N1E-115 neuroblastoma cells. *Am J Physiol* 1995;269:C979–C985.
84. Gilon P, Obie JF, Bian X, Bird GSJ, Putney JW Jr: Role of cyclic GMP in the control of capacitative Ca^{2+} entry in rat pancreatic acinar cells. *Biochem J* 1995;311:649–656.
85. Bian X, Bird GSJ, Putney JW Jr: cGMP is not required for capacitative Ca^{2+} entry in Jurkat T-lymphocytes. *Cell Calcium* 1996;19:351–354.
86. Fasolato C, Hoth M, Penner R: A GTP-dependent step in the activation mechanism of capacitative calcium influx. *J Biol Chem* 1993;268:20737–20740.
87. Bird GSJ, Putney JW Jr: Inhibition of thapsigargin-induced calcium entry by microinjected guanine nucleotide analogues. *J Biol Chem* 1993;268:21486–21488.
88. Jaconi MEE, Lew DP, Monod A, Krause K-H: The regulation of store-dependent Ca^{2+} influx in HL-60 granulocytes involves GTP-sensitive elements. *J Biol Chem* 1993;268:26075–26078.
89. Somasundaram B, Norman JC, Mahaut-Smith MP: Primaquine, an inhibitor of vesicular transport, blocks the calcium-release-activated current in rat megakaryocytes. *Biochem J* 1995;309:725–729.
90. Somasundaram B, Mahaut-Smith MP, Floto RA: Temperature-dependent block of capacitative Ca^{2+} influx in the human leukemic cell line KU-812. *J Biol Chem* 1996;271:26096–26104.
91. Irvine RF: 'Quantal' Ca^{2+} release and the control of Ca^{2+} entry by inositol phosphates—a possible mechanism. *FEBS Lett* 1990;263:5–9.
92. Petersen CCH, Berridge MJ: Capacitative calcium entry is colocalised with calcium release in *Xenopus* oocytes: evidence against a highly diffusible calcium influx factor. *Pfluegers Arch* 1996; 432:286–292.
93. DeLisle S, Blondel O, Longo FJ, Schnabel WE, Bell GI, Welsh MJ: Expression of inositol 1,4,5-trisphosphate receptors changes the Ca^{2+} signal of *Xenopus* oocytes. *Am J Physiol* 1996;270: C1255–C1261.
94. Bennett DL, Petersen CCH, Cheek TR: Cracking I_{CRAC} in the eye. *Curr Biol* 1995;5:1225–1228.
95. Wes PD, Chevesich J, Jeromin A, Rosenberg C, Stetten G, Montell C: TRPC1, a human homolog of a *Drosophila* store-operated channel. *Proc Natl Acad Sci U S A* 1995;92:9652–9656.
96. Gamberucci A, Innocenti B, Fulceri R et al: Modulation of Ca^{2+} influx dependent on store depletion by intracellular adenine–guanine nucleotide levels. *J Biol Chem* 1994;269:23597–23602.
97. Marriott I, Mason MJ: ATP depletion inhibits capacitative Ca^{2+} entry in rat thymic lymphocytes. *Am J Physiol* 1995;269:C766–C774.
98. Innocenti B, Pozzan T, Fasolato C: Intracellular ADP modulates the Ca^{2+} release-activated Ca^{2+} current in a temperature- and Ca^{2+}-dependent way. *J Biol Chem* 1996;271:8582–8587.
99. Malayev A, Nelson DJ: Extracellular pH modulates the Ca^{2+} current activated by depletion of intracellular Ca^{2+} stores in human macrophages. *J Membr Biol* 1995;146:101–111.
100. Neher E: Concentration profiles of intracellular calcium in the presence of a diffusible chelator. *Exp Brain Res* 1986;14:80–96.
101. Huber A, Sander P, Gobert A, Bähner M, Hermann R, Paulsen R: The transient receptor potential protein (Trp), a putative store-operated Ca^{2+} channel essential for phosphoinositide-mediated photoreception, forms a signaling complex with NorpA, InaC and InaD. *EMBO J* 1996;15:7036–7045.
102. Chevesich J, Kreuz AJ, Montell C: Requirement for the PDZ domain protein, INAD, for localization of the TRP store-operated channel to a signaling complex. *Neuron* 1997;18:95–105.
103. Lewis RS, Dolmetsch RE, Zweifach A: Positive and negative regulation of depletion-activated calcium channels by calcium. In: Clapham DE, Ehrlich BE, eds. *Organellar ion channels and transporters.* New York: Rockefeller University Press, *Soc. Gen Physiol Ser.* 1996;51:241–254.
104. Vaca L: Calmodulin inhibits calcium influx current in vascular endothelium. *FEBS Lett* 1996;390: 289–293.
105. Foskett JK, Roifman CM, Wong D: Activation of calcium oscillations by thapsigargin in parotid acinar cells. *J Biol Chem* 1991;266:2778–2782.

106. Foskett JK, Wong DCP: $[Ca]_i$ inhibition of Ca^{2+} release-activated Ca^{2+} influx underlies agonist- and thapsigargin-induced $[Ca^{2+}]_i$ oscillations in salivary acinar cells. *J Biol Chem* 1994;269:31525–31532.
107. Missiaen L, De Smedt H, Pary JB, Oike M, Casteels R: Kinetics of empty store-activated Ca^{2+} influx in HeLa cells. *J Biol Chem* 1994;269:5817–5823.
108. Dolmetsch R, Lewis RS: Signaling between intracellular Ca^{2+} stores and depletion-activated Ca^{2+} channels generates $[Ca^{2+}]_i$ oscillations in T lymphocytes. *J Gen Physiol* 1994;103:365–388.
109. Ali H, Maeyama K, Sagi-Eisenberg R, Beaven MA: Antigen and thapsigargin promote influx of Ca^{2+} in rat basophilic RBL-2H3 cells by ostensibly similar mechanisms that allow filling of inositol 1,4,5-trisphosphate-sensitive and mitochondrial Ca^{2+} stores. *Biochem J* 1994;304:431–440.
110. Marriott I, Mason MJ: Evidence for a phorbol ester-insensitive phosphorylation step in capacitative calcium entry in rat thymic lymphocytes. *J Biol Chem* 1996;271:26732–26738.
111. Clapham DE: TRP is cracked but is CRAC TRP? *Neuron* 1996;16:1069–1072.
112. Friel DD: TRP: its role in phototransduction and store-operated Ca^{2+} entry. *Cell* 1996;85:617–619.
113. Montell C, Rubin GM: Molecular characterization of the *Drosophila trp* locus: a putative integral membrane protein required for phototransduction. *Neuron* 1989;2:1313–1323.
114. Ranganathan R, Bacskai BJ, Tsien RY, Zuker CS: Cytosolic calcium transients: spatial localization and role in *Drosophila* photoreceptor cell function. *Neuron* 1994;13:837–848.
115. Hardie RC: Excitation of *Drosophila* photoreceptors by BAPTA and ionomycin: evidence for capacitative Ca^{2+} entry? *Cell Calcium* 1996;20:315–327.
116. Phillips AM, Bull A, Kelly LE: Identification of a *Drosophila* gene encoding a calmodulin-binding protein with homology to the *trp* phototransduction gene. *Neuron* 1992;8:631–642.
117. Gillo B, Chorna I, Cohen H et al: Coexpression of *Drosophila* TRP and TRP-like proteins in *Xenopus* oocytes reconstitutes capacitative Ca^{2+} entry. *Proc Natl Acad Sci U S A* 1996;93:14146–14151.
118. Sinkins WG, Vaca L, Hu Y, Kunze DL, Schilling WP: The COOH-terminal domain of *Drosophila* TRP channels confers thapsigargin sensitivity. *J Biol Chem* 1996;271:2955–2960.
119. Zhu X, Chu PB, Peyton M, Birnbaumer L: Molecular cloning of a widely expressed human homologue for the *Drosophila trp* gene. *FEBS Lett* 1995;373:193–198.
120. Zhu X, Jiang M, Peyton M et al: *trp,* A novel mammalian gene family essential for agonist-activated capacitative Ca^{2+} entry. *Cell* 1996;85:661–671.
121. Petersen CCH, Berridge MJ, Borgese MF, Bennett DL: Putative capacitative calcium entry channels: expression of *Drosophila trp* and evidence for the existence of vertebrate homologues. *Biochem J* 1995;311:41–44.
122. Christian EP, Spence KT, Togo JA, Dargis PG, Warawa E: Extracellular site for econazole-mediated block of Ca^{2+} release-activated Ca^{2+} current (I_{CRAC}) in T lymphocytes. *Br J Pharmacol* 1996;119: 647–654.
123. Partiseti M: Régulation de l'expression et fonctions des conductances potassium et calcium des lymphocytes au cours du cycle céllulaire. Ph.D. thesis, Université Pierre et Marie Curie, Paris, 1994.
124. Somasundaram B, Mahaut-Smith MP: Three cation influx currents activated by purinergic receptor stimulation in rat megakaryocytes. *J Physiol (Lond)* 1994;480:225–231.

Subject Index

R

S

T

CARDIFF UNIVERSITY
PRIFYSGOL CAERDYDD